N83-14687

JPL PUBLICATION 82-69

Solar Cell Radiation Handbook

Third Edition

H.Y. Tada
J.R. Carter, Jr.
TRW Systems Group

B.E. Anspaugh
R.G. Downing
Jet Propulsion Laboratory

November 1, 1982

National Aeronautics and
Space Administration

Jet Propulsion Laboratory
California Institute of Technology
Pasadena, California

The research described in this publication was carried out by the Jet Propulsion Laboratory, California Institute of Technology, under contract with the National Aeronautics and Space Administration.

Reference to any specific commercial product, process, or service by trade name or manufacturer does not necessarily constitute an endorsement by the United States Government or the Jet Propulsion Laboratory, California Institute of Technology.

TABLE OF CONTENTS

		Page
1.0	THEORY OF THE SILICON SOLAR CELL	1-1
1.1	Semiconductor Theory	1-6
1.1.1	Thermal Equilibrium Relationships and Excess Densities	1-6
1.1.2	Carrier Transport	1-10
1.2	The P-N Junction	1-12
1.3	Silicon Solar Cell Theory	1-19
1.4	Solar Cell Coatings and Contacts	1-25
1.5	Improvement of Solar Cell Efficiency	1-30
1.5.1	Considerations for Photovoltaic Materials	1-31
1.5.2	Shallow Junction Solar Cells	1-33
1.5.3	Back Surface Field (BSF) or (P+) Solar Cells	1-34
1.5.4	Textured Surface Solar Cells	1-36
1.5.5	Vertical Junction (VJ) Solar Cells	1-40
1.5.6	Back Surface Reflector (BSR) Solar Cells	1-41
	REFERENCES	1-42
2.0	INSTRUMENTATION TECHNIQUES FOR MEASUREMENT OF SOLAR CELL PARAMETERS	2-1
2.1	Light Sources and Solar Simulators	2-1
2.2	Current-Voltage Characteristics	2-8
2.3	Spectral Response Measurements	2-9
2.4	Irradiation Methods	2-10
2.5	Diffusion Length Measurement	2-11
	REFERENCES	2-16

TABLE OF CONTENTS (Continued)

			Page
3.0	RADIATION EFFECTS	3-1
	3.1	The Theory of Radiation Damage	3-1
		3.1.1 Ionization	3-2
		3.1.2 Atomic Displacement	3-6
		3.1.2.1 Electron Displacement Damage	3-8
		3.1.2.2 Proton Displacement Damage	3-12
		3.1.2.3 Neutron Displacement Damage	3-13
	3.2	Theory of Silicon Solar Cell Damage	3-17
	3.3	The Concept of Damage Equivalence	3-21
	3.4	1 MeV Electron Irradiation of Silicon Solar Cells	3-24
	3.5	Effect of Electron Energy on Solar Cell Degradation	3-25
	3.6	Effect of Proton Energy on Solar Cell Degradation	3-26
	3.7	Junction Effects of Low Energy Protons	3-33
	3.8	Effects of Neutron and Gamma Radiation on Solar Cells	3-36
	3.9	Lithium Doped Solar Cells	3-41
	3.10	Annealing of Irradiated Solar Cells	3-42
	3.11	Effect of Irradiation on Solar Cell Temperature and Illumination Intensity Dependences	3-44
	3.12	Radiation Effects on Shielding Materials	3-52
	3.13	Solar Cell Output vs. 1 MeV Electron Irradiation	3-58
		REFERENCES	3-153
4.0	RELATIVE DAMAGE COEFFICIENTS FOR SPACE RADIATION	4-1
	4.1	Geometrical Aspects of Radiation Fluences	4-1
	4.2	Effect of Shielding on Radiation	4-3
	4.3	Electron Space Radiation Effects	4-4
	4.4	Proton Space Radiation Effects	4-10
	4.5	Alpha Particle Space Radiation Effects	4-19
	4.6	Alternative Approaches	4-21
		REFERENCES	4-23

TABLE OF CONTENTS (Continued)

		Page
5.0	THE SPACE RADIATION ENVIRONMENT.	5-1
5.1	Geomagnetically Trapped Radiation	5-1
	5.1.1 Trapped Protons	5-6
	5.1.2 Trapped Electrons	5-6
5.2	Orbital Integration	5-9
5.3	Cosmic-Ray (Galactic Cosmic-Ray) Radiation	5-10
5.4	Solar Flare (Solar Cosmic-Ray) Radiation	5-11
	REFERENCES	5-24
6.0	SOLAR ARRAY DEGRADATION CALCULATIONS	6-1
6.1	General Procedure, Equivalent Fluence	6-1
6.2	Effect of Reduced Light Transmission on Solar Cell Response	6-5
6.3	Rough Degradation Calculations	6-5
6.4	Computer Calculated Equivalent Fluence	6-8
6.5	Solar Array Degradation	6-12
	REFERENCES	6-53
7.0	FLIGHT DATA.	7-1
7.1	Flight Data at Synchronous Orbit	7-3
	7.1.1 Solar Array Performance Data at Synchronous Orbit	7-3
	7.1.2 Flight Experiments at Synchronous Orbits	7-6
	7.1.2.1 LES-6	7-6
	7.1.2.2 ATS-5	7-9
	7.1.2.3 ATS-6	7-13
7.2	Flight Data at Other Than Synchronous Orbits	7-15
	7.2.1 Solar Array Performance at Other Than Synchronous Orbits	7-15

TABLE OF CONTENTS (Continued)

		Page
7.2.2	Flight Experiments at Other Than Synchronous Orbits	7-15
	7.2.2.1 ERS 6	7-15
	7.2.2.2 NTS-1 (Timation III)	7-15
	7.2.2.3 NTS-2	7-18
REFERENCES		7-21
APPENDIX A:	SHIELDING THICKNESS CONVERSIONS	A-1
APPENDIX B:	CONSTANTS, PROPERTIES AND VALUES	B-1
	REFERENCES	B-8
APPENDIX C:	SOLAR CELL TYPES	C-1
APPENDIX D:	COMPUTER PROGRAM, EQFRUX	D-1
	REFERENCES	D-5

FIGURES

		Page
1.1	Carrier Density in Illuminated Silicon	1-2
1.2	Carrier Concentrations in an Illuminated Solar Cell, Short Circuited	1-13
1.3	Solar Cell Equivalent Circuit Model	1-17
1.4	Typical Dark Solar Cell Current-Voltage Characteristic, Forward Biased	1-18
1.5	Calculated Silicon Solar Cell Spectral Response	1-22
1.6	Development of a Current-Voltage Characteristic for an Illuminated Silicon Solar Cell	1-24
1.7	Temperature-Dependent Maximum Efficiency as a Function of Energy Gap for a Few Photovoltaic Materials	1-32
1.8	(a) Textured Surface Solar Cell, (b) Vertical Junction Solar Cell	1-38
1.9	Spectral Responses of Textured Surface Cell and a Comparable Flat Surface Cell	1-39
2.1	Spectral Irradiance of Spectrolab X-25 Mark II Solar Simulator	2-4
2.2	Spectral Irradiance of ASEC Tungsten/Xenon Solar Simulator	2-5
2.3	Spectral Irradiance of Spectrolab Large Area Pulsed Solar Simulator (LAPSS)	2-6
2.4	Minority Carrier Diffusion Length vs Short Circuit Current Density of Conventional n/p Silicon Solar Cells	2-15
3.1	Stopping Power and Range Curves for Electrons in Silicon	3-4
3.2	Stopping Power and Range Curves for Protons in Silicon	3-5
3.3	Atomistic Models of Radiation Defects in Silicon	3-10
3.4	Energy-Dependent Displacement in Proton-Irradiated Silicon	3-14
3.5	Variation of Solar Cell Diffusion Length with Fluence for Various Radiations	3-22
3.6	Variation of Solar Cell Short Circuit Current Density with Fluence for Various Radiations	3-23
3.7	Electron Energy Dependence of K_L Values for n/p Silicon Solar Cells	3-27
3.8	Electron Energy Dependence of Φ_c^{-1} Values for n/p Silicon Solar Cells	3-28

FIGURES (Continued)

		Page
3.9	Comparison of Relative Damage Coefficients, Proton-Irradiated n/p Silicon Solar Cells.	3-30
3.10	Relative Damage Coefficients for Proton-Irradiated n/p Silicon Solar Cells.	3-32
3.11	Atomic Displacements as a Function of Depth for a 3 MeV Proton in Silicon	3-35
3.12	Low Energy Proton Junction Damage, 0.250 MeV Protons, 3×10^{13} p/cm^2, Partially Shielded n/p Solar Cell.	3-37
3.13	Neutron-Induced Change in n/p Silicon Solar Cells.	3-39
3.14	Recovered Power Output of Irradiated Conventional and Lithium-Doped Solar Cells	3-43
3.15	I_{sc} Temperature Coefficient vs 1 MeV Electron Fluence.	3-47
3.16	V_{oc} Temperature Coefficient vs 1 MeV Electron Fluence.	3-48
3.17	P_{max} Temperature Coefficient vs 1 MeV Electron Fluence	3-49
3.18	V_{mp} Temperature Coefficient vs 1 MeV Electron Fluence.	3-50
3.19	Variation of Coverglass Transmittance with Absorbed Dose	3-56

2 Ohm-cm n/p Conventional and Shallow Junction Silicon Cells

3.20	I_{sc} vs 1 Mev Electron Fluence.	3-61
3.21	V_{oc} vs 1 MeV Electron Fluence.	3-62
3.22	P_{max} vs 1 MeV Electron Fluence	3-63
3.23	V_{mp} vs 1 MeV Electron Fluence	3-64
3.24	I_{mp} vs 1 MeV Electron Fluence	3-65
3.25	Normalized I_{sc} vs 1 MeV Electron Fluence	3-66
3.26	Normalized V_{oc} vs 1 MeV Electron Fluence	3-67
3.27	Normalized P_{max} vs 1 MeV Electron Fluence.	3-68
3.28	Normalized V_{mp} vs 1 MeV Electron Fluence	3-69
3.29	Normalized I_{mp} vs 1 MeV Electron Fluence	3-70

10 Ohm-cm n/p Conventional and Shallow Junction Silicon Cells

3.30	I_{sc} vs 1 MeV Electron Fluence.	3-71
3.31	V_{oc} vs 1 MeV Electron Fluence.	3-72
3.32	P_{max} vs 1 MeV Electron Fluence	3-73
3.33	V_{mp} vs 1 MeV Electron Fluence.	3-74

FIGURES (Continued)

		Page
3.34	I_{mp} vs 1 MeV Electron Fluence	3-75
3.35	Normalized I_{sc} vs 1 MeV Electron Fluence	3-76
3.36	Normalized V_{oc} vs 1 MeV Electron Fluence	3-77
3.37	Normalized P_{max} vs 1 MeV Electron Fluence	3-78
3.38	Normalized V_{mp} vs 1 MeV Electron Fluence	3-79
3.39	Normalized I_{mp} vs 1 MeV Electron Fluence	3-80

10 Ohm-cm n/p Back Surface Field Silicon Cells (+TEX)

3.40	I_{sc} vs 1 MeV Electron Fluence.	3-81
3.41	V_{oc} vs 1 MeV Electron Fluence.	3-82
3.42	P_{max} vs 1 MeV Electron Fluence	3-83
3.43	V_{mp} vs 1 MeV Electron Fluence.	3-84
3.44	I_{mp} vs 1 MeV Electron Fluence.	3-85
3.45	Normalized I_{sc} vs 1 MeV Electron Fluence	3-86
3.46	Normalized V_{oc} vs 1 MeV Electron Fluence	3-87
3.47	Normalized P_{max} vs 1 MeV Electron Fluence	3-88
3.48	Normalized V_{mp} vs 1 MeV Electron Fluence	3-89
3.49	Normalized I_{mp} vs 1 MeV Electron Fluence	3-90

2 Ohm-cm n/p Silicon Cells with BSR

3.50	I_{sc} vs 1 MeV Electron Fluence.	3-91
3.51	V_{oc} vs 1 MeV Electron Fluence.	3-92
3.52	P_{max} vs 1 MeV Electron Fluence	3-93
3.53	V_{mp} vs 1 MeV Electron Fluence.	3-94
3.54	I_{mp} vs 1 MeV Electron Fluence.	3-95
3.55	Normalized I_{sc} vs 1 MeV Electron Fluence	3-96
3.56	Normalized V_{oc} vs 1 MeV Electron Fluence	3-97
3.57	Normalized P_{max} vs 1 MeV Electron Fluence.	3-98
3.58	Normalized V_{mp} vs 1 MeV Electron Fluence	3-99
3.59	Normalized I_{mp} vs 1 MeV Electron Fluence	3-100

FIGURES (Continued)

Page

2 Ohm-cm n/p Back Surface Field Silicon Cells with BSR

3.60	I_{sc} vs 1 MeV Electron Fluence.	3-101
3.61	V_{oc} vs 1 MeV Electron Fluence.	3-102
3.62	P_{max} vs 1 MeV Electron Fluence	3-103
3.63	V_{mp} vs 1 MeV Electron Fluence.	3-104
3.64	I_{mp} vs 1 MeV Electron Fluence.	3-105
3.65	Normalized I_{sc} vs 1 MeV Electron Fluence	3-106
3.66	Normalized V_{oc} vs 1 MeV Electron Fluence	3-107
3.67	Normalized P_{max} vs 1 MeV Electron Fluence.	3-108
3.68	Normalized V_{mp} vs 1 MeV Electron Fluence	3-109
3.69	Normalized I_{mp} vs 1 MeV Electron Fluence	3-110

10 Ohm-cm n/p Silicon Cells with BSR

3.70	I_{sc} vs 1 MeV Electron Fluence.	3-111
3.71	V_{oc} vs 1 MeV Electron Fluence.	3-112
3.72	P_{max} vs 1 MeV Electron Fluence	3-113
3.73	V_{mp} vs 1 MeV Electron Fluence.	3-114
3.74	I_{mp} vs 1 MeV Electron Fluence.	3-115
3.75	Normalized I_{sc} vs 1 MeV Electron Fluence	3-116
3.76	Normalized V_{oc} vs 1 MeV Electron Fluence	3-117
3.77	Normalized P_{max} vs 1 MeV Electron Fluence.	3-118
3.78	Normalized V_{mp} vs 1 MeV Electron Fluence	3-119
3.79	Normalized I_{mp} vs 1 MeV Electron Fluence	3-120

10 Ohm-cm n/p Back Surface Field Thin Silicon Cells with BSR

3.80	I_{sc} vs 1 MeV Electron Fluence.	3-121
3.81	V_{oc} vs 1 MeV Electron Fluence.	3-122
3.82	P_{max} vs 1 MeV Electron Fluence	3-123
3.83	V_{mp} vs 1 MeV Electron Fluence.	3-124
3.84	I_{mp} vs 1 MeV Electron Fluence.	3-125
3.85	Normalized I_{sc} vs 1 MeV Electron Fluence	3-126
3.86	Normalized V_{oc} vs 1 MeV Electron Fluence	3-127

FIGURES (Continued).

Page

3.87	Normalized P_{max} vs 1 MeV Electron Fluence.	3-128
3.88	Normalized V_{mp} vs 1 MeV Electron Fluence	3-129
3.89	Normalized I_{mp} vs 1 MeV Electron Fluence	3-130

<u>10 Ohm-cm n/p Back Surface Field Silicon Cells with BSR</u>

3.90	I_{sc} vs 1 MeV Electron Fluence.	3-131
3.91	V_{oc} vs 1 MeV Electron Fluence.	3-132
3.92	P_{max} vs 1 MeV Electron Fluence	3-133
3.93	V_{mp} vs 1 MeV Electron Fluence.	3-134
3.94	I_{mp} vs 1 MeV Electron Fluence.	3-135
3.95	Normalized I_{sc} vs 1 MeV Electron Fluence	3-136
3.96	Normalized V_{oc} vs 1 MeV Electron Fluence	3-137
3.97	Normalized P_{max} vs 1 MeV Electron Fluence.	3-138
3.98	Normalized V_{mp} vs 1 MeV Electron Fluence	3-139
3.99	Normalized I_{mp} vs 1 MeV Electron Fluence	3-140

<u>10 Ohm-cm n/p Textured, 2 Ohm-cm Vertical Junction, and p/n AlGaAs Cells</u>

3.100	I_{sc} vs 1 MeV Electron Fluence.	3-141
3.101	V_{oc} vs 1 MeV Electron Fluence.	3-142
3.102	V_{oc} (AlGaAs) vs 1 MeV Electron Fluence	3-143
3.103	P_{max} vs 1 MeV Electron Fluence	3-144
3.104	V_{mp} vs 1 MeV Electron Fluence.	3-145
3.105	V_{mp} (AlGaAs) vs 1 MeV Electron Fluence	3-146
3.106	I_{mp} vs 1 MeV Electron Fluence.	3-147
3.107	Normalized I_{sc} vs 1 MeV Electron Fluence	3-148
3.108	Normalized V_{oc} vs 1 MeV Electron Fluence	3-149
3.109	Normalized P_{max} vs 1 MeV Electron Fluence.	3-150
3.110	Normalized V_{mp} vs 1 MeV Electron Fluence	3-151
3.111	Normalized I_{mp} vs 1 MeV Electron Fluence	3-152

FIGURES (Continued)

		Page
4.1	Relative Damage Coefficients for Space Electron Irradiation of Shielded n/p Silicon Solar Cells.	4-5
4.2	Absorbed Dose Per Unit Fluence of Space Electrons for Various Depths in Planar Fused Silica Shielding	4-8
4.3	Relative Damage Coefficients for Space Proton Irradiation of Shielded n/p Silicon Solar Cells (Based on I_{sc})	4-13
4.4	Relative Damage Coefficients for Space Proton Irradiation of Shielded n/p Silicon Solar Cells (Based on P_{max} or V_{oc}).	4-14
4.5	Absorbed Dose Per Unit Fluence of Space Proton Irradiation for Various Depths in Planar Fused Silica Shielding	4-17
4.6	Relative Damage Coefficients for Space Alpha Particle Irradiation of Shielded n/p Silicon Solar Cells (Based on P_{max} or V_{oc})	4-20
5.1	Regions of the Magnetosphere Shown in the Noon-Midnight Meridian Plane	5-3
5.2	Charged Particle Distribution in the Magnetosphere	5-5
5.3	Outer Zone Electron Fluxes	5-8
5.4	Superposition of Cycles 18, 19, and 20	5-12
5.5	Observed and One-Year-Ahead Predicted Smoothed Sunspot Numbers.	5-13
5.6	Changes of the Interplanetary Magnetic Field Regime Model with Time	5-15
5.7	Solar Flare Proton Environment of Solar Cycle 20	5-18
5.8	Predicted Smoothed Sunspot Number for Solar Cycle 21	5-19
5.9	Solar Flare Proton Environment in a 200 nmi Circular Orbit Due to a Class Three Flare Event on July 18, 1961	5-22
7.1	Performance of Two Satellite Solar Arrays in Synchronous Orbit During the August 1972 Solar Flares	7-7
7.2	LES-6 Flight Data for 10 Ohm-cm Cell No. A3 with 6 mil 7940 Coverglass	7-10
7.3	Degradation of Solar Cell Maximum Power vs. Time in Synchronous Orbit, ATS-5 Experimental Cells.	7-12

FIGURES (Continued)

		Page
B-1	Refractive Index of Silicon	B-2
B-2	Extinction Coefficient of Silicon	B-2
B-3	Absorption Coefficient of Single Crystal Silicon at at 77 and 300 K	B-3
B-4	Resistivity of Silicon at 300 K as a Function of Acceptor or Donor Concentration	B-4

TABLES

		Page
1.1	Antireflection Coatings	1-28
2.1	Solar Spectral Irradiance - Proposed Standard Curve	2-2
3.1	Silicon Displacement Parameters, Various Electron Energies	3-9
3.2	Radiation Effects on Shielding Materials	3-53
3.3	Test Cell Descriptions	3-60
4.1	Electron Damage Coefficients	4-6
4.2	Electron Stopping Power, Rad(SiO_2)/Unit Omnidirectional Flux	4-9
4.3	Proton Damage Coefficients for I_{sc}	4-15
4.4	Proton Damage Coefficients for V_{oc} and P_{max}	4-16
4.5	Proton Stopping Power, Rad(SiO_2)/Unit Omnidirectional Flux	4-18
5.1	Observed and Predicted Unattenuated Interplanetary Annual Integral Solar Proton Fluence	5-20
6.1	Summary of Equivalent Fluence Contributions	6-4
6.2	Manual Calculation of Equivalent Fluence (AP8MAX Protons) V_{oc} and P_{max} Circular Orbit 450 nmi (833 km), 90 Degree Inclination	6-15
6.3	Manual Calculation of Equivalent Fluence (AP8MAX Protons) I_{sc} Circular Orbit 450 nmi (833 km), 90 Degree Inclination	6-16
6.4	Manual Calculation of Equivalent Fluence (AE6MAX Electrons) Circular Orbit 450 nmi (833 km), 90 Degree Inclination	6-17
6.5	Summary of Data in Tables 6.6 through 6.39	6-18

ANNUAL EQUIVALENT 1 MeV ELECTRON FLUENCE FROM TRAPPED ELECTRONS (INFINITE BACKSHIELDING)

0° Inclination

6.6	Due to Trapped Electrons	6-19
6.7	Due to Trapped Protons (V_{oc}, P_{max})	6-20
6.8	Due to Trapped Protons (I_{sc})	6-21

TABLES (continued)

Page

10° Inclination

6.9	Due to Trapped Electrons................	6-22
6.10	Due to Trapped Protons (V_{OC}, P_{max})...........	6-23
6.11	Due to Trapped Protons (I_{SC})..............	6-24

20° Inclination

6.12	Due to Trapped Electrons................	6-25
6.13	Due to Trapped Protons (V_{OC}, P_{max})...........	6-26
6.14	Due to Trapped Protons (I_{SC})..............	6-27

30° Inclination

6.15	Due to Trapped Electrons................	6-28
6.16	Due to Trapped Protons (V_{OC}, P_{max})...........	6-29
6.17	Due to Trapped Protons (I_{SC})..............	6-30

40° Inclination

6.18	Due to Trapped Electrons................	6-31
6.19	Due to Trapped Protons (V_{OC}, P_{max})...........	6-32
6.20	Due to Trapped Protons (I_{SC})..............	6-33

50° Inclination

6.21	Due to Trapped Electrons................	6-34
6.22	Due to Trapped Protons (V_{OC}, P_{max})...........	6-35
6.23	Due to Trapped Protons (I_{SC})..............	6-36

60° Inclination

6.24	Due to Trapped Electron	6-37
6.25	Due to Trapped Proton (V_{OC}, P_{max})	6-38
6.26	Due to Trapped Proton (I_{SC})	6-39

TABLES (Continued)

Page

70° Inclination

6.27	Due to Trapped Electrons.	6-40
6.28	Due to Trapped Protons (V_{oc}, P_{max}).	6-41
6.29	Due to Trapped Protons (I_{sc}).	6-42

80° Inclination

6.30	Due to Trapped Electrons	6-43
6.31	Due to Trapped Protons (V_{oc}, P_{max})	6-44
6.32	Due to Trapped Protons (I_{sc})	6-45

90° Inclination

6.33	Due to Trapped Electrons	6-46
6.34	Due to Trapped Protons (V_{oc}, P_{max}).	6-47
6.35	Due to Trapped Protons (I_{sc}).	6-48

ANNUAL EQUIVALENT 1 MeV ELECTRON FLUENCE FROM TRAPPED ELECTRONS AT SYNCHRONOUS ALTITUDE VS LONGITUDE (INFINITE BACKSHIELDING)

0° Inclination

6.36	Due to Trapped Electrons	6-49
6.37	Due to Trapped Protons (V_{oc}, P_{max})	6-50
6.38	Due to Trapped Protons (I_{sc})	6-51
6.39	Equivalent 1 MeV Electron Fluence for Solar Flare Protons Based on Fluences in Table 5.1	6-52
7.1	Synchronous Orbit Solar Cell Array Degradation	7-4
7.2	Percentage Degradation of Predicted and Observed ATS-6 Solar Cell Experiment Output	7-14
7.3	Solar Cell Array Degradation, Various Circular Orbits.	7-16
A.1	Shielding Thickness Conversion	A-1
B.2	Solar Cell Parameters for Si, 300K	B-7
C.1	Solar Cell Types	C-1
C.2	Approximate Performance Increases Due to Processing Variable Change.	C-2

PREFACE TO THE THIRD EDITION

The purpose of this second revision of the handbook is to update the data published in the first edition in 1973 and the first revision in 1977. Since 1977, new techniques have been found to improve solar cell performance. These are discussed and the data from the new solar cells added to the collection. Also, considerable work has taken place in updating the models which describe the electrons and protons trapped in the Van Allen belts. Although the models used to recompute the Tables in Chapter 6 are the latest available, we understand that new models to describe the trapped electrons will soon be available which will replace AE5 and AE17.

In Chapter 1 we have added an introductory section which gives the reader a gentler boost into the area of semiconductor and solar cell theory. We have updated the discussions of new solar cell technology and added sections to describe the vertical junction and back surface reflector cells. The concepts of band gap narrowing and Auger recombination are introduced since they may be limiting factors in realizing ultimate solar cell efficiency as historically calculated. Detailed treatments are beyond the scope of this book, but the references can give the reader a start in the right direction should he want to pursue these matters. Chapter 2 has received only minor attention, with the main changes being replacement of solar simulator irradiance data with up-to-date data, deletion of the statistics section, and a discussion of how to measure diffusion lengths under high injection level. The solar cell radiation curves in Chapter 3 have been totally supplanted with newer data on the latest solar cells available. This chapter was also reordered for self consistency, and the proton displacement theory was expanded slightly. In Chapter 4, we have totally rewritten the equations in order to clarify them. The rewritten equations were used to recompute the damage coefficients tabulated in Chapter 4. The results were so close to the results published in the 1977 edition, that the Tables were left unchanged. Chapter 5 has been moderately changed to reflect the existence of new models for the Earth's trapped radiation environment, including the decay of the Starfish electrons. The Tables of Chapter 6 have all been recomputed, now using finer inclination increments of 10° and the newer radiation models. Additional calculations were made and presented in Chapter 6 which show how the radiation can become more severe at low-altitude orbits during periods of solar minimum. These Tables differ significantly from the earlier tabulations, and this is entirely due to the use of the new models for geomagnetically trapped radiation. Chapter 7, which compares flight data to predicted solar cell degradation, has been updated with as much flight data as we could find. We again observe that nearly all solar cell flight experiments seem to be plagued with misfortune, and almost no self-consistent data exists for other than synchronous orbits. All data examined which did not originate from flight experiments or flight solar panels having operational problems were found to be predictable by the equivalent fluence methods presented in this handbook.

Finally, we would like to acknowledge the immense aid received in producing this revision. Among the principal contributors are Bob Weiss and Tetsuo Miyahira who performed the electron irradiations and took thousands of I-V curves. Randy Webster contributed many programming hours in producing the Tables in Chapter 6. Mike Hurick and Neil Divine were the principal architects in revising Chapter 5. E. G. Stassinopoulos of Goddard Space Flight Center was very generous in contributing radiation models, computer programs, and a considerable amount of time in helping us learn to use them. Oldwig von Roos contributed his considerable theoretical skills in helping us rewrite Chapter 1. Pat Payne and Jim Albeck of Spectrolab helped us prepare our radiation matrix for the electron irradiations, then carefully supervised production and measurement of the solar cells to see that we received truly representative samples. Ken Ling and Peter Iles of ASEC performed the same service for us at their facility. Sue Hofmann deserves special commendation for her patience in typing the many revisions of the manuscript.

ABSTRACT

This handbook is intended to furnish the reader with the necessary tools to permit him to predict the degradation of solar cell electrical performance in any given space radiation environment. It begins with an introduction to solar cell theory, describing how cells are manufactured and how they are modeled mathematically. The interaction of energetic charged particle radiation with solar cells is discussed in detail and the concept of 1 MeV equivalent electron fluence is introduced. The space radiation environment is described and methods of calculating equivalent fluences for the space environment are developed. A computer program was written to perform the equivalent fluence calculations and a Fortran listing of the program is included. Finally, an extensive body of data detailing the degradation of solar cell electrical parameters as a function of 1 MeV electron fluence is presented.

CHAPTER 1

1.0 THEORY OF THE SILICON SOLAR CELL

The silicon solar cell, developed at Bell Telephone Laboratories in the early 1950's, is a semiconductor device capable of directly converting light energy to electricity. Some elementary concepts of how this device works and how particle radiation can act to decrease its efficiency will be introduced in this section and expanded on in the remainder of Chapter 1.

Consider a thin slice of silicon with the solar spectrum falling normally on one surface. As the light enters the silicon it will collide with the atomic electrons, losing a photon at each collision, and continue to lose its energy as it transfers energy to the electrons. Blue light of wavelength 0.35 microns will be 99% absorbed before it has traveled 0.2 microns into the slice. Light of 0.46 microns wavelength, at the peak of the solar spectrum, will travel 2.0 microns and light of wavelength 0.94 microns will travel 200 microns before losing 99% of its energy. The net effect of absorbing all wavelengths of the solar spectrum, neglecting reflections, is illustrated in Figure 1.1a. This Figure plots the number of electrons which have received energy from solar spectrum photons as a function of distance from the front silicon surface. These electrons have all been boosted up into the conduction band leaving a hole behind in the valence band. These pairs are commonly referred to as electron-hole pairs. If no other structure is introduced into this silicon slice, the electrons will simply recombine with the holes and there will be no net effect of this absorption process other than heating up the silicon.

The solar cell introduces an additional feature to separate and collect the electron-hole pairs before they recombine. This feature is the presence of an internal electric field which is produced in the cell by a p-n junction. This junction is typically produced by taking a wafer of either p-type or n-type silicon and diffusing the opposite type of dopant into the surface at high temperature. This junction is formed very near the front surface by diffusing n-type dopant into p-type silicon to make a n/p junction (or by diffusing p-type dopant into n-type silicon to make a p/n junction).

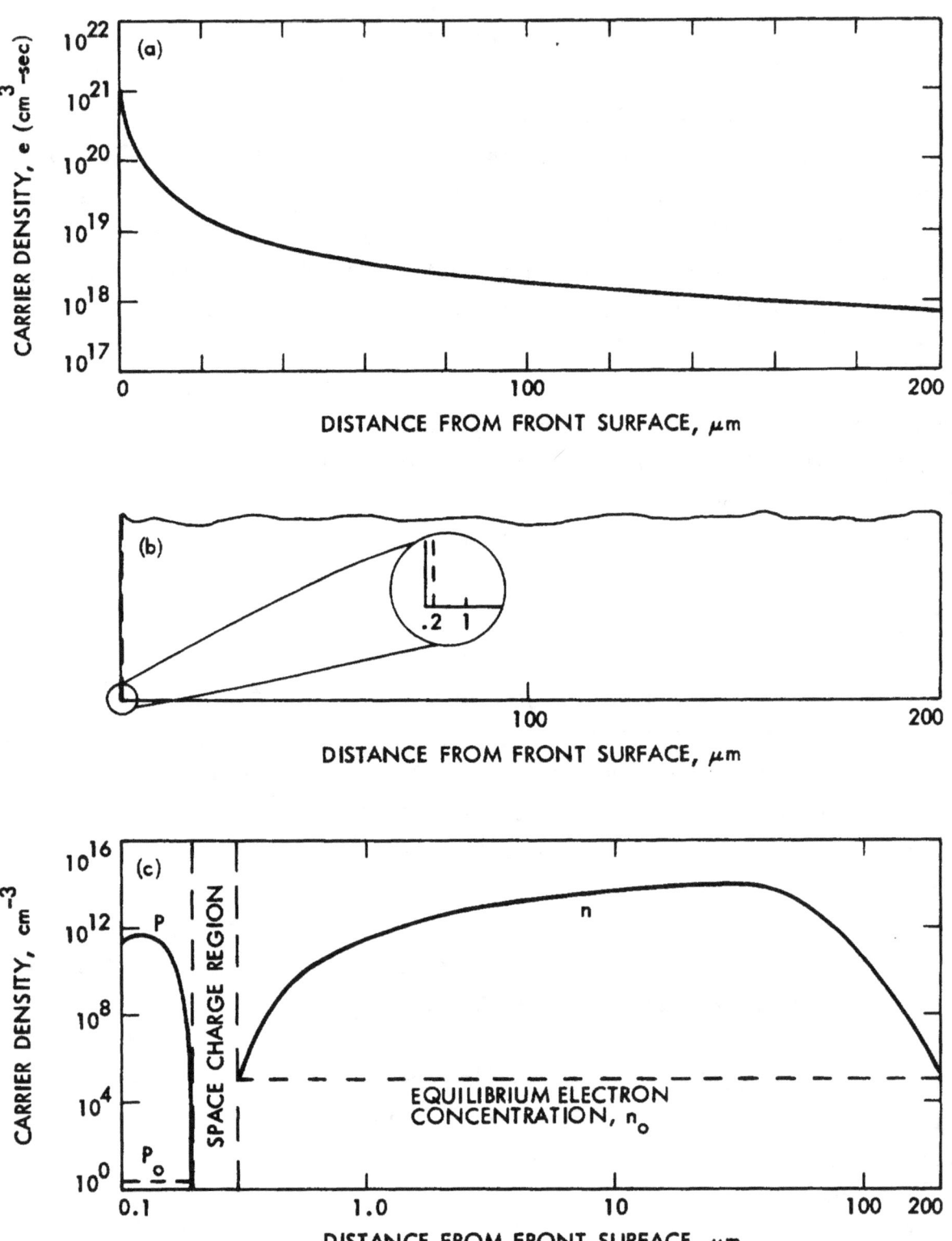

Figure 1.1 Carrier Density in Illuminated Silicon:
(a) Illuminated Silicon Slice, (b) Location of Junction, (c) Illuminated Silicon Solar Cell

The presence of such a junction in our 200 micron thick silicon slice is drawn to scale in Figure 1.1b. The internal electric field is formed because the excess electrons in the resulting n-region will move over into the existing p-type region and holes from the p-region will move to the n-region. The resulting charge separation produces a field which is strong enough to prevent further net charge movement. Electrical contacts added to the front and the back of our slice of silicon complete the formation of an elementary solar cell. For fabrication convenience, early solar cells were of the p/n type. Contemporary silicon solar cells, however, are almost exclusively of the n/p type due to their superior radiation resistance. For purposes of simplicity, therefore, only the n/p type of silicon solar cell will be discussed here although the p/n cell operates in a similar fashion.

The net effect of the junction is to produce an electric field with a polarity which accelerates electrons toward the front surface and holes toward the rear surface in an n/p solar cell. The field exists only in a very narrow region (approximately 0.1 to 0.5 microns) near the junction. The region where this field exists is called the space charge or depletion region. Any electron which now enters this region will be discharged into the n-region of the cell where it will be a majority carrier. This means that there are considerably fewer carriers of opposite sign with which it can recombine and it has a very high probability of reaching the front electrical contact. The same thing happens to the holes produced in the n-region which enter the space charge region except that they will be accelerated in the opposite direction and will be collected at the rear contact. Figure 1.1c, drawn with a logarithmic thickness scale, illustrates how the electron-hole carrier concentration has been markedly changed in this same illuminated slice of silicon because of the addition of the junction.

Effectively, then, every minority carrier which reaches the space charge region of the solar cell will reach the cell contacts and be delivered to the external electrical load. This action causes the minority carrier concentration at the edges of the space charge region to be nearly zero (more accurately it equals the equilibrium concentration). As shown in Figure 1.1c, the concentration rises very steeply just outside the space charge region. In fact, as we will see later, almost the entire electrical current produced by the cell is determined by the slope of the carrier concentration at the edge of the space charge region.

Since the space charge region is very thin, only a small fraction of the hole-electron pairs produced in the cell are produced in this region. Most of the hole-electron pairs are produced in the bulk material where they are not directly affected by the junction field. In the absence of an electric field, they move in a random-walk diffusion process until they either recombine with a majority carrier or move to the space charge region where they will be swept up.

Considering the p-region (where electrons are the minority carriers), the electrons which do not reach the junction will recombine after living for a time τ, called the lifetime. More precisely, after a time τ has passed, 1/e of the electrons will remain. During this time they are able to travel a distance L, called the diffusion length which is related to by $L = \sqrt{D\tau}$, where D is the diffusion constant for electrons in p-type material. Since the recombination of minority carriers decreases the useful output of the device, care is taken to minimize recombination (increase τ and L) and therefore increase the probability of their reaching the junction. Minority carrier diffusion lengths in today's solar cells are typically about 200 microns. Minority carriers produced deep in a solar cell of 12 mils (300 microns) thickness must live a time long enough to travel more than one diffusion length to reach the junction. Any impurities, disruption, or defects in the crystalline structure act to decrease this diffusion length so the cell will be unresponsive to the red portion of the solar spectrum.

The effect of electron, proton, neutron, or gamma ray irradiation is to produce defect sites in the solar cell which decrease the cell's diffusion length and therefore decrease the electrical output. After irradiation, the cell's response to red light decreases markedly. Radiation does not alter the blue response of the cell too much because the blue light produces its hole-electron pairs in regions of the cell very close to the junction. All blue-light-induced carriers are produced either in the space charge region or else so close to it that they have a high probability of reaching the junction. For this reason, any technique which enhances the blue response of a solar cell is very effective in making it into a radiation resistant cell. An example of a blue-enhancing technique is to make the junction shallower as was done in the violet cell. Minority carriers generated in the n-type region where lifetimes are short now have less distance to travel to reach the junction. Also since the space charge region is moved nearer the surface where the light-generated carrier density increases rapidly, more carriers will be produced directly in or near the space charge region. These carriers will continue to find their way to the junction no matter how disordered the bulk of the cell becomes when the cell is irradiated with high energy particles. Textured surface cells and vertical junction cells take advantage of structure modifications to the silicon surface to enhance production of hole-electron pairs near the junction. Such cells, built with shallow junctions and back surface reflectors, have been very effective as radiation resistant cells.

Figure 1.1c illustrates another aspect of solar cell performance which has received considerable attention in the past few years. The carrier concentration at the rear surface of the cell is seen to drop off rapidly. This is because minority carriers have a very short lifetime at the ohmic contact and electrons which diffuse to this surface are immediately lost by recombination and produce no useful power. This loss can be lessened if the carriers can be turned around and headed toward the junction. The addition of a thin layer just in front of the rear contact, doped with a p-type impurity to a much higher level than the bulk material (p+ layer) performs this function. The resulting electric field is of such a polarity that the electrons are repelled back toward the junction. Such p+ layers are commonly referred to as back surface fields. These fields are most effective for cells which are thinner than about a diffusion length. Consequently, at beginning-of-life, these fields are more effective for cells no thicker than approximately 0.02 to 0.025 cm (200 to 250 microns). After exposure to radiation has substantially decreased the diffusion length, the back surface field effectiveness will be greatly decreased, so these cells are not usually a good choice for missions which will encounter high radiation environments.

The detailed way in which the types of radiation found in outer space degrade the performance of solar cells is the subject of this book. Primarily, the effect of radiation is to decrease minority carrier diffusion length in the bulk or p-type material. Our goal here is to predict the performance of all types of solar cells in current use after they have been exposed to any spectrum of electron and proton irradiation. The rest of this chapter is devoted to a more detailed treatment of basic semiconductor and solar cell theory.

1.1 Semiconductor Theory

1.1.1 Thermal Equilibrium Relationships and Excess Densities

Semiconductors are a class of materials which have electrical properties and physical characteristics intermediate between metals and dielectrics. An important characteristic of semiconductor materials is bipolar conduction, where charge transport may occur by conduction band electrons or through empty energy states in the valence band which behave electrically like positively charged electrons and are referred to as holes. The equilibrium concentrations of conduction electrons and holes in silicon are determined from thermal considerations (law of mass action) by the following expression:

$$n_o p_o = n_i^2 = 3.62 \times 10^{31} \, T^3 \exp(-E_G/kT) \quad (1.1.1)$$

$$= 2.2 \times 10^{20} \text{ cm}^{-6}, \text{ for } T = 300 \text{ K}$$

where
- n_o = the equilibrium concentration of conduction electrons (cm^{-3})
- p_o = the equilibrium concentration of holes (cm^{-3})
- n_i = intrinsic carrier concentration (cm^{-3})
- E_G = bandgap energy (1.11 eV in Si at 300 K)
- T = temperature (K)
- k = Boltzmann constant (8.6171 × 10^{-5} eV/K)

For a highly purified semiconductor, the principal source of charge carriers is thermal excitation of electrons from the valence band to the conduction band, and the concentration of conduction electrons will equal the concentration of holes. This state, in which the electrical properties of a semiconductor are not modified by impurities, may be referred to as intrinsic. The electron and hole concentrations, n_i, in intrinsic silicon, for example, are equal to 1.5×10^{10} cm^{-3} at room temperature.

When elements from Columns III and V of the periodic table occur in substitutional solid solution in silicon, they can be thermally ionized. In the case of Column V elements, such as phosphorus or arsenic, the ionization results in an electron in the conduction band and a positively-charged donor impurity atom in the silicon lattice. Impurities from Column III, such as boron, undergo ionization in silicon by accepting a thermally ionized electron from the valence band. This process creates a hole in the valence band and a negatively-charged acceptor impurity ion. The activation energies for these donor and acceptor atoms in silicon are approximately 0.05 eV. For this reason, these equilibrium processes go to completion at temperatures near 300 K ($kT \cong 0.026$ eV), and the commonly-used Column III and V impurities in silicon can be considered to be completely ionized at room temperature.

If significant quantities of conduction electrons or holes are produced by the addition of impurities, as described above, the semiconductor may be classed as extrinsic. Extrinsic semiconductors are referred to as n-type (i.e., negative type) if the equilibrium concentration of conduction electrons exceeds the intrinsic carrier concentration. When the equilibrium concentration of holes exceeds the intrinsic carrier concentration of a semiconductor, it is referred to as a p-type (i.e., positive type). The product of the equilibrium conduction electron and hole concentrations in extrinsic semiconductors remains constant as described by equation (1.1.1). Thus, boron-doped, p-type, extrinsic silicon with a resistivity of 10 ohm-cm and a hole concentration of 1.4×10^{15} cm^{-3} must also have a conduction electron concentration of 1.6×10^{5} cm^{-3}. In this case, the holes are referred to a majority carriers and the conduction electrons as minority carriers.

The concept of Fermi level may also be used to describe several aspects of semiconductor theory. The Fermi level of a material is defined as that electron energy at which the probability of occupancy is equal to 1/2. The Fermi level is at the center of the forbidden band when silicon is intrinsic. In an n-type semiconductor, the Fermi level is above the center of the forbidden band. In a p-type semiconductor the Fermi level is below the center of the forbidden band.

Concentrations of conduction electrons and holes in excess of thermal equilibrium values can be introduced in a semiconductor by electrical processes, by the absorption of electromagnetic radiation, or in the process of stopping high energy particulate radiation. The total instantaneous concentration of carriers during an excitation process can be expressed as follows:

$$p(t) = p_0 + p'(t) \quad (1.1.2)$$
$$n(t) = n_0 + n'(t) \quad (1.1.3)$$

where $p'(t)$ and $n'(t)$ are the instantaneous excess hole and electron concentrations, which in the general case will be functions of time. The absorption of electromagnetic radiation in silicon, referred to as the optical injection of carriers, is fundamental to the operation of the solar cell. In the absorption process, an electron-hole pair is created for each photon of light absorbed. The densities of excess electrons and holes created in this manner obey the following equations:

$$\frac{dp(t)}{dt} = g_{ext} + g_{th} - r, \quad (1.1.4)$$

$$\frac{dn(t)}{dt} = g_{ext} + g_{th} - r, \quad (1.1.5)$$

where g_{ext} represents the excitation rate per unit volume due to an external cause, g_{th} is the thermal generation rate, and r is the total recombination rate. If the net rate of recombination, u, is defined,

$$u = r - g_{th}, \quad (1.1.6)$$

then for the case of holes, for example,

$$\frac{dp(t)}{dt} = g_{ext} - u \qquad (1.1.7)$$

It has been found for n-type semiconductors, for the case of small excess carrier densities (or at "low injection level", defined by $p'(t) << n_0$), that a good approximation for u is,

$$u = (p_n - p_{no})/\tau_p = p'(t)/\tau_p \qquad (1.1.8)$$

where $p'(t)$ is defined by equation (1.1.2), τ_p is the lifetime of a hole, and p_n is the concentration of holes in n-type material. The implication of this can be seen if the above expression for the time derivative of $p(t)$ is integrated, for the case of $g_{ext} = 0$, with the initial condition $p_n(0) = p_{no}$. The result is,

$$p'_n(t) = p_n(0) e^{-t/\tau_p}, \quad t > 0, \qquad (1.1.9)$$

and the lifetime is now seen to be the decay time constant governing the recombination of excess holes in n-material if the external source is removed at $t = 0$.

An explicit expression for the lifetime, τ_p, has been developed by Hall,[1.1] and Shockley and Read;[1.2] it is given by the expression, for holes,

$$\tau_p \propto (\sigma_p V_{th} N_t)^{-1}, \qquad (1.1.10)$$

where σ_p is the cross-section for capture of a hole by what Shockley and Read have termed a recombination center, V_{th} is the thermal velocity of an excess carrier and is about 10^7 cm/sec, and N_t is the density of the recombination centers. These centers, it has been determined, are responsible for the recombination of excess carriers, whether injected electrically or by electromagnetic, or particle radiation. The creation of additional centers of this type resulting from the high energy radiation in producing lattice displacements and vacancies severely shortens the carrier lifetime as will be discussed in more detail below.

1.1.2 Carrier Transport

Current flow or charge transport can occur by either of two mechanisms in semiconductors. The drift of charged carriers in an electric field is observed in semiconductors as well as metals. The drift current for the case of holes in a p-type semiconductor can be described as follows:

$$J_p = q\, p\, \mu_p\, E, \qquad (1.1.11)$$

where
- J_p = hole current density (amperes/cm^2)
- q = hole charge (coulomb)
- p = hole concentration (cm^{-3})
- μ_p = hole mobility (cm^2/volt sec)
- E = electric field (volts/cm)

The coefficients of the electric field (E) in the above expression are related to the resistivity of the material in the following manner:

$$\rho\,(\text{ohm-cm}) = \frac{1}{q\, p\, \mu_p} \qquad (1.1.12)$$

Similar expressions can be written to describe conduction electron flow and combined expressions can be used if minority carrier conduction is significant.

The second mechanism for charge transport in semiconductors is carrier diffusion. This process results from the random thermal movement of particles which exist in a concentration gradient. Such diffusion is analogous to flow of heat due to thermal gradients and the diffusion of atoms and molecules. When a gradient in the concentration of holes exists in a semiconductor, a hole flux will flow opposite to the gradient. The hole current density, for a one-dimensional geometry, is given by the following expression:

$$J_p = -q\, D_p\, \frac{dp}{dx}, \qquad (1.1.13)$$

where
- J_p = hole current density (A/cm^2)
- D_p = hole diffusion constant (cm^2/sec)
- $\dfrac{dp}{dx}$ = gradient of hole concentration.

When both mechanisms contribute to the hole flow, the following equation describes the current density:

$$J_p = q\left(p\,\mu_p\, E - D_p \frac{dp}{dx}\right). \tag{1.1.14}$$

A similar expression can be written for the conduction electron current density as follows:

$$J_n = q\left(p\,\mu_n\, E + D_n \frac{dn}{dx}\right). \tag{1.1.15}$$

The total current density is given by the sum of equations (1.1.14) and (1.1.15).

The basic equation governing the behavior of charge carriers in time and space is the time-dependent continuity equation. This equation sums the effects of the divergence of current, carrier recombination, and carrier generation. For the non-equilibrium steady state case, the total carrier concentrations (n and p) remain constant, and $\dfrac{dn}{dt}$ and $\dfrac{dp}{dt}$ equal zero. In this case the one-dimensional continuity equations for electrons and holes are as follows:

$$g_{ext} - \frac{n - n_o}{\tau_n} - \frac{1}{q}\frac{d}{dx}J_n = 0 \tag{1.1.16}$$

$$g_{ext} - \frac{p - p_o}{\tau_p} - \frac{1}{q}\frac{d}{dx}J_p = 0, \tag{1.1.17}$$

where g_{ext}, introduced earlier, is the rate of generation of carriers per unit volume. If current flow occurs only by diffusion, equation (1.1.13) can be substituted into equation (1.1.17), and a similar substitution can be made in equation (1.1.16), leading to the following equations:

$$D_n \frac{d^2n}{dx^2} - \frac{n - n_o}{\tau_n} = -g_{ext} \qquad (1.1.18)$$

$$D_p \frac{d^2p}{dx^2} - \frac{p - p_o}{\tau_p} = -g_{ext} \qquad (1.1.19)$$

1.2 The P-N Junction

The current-voltage characteristic of a p-n junction is one of the factors which determine solar cell response. In this section, the general factors which determine diode characteristics will be discussed. The carrier concentrations found in a solar cell diode are shown graphically in Figure 1.2. The base or p-type region of the device has a majority carrier density (p_{po}) of approximately 10^{15} cm^{-3}. Because the product of the carrier concentrations is roughly 10^{20} cm^{-6} (See equation (1.1.1)), the minority carrier concentration is 10^5 cm^{-3}. The surface or diffused layer has a majority carrier concentration approaching 10^{20} cm^{-3}. Equilibrium considerations therefore require the minority carrier concentration of this region to be approximately unity. The concentration of holes, electrons, donors and acceptors differ greatly on different sides of the junction. The mobile charges arrange themselves in such a way that the result is a net accumulation of positive charge on the n-side and a net negative charge on the p-side of the junction region. As a result, all of the mobile charge carriers (holes and conduction electrons) are swept out of the junction region. This region is also frequently referred to as a space charge or depletion region. In a typical solar cell, the width of the n-type diffused layer is roughly 0.2 μm. The width of the space charge region is very roughly 0.1 to 0.5 μm, varies with resistivity and bias, and extends primarily into the base region.

Equation (1.1.18) can be used to determine the behavior of excess carriers in the base region of a junction. In the case of steady-state illumination with uniform generation rate g_o,

$$D_n \frac{d^2n_p}{dx^2} - \frac{n_p - n_{po}}{\tau_n} = -g_o \, , \, x > 0. \qquad (1.2.1)$$

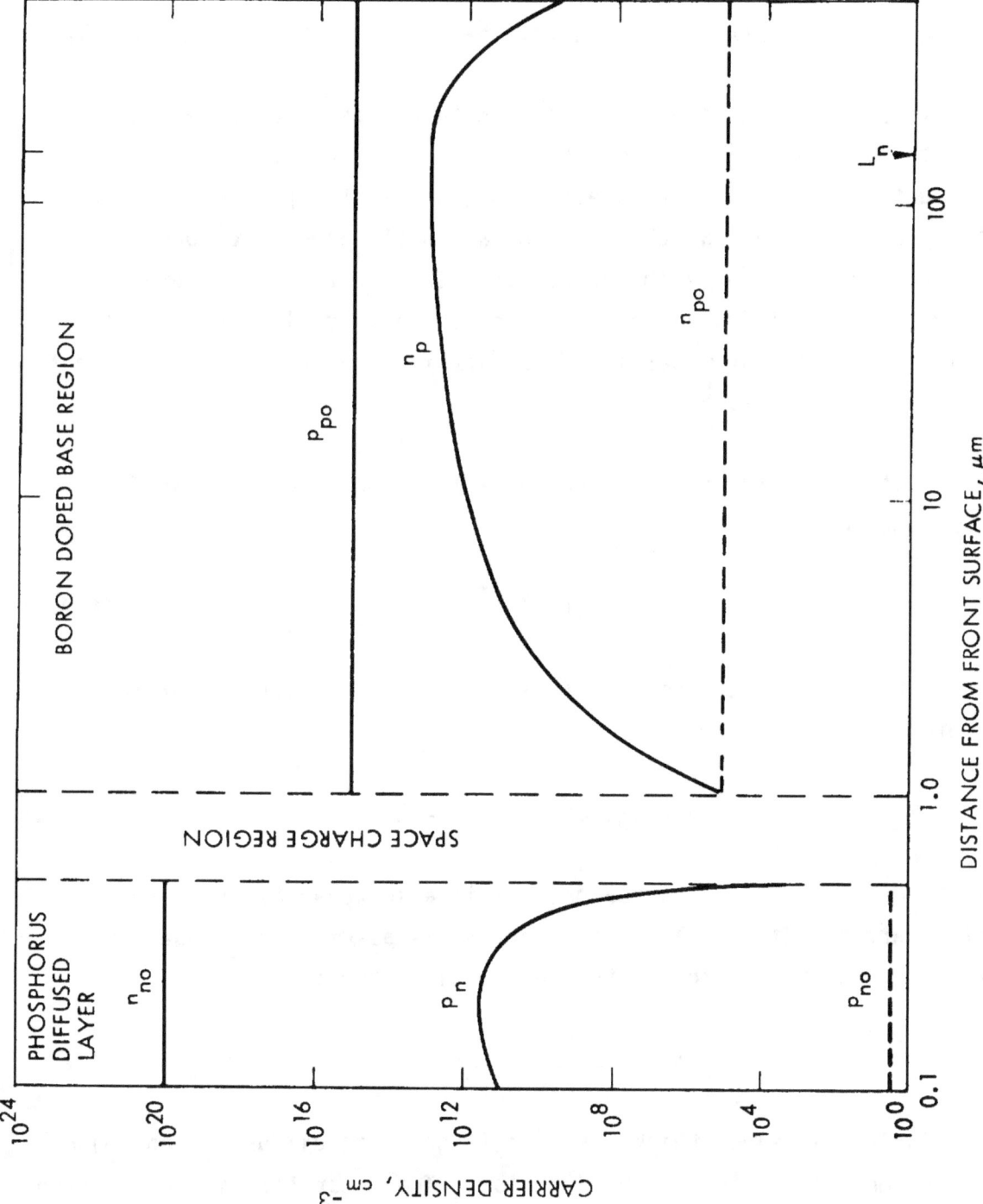

Figure 1.2 Carrier Concentrations in an Illuminated Solar Cell, Short Circuited

The solution of this equation for a semi-infinite semiconductor with the boundary condition, that at $x = 0$, $n_p = n_{po}$, is

$$n_p(x) = n_{po} + g_o \tau_n (1 - e^{-x/\sqrt{D_n \tau_n}}) \qquad (1.2.2)$$

The quantity $\sqrt{D\tau}$ has the dimensions of length and is often referred to as the diffusion length (L). The above result indicates that the steady state concentration of conduction electrons, n_p, in the p-type region will effectively approach zero at the junction and will increase exponentially with distance from the space charge region. This behavior is shown in Figure 1.2. Actual diffusion lengths found in solar cells can be as large as 200 microns. This parameter is of primary importance in the determination of the efficiency of a solar cell.

The equation for the dark current as a function of bias is as follows for a p-n junction.:

$$J = J_{01}(e^{qV/kT} - 1) \qquad (1.2.3)$$

In the case of a large forward bias ($V \gg 0$), $e^{qV/kT}$ is much larger than 1 and therefore,

$$J = J_{01} e^{qV/kT}, \quad V > 0 \qquad (1.2.4)$$

When $V \ll 0$, $J = -J_{01}$. For this reason, J_{01} is also known as the reverse saturation current. If the saturation current is assumed to be due to the diffusion of minority carriers into the junction, then:

$$J_{o1} = q \frac{D_n n_{po}}{L_n} + q \frac{D_p p_{no}}{L_p} \qquad (1.2.5)$$

valid for infinitely wide (thickness much larger than diffusion length) n- and p-layers on either side of the junction. The first term is the contribution from the p-side and the second term is from the n-side of the junction. For 1 to 10 ohm-cm solar cells the second term is 3 to 4 orders of magnitude smaller than the first term, so it can usually be ignored (see Appendix B for appropriate values for Si). Using equation (1.2.5), the calculated saturation current for a n/p, 10 ohm-cm solar cell would be roughly 10^{10} A/cm^2 at room temperature. The measured values of saturation currents found in such solar cells are considerably higher than the above value. The diffusion theory does not adequately explain the current voltage characteristics of a silicon junction diode. [1.3]

Recently it has been realized that the major cause of the discrepancy between the magnitude of the saturation current density as given by equation (1.2.5) and the actual saturation current density lies with the heavily doped n-layer. The extremely heavy doping in this layer causes a bandgap narrowing [1.4] and enhanced Auger recombination. [1.5] Bandgap narrowing is a decrease in the bandgap energy, E_G, due to many body exchange effects among the free carriers. Auger processes are three body interactions of which the most important in the n-layer are probably electron-electron-hole and electron-electron-trap interactions. Since the rate of Auger recombination processes is proportional to the square of the majority carrier concentration, this mechanism is expected to become significant in heavily doped material and thus increase the reverse saturation current by decreasing the hole lifetime, τ_p, in the heavily doped n-layer. [1.5] Bandgap narrowing also increases J_{o1} as may be seen as follows. The equilibrium concentration of holes, p_{no}, of equation (1.2.5) may be written using equation (1.1.1) as

$$p_{no} = \frac{9.77 \times 10^{38}}{N_D} \exp(-E_G/kT) \text{ at } T = 300 \text{ K}$$

A bandgap narrowing of 0.1 eV is not uncommon [1.59] and will increase the hole concentration by a factor of $\exp(0.1/0.0259) = 47.8$ at 300 K. A more detailed analysis, taking into account degeneracy and the position dependence of all relevant quantities confirms this simple picture. [1.6]

While equation (1.2.3) describes a dark current that arises from diffusion of carriers into the space charge region from its neighboring n- and p-type layers, a second contributor to dark current is the space charge region itself. The theory of the diode current-voltage relationship for this mechanism involves carrier generation and recombination through defect centers located in the space charge region. The diode or rectifier equation predicted by this theory is as follows: 1.3

$$J = J_{02}(e^{qV/2kT} - 1) \qquad (1.2.6)$$

The only difference between equations (1.2.3) and (1.2.6) is the factor of 1/2 which appears in the exponent and the form of J_{02}. The expression for J_{02} is:

$$J_{02} = \frac{q W n_i}{\tau_0} \qquad (1.2.7)$$

where W = width of space charge region
        τ_0 = carrier lifetime in space charge region
        n_i = intrinsic carrier concentration ($\approx 1.5 \times 10^{10}$ cm^{-3})

Experimental studies have shown that both the generation-recombination model and the diffusion model are necessary to describe the diode current flow at all voltages. An expression summing the currents of both models can be used to describe the current flow at all voltages. 1.3

As a result of manufacturing variations, a solar cell junction is occasionally shunted by an ohmic resistance. When the value of this shunt resistance is less than 10^4 ohms, the shunt current will dominate the diode current at forward biases of slightly less than 0.2V. The symbol for shunt resistance is R_{sh}. As a result of resistive volume elements in current paths to the diode junction, the solar cell also has a finite resistance which appears in series with the diode. This series resistance (R_s) is usually less than one ohm and will dominate the current flow through the diode at large forward biases. A model summing both of the above elements is necessary to describe the forward voltage-current characteristics of a silicon solar cell in the most general case. Such a model is shown in Figure 1.3. In Figure 1.4, a generalized current-voltage characteristic is shown for a solar cell diode using the above model. Actual solar cells will have considerable variation in the shunt and series resistance.

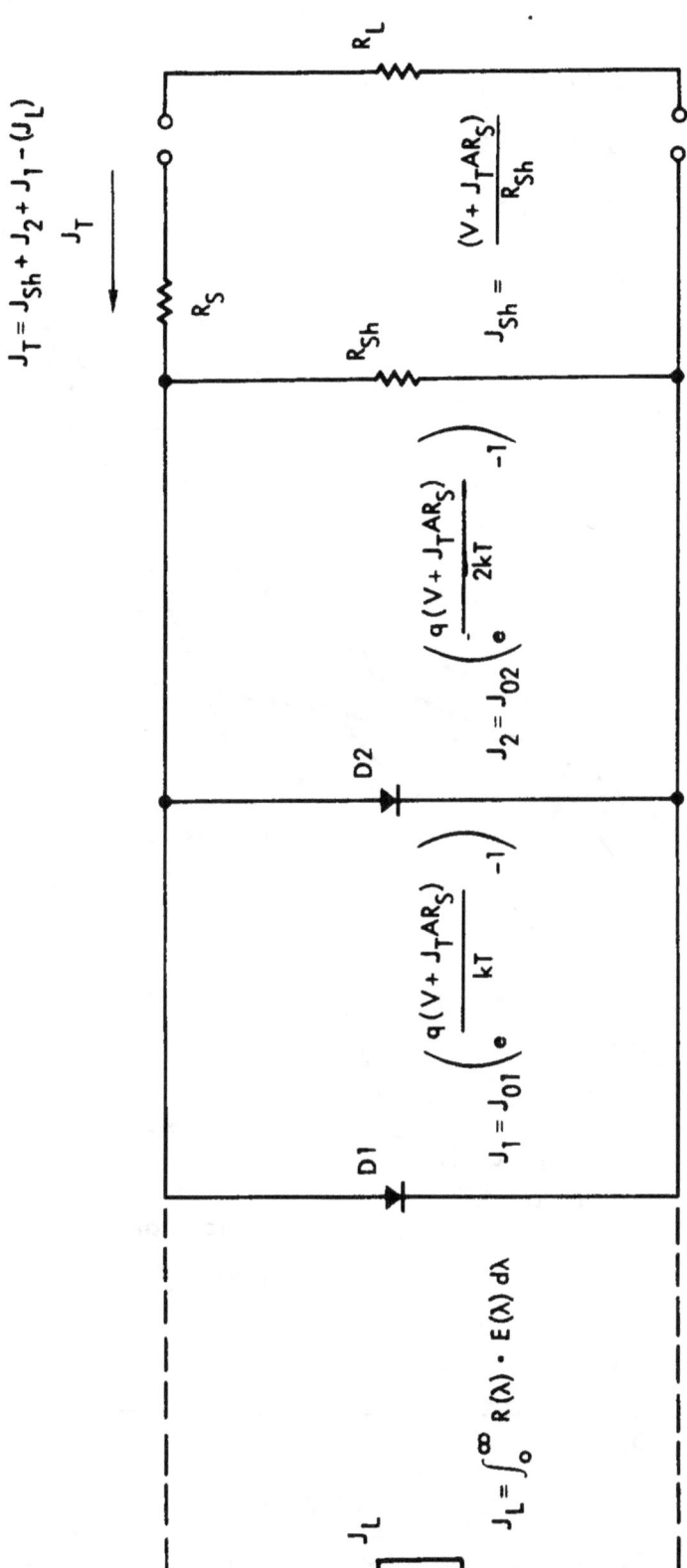

Figure 1.3 Solar Cell Equivalent Circuit Model

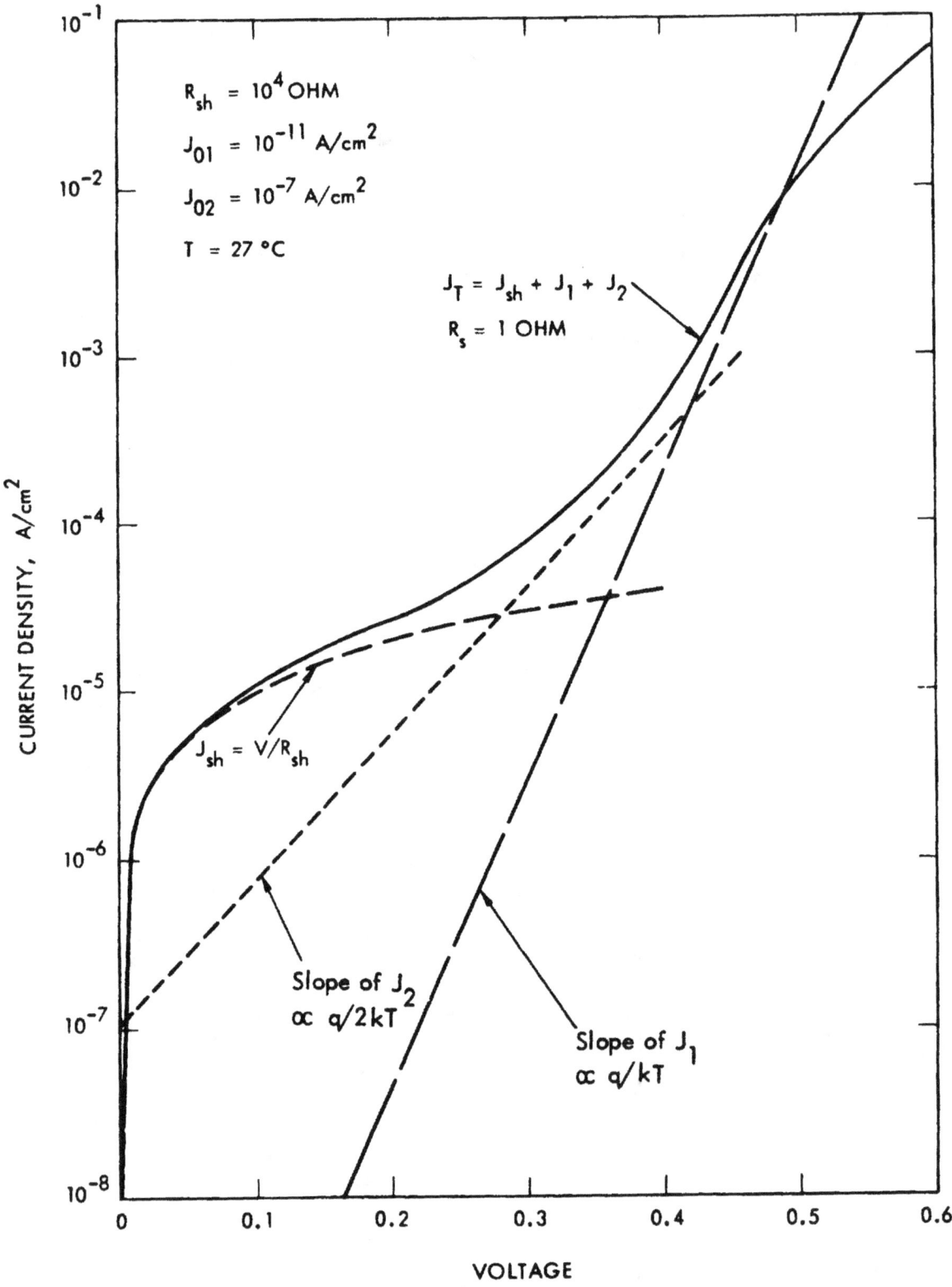

Figure 1.4 Typical Dark Solar Cell Current-Voltage Characteristic, Forward Biased

The junction space charge region of a solar cell has an associated capacitance. The capacitance of a conventional solar cell is related to the width of the space charge region in the following manner:

$$C = \frac{\epsilon A}{W} \qquad (1.2.8)$$

where C is the capacitance, A is area, $\epsilon = \kappa \epsilon_0$ is the permittivity of silicon, κ is the dielectric constant of silicon, and W is the width of the space charge region. The acceptor density in the p-type region adjacent to the space charge region can be related to the capacitance per unit area by:

$$N_a = \frac{2(V_b - V_a)C^2}{q \epsilon A^2} \qquad (1.2.9)$$

where V_a is the applied voltage, positive in forward bias, and V_b is the barrier voltage (0.6 to 0.8v depending on the cell base resistivity). The above expression assumes an abrupt or step junction which is typical of conventional solar cells. Since the acceptor or donor density is related to base resistivity, the resistivity of a solar cell may be computed by measuring cell capacitance as a function of bias.[1.7-1.9] Irvin's curves in Appendix B give the necessary N_a vs. ρ dependence.[1.10]

1.3 Silicon Solar Cell Theory

When a silicon p-n junction diode is exposed to ionizing radiation or light with a photon energy equal to or greater than the band gap of silicon, electron-hole pairs are produced in the silicon. Because of the gradient of conduction electrons (see Figure 1.2) which exists in the p-type region near the space charge region, the conduction electrons generated by the radiation diffuse to the junction. When these electrons reach the space charge region, they are accelerated by the electric field to the opposite side of the junction. A similar behavior occurs for holes generated in the n-type region of a solar cell. The diffusion flux of these generated carriers to the junction constitutes the solar cell light-generated current. Several investigators have developed general expressions for light-generated current.[1.11-1.17] These expressions are solutions of the continuity equations (1.1.18) and (1.1.19) for the case of optical carrier generation. The expression for electrons is as follows:

$$D_n \frac{d^2 n}{dx^2} - \frac{D_n(n - n_0)}{L_n^2} = \alpha N_0 (1 - R) e^{-\alpha x} \qquad (1.3.1)$$

where α = absorption coefficient for light of wavelength λ, (cm^{-1}) [1.18]
N_0 = photon flux density
R = reflection loss
x = distance from the junction
d = distance from front surface

This equation can be solved to find the minority carrier concentration gradient at the edge of the space charge region. The current density entering the space charge region can be calculated by evaluating equation (1.1.15) at the edge of the space charge region. Separate evaluations are made for the hole current from the surface layer and the electron current from the bulk region in response to monochromatic light as follows: [1.17]

Surface, Layer:

$$J_p(\lambda) = \frac{q N_0 (1 - R) \alpha L_p}{1 - \alpha^2 L_p^2} \frac{\left[\left(\alpha L_p + \frac{D_p}{S L_p}\right) \sinh \frac{a}{L_p} + \left(1 + \frac{\alpha D_p}{S}\right) \cosh \frac{a}{L_p}\right] e^{-\alpha a} - \left(1 + \frac{\alpha D_p}{S}\right)}{\sinh \frac{a}{L_p} + \frac{D_p}{S L_p} \cosh \frac{a}{L_p}} \qquad (1.3.2)$$

Bulk Response: (assuming $S = \infty$ at $d = b$)

$$J_n(\lambda) = \frac{p N_0 (1 - R) \alpha L_n}{1 - \alpha^2 L_p^2} \frac{\left(\sinh \frac{b-a}{L_n} - \alpha L \cosh \frac{b-a}{L_n}\right) e^{-\alpha a} + \alpha L_n e^{-\alpha b}}{\cosh \frac{b-a}{L_n}} \qquad (1.3.3)$$

where a = junction depth (cm)
b = cell thickness (cm)
S = surface recombination velocity (cm/sec).

Total Response:

$$J_L(\lambda) = J_n(\lambda) + J_p(\lambda) \qquad (1.3.4)$$

The above equations are written for the case of an n/p solar cell, assuming no significant drift fields are present. The cell response in A/cm^2 may be normalized to the photon flux density (N_o). In this way, the above equations describe the response of the cell in terms of amperes per photon/sec of incident light of a given wavelength. Solar cell spectral response curves are routinely measured. In these experimental measurements, the response is usually normalized to the incident optical power density (watts cm^{-2}) rather than photon density rate. The calculated response of a typical solar cell in such terms is shown in Figure 1.5.

The previous equations illustrated the role of the minority carrier diffusion length in development of the light-generated current of a solar cell. These response equations can be folded with the solar spectral irradiance and integrated to yield the light-generated solar-cell current under solar illumination (see Figure 1.5).

The light generated current can be combined with previously discussed diode rectifier equations to determine the current-voltage characteristics of an illuminated solar cell. The model for an illuminated solar cell is the same as that shown in Figure 1.3 for a dark diode, with the addition of a current source.[1.19] The current source (shown dotted in Figure 1.3) represents the light generated current. On the basis of the above model, an equation can be written to describe the cell current into an external load:

$$I = I_L - I_{D1} - I_{D2} - I_{sh} \qquad (1.3.5)$$

where
- I = cell current in external load
- I_L = light generated current
- I_{D1} = current in solar cell diode D1
- I_{D2} = current in solar cell diode D2
- I_{sh} = current in internal solar cell shunt (R_{sh})

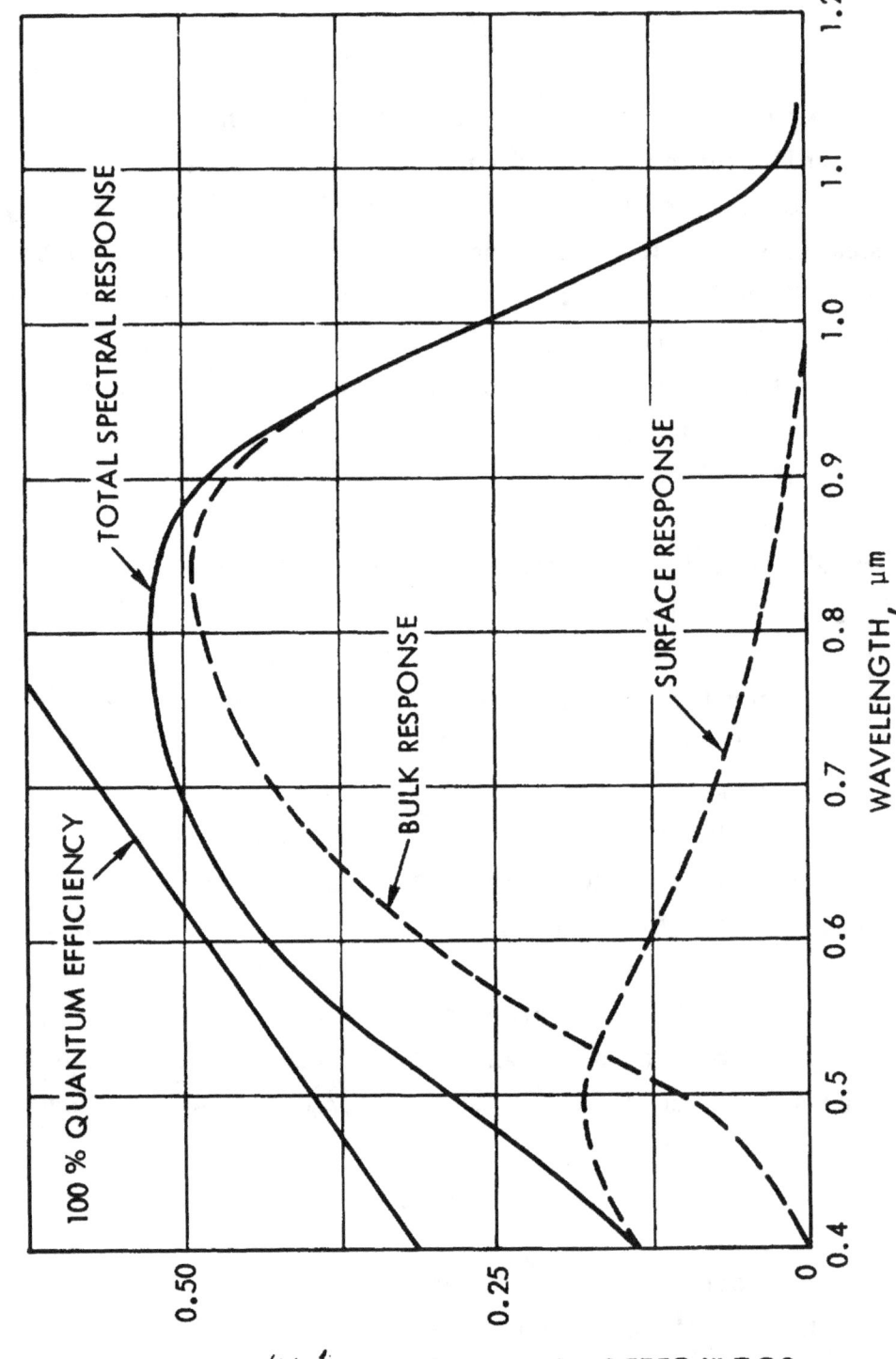

Figure 1.5 Calculated Silicon Solar Cell Spectral Response

Several observations can be made regarding the form of the above equation. The light generated current is independent of applied voltage and proportional to the intensity of the incident illumination. The development of the light generated current produces a forward bias on the solar cell diodes (D1 and D2). The light generated current (I_L) will divide between the parallel branches containing D1, D2, R_{sh} and $R_s + R_L$. The behavior of the illuminated solar cell current (I) and voltage (V) as R_L varies from zero to infinity is referred to as the I-V characteristic. This characteristic is the primary engineering tool used in evaluating solar cells. A general expression for the cell current to an external load can be obtained by substitution of equations (1.2.4.) and (1.2.6.) into equation (1.3.5). In the case of a good cell under 135 mW/cm^2 solar illumination, the current in R_{sh} can be neglected. It has been the practice to simplify the two diode currents with the following expression:

$$I_d = I_0 \left[e^{q(V+IR_s)/nkT} - 1 \right] \qquad (1.3.6)$$

where I_0 is an effective saturation current, and n is a constant, between 1 and 2. The resulting expression is often used to describe solar cell I-V characteristics:

$$I = I_L - I_0 \left[e^{q(V+IR_s)/nkT} - 1 \right] - \frac{V + I R_s}{R_{sh}} \qquad (1.3.7)$$

The development of a solar cell I-V characteristic from the light generated current and dark diode characteristic is shown graphically in Figure 1.6. An I_L value of 35 mA/cm^2 is typical of solar cells under solar illumination of 135 mW/cm^2. This I_L value is shown in Figure 1.6. In addition, the dark diode or rectifier characteristics shown in Figure 1.4 are replotted in Figure 1.6. The diode characteristics are shown with and without the series resistance. The illuminated solar cell I-V characteristic for a hypothetical cell with $R_s = 0$ is obtained by subtracting the forward current flowing in D1, D2, and R_{sh} from the light generated current I_L. When R_s is some significant quantity, the dark diode characteristic is displaced an amount ΔV before subtraction from I_L. The quantity ΔV is the voltage drop across R_s when the solar cell diode conducts a forward current equal to $+I_L$.

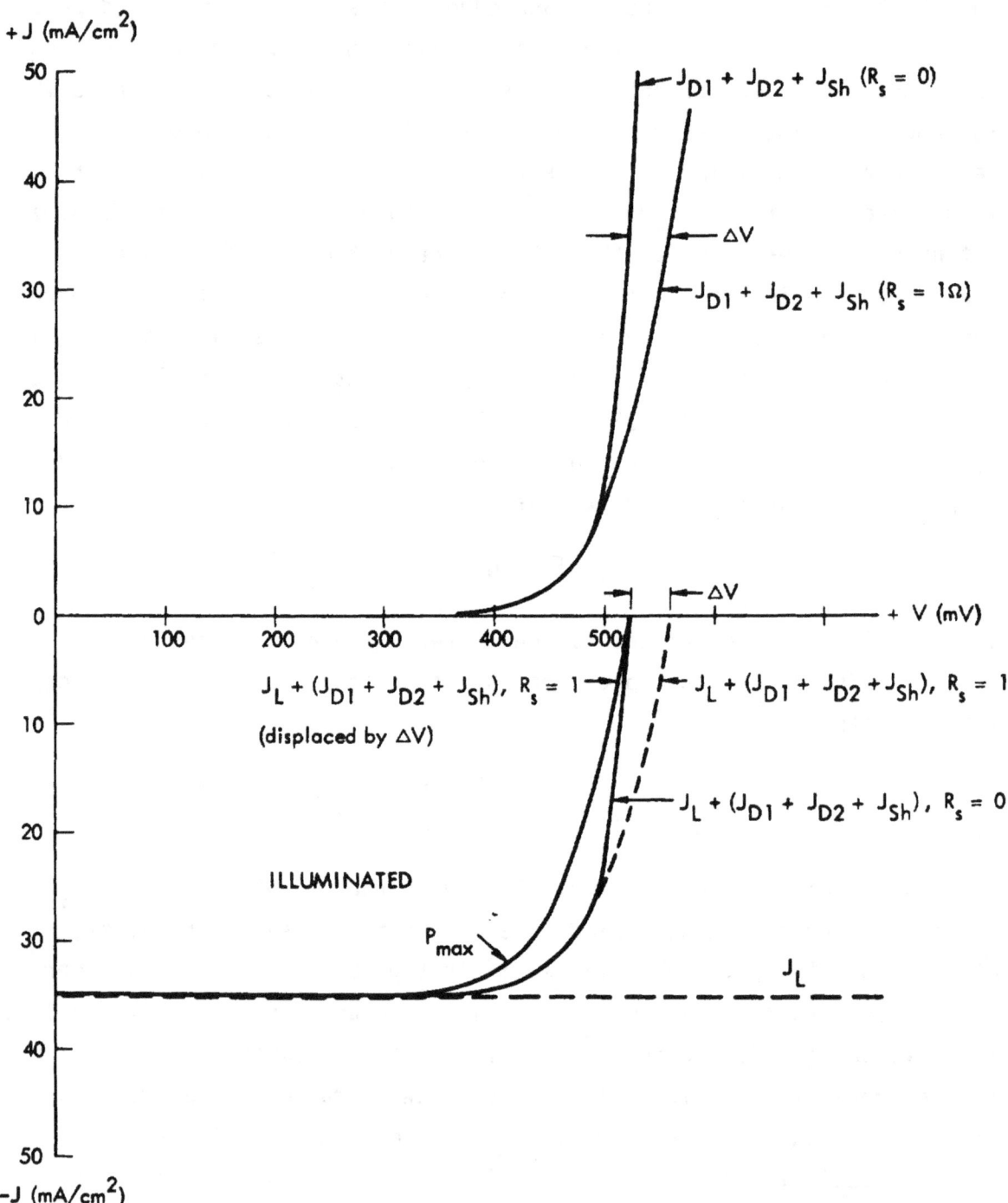

Figure 1.6 Development of a Current-Voltage Characteristic for an Illuminated Silicon Solar Cell

It should be understood that this analysis is for a solar cell at 27°C under solar illumination of 135 mW/cm^2. The quantity I_L, proportional to the light intensity, is a weak function of temperature and also a function of the spectral content of the illumination. The dark diode currents I_{D1} and I_{D2} are strong functions of temperature. Under the assumed conditions of temperature and illumination, I_{D1} (diffusion current) and R_s dominate the I-V characteristic of the solar cell. Under other conditions of temperature and illumination, the solar cell I-V characteristic may be influenced by other factors such as R_{sh} and I_{D2} (generation-recombination current).

A different set of parameters is used to describe the solar cell characteristic for engineering purposes. These are (a) short circuit current, I_{sc}, (b) open circuit voltage, V_{oc}, and (c) maximum power, P_{max}. The short circuit current is that current produced by the cell when the load resistance (R_L) approaches zero. In good solar cells, this quantity is equal to the light generated current I_L or J_L A. In cells with high or excessive internal series resistance, or in good cells at higher illumination intensities, I_{sc} will be less than the light generated current. The open circuit voltage is the voltage produced by the cell when R_L is infinite. In this load condition, all of the light generated current is consumed in forward conduction of diodes D1 and D2.

$$V_{oc} = \frac{nkT}{q} \ln\left(\frac{I_L}{I_o} + 1\right) \qquad (1.3.8)$$

A maximum in the power delivered to the load resistance occurs at some point of the solar cell I-V characteristic. The power developed under such a load is called the maximum power (P_{max}). A method of determining an analytical expression for the I-V characteristic from parameters such as I_{sc}, V_{oc}, P_{max}, and V_{mp} has been described in the literature.[1.20]

1.4 Solar Cell Coatings and Contacts

A silicon solar cell is a composite of several layers of material. The layers of n- and p-type silicon form the basic cell structure in which the current is generated. Additional practical problems are involved in maximizing the light entering the silicon and providing a low resistance path for collection of the generated current from the solar cell. When the light passes from one medium to another medium which has a different index of refraction, some light is reflected. The amount of light reflected can be determined from the following relationship when the second medium is absorbing.[1.21]

$$R = \frac{(n_1 - n_2)^2 + k_2^2}{(n_1 + n_2)^2 + k_2^2} \quad (1.4.1)$$

where R = reflectivity (fraction of normal incidence light intensity reflected)
n_1 = index of refraction, medium 1
n_2 = index of refraction, medium 2
k_2 = extinction coefficient, medium 2

The extinction coefficient, k, is the imaginary part of the index of refraction, where $k = \frac{\alpha \lambda}{4\pi}$, α is the absorption coefficient and λ is the wavelength. The above relationship holds only for normal incidence light. The more general case of light incident at an arbitrary angle θ from the normal is determined by Fresnel's equation.[1.22] Silicon has a high index of refraction (between 3.5 to 6.9 in the optical region).[1.23, 1.24] (See Appendix B). Reflection losses of incident light at an air-silicon interface are quite significant (about 30% in the long wavelength region, 71% at 0.275 µm, and 62% at 0.3 µm). The use of an antireflection (AR) coating, a surface layer with an intermediate index of refraction, will reduce the reflection loss.

The reflectivity in the presence of intermediate layers has an optimum effect at film thickness of one quarter wavelength, $(\lambda_0/4)$, where the thickness of a nonabsorbing coating d_1 satisfies $n_1 d_1 = (2j + 1)(\lambda_0/4)$ and j is an integer.[1.25, 1.26] The reflectivity is minimum when the index of refraction for the intermediate layer is

$$n_1^2 = n_0 n_2 \quad (1.4.2)$$

where n_0 = ambient index of refraction

Since n_0 is equal to 1 for air, the optimum index of refraction for an antireflection coating at an air-silicon interface is approximately 1.9. Silicon monoxide (SiO), with an index of refraction in the range of 1.8 to 1.9 was therefore most often used in the past as an AR coating on solar cells to minimize reflectance. The SiO has some absorption loss in the visible region.

Lower average reflectivity over a wider wavelength range can be obtained by using two AR coatings instead of one.[1.25, 1.27-1.28] In a practical space environment application, the solar cell is always covered by glass to shield against radiation and to raise the effective emissivity for better thermal control. This constitutes the double-layer system. It turns out that the reflection for a two-layer system has either a minimum or a local maximum for a quarter wavelength optical coating; i.e., $n_1 d_2 = n_2 d_2 = \lambda_0/4$.[1.25] The reflectance approaches zero if $(n_2/n_1)^2 = n_3/n_0$, where n_3 is now the index of refraction of silicon, and the average reflectance is lower over a broader wavelength range than for a single-layer coating. Thus, the coverglass and the adhesive must be considered as a part of an AR coating system. Since the adhesives have n values of approximately 1.4, the previous equation reveals that an AR coating with n = 2.2 - 2.4 would be optimum for a solar cell to be used with a coverglass.[1.27, 1.29-1.31] The addition of a MgF_2 coating (n = 1.38) to the coverglass is also used to give an even better match to air or vacuum.

Titanium oxide (TiO_x, n = 2.20) has both a higher refraction index and less absorption than silicon monoxide (n = 1.90), and is a better choice for this double-layer system.[1.32-1.36] Both of these materials, however, exhibit stronger absorption in the shorter wavelength region (0.4 μm), and are thus not suitable to a cell with high spectral response in this wavelength region. Tantalum pentoxide[1.37-1.40] (Ta_2O_5) has a high refractive index (n = 2.15 to 2.26) with less absorption in the shorter wavelength region than the above two, and is suitable for use on modern solar cells covered with quartz coverglasses.[1.38]

Recently, cells with two layer AR coatings (dual AR or MLAR coatings) have been put into routine production on solar cell manufacturing lines. On these cells, a layer of TiO_x followed by a layer of Al_2O_3 is deposited on the cells. These layers, in conjunction with the coverglass adhesive, coverglass, and MgF_2 form a system which gives a short circuit current increase of 4 to 5% over cells with only a single Ta_2O_5 coating.

Many properties of AR coatings vary greatly with the fabrication technique and conditions. The transparency, refractive index, and absorption are all related to the deposition rate, substrate temperature, oxygen pressure in the evaporation chamber, and film thickness, as well as defects formed during the processes. For example, Revesz reported that chemical vapor deposition of Ta_2O_5 AR film on the violet cell showed far better optical properties than sputtered Ta_2O_5 films. [1.37]

A summary of the AR coatings commonly used on solar cells and their indices of refraction is given in Table 1.1.

Table 1.1 Antireflection Coatings

Material	Index of Refraction
Silicon	3.8
Ta_2O_5	2.15-2.26
TiO_x	2.15-2.20
SiO	1.8-1.9
Al_2O_3	1.6
Coverglass	1.47
Adhesive	1.4
MgF_2	1.38

The contacts of current commercial solar cells are formed by evaporating titanium, palladium, and silver metal on the entire back surface and in a contact pattern on the front surface. The total thickness of this evaporated metallization is approximately 5 m. After the metallization, the cells are usually solder dipped. The solder thickness may vary between 10 and 80 m (0.4 - 3 mils). One of the primary considerations in the selection of the contact is the electrical behavior of the metal-semiconductor interface. In general, such interfaces should be ohmic with little or no contact resistance or Schottky barriers. A Schottky barrier has a current voltage characteristic of the same form as that for a p-n junction. The saturation current for a Schottky barrier is as follows: 1.41

$$I_0 = A T^2 \exp - (\phi_B/kT) \qquad (1.4.3)$$

where A = effective Richardson constant (A cm^{-2} K^{-2})
ϕ_B = effective barrier height (eV)

The quantity A is approximately 100 A cm^{-2} K^{-2} and ϕ_B is approximately 0.50 (eV) for most metals in contact with p-type silicon. The saturation current (I_0) at room temperature (T = 300 K) will be between 10^{-2} and 10^{-1} A cm^{-2}. The effect that the Schottky barrier has on the solar cell will be related to the forward resistance of the barrier. Since the form of the barrier current-voltage characteristic is:

$$I = I_0(e^{qV/kT} - 1) \qquad (1.4.4)$$

The dynamic impedance of the junction is as follows:

$$\frac{dV}{dI} = \frac{kT/q}{I_0} e^{-qV/kT} \qquad (1.4.5)$$

It can be seen that the impedance of this barrier is inversely proportional to the saturation current. Since the saturation current at room temperature is very high, the impedance of a Schottky barrier is very low. If the barrier potential (ϕ_B) for a particular metal on silicon is low enough, the barrier I-V characteristic will approach low resistance ohmic behavior. This is the case for a titanium layer on p-type silicon at room temperature. At low temperatures the saturation current of such a Schottky barrier is reduced and the diode characteristics become more significant. In this case, the Schottky barrier adds a nonlinear voltage drop to the solar cell model in series with R_s. This problem has received considerable attention in the literature. 1.42-1.47 The problem associated with non-ohmic contacts can be reduced by producing a heavily doped (p+) layer on the silicon interface. In such cases, the space charge region associated with the Schottky barrier is generally reduced. Quantum mechanical tunneling of the space charge region dominates the behavior of such thin barriers, and provides a highly conductive metal-semiconductor interface. The solar cell front contact is applied to a silicon interface which is very heavily doped with phosphorous. Above a dopant concentration of 3.2×10^{18} cm^{-3}, n-type silicon undergoes a metal-insulator transition.$^{1.33}$ Above 2×10^{19} cm^{-3} the impurity band has merged with the conduction band and silicon becomes indistinguishable from a low resistivity metal like Ag as far as its conductivity is concerned. Therefore metallization of the front surface provides ohmic contacts without formation of a Schottky barrier.

1.5 Improvement of Solar Cell Efficiency

For the improvement of solar cell efficiency, certain variables affecting the output must be considered: (a) physical properties inherently associated with materials such as band gap and absorption coefficient, (b) configuration geometry such as junction depth or contact patterns, and (c) physical parameters or properties such as impurity concentration which can be modified by manufacturing processes.

The choice of material is important in that the physical properties, such as absorption coefficient or energy gap, are suitable for efficient photovoltaic action. Junction depth (thickness of diffused layer) and cell thickness also affect the solar cell output as expressed in equations (1.3.2) and (1.3.3). Textured cells with tetrahedral surface structures belong to the second category. Bulk (base) resistivity and base material type (n or p type) can be manipulated by the amount and type of dopant. Fabrication technique and configuration at the front and back contacts change not only the series resistance but also the surface recombination velocity. For example, the back surface field (BSF) or p+ cell*, produced by introducing an impurity gradient near the back contact, reduce the recombination velocity at the rear surface to essentially zero. These variables, together with AR coatings, are examples of improvements that can be made with new materials, and belong to the third category.

---

* The p+ is a symbol to identify a much higher than normal concentration of p-type impurity in the base region, approximately $10^{18} \sim 10^{19}$ cm^{-3} as compared to the normal concentration of $10^{15} \sim 10^{16}$ cm^3.

These aspects are briefly discussed in the following sections with particular emphasis on five types of new technology cells, namely the violet cell, the BSF cell, the textured surface (or black) cell, the vertical junction cell, and the back surface reflector (BSR) cell.

1.5.1 Considerations for Photovoltaic Materials. [1.12, 1.48]

For a photovoltaic effect, the material has to absorb a photon or ionizing radiation energy to create excess carriers. Suitable materials for achieving a photovoltaic effect are therefore inherently limited to those with an energy gap slightly less than the energy of the photon radiation under consideration. The material thickness required for complete photon absorption is governed by the magnitude of the absorption coefficient and its change as a function of increasing photon energy. Those materials with a large $\alpha(\lambda)$ and a steep increase in absorption coefficient with respect to photon energy do not require a thick base material for complete absorption of sunlight, and hence are suitable for use in a thin-film cell; while those with a gradual absorption coefficient increase or low $\alpha(\lambda)$ require a greater thickness. Materials like silicon and gallium phosphide belong to the latter type while many group II-VI and III-V compounds such as gallium arsenide belong to the former.

For the charge separation mechanism, an electrostatic potential is created by a metal-semiconductor junction (Schottky barrier) or by a p-n junction. The latter falls into two types: one is the homojunction, made from a single semiconductor such as those in group IV compounds (Ge, Si, C) and group III-V compounds (GaAs, InP, AlSb, etc.): and another the heterojunction, consisting of two different and distinct semiconductors separated by the junction, such as CuS-CdS or group II-VI compounds (CdS, ZnS, CdSe, etc.). Theoretical maximum efficiency under the solar spectrum is plotted against the energy gaps for a few photovoltaic materials, in Figure 1.7. [1.48] Interestingly, the output of every material monotonically decreases with increasing temperature, but the rates are different (Figure 1.7). The maximum power of silicon is much less than that of GaAs at higher temperature. This is the reason why GaAs is expected to be a better solar cell material than Si in high temperature applications such as for solar concentrators. [1.49]

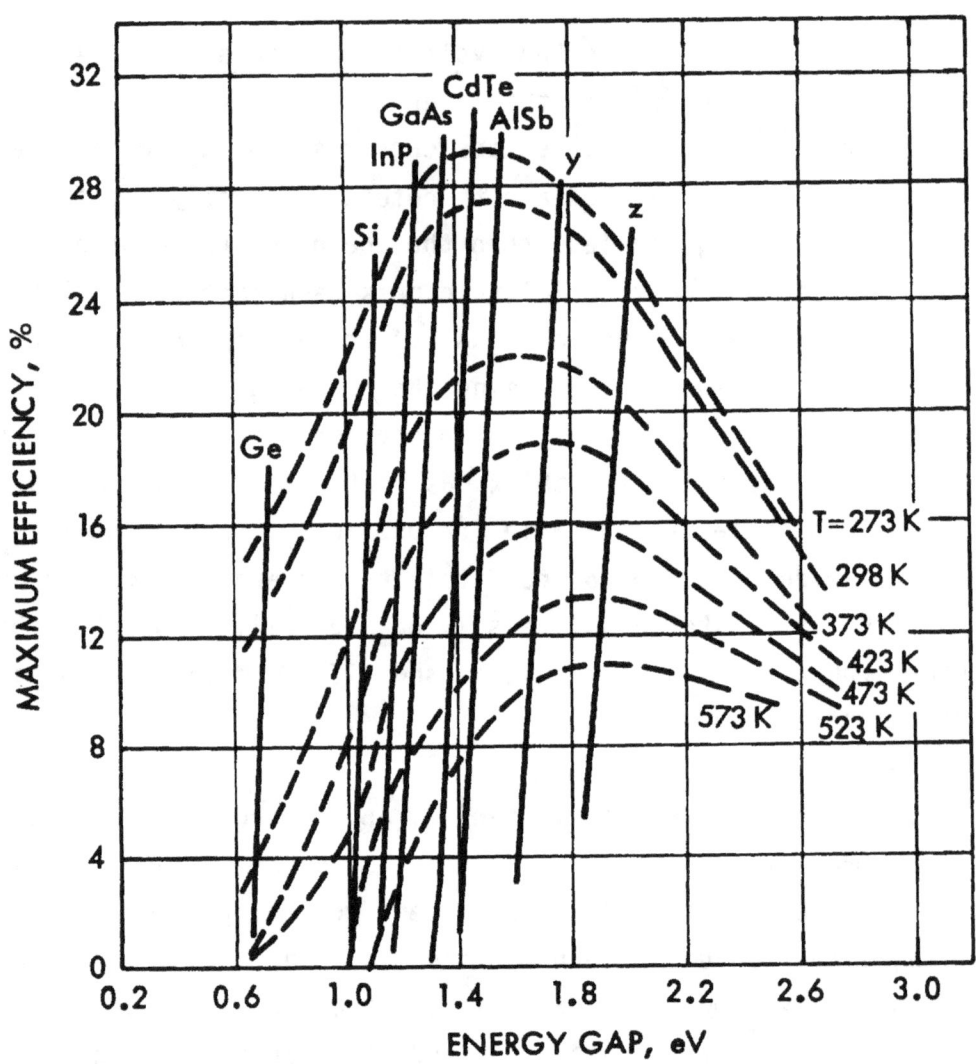

Figure 1.7 Temperature-Dependent Maximum Efficiency as a Function of Energy Gap for a Few Photovoltaic Materials [1.48]

1.5.2 Shallow Junction Solar Cells.

Since the solar irradiance is abundant in the blue and ultraviolet regions, an increased spectral response in these regions is essential for a higher efficiency. A blue-rich solar spectrum creates a heavier carrier concentration near the front surface of the cell than in the bulk. Therefore, if a junction is placed near or in the middle of this heavy carrier concentration by making the junction as shallow as possible, more carriers will be collected before recombination than in the cell with a deep junction. Since a shallow junction introduces a greater sheet resistance in the diffused layer, thereby increasing the potential drop there, an improved carrier collection mechanism is needed to increase overall efficiency.

The sheet resistance can be decreased by increasing the number of grid lines (decreased distance between grids). However, increasing the number of grid lines increases the shadowing effect on the cell. The grid lines can be made thinner to reduce shadowing, but this, in turn, increases the voltage drop in the grid. Hence, a careful optimization of all the parameters is required to ensure the best blue response.

Lindmayer, et al.,[1.36, 1.38] developed an improved diffusion technology which addressed the above problems of making practical shallow junction solar cells. In their cell, which they called the "violet cell", they diffused a junction which was less than 0.2 microns deep in contrast to the usual cells, which had junctions 0.3 to 0.5 microns deep. They were able to decrease the sheet resistance by using closely spaced grid lines (10 to 30 grids/cm), yet so narrow as to only cover 6 to 7% of the total surface area. Older cells used approximately 3 grids/cm and covered about 10% of the surface area. The series resistance of 2 x 2 cm cell was reduced to a value of about 0.05 ohm from the previously common values of 0.2 to 0.25 ohms.[1.50] Since the enhanced blue response of the cell would be useless if the old antireflection (AR) coatings which absorbed in the blue were used, tantalum pentoxide (Ta_2O_5) AR coatings were developed which absorbed less blue light and had a high index of refraction as well. (See Section 1.4).[1.37]

With the above improvements, the spectral response in the blue and ultraviolet regions is greatly enhanced, hence the name "violet" cell. This technique of producing shallow junction cells has been so successful it has now become the industry standard and the older "conventional" cells with deep junctions have become obsolete. The short circuit current for AM0 sunlight is about 40 mA/cm^2 as compared with about 35 mA/cm^2 for older cells. AM0 efficiencies of 14 to 14.5% are reported for 2 ohm-cm cells.[1.36]

1.5.3 Back Surface Field (BSF) or (P+) Solar Cells [1.51-1.60]

Back Surface Field cells are solar cells with a built-in electric field on the back surface just forward of the contact. The field polarity is such that minority carriers which would otherwise diffuse toward the back surface will be encouraged to reverse direction and diffuse toward the junction. BSF cells have been quite successful in improving the beginning of life performance of nearly all types of solar cells and in dramatically improving the output of very thin cells. Some properties of BSF cells may be summarized as follows:

> High open circuit voltage can be produced in cells made from high resistivity silicon. Open circuit voltages of 600 mV are common in 10 ohm-cm BSF cells, as compared to voltages of 550 mV attainable in non-BSF cells.

> The open circuit voltage of good BSF cells is independent of cell thickness. 10 ohm-cm cells as thin as 50 microns also have V_{oc}'s of 600 mV.

> Increases in short circuit current, fill factor, and maximum power are also achieved by the addition of the BSF. I_{sc} increases of approximately 10-15% are commonly observed as a direct result of BSF action. Increases in P_{max} range between 13 and 26% with the highest increases occurring for thinner cells.

> The large advantage of the back surface field disappears after a modest amount of irradiation. As can be seen from the curves in Chapter 3, the V_{oc} of a BSF cell degrades much faster than that of a non-BSF cell and the I_{sc} of a BSF cell degrades a little faster than the I_{sc} of a non-BSF cell. The net result is that the P_{max} degradation of a BSF cell with electron fluence is much worse than for non-BSF cells. These degradations are worse for thick cells than for thin cells.

The back surface field is produced in a silicon solar cell by forming an acceptor impurity gradient at the rear surface either by diffusing in an impurity such as boron or by alloying in an impurity such as aluminum. Mandelkorn and Lamneck, who reported making the first successful BSF cells in 1972,[1.51] used the aluminum alloying process to make their cells. The aluminum field cells are still the most commonly produced cells on today's production lines. The aluminum is applied to the rear surface either by evaporation or by screen printing a "paste" compound containing aluminum. The aluminum is then alloyed with the silicon by firing at a temperature of approximately 825 to 850°C.

The original theory of drift fields in cells was developed by Wolf[1.52] many years before BSF cells, which are a type of drift field cell, were reduced to practice. Godlewski, et al.,[1.53] proposed that the high V_{OC} obtained with the BSF cells was attributable to a decrease in reverse saturation current, I_0, which according to equation (1.3.8) will increase as the logarithm of (I_L/I_0). They investigated three possible causes for the decrease in I_0: (a) the reduction of surface recombination velocity at the rear contact using an otherwise conventional cell model, (b) the presence of a drift field, or (c) an abrupt change in the acceptor concentration (low-high junction). In this investigation, it was concluded that there was no clear choice as to which model is more appropriate for the explanation of the high voltage of the BSF cell. Studies by Brandhorst, et at.,[1.54] on BSF cells constructed by epitaxial deposition of 10 ohm-cm silicon layers onto substrates with various resistivities showed that the low-high junction model explains the variation in V_{OC} with substrate resistivity and the degradation of V_{OC} and I_{SC} with radiation fluence. These studies were in essential agreement with Mandelkorn et al.,[1.55] who made BSF cells by several different methods and reported that the V_{OC} increase was not caused simply by the mechanism of "blocking" of minority carriers at the rear surface.

Further investigation of the BSF cells by Mandelkorn and Lamneck,[1.56] in which p p+ cells were measured after removing the n+ surface layer from a n+ p p+ BSF cell, showed that a photovoltage is produced at the p p+ back junction of the cell. This phenomenon is in agreement with earlier work [1.57, 1.58] where it was found that when a sharp difference in doping concentration exists between one region and another (called a "low-high" junction by Gunn [1.58]), the difference in minority carrier concentration across the interface (or minority concentration barrier) results in a large potential difference.

Additional theoretical treatments by Fossum [1.59] and von Roos [1.60] basically corroborate the earlier work, but emphasize that both the reduction in I_0 and the barrier action at the rear surface are important. Fossum computes that in a 10 ohm-cm cell without a BSF, about 44% of the total recombination occurring in the device takes place at the back surface under short circuit conditions, while in a similar BSF cell only 2% of the total recombination takes place at the back surface. Von Roos also finds that the voltage applied across the cell is shared by the n-p and p-p+ junctions in such a manner that the dark current is diminishing while the short circuit current virtually stays the same. As a result, the voltage across the n-p junction is lowered, thus decreasing I_0 and increasing V_{oc} according to equation (1.3.8). In all these theories, the enhanced performance of the BSF cell is a function of the ratio L_n/W, where W is the thickness of the base material. When L_n is reduced to the point where it is approximately equal to W, for example by radiation, the enhanced performance due to the presence of the BSF begins to die rapidly. This is also why the BSF cell is most effective on thin cells which will maintain a high L_n/W ratio to higher values of radiation fluence.

1.5.4 Textured Surface Solar Cells [1.39, 1.40, 1.61, 1.62]

The textured surface solar cell is a cell with a configuration of very small densely packed pyramids etched into its front surface to act as light traps. This surface treatment was first successfully applied to solar cells in 1974 by Haynos et al. of Comsat Laboratories [1.39, 1.40] and has subsequently been put into practice on various solar cell production lines. [1.61] Preferential chemical etches by sodium or potassium hydroxide, or hydrazine hydrate applied to the surface of (100) orientation silicon will rapidly attack the (100) surface planes to selectively expose the <111> planes and produce the tetrahedral structure.[1.62] The pyramids have a base width and a height of from 1 to 15 microns. The included angle between opposite faces at the top of the pyramid is 70.5°. Figure 1.8a is a cross section through an idealized solar cell etched in this manner. Since incident light undergoes two reflections before complete escape, the optical reflection loss is reduced to the square of the normal reflection loss $(0.35)^2$ or about 12%. The addition of an AR coating will reduce the loss still further. These cells appear velvet black when viewed from the top because there is no reflection in this direction. The reduced reflection occurs at all wavelengths in contrast to that of planar cells where an antireflection coating can be made to minimize reflections only in selective wavelength bands. The Comsat group named this cell "Comsat Non-Reflecting" or CNR cell for short.[1.39, 1.40]

The addition of a textured surface to a solar cell produces about an 8% increase in both I_{sc} and P_{max}, with V_{oc} remaining about the same or perhaps decreasing slightly. The increased output is not only due to the decreased reflectance at the front surface, but also due to the fact that the light is refracted at the surface and enters the cell at an off-normal angle. Figure 1.8a shows that after first striking a pyramid, light enters the cell at an angle of 42° from the normal. The light which does not enter the cell is reflected and strikes a neighboring pyramid, entering the cell at an angle of 59° from the normal. The net result is that hole-electron pairs created by long wavelength light are produced much closer to the junction. This factor not only enhances the red response but also makes them quite radiation resistant. It also greatly enhances the output of very thin textured cells. The spectral response curves of Figure 1.9 illustrate the enhanced output of textured cells over the entire wavelength region.

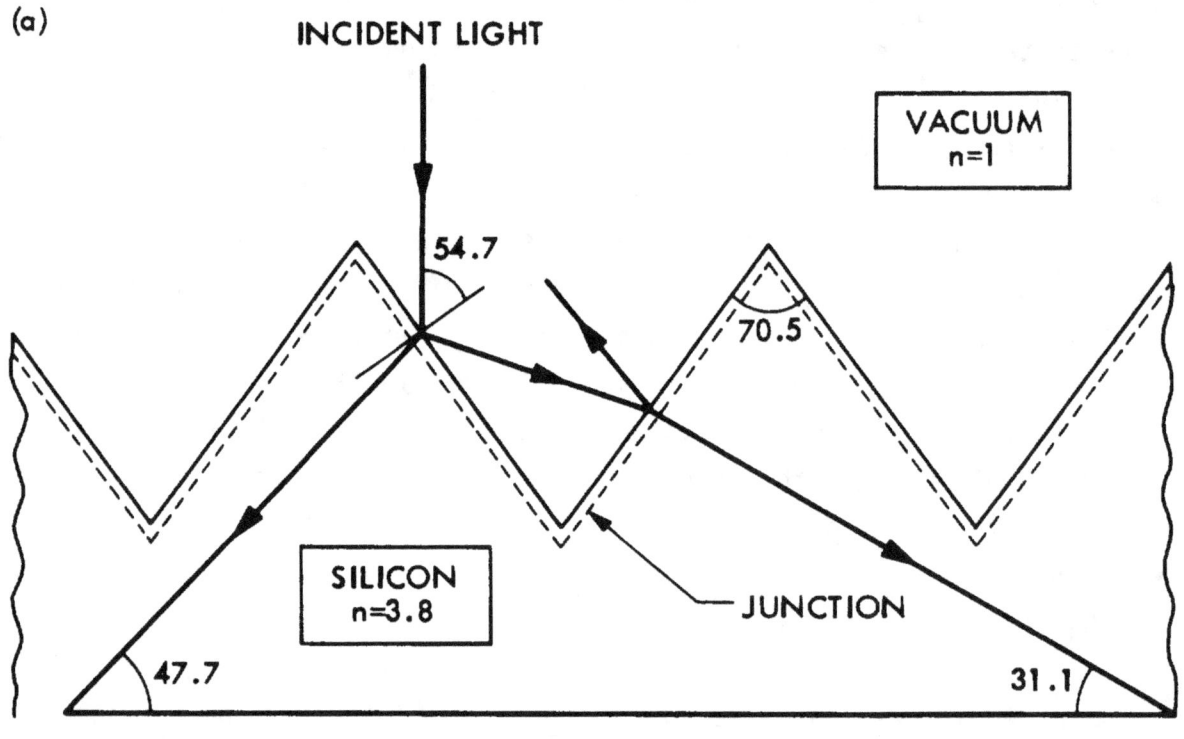
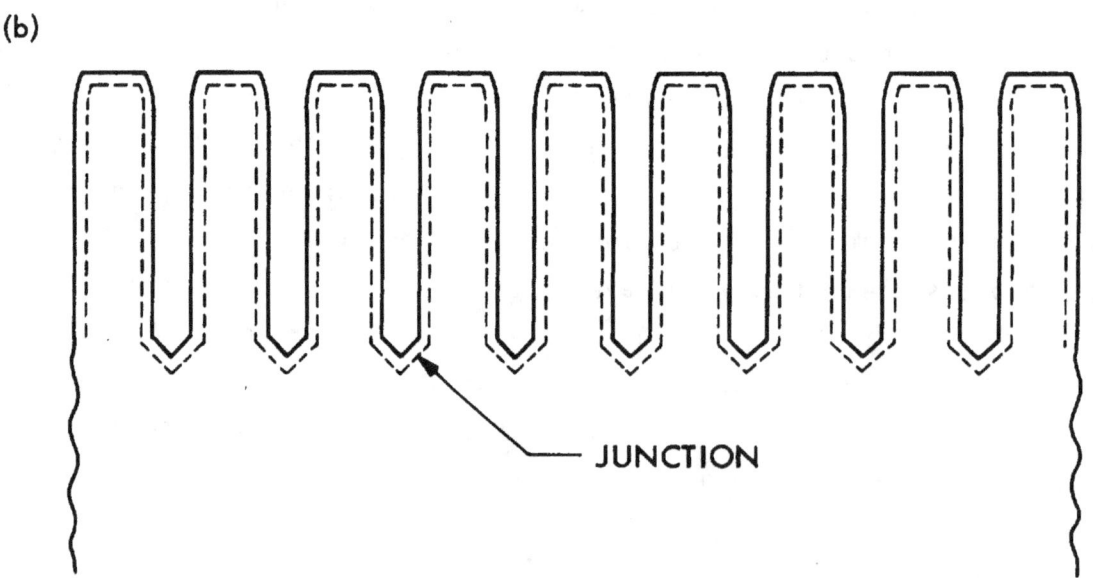

Figure 1.8 (a) Textured Surface Solar Cell,
(b) Vertical Junction Solar Cell

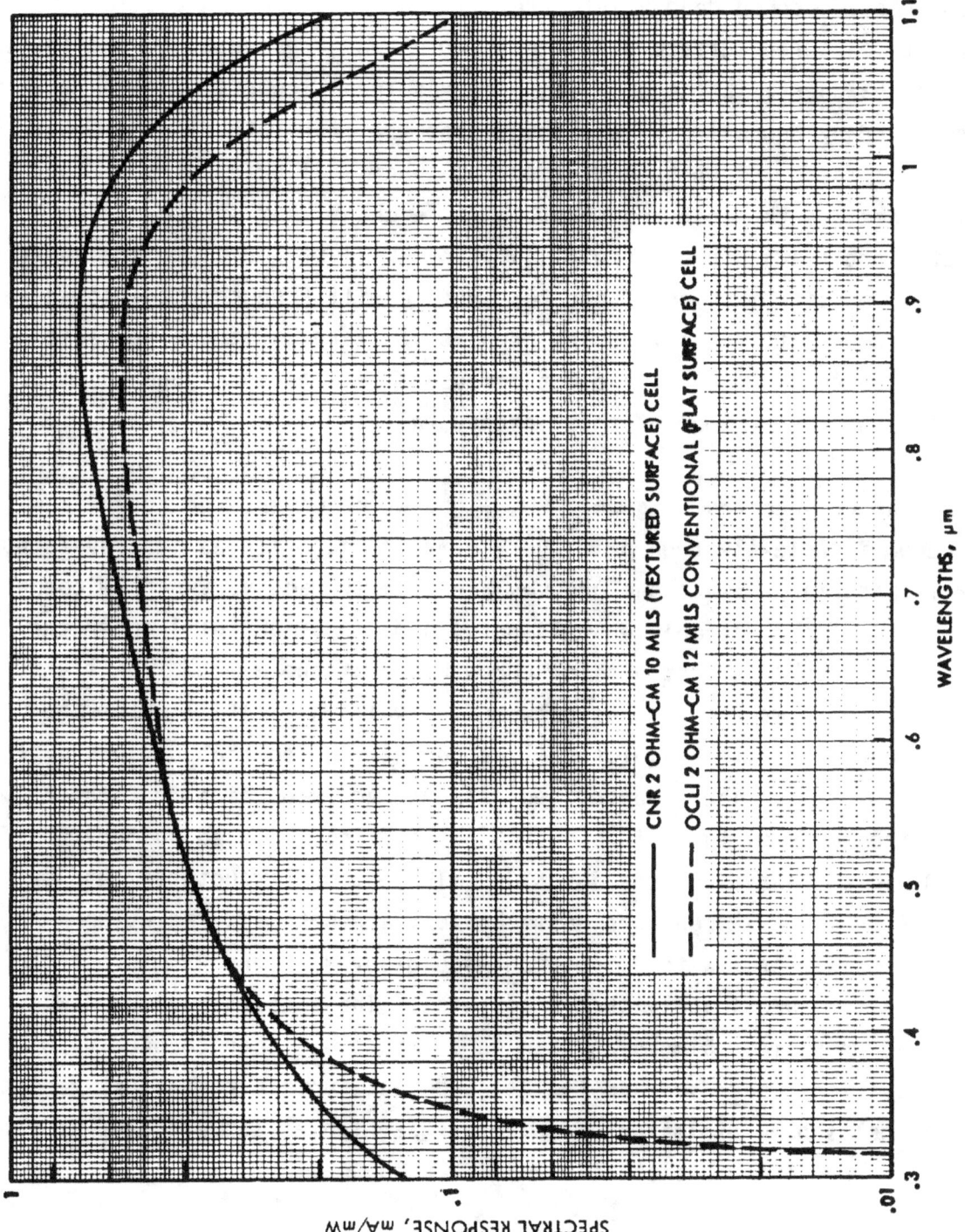

Figure 1.9 Spectral Responses of Textured Surface Cell and a Comparable Flat Surface Cell

1.5.5 Vertical Junction (VJ) Solar Cells[1.63-1.65]

The vertical junction cell is another cell structure which uses a modified front surface as shown in Figure 1.8b. These cells utilize very narrow grooves etched deeply into the silicon surface which act in a fashion similar to textured surface solar cells to enhance red response and increase radiation hardness. Etching of deep grooves in silicon is based on a technique reported by Kendall[1.63] who found that the etch rate of silicon by KOH varies by a factor of up to 400 depending on crystal orientation. The (111) planes are the most resistant to the etchant. If silicon wafers are cut along (110) planes, the (111) planes will be normal to the surface. Suitable precision masking of the top surface will then permit etching very narrow grooves into the silicon as deep as 75 microns without appreciable groove widening.[1.65] A subsequent diffusion results in a junction which follows the contour of these etched walls.

As can be seen from Figure 1.8b, light which enters the top of a wall will generate hole-electron pairs very near the junction as it travels down the wall interior. Since the grooves are usually made very narrow compared to the wall thickness, only a small portion of the light will enter the bulk silicon without first passing through one of the walls. The light that does come down the grooves between the walls will still enter the bulk silicon and initiate solar cell action similar to that of an ordinary planar solar cell. The enhanced red response and radiation resistance results from most of the hole-electron pairs being produced no more than 5 or 10 microns away from the junction. The optimum structure found by Scheinine et al.[1.65] considering mechanical strength and radiation resistance, is a cell with 2.5 micron wide grooves etched 25 microns deep, leaving walls approximately 15 microns wide. These cells have been successfully made on silicon blanks as thin as 75 microns. When coupled with dual AR coatings, back surface fields, and back surface reflectors vertical junction cells have exhibited efficiencies as high as 15% and have proven to be quite radiation resistant.

1.5.6 Back Surface Reflector (BSR) Solar Cells [1.61, 1.66]

As discussed above, the increased light trapping power of the textured cell was very effective at all wavelengths of incident light. In addition to effectively utilizing those wavelengths which produced electrical output of the cell, long wavelengths which do not have enough energy to produce hole-electron pairs in the cell are also trapped. The increased absorptivity (α = 0.93 to 0.94) causes textured cells to run so much hotter in space that they lose their performance advantage over planar cells. In order to reduce the temperature of these cells, a reflective back contact was developed with the idea of reflecting long wavelength light from the rear surface, sending it back through the cell and rejecting it out through the front surface. This contact, consisting of Al-Ti-Pd-Ag, gave 8 to 10% higher reflectivity at wavelengths between 1 and 2.5 microns and reduced the total absorptivity to values between 0.87 and 0.88.[1.66] This contact, referred to as a back surface reflector (BSR) was also found to be effective in reflecting shorter wavelength photons back through the cell so they have a second chance to produce electron-hole pairs near the junction. The increase in cell I_{sc} and P_{max} produced by the addition of the BSR to textured cells was found to be 1 to 3%.[1.61, 1.66] The BSR contact was found to produce the same kind of increased electrical performance for planar cells as it did for textured cells, so that today all cells produced for use in space incorporate this technology.

REFERENCES

1.1 R. N. Hall, "Electron-Hole Recombination in Germanium," Phys. Rev., 87, 387, 1952.

1.2 W. Shockley and W. T. Read, Jr., "Statistics of the Recombination of Hole and Electrons," Phys. Rev., 87, 835, 1952.

1.3 A. S. Grove, Physics and Technology of Semiconductor Devices, John Wiley, 1967.

1.4 G. D. Mahan, "Energy Gap in Si and Ge: Impurity Dependence," J. Appl. Phys., 51, 5, 2634, May 1980.

1.5 D. Redfield, "Mechanism of Performance Limitations in Heavily Doped Silicon Devices," Appl. Phys. Letters, 33, 6, 531, Sept. 1978.

1.6 J. G. Fossum, M. A. Shibib, and F. A. Lindholm, "The Importance of Surface Recombination and Energy-Bandgap Narrowing in p-n Junction Silicon Solar Cells," IEEE Trans. on Electron Devices, ED-26, 1294, Sept. 1979.

1.7 J. Hilibrand and R. D. Gold, "Determination of the Impurity Distribution in Junction Diodes from Capacitance-Voltage Measurements," RCA Review, 21, 245, June 1960.

1.8 N. I. Meyer and T. Guldbrandsen, "Method for Measuring Impurity Distributions in Semiconductor Crystals," Proc. IEEE, 51, 1631, Nov. 1963.

1.9 C. P. Wu, E. C. Douglas and C. W. Mueller, "Limitations of the CV Technique for Ion-Implanted Profiles," IEEE Trans. on Electron Devices, ED-22, 6, 319, June 1975.

1.10 J. C. Irvin, "Resistivity of Bulk Silicon and of Diffused Layers in Silicon," Bell System Technical Journal, 41, 387, March 1962.

1.11 L. M. Terman, "Spectral Response of Solar Cell Structures," Solid State Electronics, 2, 1, 1, 1961.

1.12 D. A. Kleinman, "Considerations on the Solar Cell," Bell System Technical Journal, 40, 81, 1961.

1.13 T. S. Moss, "Potentialities of Silicon and Gallium Arsenide Solar Batteries," Solid-State Electronics, 1, 222, 1961.

1.14 B. Dale and F. P. Smith, "Spectral Response of Solar Cells," J. Appl. Phys. 32, 7, 1377, 1961.

1.15 J. J. Loferski and J. J. Wysocki, "Spectral Response of Photovoltaic Cells," RCA Review, 22, 38, 1961.

1.16 J. W. Oliver, "Minority Carrier Diffusion Analysis in Photovoltaic Devices," TRW Report 8987-001-RU-001, 19 Feb. 1962.

1.17 H. Y. Tada, "A Survey of the Effects of Electron and Proton Radiation of Silicon Solar Cells," TM-755, Aerospace Group, Hughes Aircraft Co., NAS5-2797, June 1963.

1.18 W. C. Dash and R. Newman, "Intrinsic Optical Absorption in Single Crystal Germanium and Silicon at 77°K and 300°," Phys. Rev., $\underline{99}$, 4, 1151, 1955.

1.19 M. Wolf and H. Rauschenbach, "Series Resistance Effects on Solar Cell Measurements," Advanced Energy Conversion, $\underline{3}$, 455, 1963.

1.20 W. T. Picciano, "Determination of the Solar Cell Equation Parameters Including Series Resistance from Empirical Data," Energy Conversion, $\underline{9}$, 1, 1969.

1.21 D. E. Gray, Ed., *American Institute of Physics Handbook*, 3rd Ed., McGraw-Hill, 1972.

1.22 F. W. Sears, *Optics*, Addison, Cambridge, Mass., 1973, 1949.

1.23 H. R. Philipp and E. A. Taft, "Optical Constants of Silicon in the Region of 1 to 10 eV," Phys. Rev. $\underline{120}$, 1, 37, 1960.

1.24 Kirk-Othmer, eds., Encyclopedia of Chemical Technology, $\underline{18}$, John Wiley, New York, 1964.

1.25 E. Y. Wang, F. T. S. Yu, V. L. Simms, H. W. Brandhorst, Jr., and J. D. Broder, "Optimum Design of Antireflection Coating for Silicon Solar Cells," Conf. Rec. of the 10th IEEE Photovoltaic Specialists Conf., 168, 1973.

1.26 H. Anders, *Thin Films in Optics*, Chapter 1, Focal Press, London, 1967.

1.27 J. P. Schwartz, "Improved Silicon Solar Cell Antireflective Coatings," Conf. Rec. of the 8th IEEE Photovoltaic Specialists Conf., 173, 1970.

1.28 A. Musset and A. Thelen, *Progress in Optics* (E. Wolf, ed.), Chapter 4, North-Holland, Amsterdam, 1970.

1.29 P. M. Stella and H. Somberg, "Integrally Covered Silicon Solar Cells," Conf. Rec. of the 9th IEEE Photovoltaic Specialists Conf., 179, 1972.

1.30 V. Magee, H. G. Webb, A. D. Haigh, and R. Freestone, "Design and Practical Aspects of Maximum Efficiency Solar Cells for Satellite Applications," Conf. Rec. of the 9th IEEE Photovoltaic Specialists Conf., 6, 1972.

1.31 R. L. Crabb and A. Atzei, "Environmental Study of European Silicon Solar Cells with Improved Antireflection Coatings," Conf. Rec. of the 8th IEEE Photovoltaic Specialists Conf., 78, 1970.

1.32 A. Atzei, et al., "Improved Antireflection Coatings on Silicon Solar Cells," Solar Cells, 349, Gordon and Breach Sci. Pub., London, 1971.

1.33 N. F. Mott, Electronic Processes in Non-Crystalline Materials, 119, Clarendon Press, Oxford, 1979.

1.34 J. Roger and P. Colardelle, "Antireflecting Silicon Solar Cell with Titanium Dioxide," Conf. Rec. of the 8th IEEE Photovoltaic Specialists Conf., 84, 1970.

1.35 C. Misiano and G. Greco, "Titanium Dioxide Antireflection Coating for Silicon Solar Cells," Solar Cells, 363, Gordon and Breach Sci. Pub., London, 1971.

1.36 J. Lindmayer and J. Allison, "The Violet Cell: An Improved Silicon Solar Cell," COMSAT Technical Review, 3, 1, 1, Spring 1973.

1.37 A. G. Revesz, "Vitreous Oxide Antireflection Films in High-Efficiency Solar Cells," Conf. Rec. of the 10th IEEE Photovoltaic Specialists Conf., 180, 1973.

1.38 J. Lindmayer and J. Allison, "An Improved Silicon Solar Cell - The Violet Cell," Conf. Rec. of the 9th IEEE Photovoltaic Specialists Conf., 83, 1972.

1.39 R. A. Arndt, J. F. Allison, J. G. Haynos, and A. Meulenberg, Jr., "Optical Properties of the COMSAT Non-Reflective Cell," Conf. Rec. of the 11th IEEE Photovoltaic Specialists Conf., 40, 1975.

1.40 J. Haynos, J. Allison, R. Arndt, and A. Meulenberg, Jr., "The COMSAT Non-Reflective Silicon Solar Cell: A Second Generation Improved Cell," Proceedings of International Conf., on Photovoltaic Power Generation, Hamburg, Sept. 1974.

1.41 S. M. Sze, Physics of Semiconductor Devices, John Wiley, 1969.

1.42 C. H. Liebert, "Solar Cell Performance at Jupiter Temperature and Solar Intensity," Conf. Rec. of the 7th Photovoltaic Specialists Conf., 92, 1968.

1.43 R. J. Lambert, "Characteristics of Solar Cells at Low Temperatures," Conf. Rec. of the 7th Photovoltaic Specialists Conf., 97, 1968.

1.44 P. A. Payne and E. A. Ralph, "Low Temperature and Low Solar Intensity Characteristics of Silicon Solar Cells," Conf. Rec. of the 8th IEEE Photovoltaic Specialists Conf., 123, 1970.

1.45   J. C. Ho, et al., "Solar Cell Low Temperature, Low Solar Intesity Operation," Conf. Rec. of the 8th IEEE Photovoltaic Specialists Conf., 150, 1970.

1.46   W. Luft, "Silicon Solar Cells at Low Temperature," Conf. Rec. of the 8th IEEE Photovoltaic Specialists Conf., 161, 1970.

1.47   H. Fisher, et al., "Some Innovations in Silicon Solar Cell Technology," Solar Cells, Gordon and Breach Sci. Pub., London, 1971.

1.48   J. J. Loferski, "Principles of Photovoltaic Solar Energy Conversion," Conf. Rec. of the 10th IEEE Photovoltaic Specialists Conf., 1, 1973.

1.49   L. W. James and R. L. Moon, "GaAs Concentrator Solar Cell," Conf. Rec. of the 11th IEEE Photovoltaic Specialists Conf., 402, 1975.

1.50.   H. H. Hovel, Solar Cells, Semiconductors and Semimetals, 11, Academic Press, New York, 1975.

1.51   J. Mandelkorn and J. H. Lamneck, Jr., "Simplified Fabrication of Back Surface Electric Field Silicon Cells and Novel Characteristics of Such Cells," Conf. Rec. of the 9th IEEE Photovoltaic Specialists Conf., 66, 1972.

1.52   M. Wolf, "Drift Fields on Photovoltaic Solar Energy Converter Cells," Proc. IEEE, 51, 5, 674, May 1963.

1.53   M. P. Godlewski, C. R. Baraona, and H. W. Brandhorst, Jr., "Low-High Junction Theory Applied to Solar Cells," Conf. Rec. of the 10th IEEE Photovoltaic Specialists Conf., 40, 1973.

1.54   H. W. Brandhorst, Jr., C. R. Baraona, and C. K. Swartz, "Performance of Epitaxial Back Surface Field Cells," Conf. Rec. of the 10th IEEE Photovoltaic Specialists Conf., 212, 1973.

1.55   J. Mandelkorn, J. H. Lamneck, and L. R. Scudder, "Design, Fabrication and Characteristics of New Types of Back Surface Field Cells," Conf. Rec. of the 10th IEEE Photovoltaic Specialists Conf., 207, 1973.

1.56   J. Mandelkorn and J. H. Lamneck, "Advances in the Theory and Application of BSF Cells," Conf. Rec. of the 11th IEEE Photovoltaic Specialists Conf., 36, 1975.

1.57   M. Becker and H. Y. Fan, "Photovoltaic Effect of p-n Junction in Germanium," Phys. Rev. 78, 302, 1950.

1.58   J. B. Gunn, "On Carrier Accumulation and Properties of Certain Semiconductor Junctions," J. Electron. Control, 4, 17, 1958.

1.59 J. G. Fossum, "Physical Operation of Back-Surface-Field Silicon Solar Cells," IEEE Trans. on Electron Devices, ED-24, 4, 332, April 1977.

1.60 O. von Roos, "A Simple Theory of Back-Surface-Field (BSF) Solar Cells," J. Appl. Phys., 50, 5371, Aug. 1979.

1.61 C. F. Gay, "Development of High Efficiency, Radiation Tolerant, Thin Solar Cells," JPL Contract No. 954600, Final Report, Oct. 1977.

1.62 C. R. Baraona and H. W. Brandhorst, Jr., "V-Grooved Silicon Solar Cells," Conf. Rec. of the 11th IEEE Photovoltaic Specialists Conf., 44, 1975.

1.63 D. L. Kendall, "On Etching Very Narrow Grooves in Silicon," App. Phys. Letters, 26, 4, 195, Feb. 1975.

1.64 W. W. Lloyd, "Fabrication of an Improved Vertical Multijunction Solar Cell," Conf. Rec. of the 11th IEEE Photovoltaic Specialists Conf., 349, 1975.

1.65 A. L. Scheinine, J. H. Wohlgemuth, and E. Sparks, "Silicon Solar Cell Optimization," Technical Report AFWAL-TR-81-2052, June 1981.

1.66 P. A. Payne and R. L. Oliver, "Impound Helios Cell Output," Conf. Rec. of the 12th IEEE Photovoltaic Specialists Conf., 595, 1976.

CHAPTER 2

2.0 INSTRUMENTATION TECHNIQUES FOR MEASUREMENT OF SOLAR CELL PARAMETERS

In this section, the various commonly used experimental methods for the analysis of radiation effects are discussed. The most commonly used measurement in the analysis of radiation effects in silicon solar cells is the current-voltage characteristic under illumination. Since solar cell response is a strong function of optical wavelength, the light source is a major variable in the evaluation of solar cell parameter changes.

2.1 Light Sources and Solar Simulators

The spectral irradiance of the sun at 1.5×10^{11} m (one AU) is of primary importance in solar cell analysis for Earth orbits. The values of solar spectral irradiance proposed by Johnson [2.1] were used extensively up until about 1970. Johnson's results indicated that the solar constant was 139.5 mW/cm^2, and also that the solar spectrum closely approximates that of a 6000 K black body. Several high-altitude measurements made in recent years have been reviewed by Thekaekara, et al. [2.2] Their findings indicated a solar constant of 135.3 ± 2.1 mW/cm^2 is a better fit to the available data. They also published a detailed spectral irradiance which is tabulated in Table 2.1. More recently Neckel and Labs [2.3] have published additional work on the value of the solar constant and spectral irradiance. They find a value of between 136.8 and 137.7 mW/cm^2 for the solar constant. Currently, the values of Thekaekara are still used for most solar cell work but the Neckel and Labs data is finding increasing acceptance. Silicon solar cell response is generally limited to the region between 0.3 and 1.2 μm. In this range, the solar power density (Thekaekara) is 104.4 mW/cm^2.

Among several solar simulation techniques, the most common method is the use of a xenon arc lamp with filters to remove undesired line spectra in the near infrared. Unfiltered xenon lamps are also used in the pulsed mode, which does not generate the undesired line spectra. Unfiltered carbon arcs are also used to simulate solar illumination with a reasonable spectral match. A close spectral match to the solar spectrum is obtained by the use of a xenon-filtered tungsten combination or filtered xenon source.

Table 2.1. Solar Spectral Irradiance - Proposed Standard Curve[2.2]

λ — WAVELENGTH IN MICROMETERS

$E(\lambda)$ — SOLAR SPECTRAL IRRADIANCE AVERAGED OVER SMALL BANDWIDTH CENTERED AT λ, IN $Wm^{-2}\mu m^{-1}$

$E(0-\lambda)$ — AREA UNDER THE SOLAR SPECTRAL IRRADIANCE CURVE IN THE WAVELENGTH RANGE 0 TO λ IN Wm^{-2}

$D(0-\lambda)$ — PERCENTAGE OF THE SOLAR CONSTANT ASSOCIATED WITH WAVELENGTHS SHORTER THAN λ

SOLAR CONSTANT = 1353 Wm^{-2}

λ	$E(\lambda)$	$E(0-\lambda)$	$D(0-\lambda)$	λ	$E(\lambda)$	$E(0-\lambda)$	$D(0-\lambda)$	λ	$E(\lambda)$	$E(0-\lambda)$	$D(0-\lambda)$
.120	.100	.005999	.00044	.525	1852	352.591	26.059	1.70	202.	1221.23	90.261
.140	.030	.007299	.00053	.530	1842	361.826	26.742	1.75	180.	1230.78	90.967
.150	.07	.00780	.00057	.535	1818	370.976	27.418	1.80	159.	1239.25	91.593
.160	.23	.00930	.00068	.540	1783	379.979	28.084	1.85	142.	1246.78	92.149
.170	.63	.01360	.00100	.545	1754	388.821	28.737	1.90	126.	1253.48	92.644
.180	1.25	.02300	.00169	.550	1725	397.519	29.380	1.95	114.	1259.48	93.088
.190	2.71	.04280	.00316	.555	1720	406.131	30.017	2.00	103.	1264.90	93.489
.200	10.7	.10985	.00811	.560	1695	414.669	30.648	2.10	90.	1274.55	94.202
.210	22.9	.27785	.02053	.565	1705	423.169	31.276	2.20	79.	1283.00	94.826
.220	57.5	.67985	.05024	.570	1712	431.711	31.907	2.30	69.	1290.40	95.373
.225	64.9	.98585	.0728	.575	1719	440.289	32.541	2.4	62.0	1296.95	95.8580
.230	66.7	1.31485	.0971	.580	1715	448.874	33.176	2.5	55.0	1302.80	96.2903
.235	59.3	1.62985	.1204	.585	1712	457.441	33.809	2.6	48.0	1307.95	96.6710
.240	63.0	1.93560	.1430	.590	1700	465.971	34.439	2.7	43.0	1312.50	97.0073
.245	72.3	2.27385	.1680	.595	1682	474.426	35.064	2.8	39.0	1316.60	97.3103
.250	70.4	2.63060	.1944	.600	1666	482.796	35.683	2.9	35.0	1320.30	97.5838
.255	104.	3.05660	.2266	.605	1647	491.079	36.295	3.0	31.0	1323.60	97.8277
.260	130.	3.65160	.2698	.610	1635	499.284	36.902	3.1	26.0	1326.45	98.0383
.265	185.	4.43910	.3280	.620	1602	515.469	38.098	3.2	22.6	1328.88	98.2179
.270	232.	5.48160	.4051	.630	1570	531.329	39.270	3.3	19.2	1330.97	98.3724
.275	204.	6.5716	.4857	.64	1544	546.899	40.421	3.4	16.6	1332.76	98.5047
.280	222.	7.6366	.5644	.65	1511	562.174	41.550	3.5	14.6	1334.32	98.6200
.285	315.	8.9791	.6636	.66	1486	577.159	42.657	3.6	13.5	1335.73	98.7238
.290	482.	10.9716	.8109	.67	1456	591.869	43.744	3.7	12.3	1337.02	98.8192
.295	584.	13.6366	1.0078	.68	1427	606.284	44.810	3.8	11.1	1338.19	98.9056
.300	514.	16.3816	1.2107	.69	1402	620.429	45.855	3.9	10.3	1339.26	98.9847
.305	603.	19.1741	1.4171	.70	1369	634.284	46.879	4.0	9.5	1340.25	99.0579
.310	689.	22.4041	1.6558	.71	1344	647.849	47.882	4.1	8.7	1341.16	99.1252
.315	764.	26.0366	1.9243	.72	1314	661.139	48.864	4.2	7.8	1341.98	99.1861
.320	830.	30.0216	2.2188	.73	1290	674.159	49.826	4.3	7.1	1342.73	99.2412
.325	975.	34.5341	2.552	.74	1260	686.909	50.769	4.4	6.50	1343.4141	99.291507
.330	1059.	39.6191	2.928	.75	1235	699.384	51.691	4.5	5.90	1344.0341	99.337331
.335	1081.	44.9591	3.323	.76	1211	711.614	52.595	4.6	5.30	1344.5941	99.378721
.340	1074.	50.3566	3.721	.77	1185	723.594	53.480	4.7	4.80	1345.0991	99.416045
.345	1069.	55.7141	4.117	.78	1159	735.314	54.346	4.8	4.50	1345.5641	99.450413
.350	1093.	61.1191	4.517	.79	1134	746.779	55.194	4.9	4.10	1345.9941	99.482195
.355	1083.	66.5591	4.919	.80	1109	757.994	56.023	5.0	3.83	1346.3906	99.511500
.360	1068.	71.9366	5.316	.81	1085	768.966	56.834	6.0	1.75	1349.1806	99.717708
.365	1132.	77.4366	5.723	.82	1060	779.694	57.627	7.0	.99	1350.5506	99.818965
.370	1181.	83.2191	6.150	.83	1036	790.174	58.401	8.0	.60	1351.3456	99.877723
.375	1157.	89.0641	6.582	.84	1013	800.419	59.158	9.0	.380	1351.8356	99.913939
.380	1120.	94.7566	7.003	.85	990	810.434	59.899	10.0	.250	1352.1506	99.937221
.385	1098.	100.3016	7.413	.86	968	820.224	60.622	11.0	.170	1352.3606	99.952742
.390	1098.	105.7916	7.819	.87	947	829.799	61.330	12.0	.120	1352.5056	99.963459
.395	1189.	111.5091	8.241	.88	926	839.164	62.022	13.0	.087	1352.6091	99.971108
.400	1429.	118.0541	8.725	.89	908	848.334	62.700	14.0	.055	1352.6801	99.976356
.405	1644.	125.7366	9.293	.90	891	857.329	63.365	15.0	.049	1352.7321	99.980199
.410	1751.	134.2241	9.920	.91	880	866.184	64.019	16.0	.038	1352.7756	99.983414
.415	1774.	143.0366	10.571	.92	869	874.929	64.665	17.0	.031	1352.8101	99.985964
.420	1747.	151.8391	11.222	.93	858	883.564	65.304	18.0	.024	1352.8376	99.987997
.425	1693.	160.4391	11.858	.94	847	892.08	65.934	19.0	.02000	1352.8596	99.989623
.430	1639.	168.7691	12.473	.95	837	900.50	66.556	20.0	.01600	1352.8776	99.990953
.435	1663.	177.0241	13.083	.96	820	908.79	67.168	25.0	.00610	1352.9328	99.995037
.440	1810.	185.7066	13.725	.97	803	916.90	67.768	30.0	.00300	1352.9556	99.996718
.445	1922.	195.0366	14.415	.98	785	924.84	68.355	35.0	.00160	1352.9671	99.997568
.450	2006.	204.8566	15.140	.99	767	932.60	68.928	40.0	.00094	1352.9734	99.998037
.455	2057.	215.0141	15.891	1.00	748	940.18	69.488	50.0	.00038	1352.9800	99.998525
.460	2066.	225.3216	16.653	1.05	668	975.58	72.105	60.0	.00019	1352.9829	99.998736
.465	2048.	235.6066	17.413	1.10	593	1007.10	74.435	80.0	.00007	1352.9855	99.998928
.470	2033.	245.8091	18.167	1.15	535	1035.30	76.519	100.0	.00003	1352.9865	99.999002
.475	2044.	256.001	18.921	1.20	485	1060.80	78.404	1000.0	.00000	1353.0000	100.000000
.480	2074.	266.296	19.681	1.25	438	1083.88	80.109				
.485	1976.	276.421	20.430	1.30	397	1104.75	81.652				
.490	1950.	286.236	21.155	1.35	358	1123.63	83.047				
.495	1960.	296.011	21.878	1.40	337	1141.00	84.331				
.500	1942.	305.766	22.599	1.45	312	1157.23	85.530				
.505	1920.	315.421	23.312	1.50	288	1172.23	86.639				
.510	1882.	324.926	24.015	1.55	267	1186.10	87.665				
.515	1833.	334.214	24.701	1.60	245	1198.90	88.611				
.520	1833.	343.379	25.379	1.65	223	1210.60	89.475				

These sources match the solar spectrum well enough that cell measurements made with them can be considered representative of measurements made in the space environment. Many other types of light sources have been used in radiation effects studies. Unfiltered and filtered incandescent tungsten sources have peak responses in the red and near infrared. Since this is the wavelength region of the solar cell response which is most changed by irradiation, use of these sources will show much more severe cell radiation degradation as compared to evaluations with a suitable solar simulator. This characteristic severely limits the use of tungsten source simulation in the evaluation of radiation effects.

Filtered xenon arc solar simulators are manufactured by Spectrolab, Oriel, Schoeffel and Xenon Corp., among others. The spectral irradiance for the Spectrolab X-25 simulator is shown in Figure 2.1.[2.4] Similar data is shown in Figure 2.2 for a combination xenon-tungsten source simulator used by Applied Solar Energy Corporation.[2.4]

An important recent development in the field of solar simulators is the use of pulsed xenon arc lamps for solar cell and solar cell array testing.[2.5, 2.6] These developments have been prompted by the need for a suitable alternative to testing large arrays in natural sunlight on the Earth's surface. In these systems, a 2 msec pulse of light is produced. Solar cell output data can be accumulated during about 1 msec of the pulse length. Electronic data handling systems are necessary to record cell or array outputs and commutate external load resistances during the light pulse. Variations in test cell current due to light intensity variations are corrected to a normalized value at the desired illumination. The high-intensity peaks in the 0.8 to 0.9 μm region of the xenon arc spectrum are not generated by pulsed operation with high current densities. By this means, it is possible to achieve a reasonably close match to the solar spectrum.[2.7] The spectral irradiance of a Spectrolab LAPSS (Large Area Pulsed Solar Simulator) is shown in Figure 2.3.[2.4] Array areas up to eight feet by eight feet can be illuminated with excellent temperature control by such simulators.

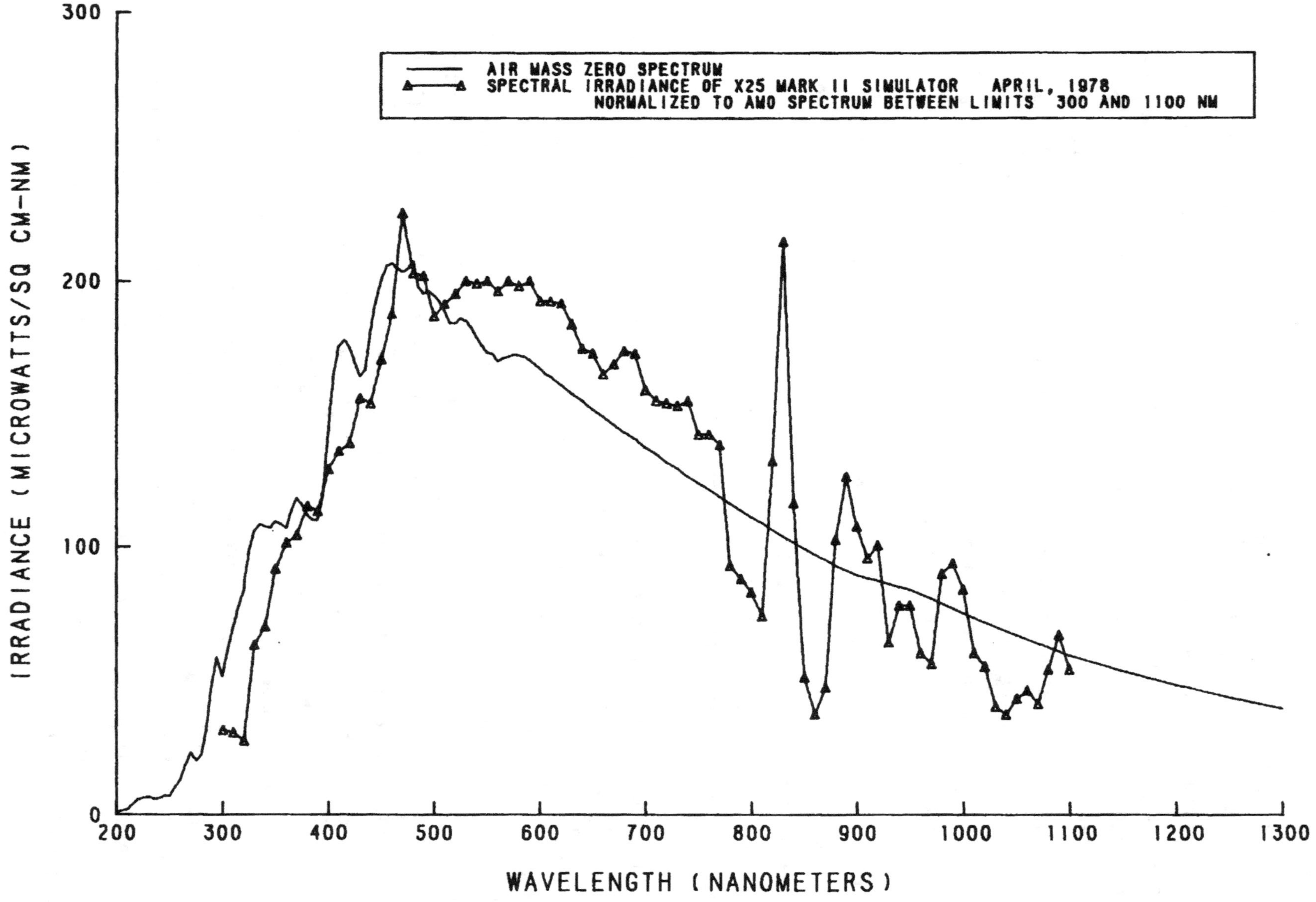

Figure 2.1 Spectral Irradiance of Spectrolab X-25 Mark II Solar Simulator

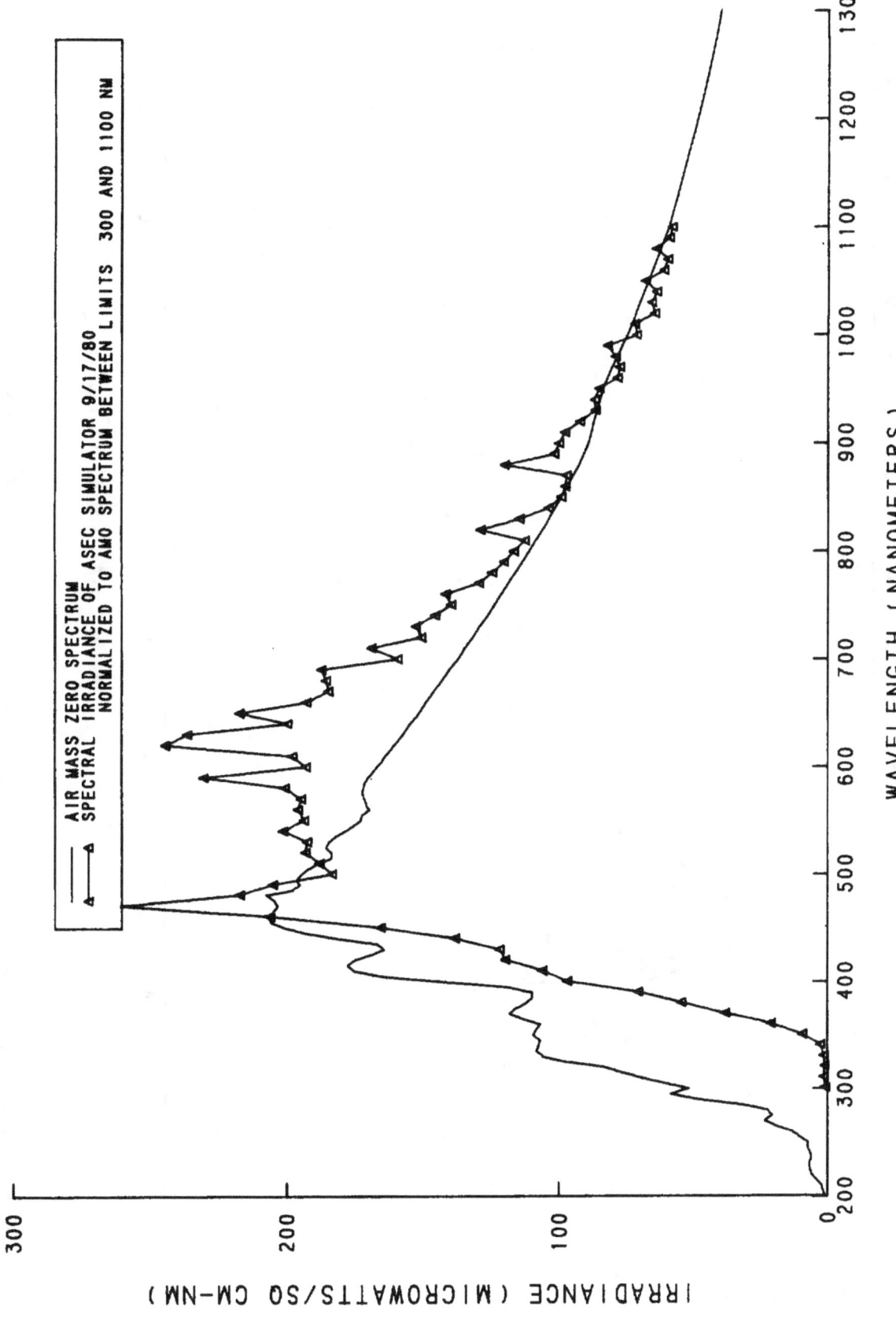

Figure 2.2 Spectral Irradiance of ASEC Tungsten/Xenon Solar Simulator

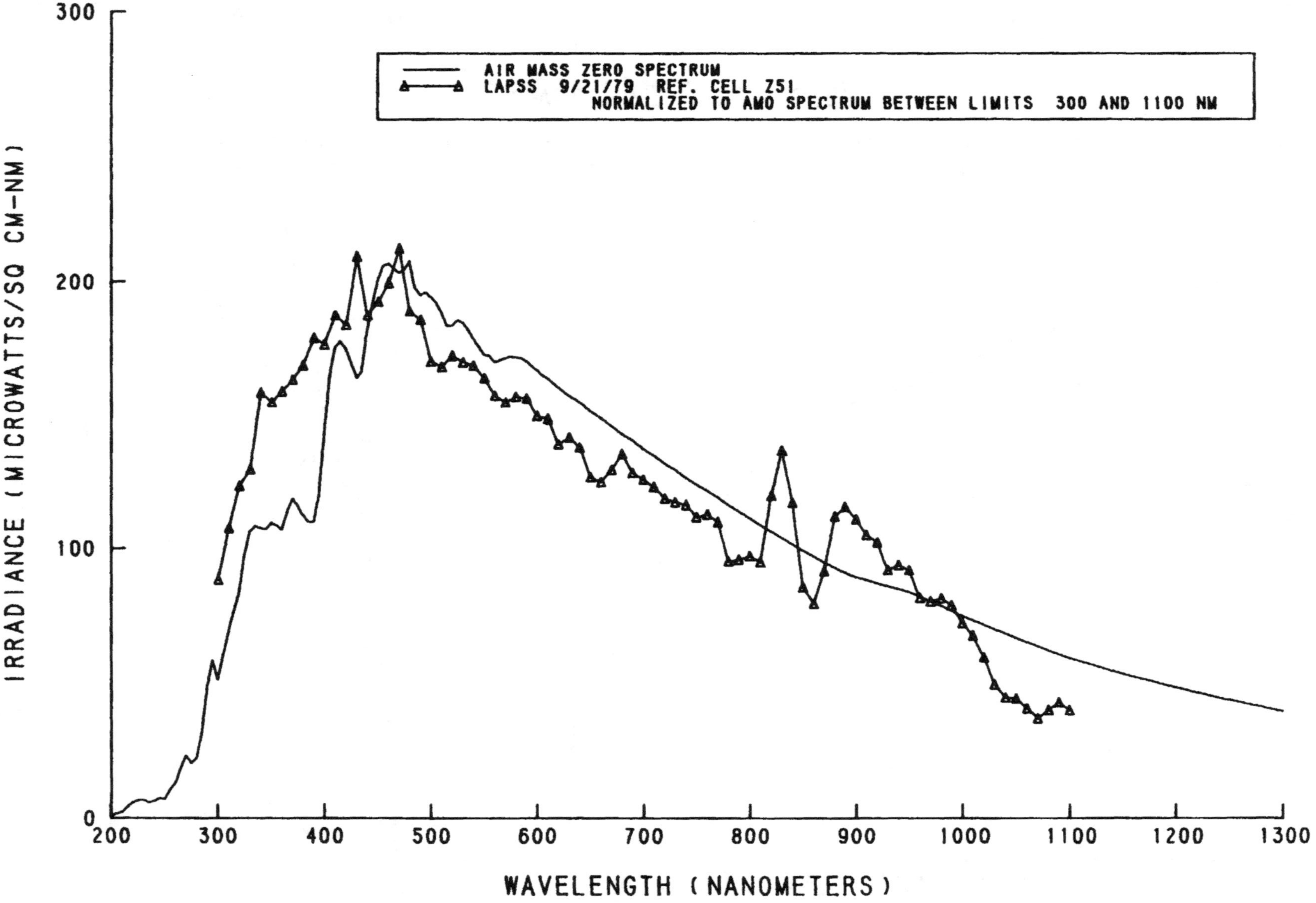

Figure 2.3 Spectral Irradiance of Spectrolab Large Area Pulsed Solar Simulator (LAPSS)

The extent to which a lack of solar spectral match affects a solar cell measurement can be estimated if the spectral intensity of the light source and the spectral response of the solar cell are known. The light-generated current of the illuminated cell can be calculated as follows:

$$I_L \ (A/cm^2) = \int R(\lambda) E(\lambda) \ d\lambda \tag{2.1.1}$$

where
$R(\lambda)$ = solar cell spectral response, A/W
$E(\lambda)$ = spectral irradiance, W/cm^2-μm
$d\lambda$ = an increment of wavelength, μm

The above equation can be used to determine the light-generated currents under solar and simulator illuminations. The generated current under solar illumination can be calculated from the generated current under simulator illumination if the spectral response of the cell and the spectral irradiance of the simulator are known. The relation is as follows:

$$I_L(\text{space}) = I_L(\text{simulator}) \frac{\int R(\lambda) \ E(\lambda)_{\text{space}} \ d\lambda}{\int R(\lambda) \ E(\lambda)_{\text{sim.}} \ d\lambda} \tag{2.1.2}$$

Solar simulator intensities are determined by the short circuit current outputs of calibrated primary or secondary standard cells. The primary standard cells commonly in use are generated by a NASA/JPL program of telemetered balloon flights.[2.8, 2.9, 2.10] Similar programs have been conducted by aircraft and high-altitude terrestrial measurements.[2.11-2.16] When the effects of atmospheric absorption are properly accounted for, the results are in good agreement with the balloon flight data.[2.17]

The availability of primary standard cells is limited; they are considered too valuable for general usage in setting simulator intensities. For this reason, secondary standard cells are calibrated for use as working standards. Palmer has reviewed the methods of generating secondary standard cells and concluded that previously proposed methods of calibration may yield poor results.[2.18] Palmer has proposed the use of alternate methods which insure that secondary standard cell calibration accuracy will approach that of primary standard cells.[2.18] A solar simulator intensity which produces a standard cell response equal to that for free space at one AU is referred to as one sun, air mass zero (AM0) or 135.3 mW/cm^2. Since the spectral output of xenon arc bulbs changes with time as does the reflectance of mirrors used in simulators, solar simulator spectral quality should be routinely monitored by means of narrow bandpass or cutoff filters with calibrated spectral response detectors.

2.2 Current-Voltage Characteristics

The measurement of solar cell current-voltage characteristics is the primary means of evaluating the device. The evaluation is made by measuring the cell voltage developed and the cell current into load resistances varying between zero and infinity. The measurement is simple in principle but attention to several practical details is necessary to insure accurate results. Solar cell response is a strong function of temperature. For this reason, the cell must be in thermal equilibrium at a known temperature during the measurement. With adequate heat sinking and cooling, cells measured under one sun irradiance at room temperature can be stabilized at 28°C. To insure that the voltages measured are representative of those developed on the cell contacts, separate probes are employed to measure cell voltage and cell current. In this way, any voltage drops which occur in the current carrying circuit due to contact and lead resistance do not cause errors in the measured cell voltage. Since connecting the cell to a variable resistive load cannot yield a true short circuit reading, the test cells are often connected to a power supply capable of sinking current. Variation of power supply voltage then changes cell load. A bipolar supply or two supplies connected with polarities in opposition are necessary to achieve short circuit current measurement. The current-voltage data is usually plotted with an X-Y recorder. The solar cell parameters such as I_{sc} and V_{oc} can be read directly with digital meters. Multiplier circuits are available which produce a voltage proportional to the product of cell voltage and current. This output may be plotted as a function of cell voltage to directly indicate the maximum power and voltage at maximum power. The cell series resistance is also determined from current-voltage characteristics at two or more different illumination levels.[2.19, 2.20]

2.3 Spectral Response Measurements

Spectral response measurements are very useful for evaluating changes in solar cells due to radiation effects. The spectral response (amps/watt) is a measure of the short-circuit current density generated by the cell under various monochromatic illuminations of a known power density. The spectral response is often reported in terms of relative units when absolute values of light intensities are not determined. Various schemes have been used to measure the spectral response of solar cells. High-resolution monochromators are used when extreme accuracy is desired. When less accuracy is needed, narrow bandpass filters can be used as sources of monochromatic light. When a monochromator is used, there are two methods to normalize the solar cell output to the light intensity. Tungsten light sources are usually used in monochromators, and the entrance slit width can be varied to control the optical power density illuminating the cell under test. In some systems, the entrance slit width can be automatically controlled to maintain a constant optical power density on the solar cell. An alternate approach is to maintain a constant slit width and allow the optical power density on the cell to vary with wavelength. This variation is then factored into the spectral response calculation.

One disadvantage of these methods of measurement is that the solar cell response is determined at very low minority carrier injection levels. Solar cells irradiated with neutrons and protons have response characteristics which are dependent upon the concentration of injected minority carriers. In such cases the cell must be illuminated with a light source similar in intensity and spectral content to the intended space environment during the spectral response evaluation. This can be achieved by chopping the monochromatic light and measuring the test cell output with a lock-in amplifier tuned to the chopper frequency.[2.21] A dc bias light may then be used to illuminate the solar cell to achieve the required injection level without directly influencing the output of the lock-in.

2.4 Irradiation Methods

The evaluation of solar cell radiation effects requires a wide range of specialized equipment and instrumentation. Charged particle accelerators are the primary sources for space radiation simulation. The range of electron energies of interest is 0.3 to 10 MeV. Electron energies of 0.3 to 3 MeV are usually obtained with Van de Graaff and Dynamitron accelerators. Higher electron energies can be reached with linear electron accelerators. Proton energies greater than 10 MeV can be obtained from cyclotrons. For lower energy protons, it is necessary to transport the proton beam and perform the irradiation in vacuum to avoid excessive energy losses. A survey of all types of accelerator facilities has been published but is currently out of date.[2.22] A more recent survey covers cyclotrons throughout the world, but does not address other types of accelerators.[2.23] Accelerators invariably produce irradiation rates which are many orders of magnitude greater than those of space environments. Real-time irradiations of solar cells have been done using beta emitting sources.[2.24, 2.25] These sources generate a spectrum of electron energies and fluxes similar to that of some space environments.

A successful experiment must include accurate knowledge of the particle energy, measurement of cross-sectional beam intensity at the irradiation area, as well as the intensity during the irradiation. Although there are several methods of accomplishing the above measurements, all can be done with a Faraday cup. A design of a Faraday cup suitable for accelerators in the 1 MeV range is shown in the literature.[2.26] The desirable characteristics of a Faraday cup are as follows:

a. Shielding thickness must exceed particle energy range.
b. A high cup length-to-diameter ratio.
c. Use of low Z (atomic number) materials to reduce secondary electron emission and bremsstrahlung production.
d. Cup must be in vacuum or potted.
e. Cup should be a reentrant cavity.
f. Cup should be screened to suppress secondary electron emission if necessary.

A Faraday cup requires a current measuring instrument which operates in the range of 10^{-10} to 10^{-6} amperes and integrates charge. Instruments of this nature are produced by Keithley Instruments, Cleveland, Ohio.

The particle energy can be determined by means of a range-energy measurement. In this measurement, an increasing thickness of absorber is introduced into a constant flux beam, while the flux of particles exiting the absorber is monitored with a Faraday cup or a radiation-degraded solar cell. If the beam is monoenergetic, a plot of cup current (or cell short circuit current) versus absorber thickness is extrapolated to zero current to yield an absorber thickness which is equal to the projected range of the mean particle energy of the beam. This technique is satisfactory for electrons and high energy protons. Since the beam current must remain constant as absorber thickness is increased, a second independent means of monitoring beam current must be available. Van de Graaff generators are equipped with a generating voltmeter which produces a dc voltage proportional to the potential difference on the accelerator tube and therefore gives the energy of the accelerated particles. A check calibration at one operating energy is sufficient to insure accurate calibration. Corrections must be made for energy loss in the accelerator exit window and in the atmosphere between the exit port and the target in determining the energy of the beam incident on the solar cell in the target plane.

2.5 Diffusion Length Measurement

The importance of minority carrier diffusion length (or lifetime) in the study of solar cells was discussed in Section 1.2. A decrease in minority carrier lifetime is the primary reason for solar cell degradation in radiation environments. An experimental technique for measuring minority carrier diffusion length using gamma ray or electron irradiation was suggested by Gremmelmaier.[2.27] This technique requires the uniform generation of electron-hole pairs throughout the active volume of the p-n junction device. Under these conditions the generated current density (J_L) is expressed as follows:

$$J_L = q\, g_0 (L_p + W + L_n) \qquad (2.5.1)$$

where q = electronic charge
g_0 = generation rate of electron-hole pairs in unit volume
L_p = hole diffusion length in an n-type layer
W = width of space charge layer
L_n = electron diffusion length in a p-type layer

Since L_p and W are usually very small compared to L_n, they may be neglected, and the measured short circuit current becomes proportional to the diffusion length in the p-type base region (L_n). The generation rate, determined for this uniform radiation, thus allows accurate determination of diffusion length from the measured short circuit current.

There are several experimental methods of uniformly injecting electron hole pairs. In addition to the use of gamma radiation, high energy electrons, high energy protons [2.28] and infrared light [2.29, 2.30] have been used to achieve uniform injection. When using 1 MeV electrons to inject carriers for this purpose, it is necessary to introduce a 0.030 cm (0.012 in.) aluminum shield immediately in front of the cell during a normal incidence front irradiation. The details of this procedure and the experimental evaluation of the generation rate have been covered by Rosenzweig.[2.28]

The experimental measurement of diffusion length by the above methods has several inherent limitations. Since the diffusion length is that distance from which 1/e of injected minority carriers will diffuse to the junction during their lifetime, the diffusion length concept involves both minority carrier lifetime and diffusion. Minority carrier lifetime, in the most general case, could vary throughout the active region of a solar cell. In practice this situation arises when solar cells are irradiated with low energy protons which do not penetrate the entire active volume of the cell. Diffusion lengths of solar cells with a nonuniform minority carrier lifetime in the active base region cannot be accurately measured by the above methods. Surface recombination at the solar cell back contact, which is strongly modified by the presence of back surface fields, can also cause errors in measured diffusion lengths. These errors are negligible for cells in which the thickness exceeds two or three times the diffusion length. Corrections for varying back surface recombination velocities and for cases where the cell thickness is not greater than L have been covered in a review paper by Reynolds and Meulenberg.[2.31]

The measurement of diffusion length by the above methods also assumes the external cell current generated is collected entirely by diffusion of excess minority carriers to the junction. Some solar cell structure designs utilize "drift fields." In such cases, excess minority carrier collection is aided by the presence of an electric field in the base region; and the short circuit current under conditions of uniform pair production cannot be related to diffusion length by the above equation.

An additional limitation arises if 1 MeV electron or other high energy radiations are used in the diffusion length measurement. The radiation flux must be kept low to minimize damage to the cell during the measurement. The 1 MeV electron beam current during such a measurement is approximately 10^{-9} A/cm^2. The generation rate of excess minority carriers produced by this electron flux is considerably lower than that produced by solar illumination at 135 mW/cm^2. In most cases the diffusion length or lifetime is not dependent upon the concentration of excess minority carriers. In such cases, the diffusion length measured with low levels of injected minority carriers is the same as that for a cell under one sun illumination. Silicon solar cells irradiated with protons and neutrons exhibit injection level dependence of the diffusion length, and must be illuminated with simulated solar illumination to allow accurate measurement of the diffusion length. The schemes used by Denney[2.32] for proton irradiated cells and Stofel, et al.[2.21] for neutron irradiated cells have shown that the diffusion lengths of such cells measured under approximate solar illumination are roughly two times greater than that measured under low injection level conditions. This diffusion length measurement scheme is performed in the following manner. First, each candidate cell type is irradiated with 1 MeV electrons. At various fluences during the irradiation, diffusion length and short circuit current under tungsten illumination are measured. Since tungsten illumination consists mostly of long wavelength light which penetrates deeply into silicon, the short circuit current is strongly dependent on diffusion length. In this fashion an accurate relationship between diffusion length and tungsten short circuit is generated, because electron radiation damage does not exhibit injection level effects. A typical curve showing this relationship is given in Figure 2.4. The diffusion lengths of solar cells irradiated with protons or neutrons may be determined by using such a curve by measuring their tungsten short circuit currents and reading the corresponding diffusion lengths from the curve.

Although the techniques described here all utilize the uniform generation of carriers throughout the cell's active volume, other techniques exist and may be advantageous in some circumstances. A review of several diffusion length measurement techniques applicable to solar cells together with an extensive bibliography is given by Woollam et al.[2.33]

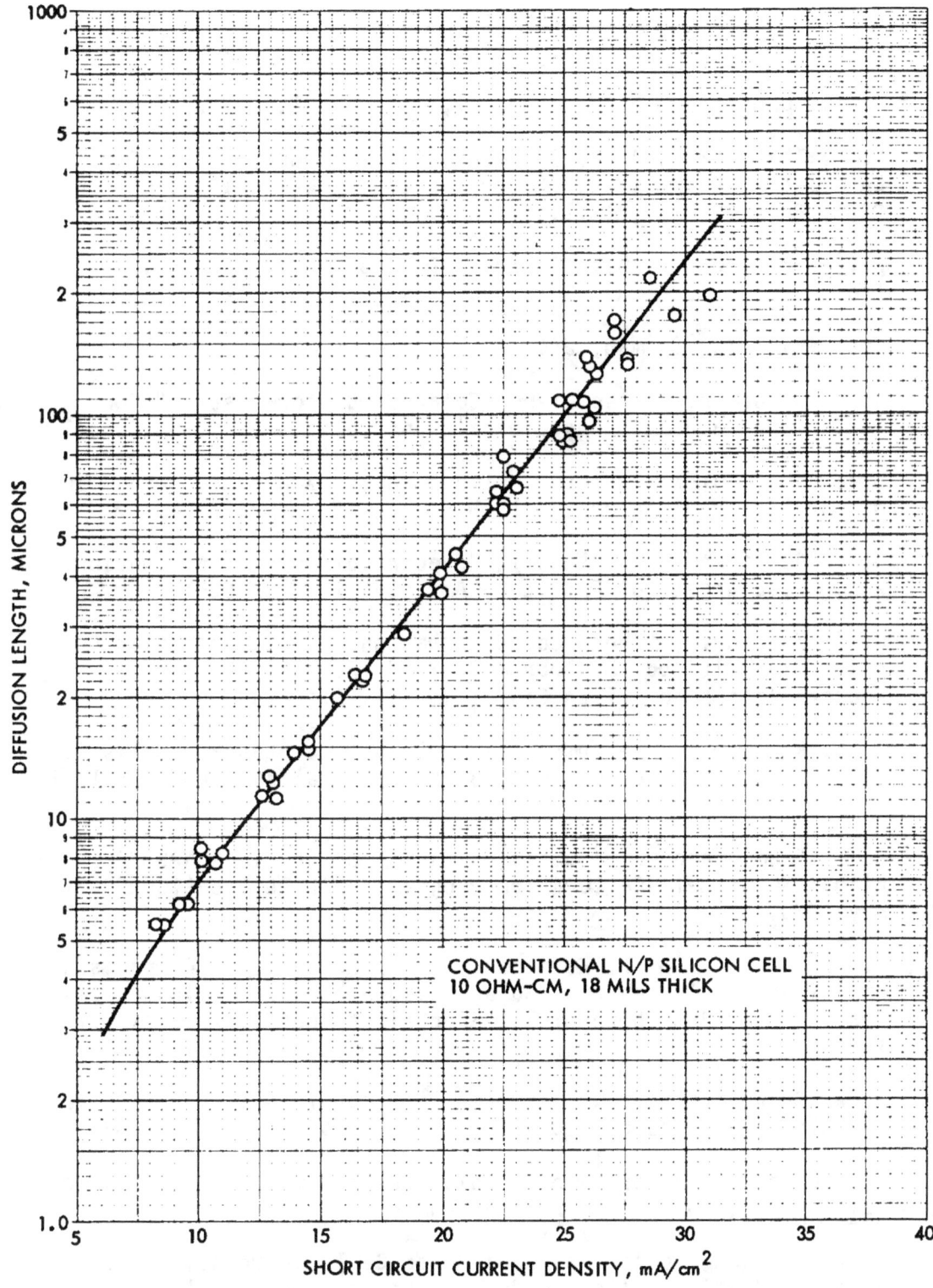

Figure 2.4 Minority Carrier Diffusion Length vs Short Circuit Current Density of Conventional n/p Silicon Solar Cells

REFERENCES

2.1 F. S. Johnson, J. of Meteorology, 11, 6, 431, 1954.

2.2 M. P. Thekaekara, A. J. Drummond, D. G. Murcray, P. R. Gast, E. G. Laue and R. C. Willson, "Solar Electromagnetic Radiation," NASA SP 8005, Revised, May, 1971.

2.3 H. Neckel and D. Labs, "Improved Data of Solar Spectral Irradiance from 0.33 to 1.25 m," Solar Physics, 74, 231, 1981.

2.4 Courtesy of C. Seaman, JPL.

2.5 R. W. Opjorden, "Pulsed Xenon Solar Simulator System," Conf. Rec. of the 8th IEEE Photovoltaic Specialists Conf., 312, 1970.

2.6 D. Creed, "Solar Simulator Using a Pulsed Xenon Arc Tube," Institute of Environmental Sciences, 1969 Proc., 363, April 1969.

2.7 J. H. Goncz and P. B. Newell, "Spectra of Pulsed and Continuous Xenon Discharge," J. Opt. Soc. Am., 56, 1, 87, 1966.

2.8 J. A. Zoutendyk, "The Space Calibration of Standard Solar Cells Using High-Altitude Balloon Flights," Proc. of the 4th Photovoltaic Specialists Conf., Vol II, C-4, 1964.

2.9 D. W. Ritchie, "Development of Photovoltaic Standard Cells for NASA," Proc. of the 4th Photovoltaic Specialists Conf., Vol. II, C-5, 1964.

2.10 C. H. Seaman and R. S. Weiss, "Results of the 1979 NASA/JPL Balloon Flight Solar Cell Calibration Program," JPL Publication 80-31, May 1980.

2.11 F. J. McKendry, H. W. Kuzminski and C. P. Hadley, "Comparison of Flight and Terrestrial Solar Measurements on Silicon Cells," Proc. of the 4th Photovoltaic Specialists Conf., Vol. II, C-3, 1964.

2.12 H. W. Brandhorst, Jr., "Airplane Testing of Solar Cells," Proc. of the 4th Photovoltaic Specialists Conf., Vol. II, C-2, 1964.

2.13 M. Audibert, "Calibration of Solar Cells for Space Applications," Solar Cells, Gordon and Breach, London, 1971.

2.14 M. W. Walkden, "Calibration of Solar Cells in Uncollimated Sunlight," Solar Cells, Gordon and Breach, London, 1971.

2.15 N. L. Thomas and D. M. Chisel, "Space Calibration of Standard Solar Cells Using High Altitude Sounding Rockets," Conf. Rec. of the 11th Photovoltaic Specialists Conf., 237, 1975.

2.16 N. L. Thomas and F. W. Sarles, Jr., "High Altitude Calibration of Thirty-Three Silicon and Gallium Arsenide Solar Cells on a Sounding Rocket," Conf. Rec. of the 12th IEEE Photovoltaic Specialists Conf., 560, 1976.

2.17 H. W. Brandhorst, Jr., "Calibration of Solar Cells Using High-Altitude Aircraft," Solar Cells, Gordon and Breach, London, 1971.

2.18 J. M. Palmer, "Solar Cell Standardization: A Practical Approach to Secondary Standard Generation," Solar Cells, Gordon and Breach, London, 1971.

2.19 R. J. Hardy, " Theoretical Analysis of the Series Resistance of a Solar Cell," Solid-State Electronics, 10, 765, 1967.

2.20 M. S. Imamura and J. I. Portscheller, "An Evaluation of the Methods of Determining Solar Cell Series Resistance," Conf. Rec. of the 8th IEEE Photovoltaic Specialists Conf., 102, 1970.

2.21 E. J. Stofel, et al., "Neutron Damage to Silicon Solar Cells," IEEE Trans. on Nuc. Sci., NS-16, 5, 97, 1969.

2.22 D. J. Hamman and W. H. Veazie, Jr., "Survey of Particle Accelerators," REIC Report No. 31, Part II, Sept. 1964.

2.23 J. A. Martin, "Cyclotrons - 1978," IEEE Trans. on Nuclear Science, NS-26, 2, 2443, April 1979.

2.24 D. L. Reynard and D. G. Peterson, "Results of a Real-Time Irradiation of Lithium P/N and Conventional N/P Silicon Solar Cells," Conf. Rec. of the 9th IEEE Photovoltaic Specialists Conf., 303, May 1972.

2.25 M. C. Whiffen and E. B. Trent, "The Effects of Radiation on Lithium Doped Solar Cells," Lockheed-Georgia Co., Report #ER-11150, July 1971.

2.26 F. M. Smits, et al., "Report of Solar Cell Work at Bell Telephone Laboratories," Proceedings of Solar Working Group Conf., Vol. I, 9-1, Feb. 1962.

2.27 R. Gremmelmaier, "Irradiation of P-N Junctions with Gamma Rays: A Method for Measuring Diffusion Lengths," Proc. I.R.E., 46, 6, 1045, 1958.

2.28 W. Rosensweig, "Diffusion Length Measurement by Means of Ionizing Radiation," Bell Systems Tech. J. 41, 1573, 1962.

2.29 J. R. Bilinski, et al., "Proton-Neutron Damage Equivalence in Si and Ge Semiconductors," IEEE Trans. on Nuc. Science, NS-10, 5, 71, 1963.

2.30 B. L. Gregory, "Minority Carrier Recombination in Neutron Irradiated Silicon," IEEE Trans. on Nuc. Science, <u>NS-16</u>, 6, 53, 1969.

2.31 J. H. Reynolds and A. Meulenberg, Jr., "Measurement of Diffusion Length in Solar Cells," J. Appl. Phys., 45, 6, 2582, 1974.

2.32 J. M. Denney and R. G. Downing, "Proton Radiation Damage in Silicon Solar Cells," TRW Systems Report No. 8653-6026-KU-000, July 1963.

2.33 J. A. Woollam, A. A. Khan, R. J. Soukup and A. M. Hermann, "Diffusion Length Measurements in Solar Cells - An Analysis and Comparison of Techniques," NASA Lewis 1982 SPRAT Conf., To be published.

CHAPTER 3

3.0 RADIATION EFFECTS

The behavior of solar cells in a radiation environment can be described in terms of the changes in the engineering output parameters of the devices. This approach limits the understanding of the physical changes which occur in the device. Since other environmental factors may need consideration, an understanding of a physical model provides a basis for estimates of the behavior in a complex environment. In addition, solar arrays of the future will become more complex and may utilize materials which are affected by different aspects of radiation damage. For these reasons, the engineer should be aware of the process by which radiation interacts with matter, and understand the physical models which describe the processes.

3.1 The Theory of Radiation Damage

The radiation usually of interest in the study of degradation of materials and devices consists of energetic or fast massive particles (i.e., electrons, protons, neutrons, or ions). The origin of these particles may be particle accelerators, the natural space radiation environment, nuclear reactions, or secondary mechanisms such as Compton electrons produced by gamma rays. Because they have mass, energy and possibly charge, these particles or other particles generated by them can interact in several ways with materials. The dominant interactions are:

a. **Inelastic Collisions with Atomic Electrons.** Inelastic collisions with bound atomic electrons are usually the predominant mechanism by which an energetic charged particle loses kinetic energy in an absorber. In such collisions, electrons experience a transition to an excited state (excitation) or to an unbound state (ionization).

b. **Elastic Collisions with Atomic Nuclei.** Energetic charged particles may have coulombic interactions with the positive charge of the atomic nucleus through Rutherford scattering. In some cases the amount of energy transferred to the atom will displace it from its position in a crystalline lattice.

This energetic displaced atom may in turn undergo similar collisions with other atoms of the material. Energetic particles may also interact directly by a hard sphere collision with the nucleus. The probability of this type of event is usually less than that for Rutherford scattering, except at higher energies. If sufficient energy is transferred to displace an atom from its lattice site, that atom will probably be energetic enough to displace many other atoms.

c. <u>Inelastic Collisions with Atomic Nuclei</u>. This general category of interaction includes several processes which are important in radiation damage studies. Highly energetic protons undergo inelastic collisions with the atomic nucleus. In this process the energetic proton interacts with the nucleus and leaves the nucleus in an excited or activated state. The excited nucleus emits energetic nucleons and the recoiling nucleus is displaced from its lattice site. This recoiling nucleus in turn causes more displacements. This process is also referred to as spallation. Collisions between neutrons of thermal energy and nuclei can also be included in this group. However, these interactions are of little importance in solar array degradation.

The major types of radiation damage phenomena in solids which are of interest to the solar array designer are ionization and atomic displacement. It is important to classify an effect into one of these two categories, if possible, because the general behavior of each phenomenon has been characterized to a large extent.

3.1.1 Ionization

Ionization occurs when orbital electrons are removed from an atom or molecule in gases, liquids or solids. The measure of the intensity of ionizing radiation is the roentgen. This unit is defined by a charge generation of 2.58×10^{-4} coulomb/kilogram of air. The measure of the absorbed dose in any material of interest is usually defined in terms of absorbed energy per unit mass. The accepted unit of absorbed dose is the rad (100 ergs/gm or 0.01 joules/kg). A recently adopted SI unit of absorbed dose is the gray (Gy) defined to be 1 joule/kg.

Through the use of the concept of absorbed dose, various radiation exposures can be reduced to absorbed dose units which will reflect the degree of ionization damage in the material of interest. This concept can be applied to electron, gamma and X-ray radiations of all energies. For electrons, the absorbed dose may be computed from the incident fluence

$$\text{Dose (rad)} = 1.6 \times 10^{-8} \frac{dE}{dx} \left(\frac{\text{MeV-cm}^2}{\text{gm}}\right) \Phi \left(\frac{e}{\text{cm}^2}\right) \qquad (3.1.1)$$

where $\frac{dE}{dx}$ is the electron stopping power in the material of interest. In this manner, the effects of an exposure to fluxes of trapped electrons of various energies in space can be reduced to an absorbed dose. In general, this practice is also applicable to proton irradiations; however, some caution must be exercised. In some types of materials, the effects of the ionization caused by heavy particles are confined to the vicinity of the particle track. If homogeneous ionization is produced by protons in the absorber material of interest, one can convert proton fluences to absorbed doses and sum them with doses from other radiations.

The variations of stopping power and range for electrons and protons of various energies can be seen in Figures 3.1 and 3.2. The data presented are for silicon and have been normalized for density. The stopping power and range of a fast particle are not strong functions of the atomic number of the absorber material. For this reason, the data in Figures 3.1 and 3.2 can be used for materials with a similar atomic number with a negligible error.

Radiation may affect solar cell array materials by several ionization-related effects. The reduction of transmittance in solar cell coverglasses is an important effect of ionizing radiation. The darkening is caused by the formation of color centers in glass or oxide materials. The color centers form when ionizing radiation excites an orbital electron to the conduction band. These electrons become trapped by impurity atoms in the oxide to form charged defect complexes which can be relatively stable at room temperature.

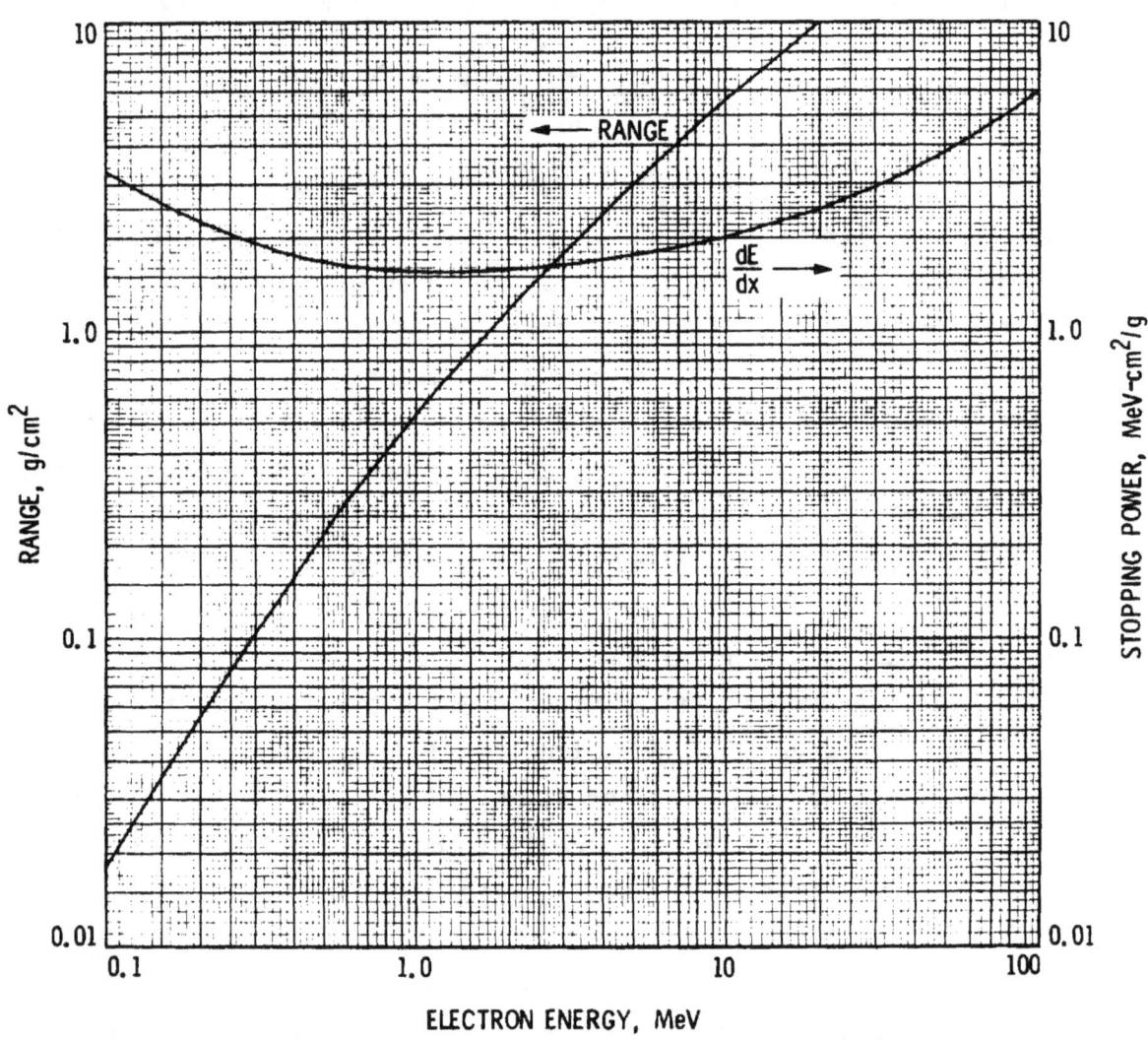

Figure 3.1 Stopping Power and Range Curves for Electrons in Silicon.
Reference: Berger and Seltzer, NASA SP-3036, 1966

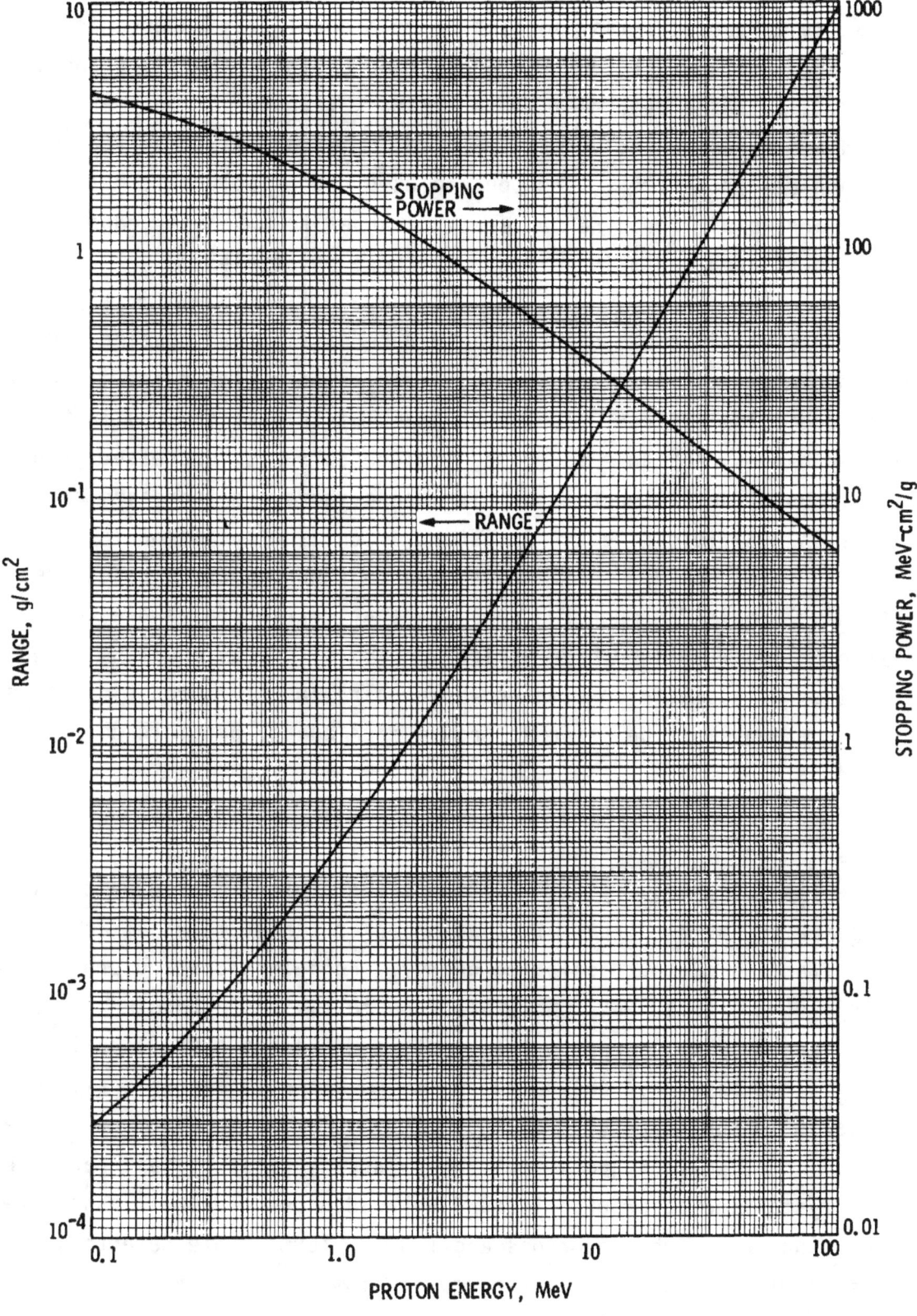

Figure 3.2 Stopping Power and Range Curves for Protons in Silicon.
Reference: Janni, AFWL-TR-65-150, 1966

Radiation produces many ionization-related effects in organic materials. These changes all result from the production of ions, free electrons, and free radicals. As a result of these actions, transparent polymers are darkened and crosslinking between main-chain members may drastically alter the mechanical properties. The contemplated use of polymeric materials in solar arrays will require the array designer to have knowledge of the ionization-related radiation effects in those materials.

The use of silicon dioxide as a surface passivation coating and dielectric material in silicon devices results in a wide range of ionization-related radiation effects. The development of trapped charges in the silicon dioxides can cause increased leakage currents, decreased gain, and surface channel development in bipolar transistors and increased threshold voltages in MOS field effect transistors. Ionizing radiation in silicon excites the electrons of the valence band to the conduction band, creating electron-hole pairs in much the same way that carrier pairs are generated by visible light. Although an optical photon of energy equal to or greater than 1.1 eV will create an electron-hole pair, roughly three times this amount of energy must be absorbed from a high energy particle to produce the same carriers.

3.1.2 Atomic Displacements

The loss of energy by fast electrons and protons caused by collision processes with the electrons of an absorber or target material accounts for a large fraction of the dissipated energy. For electrons and protons in the range of 0.1 to 10 MeV, these electron collisions determine the particle range in an absorber. Despite this fact, a different type of collision process is the basis for the damage which permanently degrades silicon solar cells in the space environment. The basis for this damage is the displacement of silicon atoms from their lattice sites by fast particles in the crystalline absorber. These displaced atoms and their associated vacancies undergo other reactions and finally form stable defects which produce significant changes in the equilibrium carrier concentrations and the minority carrier lifetime.

The displacement of an atom from a lattice site requires a certain minimum energy similar to that of other atomic movements. The energy of sublimation for a silicon atom is 4.9 eV. The energy for the formation of a vacancy in the silicon lattice is 2.3 eV. The displacement of an atom involves the formation of a vacancy, the formation of an interstitial atom and other electronic and phonon losses. It is reasonable to expect that the energy of displacement is several times larger than the energy of formation for a vacancy. Seitz has estimated that the displacement energy is roughly four times the sublimation energy.[3.1] Electron threshold energies of 145 keV and 125 keV have been reported by various investigators.[3.2 - 3.4] The following equation relates the electron threshold energy to the displacement energy.[3.1]

$$E_d = 2 \frac{m_e E_t}{M \, m_e c^2} (E_t + 2 m_e c^2) \qquad (3.1.2)$$

where
E_d = displacement energy (MeV)
E_t = threshold energy (MeV)
m_e = electron mass (1/1836)
M = atomic weight, Si (28)
$m_e c^2$ = electronic mass-energy equivalence (0.511 MeV)

The reported threshold energies indicate displacement energies of 12.9 eV or 11.0 eV, respectively.

Although proton threshold energies have not been determined, they can be calculated from the classical form of the above equation:

$$E_d = \left(\frac{4 M M_p}{(M_p + M)^2} \right) E_t \qquad (3.1.3)$$

where M_p = proton mass. The above values of displacement energies indicate proton or neutron thresholds of 97.5 or 82.5 eV in silicon. Since particles below the threshold energies cannot produce displacement damage, the space environment energy spectra are effectively cut off below these values.

For particles above the threshold energy, the probability of an atomic displacement can be described in terms of a displacement cross section. Using this concept, the number of displacements can be estimated from:

$$N_d = n_a \sigma \bar{\nu} \Phi \qquad (3.1.4)$$

where
- N_d = number of displacements per unit volume
- n_a = number of atoms per unit volume of absorber (5×10^{22} silicon atoms/cm^3)
- σ = displacement cross section (cm^2)
- $\bar{\nu}$ = average displacements per primary displacement
- Φ = radiation fluence (particles/cm^2)

3.1.2.1 Electron Displacement Damage

The displacement cross sections for fast electrons of various energies can be calculated from the relativistic generalization of the Rutherford scattering cross section equation.[3.1] For silicon, the calculated displacement cross section for 1 MeV electrons is about 68×10^{-24} cm^2 and increases only 10% for electron energies of 5 MeV and greater. The electron displaced silicon atom may receive enough energy to in turn displace other silicon atoms. The mechanism for these secondary displacements is Rutherford interactions for silicon atoms of energies greater than 10^3 eV and hard sphere collisions for lower energy atoms. Although different theories of the production of secondary displacement have been presented, their results are very similar. Using the model of Kinchin and Pease,[3.5] the average number of displacements in silicon is 1.53 for a 1 MeV electron. The electron energy variation of the various parameters is shown in Table 3.1.

Table 3.1
Silicon Displacement Parameters, Various Electron Energies

Electron Energy (Mev)	σ (10^{-24} cm^2)	$\bar{\nu}$	$\sigma\bar{\nu}$ (10^{-24} cm^2)	$n_a\sigma\bar{\nu}$ (cm^{-1})	$\Delta N_d/\Delta\Phi$
1	68	1.53	104	5.2	1.0
2	73	2.00	146	7.3	1.4
5	77	2.76	212	10.6	2.0
10	77	3.39	261	13.0	2.5
20	77	4.09	314	15.7	3.0
40	77	4.74	363	18.2	3.5

The direct result of the radiation is the production of vacant lattice sites (vacancies) and silicon atoms which come to rest in the interstices of the crystal lattice (interstitials). The distribution of vacancies will not be uniform, because the vacancies from secondary displacements will lie relatively close to the associated primary vacancy.

The experimental studies must be reviewed to gain a more complete model of displacement damage in silicon. Vacancies and interstitials are particularly mobile and unstable at room temperature. In n-type silicon, it has been shown that vacancies react with oxygen impurities to form close coupled vacancy-oxygen pairs (V-O) [3.6-3.9] (see Figure 3.3), and with impurity donor atoms, such as phosphorus and arsenic, to form close coupled vacancy-donor pairs (V-P, V-As) [3.6, 3.10] (see Figure 3.3). Both defects are electrically active and can become negatively charged by accepting an electron from the conduction band. The acceptor energy levels of the V-O and V-P pairs are 0.17 eV and 0.4 eV below the bottom of the conduction band. [3.11, 3.12] These defects are recombination centers and their formation during electron irradiation of n-type silicon reduces the minority carrier lifetime. [3.13] Since these defects are formed from single

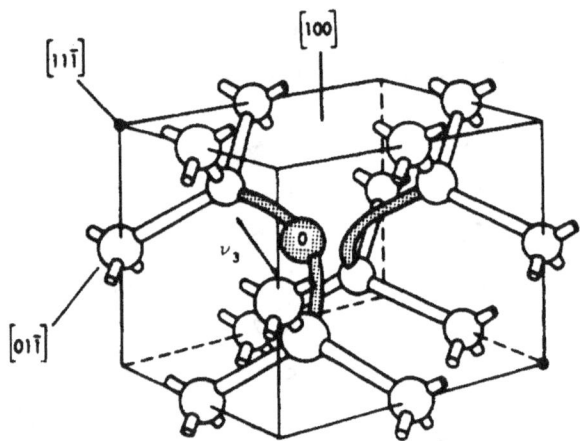

A model of the oxygen-vacancy complex or the Si-A center.

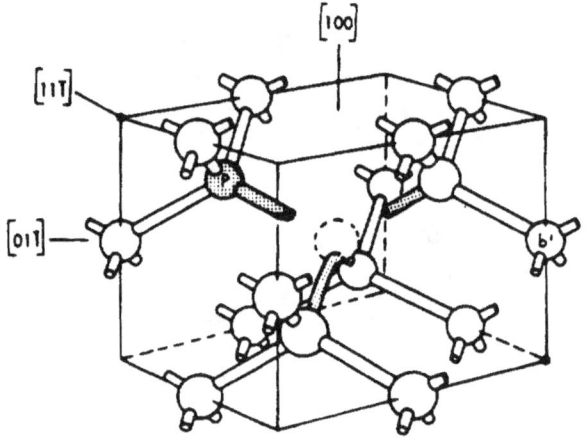

A model of the phosphorus-vacancy complex or the Si-E center.

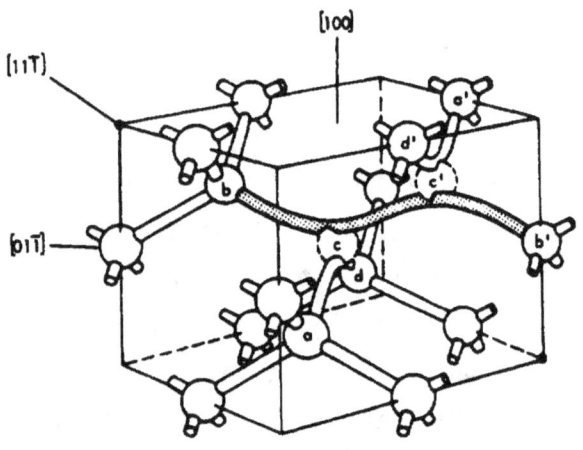

Model of the divacancy.

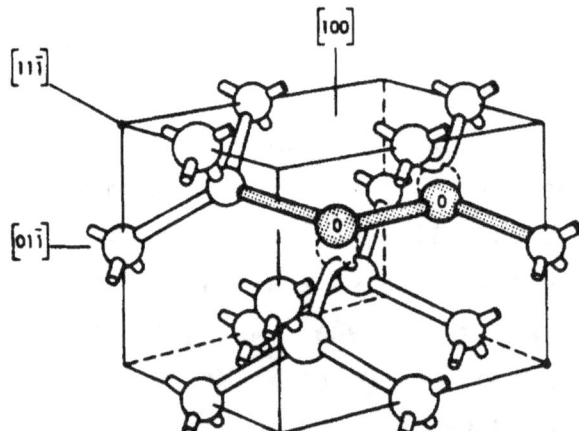

RCA tentative model for the K-center.

Figure 3.3 Atomistic Models of Radiation Defects in Silicon. Reference: M. M. Sokoloski, "Structure and Kinetics of Defects in Silicon" NASA TN D-4154, 1967

vacancies, considerations of mass action indicate that the formation of these defects might have the same variation with incident electron energy as that for the formation of single vacancies ($\sigma \bar{\nu}$). This relationship has been verified experimentally.[3.14] The V-P pair anneals rapidly near 150°C[3.15] and the V-O pair anneals rapidly near 350°C.[3.9] The introduction rates (change in defect concentration per unit fluence) for these defects are in the range of 0.1 to 0.3 cm^{-1} for 1 MeV electrons. Since the calculated displacement rate is 5.2 cm^{-1}, it appears that many of the vacancies are involved in other reactions at room temperature, such as recombination with interstitial atoms.

The electron irradiation of p-type silicon at room temperature results in a defect structure with net donor characteristics.[3.11, 3.12] This defect can donate an electron to (i.e., accept a hole from) the valence band. The energy level of this donor defect is located 0.27 to 0.30 eV above the top of the valence band. The room-temperature introduction rate of this defect in silicon by 1 MeV electrons is roughly 0.03 cm^{-1}. This value is considerably lower than those of defects found in n-type silicon. In addition, the introduction rate of this defect by 10 MeV electrons is about 16.5 times greater than that for 1 MeV electrons.[3.14] Since the single displacement rate increases by only a factor of 2.5 with that electron energy increase, this defect appears to involve a more complex structure. It has been shown that defects involving the coupling of more than one vacancy will result in defects with introduction rates which increase more rapidly with electron energy than does the displacement rate.[3.16, 3.17] Two defect structures (divacancy[3.18, 3.19] and "K" center[3.20]), which have been studied by electron spin resonance techniques, may explain this behavior. These defects, shown in Figure 3.3, involve the coupling of two vacancies in each defect. Several attempts to determine the dominant recombination center in electron irradiated p-type silicon have yielded conflicting results.[3.11, 3.12, 3.21, 3.22] Experiments by Gorodetskii et al. have indicated that a defect with an energy level in the range of 0.27 ± 0.02 eV above the top of the valence band controls recombination in electron irradiated p-type silicon.[3.22] This conclusion is consistent with the known energy dependence of p-type silicon in that the diffusion length damage coefficient has been shown to vary with electron energy 3.23 in the same manner as the introduction rate of the $E_v + 0.3$ eV level defect. 3.12

3.1.2.2 Proton Displacement Damage

The production of displacement damage in silicon by energetic protons is considerably different because the displacement cross sections are several orders of magnitude larger than those for fast electrons and vary rapidly with proton energy. The displacement cross section for protons in silicon is 3.1

$$\sigma = 4\pi a_0^2 \left[\frac{M_p Z_p^2 Z_{Si}^2 E_R^2}{M\, E\, E_d} \right]$$

where
a_0 = Bohr radius (5.3×10^{-9} cm)
E_R = Rydberg energy for hydrogen (13.6 eV)

which yields a cross section for 1 MeV protons in silicon of 3.5×10^{-20} cm^2.

Kinchin and Pease 3.5 give the following relationship for the average total number of displaced atoms produced for every primary knock-on including the primary knock-on:

$$\bar{\nu} = \frac{1}{2}\left[1 + \ln(\Lambda E/2E_d)\right]$$

where
$$\Lambda = \frac{4 M M_p}{(M + M_p)^2}$$

The average number of atomic displacements, , resulting from such a primary displacement caused by a 1 MeV proton is 4.8. Using equation (3.1.4) the displacement rate is found to be 8500 cm^{-1} per proton for 1 MeV protons in silicon. The range of a 1 MeV proton in silicon is only 17.5 m; therefore its energy and displacement rate will change rapidly after it enters a silicon crystal. The variation of the displacement rate with proton energy has been calculated by several authors. These results are shown in Figure 3.4. [3.24-3.28] Although there are some differences in results, the displacement rate is proportional to (lnE/E) for protons of energies between 1 and 10 MeV. Above 10 MeV, the various models differ as to the relative influence of Rutherford scattering, nuclear scattering, and inelastic processes of spallation. Experimentally measured defect introduction rates for proton irradiation of silicon are less than one tenth of the calculated displacement rates. The defect energy levels in proton irradiated silicon are in some respects similar to those previously discussed for electron irradiated silicon. [3.29-3.31] The proton damage, however, is highly inhomogeneous because the numerous secondary displacements occur near the site of the primary displacement.

In addition to the displacement rates discussed above, Kinchin and Pease [3.5] have computed the total number of displacements, $N_{td}(E)$, a projectile will produce in a material as it enters with energy E and comes to rest in the material. For the specific case of protons in silicon their formula reduces to:

$$N_{td} = \frac{P(10^6 E - 567.3) + 283.6}{12.9}$$

where

$$P = \frac{6.81 \times 10^{-5}[1 + \ln(5162\ E)]}{\ln(12.82\ E)}$$

and

E = Proton energy (MeV) $E \gtrsim 0.1$ MeV

Bulgakov and Kumakhov [3.26] also give a relation for the total number of displacements which has a wider range of validity but is more complex.

3.1.2.3 Neutron Displacement Damage

Neutron displacement damage in silicon is characterized by two important differences. The silicon displacement cross section for a 1 MeV neutron is 2.4×10^{-24} cm^2. This value is well below those for 1 MeV protons and 1 MeV electrons. For this reason, the number of primary displaced silicon atoms will be relatively small. The second difference involves the amount of energy transferred to the displaced silicon atom by the neutron. Since the 1 MeV

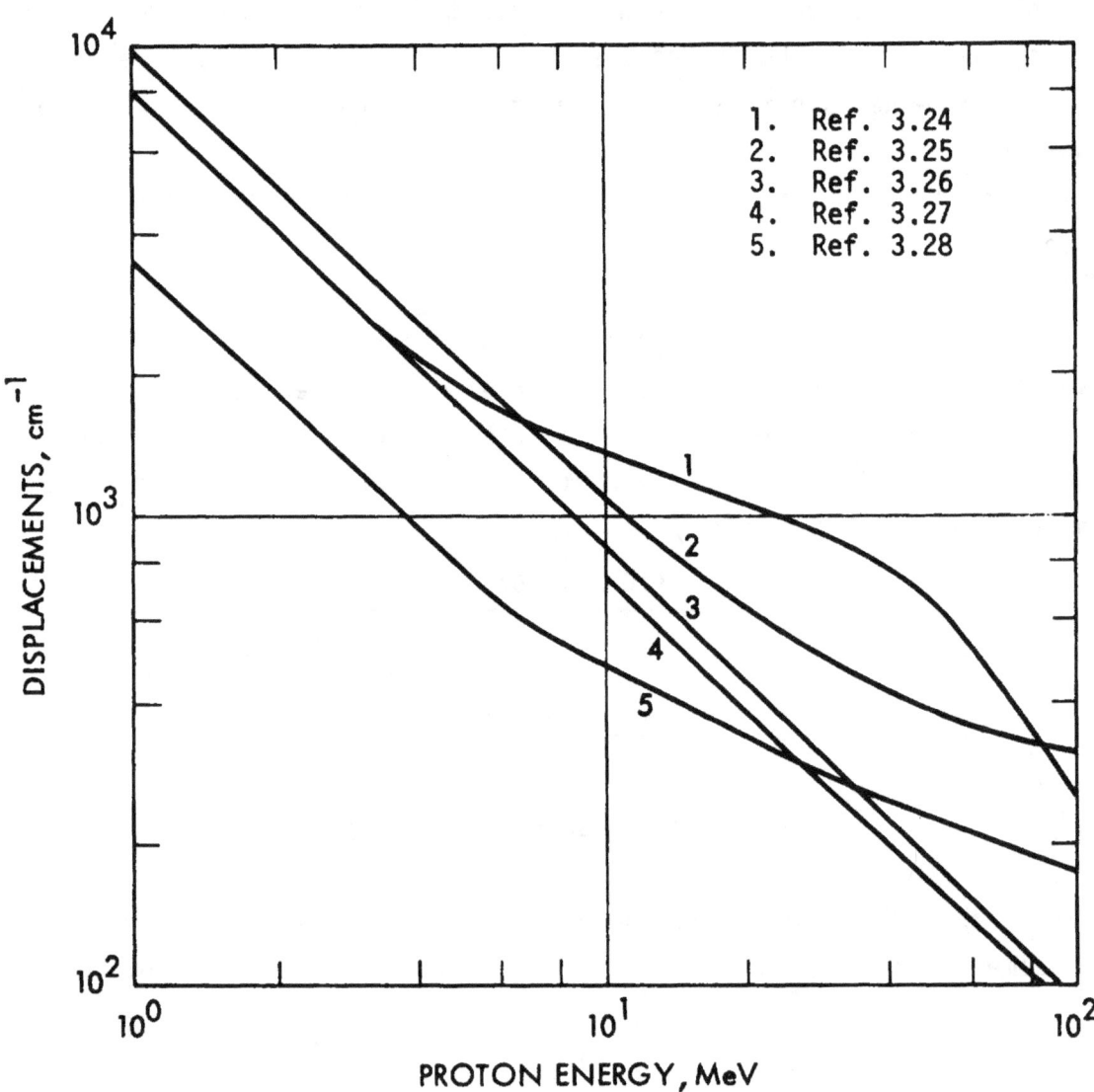

Figure 3.4 Energy-Dependent Displacement in Proton-Irradiated Silicon

neutron-silicon interaction is a hard sphere rather than coulombic collision, an average of about 70 keV is transferred to the recoiling silicon atoms. The subsequent secondary collisions between silicon atoms will displace about 1500 silicon atoms. This displacement damage will be clustered near the site of the primary displacement. Theoretical models of the neutron damage indicated that the high concentration of electrically active defects in the cluster causes the center of the cluster to behave as intrinsic silicon.[3.32] This intrinsic silicon core is separated from the bulk silicon by a layer of space charge. Extensions of this model have been used to explain the majority carrier removal and minority carrier recombination behavior of neutron irradiated silicon.[3.33-3.35]

Considerable work has been performed in recent years to study radiation-induced defect structures using deep-level transient spectroscopy (DLTS), electron paramagnetic resonance (EPR) and capacitance-voltage measurements.[3.36-3.42] Although these studies appear to be sensitive mostly to majority carrier traps, they have produced some results which may have some bearing on solar cell performance. Weinberg and Swartz[3.37] have shown that the slight annealing of boron-doped, electron-irradiated silicon is due to the disappearance of a boron-related defect as the temperature is raised to 150° C. The disappearance of this defect is followed by a reverse anneal attributed to the growth of a boron-oxygen-vacancy defect (or possibly a silicon di-interstitial). Other defects which are possibly related to carbon do not anneal until temperatures reach 400° C. These types of studies are continuing in an effort to accurately identify the defects introduced by processing (e.g., using dopants such as gallium or aluminum rather than boron) or by irradiation.[3.38-3.40] These studies may have practical applicability to solar cell design if a specific impurity could be identified as a major defect center constituent in silicon and, following its identification, a method of removing it from silicon could be devised.

The main importance of the displacement defects produced by the irradiation of silicon solar cells is in their effect on the minority carrier lifetime of the silicon. In particular, the lifetime in the bulk p-type silicon of an n/p solar cell is the major radiation sensitive parameter. Since minority carrier lifetimes are inversely proportional to the recombination rates, the reciprocal lifetime contributions caused by various sets of recombination centers can be added to determine the inverse of the lifetime as follows:

$$\frac{1}{\tau} = \frac{1}{\tau_0} + \frac{1}{\tau_e} + \frac{1}{\tau_p} + \ldots \qquad (3.1.5)$$

where τ = minority carrier lifetime
 τ_0 = minority carrier lifetime before irradiation
 τ_e = minority carrier lifetime due to electron irradiation
 τ_p = minority carrier lifetime due to proton irradiation

One of the most commonly used analytical tools for the determination of the particle type and energy dependence of degradation in silicon solar cells has been developed from the basic relationship for lifetime degradation:

$$\frac{1}{\tau} = \frac{1}{\tau_0} + K_\tau \Phi \qquad (3.1.6)$$

where τ = final minority carrier lifetime
 τ_0 = initial minority carrier lifetime
 Φ = irradiation fluence
 K_τ = damage coefficient (lifetime)

Minority carrier diffusion length is a more applicable and more easily determined parameter for solar cell analysis than minority carrier lifetime. Using $L^2 = D\tau$, the above expression becomes:

$$\frac{1}{L^2} = \frac{1}{L_0^2} + K_L \Phi \qquad (3.1.7)$$

where L = final minority carrier diffusion length
 L_0 = initial minority carrier diffusion length
 Φ = particle fluence
 K_L = damage coefficient (diffusion length)
   = K_τ/D

When the fluence is sufficiently high so that $L \ll L_0$ we have:

$$K_L = 1/L^2 \Phi \qquad (3.1.8)$$

If a plot of $\ln L$ vs. $\ln \Phi$ exhibits a $-1/2$ slope, the damage coefficient, K_L, can be used to uniquely define the particle type and energy dependence of silicon solar cell degradation.

The minority carrier lifetime or diffusion length in an irradiated solar cell may be a function of the concentration of excess or nonequilibrium minority carriers present in the semiconductor. In solar cells, this behavior is referred to as injection level dependence. This behavior is usually associated with solar cells damaged with high energy protons or neutrons. Gregory [3.33] has shown that the injection level dependence of lifetime in neutron-irradiated solar cells does not follow classical predictions and has proposed a model based on the behavior of clustered damage. The methods of measuring minority carrier lifetime or diffusion length often involve the injection of excess minority carrier concentrations which are many orders of magnitude smaller than those found in solar cells operating in space. Such low injection level methods are inadequate for the generation of data for the prediction of proton- and neutron-irradiated solar cell performance in space. One method of measuring diffusion lengths at high injection level is discussed in Section 2.5.

3.2 Theory of Silicon Solar Cell Damage

The basic solar cell equations can be used to describe the changes which occur during irradiation. This method would require data regarding the changes in the light generated current, series resistance, shunt resistance, and the basic diode parameters of saturation current and diode quality factor. Although such a method would be a logical analysis, most investigations have not reported enough data to determine the variations in the above parameters. The usual practice in the study of solar cell damage has been to reduce the experimental data in terms of changes in the cell short circuit current, open circuit voltage, and maximum power.

It is also possible to characterize solar cell damage in terms of the changes in the minority carrier diffusion length. Since the diffusion length can be measured experimentally and is a measure of the amount of displacement damage in the base of the solar cell, this method has been widely used. There are several practical and fundamental limitations to this scheme. The most serious limitations involve the evaluation of low energy proton damage in terms of diffusion length. Very low energy protons do considerable displacement damage within the junction space charge region of a solar cell. This nonuniform damage increases the diode saturation current (I_o) and quality factor (n) by mechanisms which are not related to minority carrier diffusion. This damage can cause serious reduction in solar cell V_{oc} without changing the cell diffusion length. In addition, the relation between diffusion length and the solar cell output parameters is not well defined, diffusion length is more difficult to measure than cell output parameters (particularly in the case of proton irradiated cells) and accurate measurement of diffusion length of thin or drift field cells is extremely difficult. Because of these problems, methods have been evolved to evaluate solar cell radiation effects in terms of common engineering output parameters. Experience has shown that the variation of common solar cell output parameters during irradiation can be described as shown for I_{sc} in the following case:

$$I_{sc} = I_{sco} - C \log\left(1 + \frac{\Phi}{\Phi_x}\right) \qquad (3.2.1)$$

The Φ_x represents the radiation fluence at which I_{sc} starts to change to a linear function of the logarithm of the fluence. The constant C represents the decrease in I_{sc} per decade in radiation fluence in the logarithmic region. Although the above relationship is empirical, there is some theoretical justification for the expression. Several observers have reported that the relation between the solar cell short circuit current and the diffusion length is as follows: [3.43, 3.44]

$$I_{sc} = A \ln L + B \qquad (3.2.2)$$

The constants A and B are dependent upon the spectral content and intensity of the light source used to measure I_{SC}. Tada has shown that the above expression is theoretically valid over a wide range of diffusion lengths for tungsten illumination and to a lesser range under solar illumination.[3.45] A previously discussed relation, equation (3.1.7) can be transformed as follows:

$$L = \left(K_L \Phi + \frac{1}{L_0^2}\right)^{-1/2} \qquad (3.2.3)$$

and substituted in equation (3.2.2.). The resulting expression

$$I_{SC} = B - \frac{A}{2} \ln\left(K_L \Phi + \frac{1}{L_0^2}\right) \qquad (3.2.4)$$

has the same form as equation (3.2.1).

The variation of solar cell V_{OC} during irradiation also may be empirically characterized by an expression similar to equation (3.2.1).

$$V_{OC} = V_{OCO} - C' \log\left(1 + \frac{\Phi}{\Phi_X}\right) \qquad (3.2.5)$$

In general, the open circuit voltage of a silicon solar cell can be represented by the following equation which was discussed in Chapter 1:

$$V_{OC} = \frac{kT}{q} \ln\left(\frac{I_{SC}}{I_0} + 1\right) \qquad (3.2.6)$$

In using this expression, it is assumed that the saturation current (I_0) is dominated by the diffusion component. In such cases the saturation current density is given by equation (1.2.5). If this expression is combined with equation (3.2.3), the following expression for the saturation current as a function of radiation fluence is obtained:

$$I_0 = q\, D_n\, n_p\, S \left(K_L \Phi + \frac{1}{L_0^2}\right)^{1/2} \qquad (3.2.7)$$

where S is the cell area. Equations (3.2.4) and (3.2.7) can be substituted into equation (3.2.6) to obtain the following expression:

$$V_{oc} = \frac{kT}{q} \ln \left[\frac{B - \frac{A}{2} \ln \left(K_L \Phi + \frac{1}{L_0^2} \right)}{q \, D_n \, n_p \left(K_L \Phi + \frac{1}{L_0^2} \right)^{1/2}} \right] \qquad (3.2.8)$$

The radiation fluence term (Φ) appears twice in the above expression. The fluence term in the numerator will have a much lesser effect on V_{oc} than that in the denominator because it varies as the logarithm of the fluence rather than as the square root of the fluence. It appears therefore that the V_{oc} variation with radiation fluence is dominated by the denominator of equation (3.2.8) and can be approximated by equation (3.2.5).

The maximum power (P_{max}) of a solar cell can be represented as the product of I_{sc}, V_{oc}, and a constant as follows:

$$P_{max} = F \, I_{sc} \, V_{oc} \qquad (3.2.9)$$

where F is the form (or fill) factor. The fill factor, F, is relatively insensitive to electron radiation which penetrates uniformly through a solar cell. In this case, the variation of P_{max} with irradiation is the same as that for the product of I_{sc} and V_{oc}. Equations (3.2.1) and (3.2.5) can be substituted into (3.2.9) and the resulting expression approaches the form of:

$$P_{max} = P_{maxo} - C'' \log \left(1 + \frac{\Phi}{\Phi_x} \right) \qquad (3.2.10)$$

Expressions of this form are found to closely describe the variation of P_{max} during irradiation.

3.3 The Concept of Damage Equivalence

The wide range of electron and proton energies present in the space environment necessitates some method of describing the effects of various types of radiation in terms of a radiation environment which can be produced under laboratory conditions. Since the changes in most solar cell parameters due to irradiation are in some way related to the minority carrier diffusion length, it is possible to determine an equivalent damage based upon this parameter. In Figure 3.5, the diffusion length changes are shown for 10 ohm-cm, n/p silicon solar cells which have been subjected to several different types of irradiation. The results are described by equation (3.2.3) where the constant K_L is dependent upon the radiation type.

The concept of damage equivalence can alternatively be based on common solar cell parameters. The variation of short circuit current density for 10 ohm-cm n/p solar cells irradiated in various environments is shown in Figure 3.6. The I_{SC} variation in each environment is described by equation (3.2.1). In this case two constants, C and Φ_x, are required to describe the changes in I_{SC}. Experience has shown that the constant C, under solar simulator illumination, does not vary greatly for different radiation environments. For electron irradiations in the 1 MeV and greater range, C is approximately 4.5 to 5.5 mA/cm^2-decade. For proton and neutron irradiations, C approaches 6 to 7 mA/cm^2-decade. For solar cells with the same starting I_{SC}, the constant Φ_x is a measure of the damage effectiveness of different radiation environments. The constant Φ_x for a particular radiation can be determined graphically on a semi-log plot at the intersection of the starting I_{SC} and the extrapolation of the linear degradation region.

Since the value of Φ_x is dependent upon the starting I_{SC} value, it is not a good practical measure for relative damage effectiveness. It has been the practice to define an arbitrary constant referred to as the critical fluence (Φ_c). One method of defining this value is that fluence which degrades a solar cell parameter 25% below its unirradiated state. Such a parameter is valid only when comparing cells with similar initial parameters. To eliminate this problem, critical fluence may be defined

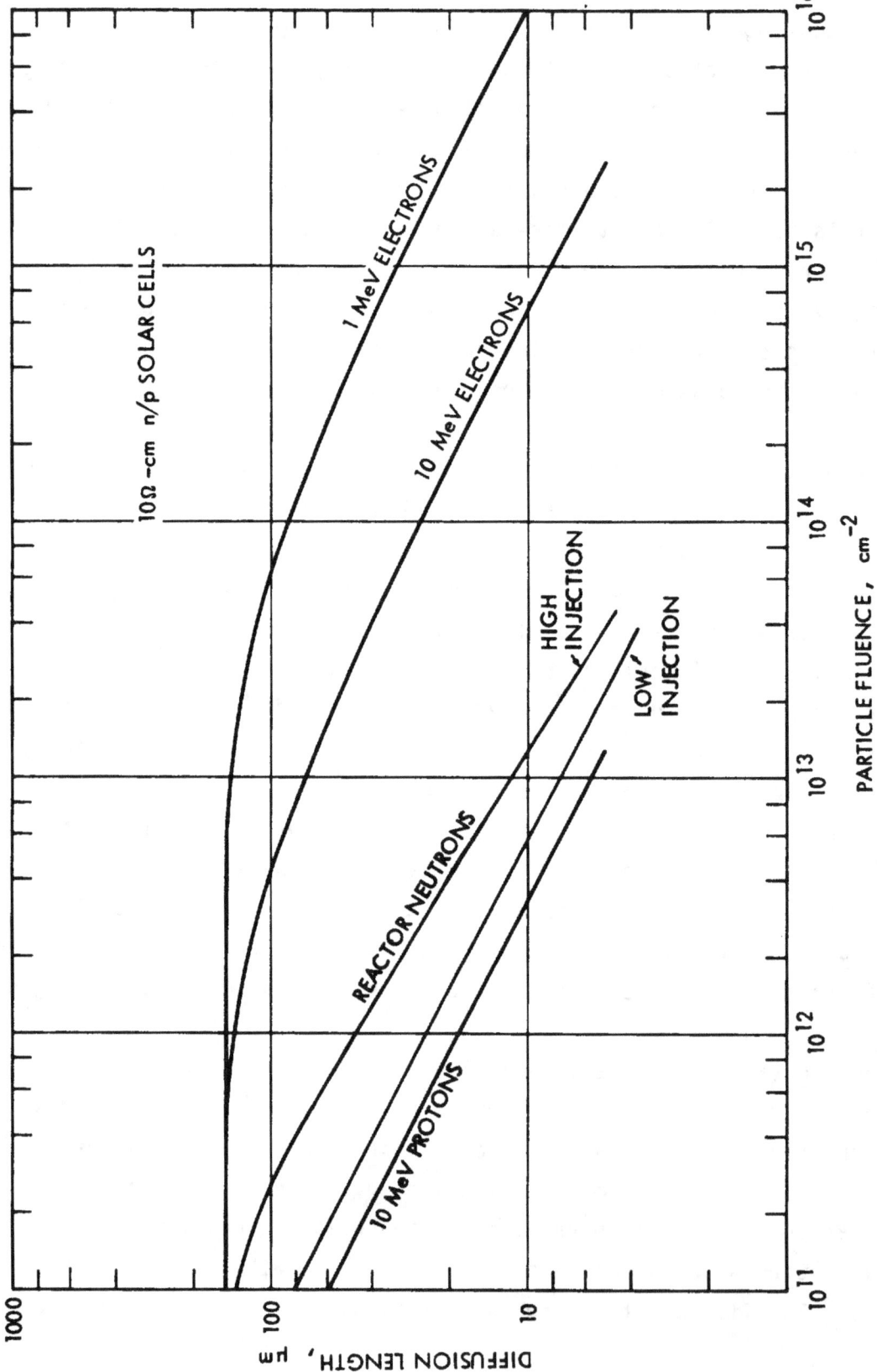

Figure 3.5 Variation of Solar Cell Diffusion Length with Fluence for Various Radiations

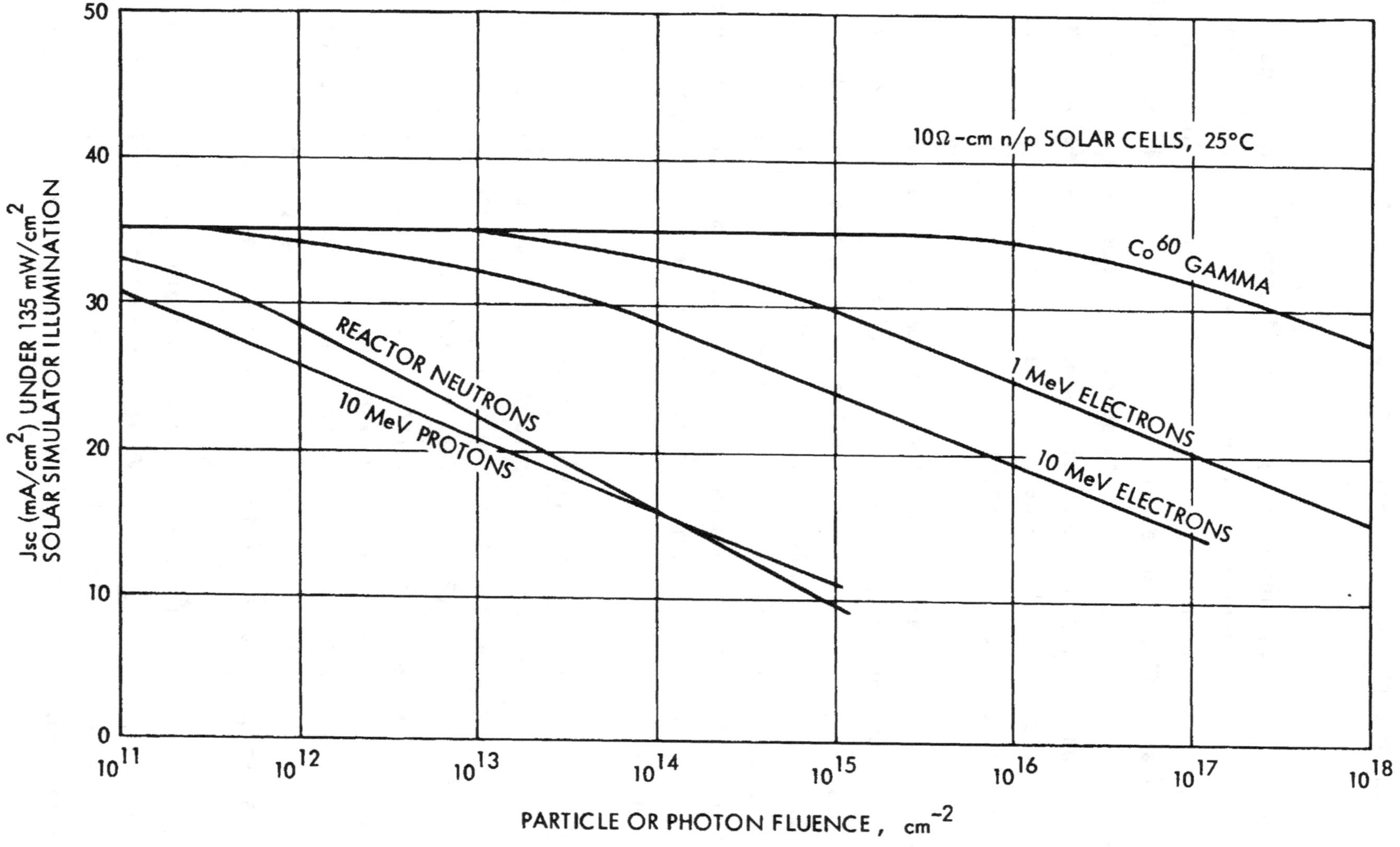

Figure 3.6 Variation of Solar Cell Short Circuit Current Density with Fluence for Various Radiations

alternatively as that fluence which will degrade a cell parameter to a certain value.

By use of the critical fluence (Φ_C) or the diffusion length damage coefficient (K_L), it is possible to construct a model in which the various components of a combined radiation environment can be described in terms of a damage equivalent fluence of a selected monoenergetic particle. 1 MeV electrons are a common and significant component of space radiation and can be produced conveniently in a test environment. For this reason, 1 MeV electron fluence has been used as a basis of the damage equivalent fluences which describe silicon solar cell degradation.

The use of the damage equivalent fluence scheme involves two separate problems. The first problem is to adequately describe the degradation of an unshielded silicon solar cell under 1 MeV electron irradiation under laboratory conditions (i.e., normal incidence). The second problem is to reduce the effect of the space radiation environment (i.e., continuous energy spectra of electrons and protons, isotropic incidence) on a shielded silicon solar cell to a damage equivalent fluence of 1 MeV electrons under laboratory conditions.

3.4 1 MeV Electron Irradiation of Silicon Solar Cells

The effects of 1 MeV electron laboratory irradiation of solar cells are reviewed and discussed in this section. Data will be presented in Section 3.13 which will form the basis for estimating solar cell performance, after the space radiation environment is reduced to a damage equivalent 1 MeV fluence. A very large volume of work has been reported concerning the effects of 1 MeV electron irradiation on silicon solar cells. However, this section considers only solar simulator data and is also limited to the types of solar cells currently in common use on spacecraft.

Currently, n/p solar cells are in use as a primary power source on nearly all earth orbiting satellites. Variations in base resistivity and cell thickness cause significant differences in the response to 1 MeV electron irradiation.[3.23, 3.46, 3.53, 3.54] Other variables such as the irradiation temperature in the range of 200 to 370 K,[3.47] and p-type base dopant (boron vs. aluminum) have been shown to have little or no effect on the solar cell response to radiation.[3.48-3.51]

The variation of n/p solar cell response with base resistivity has been studied and reported for the range of 1 to 20 ohm-cm.[3.23, 3.52] Current n/p solar cell usage is confined to the ranges of 1 to 3 ohm-cm and 7-13 ohm-cm. Cells in the base resistivity range of 1-3 ohm-cm have greater initial maximum power output than cells in the 7 to 13 ohm-cm range. The radiation hardness of n/p cells in the 7 to 13 ohm-cm range is greater than that of the 1 to 3 ohm-cm range, when the hardness is determined by parameters such as the critical fluence (Φ_c) or diffusion length damage coefficient (K_L). As a result, 10 ohm-cm cells have greater maximum power output after a certain radiation level is reached. This crossover point depends upon cell thickness but is approximately 1×10^{14} 1 MeV electons per cm^2.

Solar cell thickness has been shown to have a strong effect on the output parameters of irradiated cells.[3.46] Cell thickness does not affect measures of inherent hardness such as the critical fluence (if properly determined) or the diffusion length damage coefficient. The thickness does, however, significantly affect the cell output parameters during the initial or low fluence stage of an irradiation. JPL data showing output parameters (I_{sc}, V_{oc}, P_{max}, V_{mp}, and I_{mp}) as a function of electron fluence (1 MeV) are shown at the end of this chapter. The cells discussed here, which are thicker than 200 microns, are available on a production basis from solar cell manufacturers. The 50 and 100 micron cells are custom made, but the data are included since they are expected to be representative of production line cells in the near future. Temperature of the cells during measurement was 28°C.

3.5 Effect of Electron Energy on Solar Cell Degradation

The concept of damage equivalent 1 MeV electron fluence requires some method of evaluating the damage effectiveness of electrons of various energies. This effectiveness can be measured by the diffusion length damage constant (K_L) or solar cell critical fluence (Φ_c) for various electron energies. Experimental data have been reported for the electron energy range of 1 to 3 MeV [3.55] and from 0.6 to 40 MeV. [3.23] The results of these studies are in essential agreement and the results of Reference 3.23 are shown in Figure 3.7 (K_L) and Figure 3.8 (Φ_C). In this case Φ_C is defined as that fluence which degrades I_{sc} to 19 mA/cm^2 under 100 mW/cm^2 of tungsten light. In both figures, data are shown for cells of various resistivities. The short circuit current is directly related to the minority carrier diffusion length in the base region. Some important observations can be made from these data. The relative variations of the K_L and Φ_C^{-1} with electron energy are identical. The relative variations of both parameters with cell base resistivity are also identical. On the basis of the experimental data, one can therefore define a relative damage effectiveness for each electron energy which will be a measure of the ratio of that electron fluence at a given energy to the 1 MeV electron fluence necessary to degrade an n/p solar cell to the same output parameter value. For instance, if a given 10 MeV electron fluence degrades a solar cell to a certain state of damage, then a 1 MeV electron fluence 16.5 times that of the 10 MeV electron fluence would be required to degrade the same cell to the same state. This relationship will hold regardless of whether 2 or 10 ohm-cm resistivity cells are under consideration.

Wysocki reported data at 0.8 and 5.8 MeV which indicated that the relative electron damage constant increased more rapidly with energy. [3.56] Gorodetskii, et al., [3.57] reported data in rough agreement with References 3.23 and 3.55 below 2 MeV, but indicate a much slower rise above that energy. More recent studies by Bernard, et al., [3.44] and Lesbre [3.50] indicate good agreement with the results in References 3.23 and 3.55 up to 3 MeV and 4.5 MeV, respectively.

3.6 Effect of Proton Energy on Solar Cell Degradation

The concept of damage equivalent 1 MeV electron fluence can be extended to the effects of proton irradiation. The problem is more complex in the proton case, because the range of protons below 5 MeV is less than the usual solar cell thickness. For this reason, low energy protons produce non-uniform damage. This situation is further complicated by the fact that the

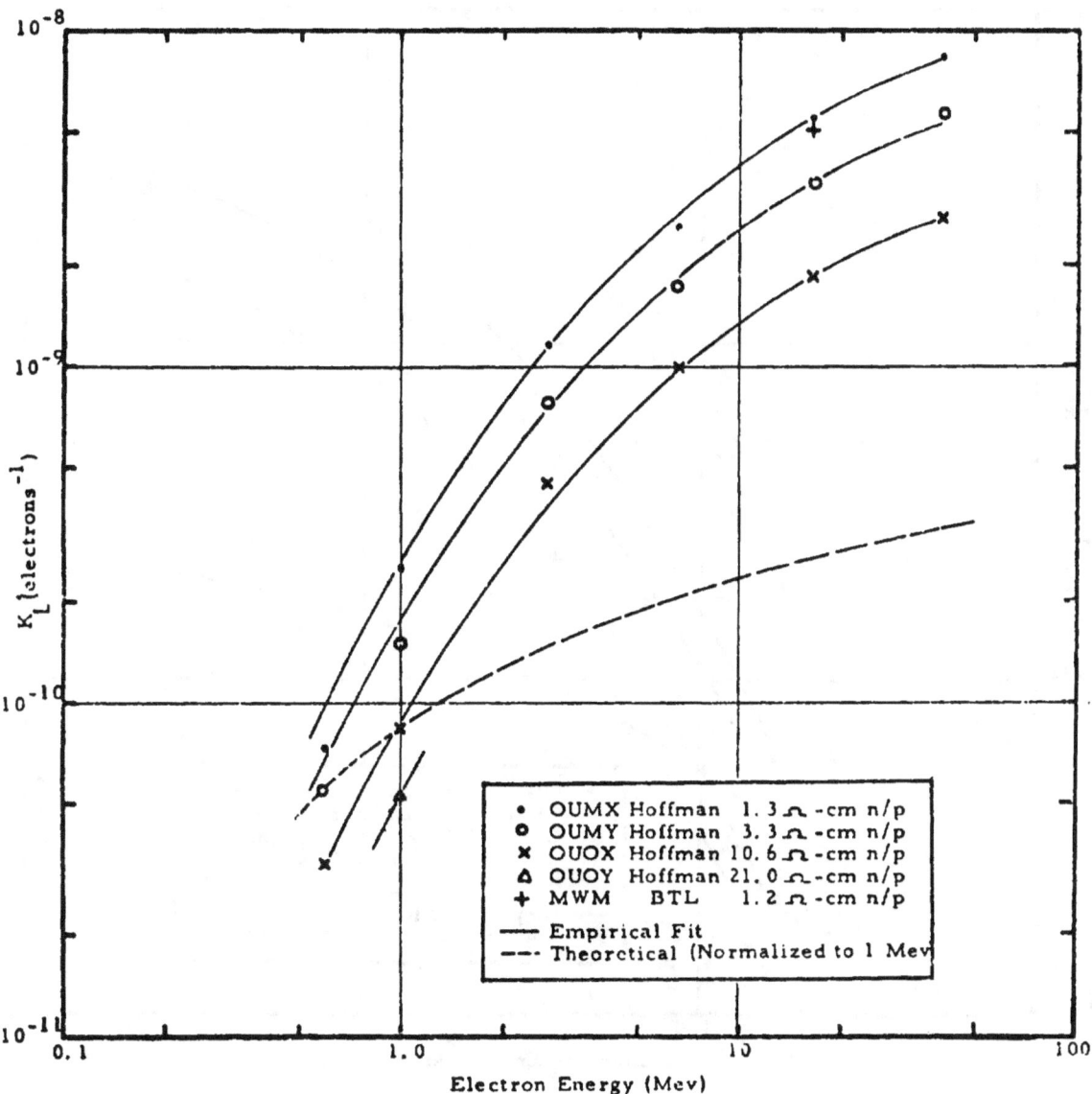

Figure 3.7 Electron Energy Dependence of K_L Values for n/p Silicon Solar Cells [3.23]

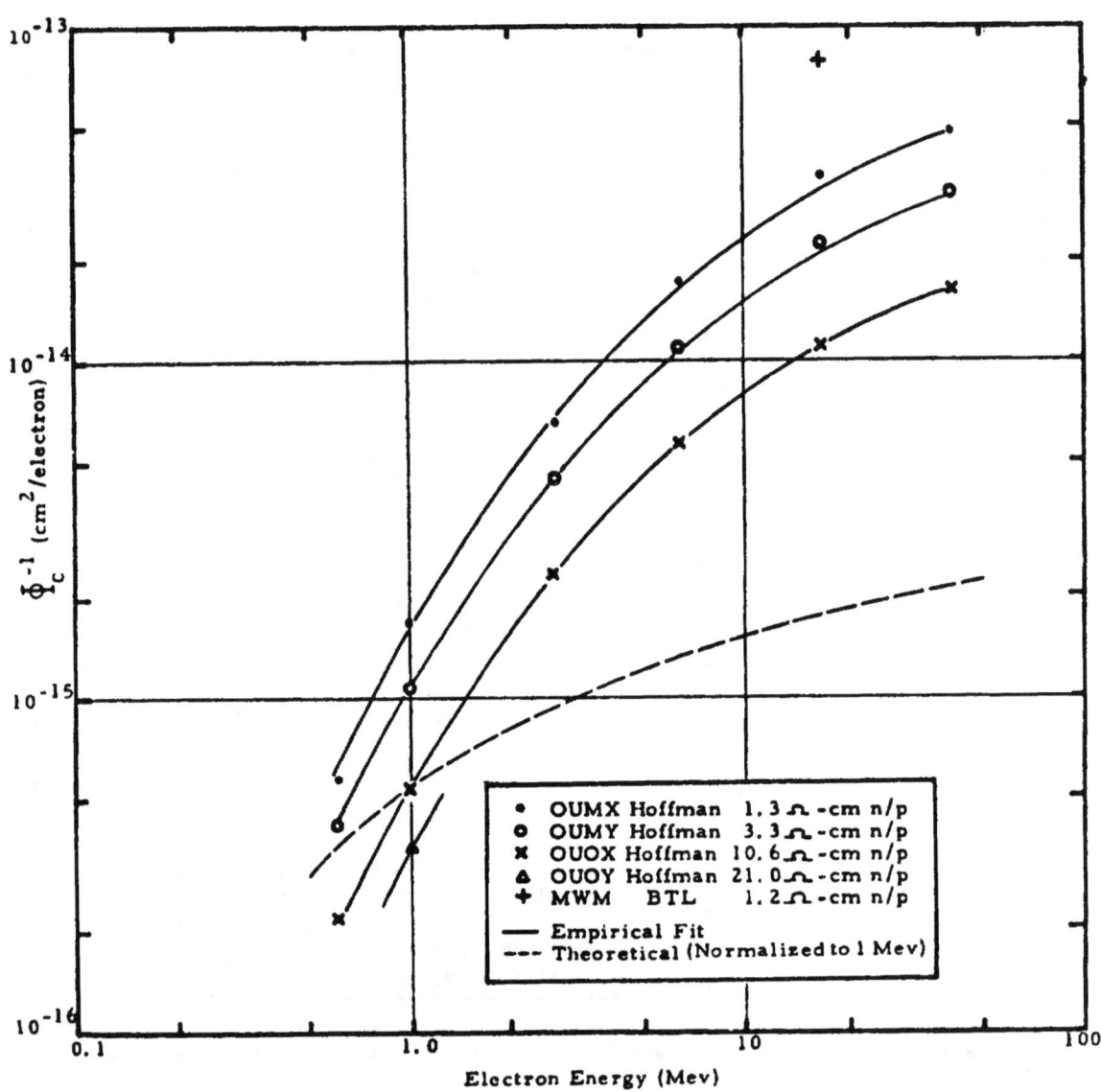

Figure 3.8 Electron Energy Dependence of Φ_c^{-1} Values for n/p Silicon Solar Cells [3.23]

damage produced per unit path length increases as the proton energy decreases. As a result, when a low energy proton is stopped in a solar cell, a large amount of damage is concentrated at the end of the proton track.

When radiation damage is uniform throughout a solar cell, the relative effectiveness of various energy particles is the same when measured by the diffusion length damage coefficients, or critical fluences determined by cell parameters such as I_{sc}, V_{oc}, or P_{max}. This is demonstrated by the data of Figures 3.7 and 3.8. In the case of protons with energies greater than 5 MeV, the damage to solar cells is relatively uniform. In this high energy range, the general concept of equivalency is directly applicable. At lower proton energies, the general concept of equivalency is not applicable; however, it can be used in a restricted manner as discussed below.

Early experimental studies of the variation of damage in n/p silicon solar cells with higher proton energies indicated conflicting results. The results reported by workers at BTL [3.58] and TRW [3.59] are shown in Figure 3.9 in normalized form. The major difference involves the behavior of the damage constant at proton energies greater than 10 MeV. Recent experimental investigations have confirmed that the variation of damage in this proton energy range is very small. [3.60 - 3.62] The results of these recent investigations are also shown in Figure 3.9.

The degradation of n/p solar cells irradiated with protons of energies below 3 MeV is more complex because of the nonuniform nature of the damage. Several experimental studies of low energy proton effects on unshielded solar cells have been reported in the literature. [3.48, 3.63-3.69] Although there are some differences in the reported results, a few general observations can be made. Protons in the energy range from 1.5 to 3 MeV produce a maximum in relative radiation damage in silicon solar cells. The relative damage to silicon solar cell V_{oc} and P_{max} due to low energy protons is more severe than that exhibited by I_{sc}.

Proton damage in silicon solar cells can be normalized to the damage produced by protons of one energy. The proton energy employed for normalization of relative damage should be close to that producing maximum damage

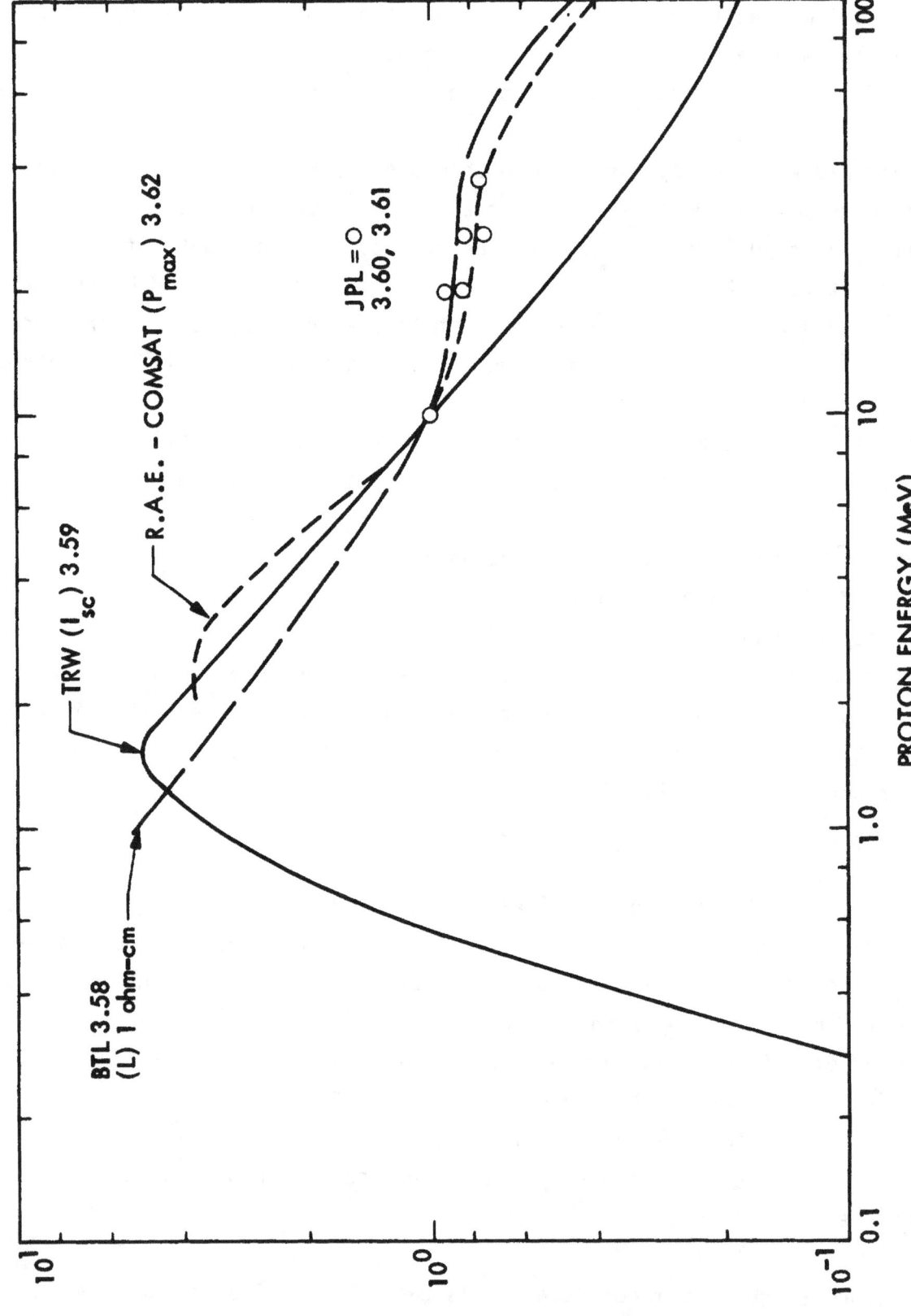

Figure 3.9 Comparison of Relative Damage Coefficients, Proton-Irradiated n/p Silicon Solar Cells

in space environments, produce relatively uniform damage, and be available for laboratory evaluations. The use of 10 MeV proton damage is based on a compromise of the above requirements. The results of several studies of proton damage have been summarized in terms of relative silicon solar cell damage as a function of proton energy.[3.48, 3.60-3.63] These relative damage results, normalized to 10 MeV proton damage, are shown in Figure 3.10. The results in Figure 3.10 have been shown to hold for both 10 ohm-cm and 2 ohm-cm solar cells at proton energies greater than 10 MeV.[3.60, 3.61]

It is emphasized that the results in Figure 3.10 are obtained by normal incidence laboratory irradiation of solar cells from the front side. If similar data were prepared for normal incidence rear irradiations, the result would be similar for proton energies above 10 MeV.[3.60] For cells of 200 to 300 microns thickness, the effects due to rear incidence protons with energies below 10 MeV would be much lower than shown in Figure 3.10.[3.70] The lower effectiveness occurs because rear incident low energy protons have insufficient range in silicon to cause atomic displacements in the active region of the solar cell. However, 2 MeV protons have sufficient energy to reach the junction through 50 micron thick cells. Since the much higher values of the V_{oc}, P_{max} damage coefficients for low proton energies are due to the effects they produce near the junction, it should be pointed out that these higher values should only be used when the protons are incident on the front surface of the cells. When considering low energy protons incident on the rear cell surface, such as for the case of solar panels using lightweight substrates, only the I_{sc} damage coefficients should be used.

The variation of solar cell output parameters with 10 MeV proton fluence is described by equations (3.2.1), (3.2.5) and (3.2.10) in much the same way as is done for 1 MeV electrons. The values of the constants C, C', and C" tend to be somewhat greater than those found for 1 MeV electron irradiation. These values determine the decrease in solar cell output parameter per decade of radiation fluence. The fact that these constants are somewhat different for electron and proton irradiation indicates that the

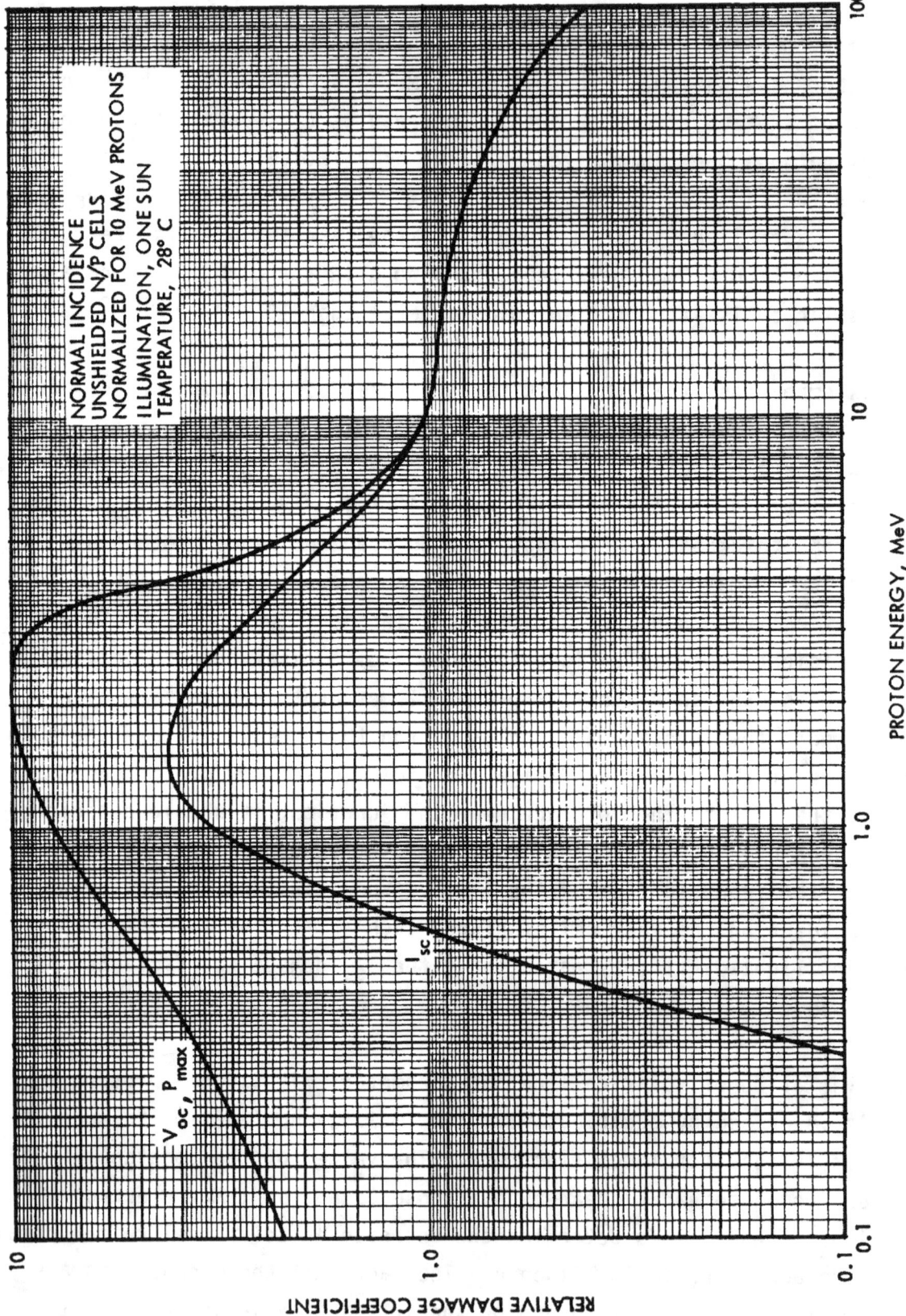

Figure 3.10 Relative Damage Coefficients for Proton-Irradiated n/p Silicon Solar Cells

concept of equivalency between the different types of radiation has limitations and is basically an approximation. This equivalence is further discussed in Chapter 6.

3.7 <u>Junction Effects of Low Energy Protons</u>

In addition to the low energy proton effects on unshielded cells discussed in the previous section, there are two aspects of low energy proton damage to be considered. These involve the effects of low energy protons on small unshielded gap areas on the front of solar cells and on unshielded backs of solar cells.

When the ATS-1 and Intelsat II-F4 satellites suddenly exhibited degradations in power output of the order of 20% in weeks to a month after launch, the importance of low energy proton damage was dramatically demonstrated. Subsequent efforts related this anomalous degradation to the bombardment of narrow exposed surface areas of the solar cells by the intense low energy proton fluence existing at synchronous altitude. The exposed areas resulted from slightly undersized or improperly applied coverglasses which exposed up to a 0.038 cm (15 mil) strip of solar cell surface. The high-intensity low energy proton fluence, though incapable of penetrating the solar cell to a depth of more than a few microns, was able to produce junction damage which would shunt the power-producing capability of the whole device. Exposed strips as narrow as 0.005 cm (2 mils) were sufficient to drastically alter the device's power-producing capability. The absence of this effect in earlier solar array system was attributed to the use of a cell-shingling type panel construction and the presence of overlapping adhesive.

The results discussed in the previous section clearly indicated that low energy proton irradiation has an inordinately greater effect upon solar cell V_{oc} and P_{max} as compared to similar irradiations with electrons or higher energy protons. The anomalous degradation of the ATS-1 and Intelsat II-F4 prompted many investigations into the effects of low energy proton irradiation on partially shielded solar cells. [3.71-3.75] Curiously, Brucker and coworkers observed and reported this degradation effect in laboratory studies several months before the launch of ATS-1. [3.71]

The results of these studies confirmed that the small unshielded areas can cause significant effects on cell power output. As a result of these studies, array manufacturers have taken measures to cover all areas of the silicon cell front surface with a coverglass and fill any gaps between the cell and coverglass with adhesive.

The changes caused by the irradiation of small unshielded areas of solar cells with low energy protons can be explained in terms of solar cell theory. It was previously mentioned that the range of low energy protons in silicon is limited to less than the cell thickness. Particles which do not penetrate the cell produce defects only to their depth of penetration. This limited penetration results in unusual effects in the case of protons because lower-energy protons produce more displacements per unit path length. The results of this behavior are shown graphically in Figure 3.11. In this figure, the calculated number of displaced silicon atoms per unit proton path is plotted as a function of depth in silicon for a 3 MeV proton (range 92.7 μm). It can be seen that the damage rises rapidly to a maximum near the end of the proton track. Every proton which is stopped in the silicon produces such a damage peak at the end of its track. Protons which enter the silicon with energies of 0.5 MeV or less produce damage which is concentrated within a few microns of the cell surface. The space charge region of a modern cell extends from 0.4 to 1 micron below the cell surface. For this reason, low energy proton displacement damage is concentrated in the junction region.

The entire solar cell junction can be considered to be an array of small parallel diodes, each having a characteristic described by the parallel combination of equations (1.2.3) and (1.2.6). Damage to only a small portion of this parallel diode array results in an increased effective leakage or saturation current for the entire array.[3.69, 3.71] In Section 1.2, the nature of the generation-recombination current was discussed. The saturation current, J_{02}, due to generation-recombination in the space charge region (equation 1.2.7) increases linearly as the carrier lifetime decreases (i.e., displacement damage increases) in the space charge region. The increased leakage current of a solar cell reduces the cell V_{oc} because

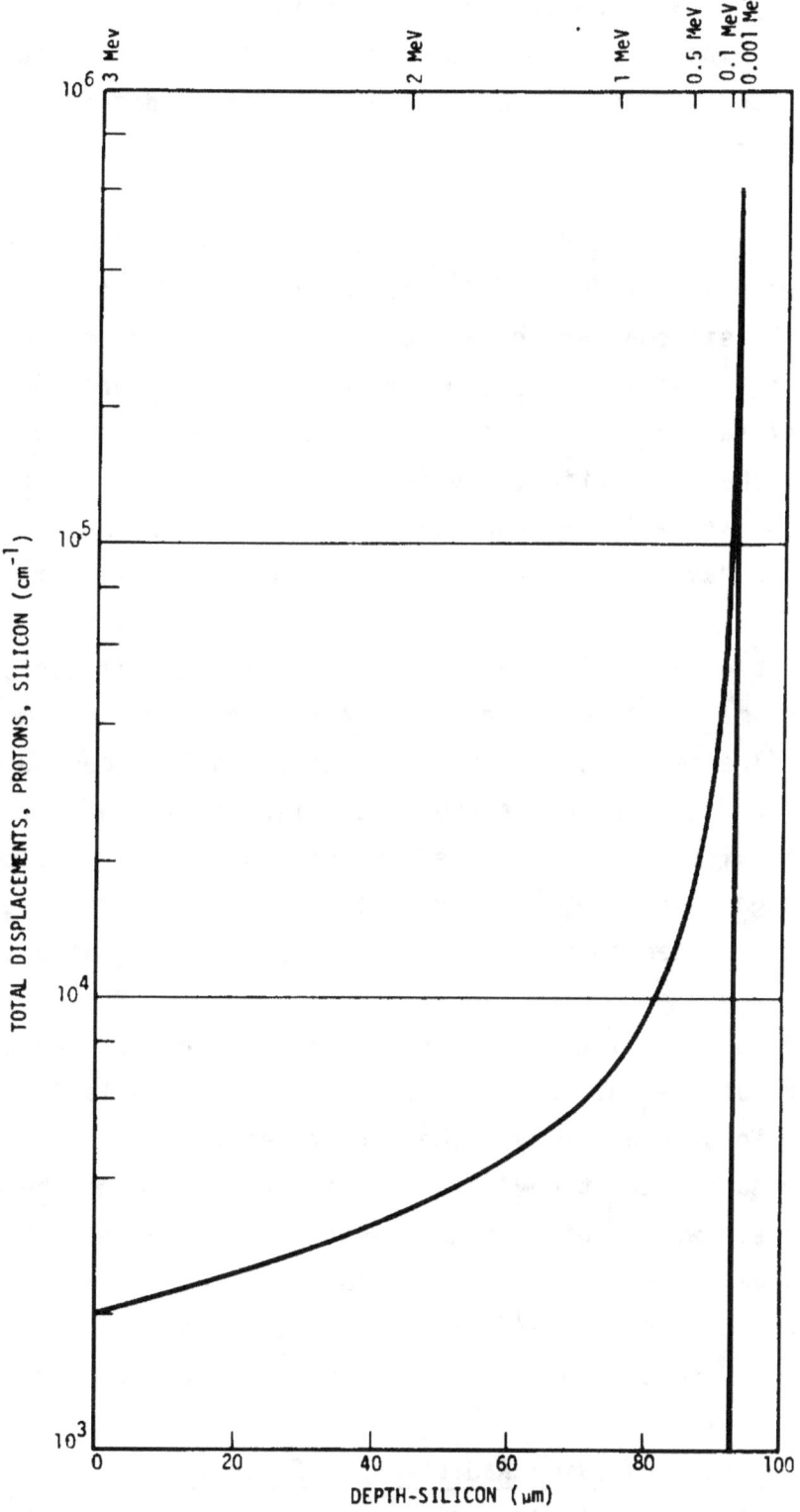

Figure 3.11 Atomic Displacements as a Function of Depth for a 3 MeV Proton in Silicon [3.26]

of the relationship of V_{oc} and I_o (junction leakage current) shown in equation (1.3.9). Since cell diode forward current (J_2) is increased at all voltages, the current available to an external load decreases, and so P_{max} will also decrease.

This effect is illustrated in Figure 3.12. A partially shielded solar cell was irradiated with 3×10^{13} p/cm^2 of 0.250 MeV protons. The protons entered the silicon through a 0.0076 cm gap between the coverglass and the metallized bus strip. The current-voltage characteristics of this cell are shown before and after irradiation. Although the I_{sc} of the cell was unaffected by the irradiation, significant degradations occurred in V_{oc} and P_{max}. Since solar cells are usually operated near the maximum power point, such changes have grave implications on in-flight performance.

It has been observed in laboratory studies that the effects of low energy protons on small unshielded areas of cells produce a maximum in degradation at a fluence of about 3×10^{13} p/cm^2. It has been suggested that the reversal of degradation is due to carrier removal effects.[3.69, 3.76] Considerable data exist regarding the effect of proton energy spectrum and busbar-coverglass gap width on the degradation.[3.73] Most reported laboratory studies have been confined to normal incidence proton irradiations.

In the past, solar cell usage has been confined to body-mounted solar cells on spinning satellites. Such applications provide a large measure of back shielding to a solar array. The requirements for increased spacecraft power and reduced weight have established trends toward the usage of oriented solar panels with minimal back shielding. Stofel has shown that low energy proton back side irradiation degrades silicon solar cells through carrier removal effects.[3.69, 3.73] The use of thin soldered back contacts or other minimal back shielding should greatly reduce these effects.

3.8 Effects of Neutron and Gamma Radiation on Solar Cells

The radiation associated with nuclear weapons degrades solar arrays in the same manner as the radiation of the space environment. Solar array designers must allow for these effects when weapon events are included in

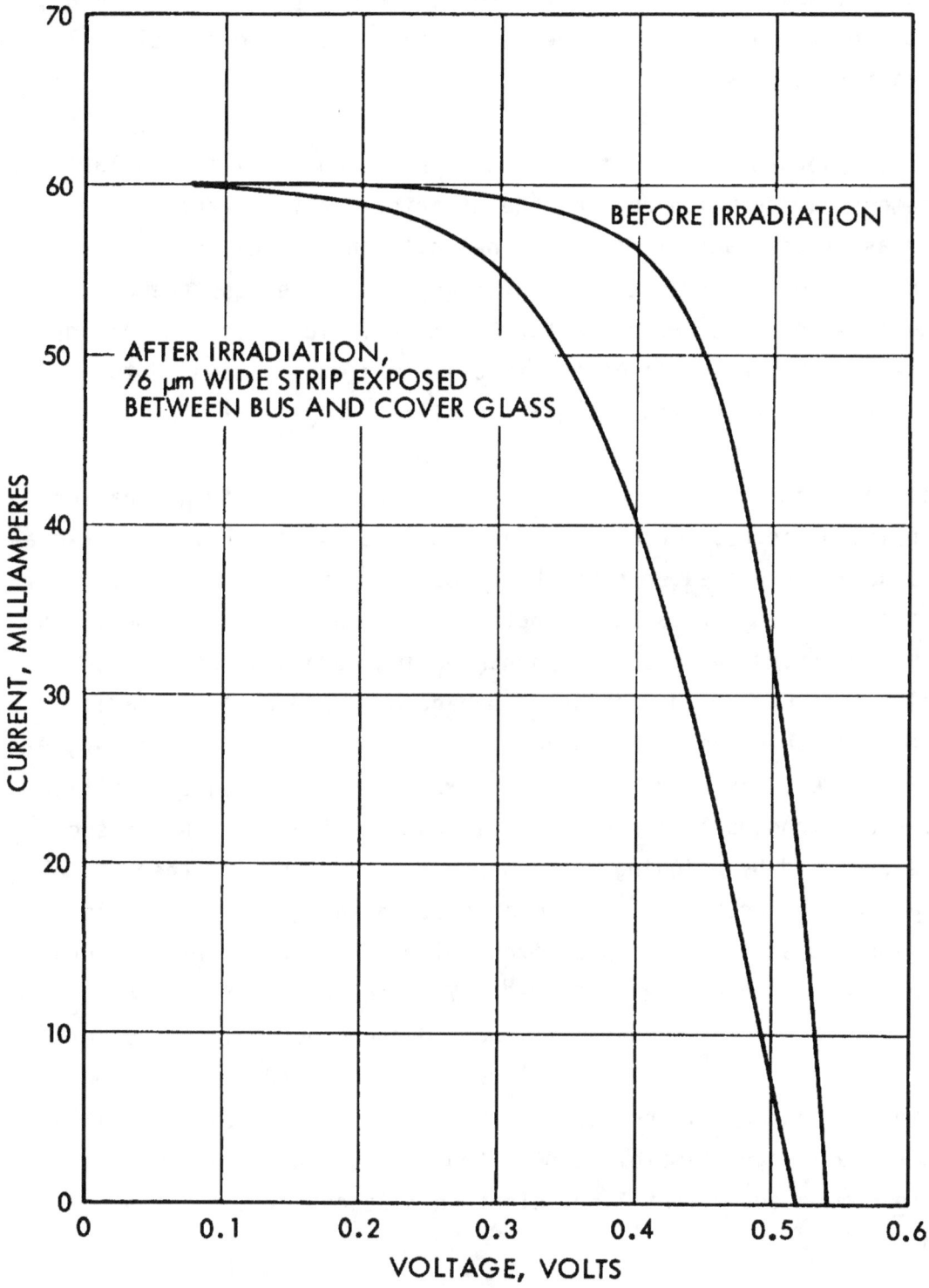

Figure 3.12 Low Energy Proton Junction Damage, 0.250 MeV Protons, 3×10^{13} p/cm^2, Partially Shielded n/p Solar Cell

the environment. The radiation from a weapon event is delivered at a much higher rate than space radiation. Because of these high radiation rates, other aspects of radiation effects become more apparent immediately following a nuclear radiation pulse.

The most important effect of neutron irradiation in silicon solar cells is displacement damage which reduces the minority carrier lifetime in the same manner as protons and electrons. When silicon devices receive neutron irradiation at room temperature, a large fraction of the displacement damage anneals within 100 seconds after the irradiation. The annealing factor is defined as the ratio of the initial (maximum) damage to the damage which remains after annealing is complete. [3.34, 3.77]

Annealing factors larger than 10 have been reported. Such behavior is not surprising, because calculated displacement rates for various radiations are usually much greater than those found experimentally. The transient annealing of neutron damage is not an important consideration in the design of solar arrays; however, the nonannealing component of neutron damage will contribute to the permanent damage produced by space radiation. This aspect of neutron damage has been studied by Brucker,[3.78] Carter,[3.79] Morris,[3.80] Stofel,[3.81] and Hicks.[3.82] Most of these studies utilized fission neutrons from nuclear reactors. If the fission spectrum of such reactors is averaged by weighing each energy component by its theoretical displacement damage factor,[3.83] the mean neutron energy is very close to 1 MeV. The degradation of n/p silicon solar cell parameters with neutron irradiation is shown in Figure 3.13.[3.81] The conversion of neutron fluences to damage equivalent 1 MeV electron fluences depends not only on output parameter but also on the degradation level. For I_{sc}, this conversion factor varies approximately from 1500 to 9000, and at the 75% degradation level, the ratio is approximately 2400. Neutron fluences may thus be converted to damage equivalent 1 MeV electron fluences by the following expression:

$$\Phi_{1 \text{ MeV } e} = 2400 \times \Phi_{1 \text{ MeV } n} \qquad (3.8.1)$$

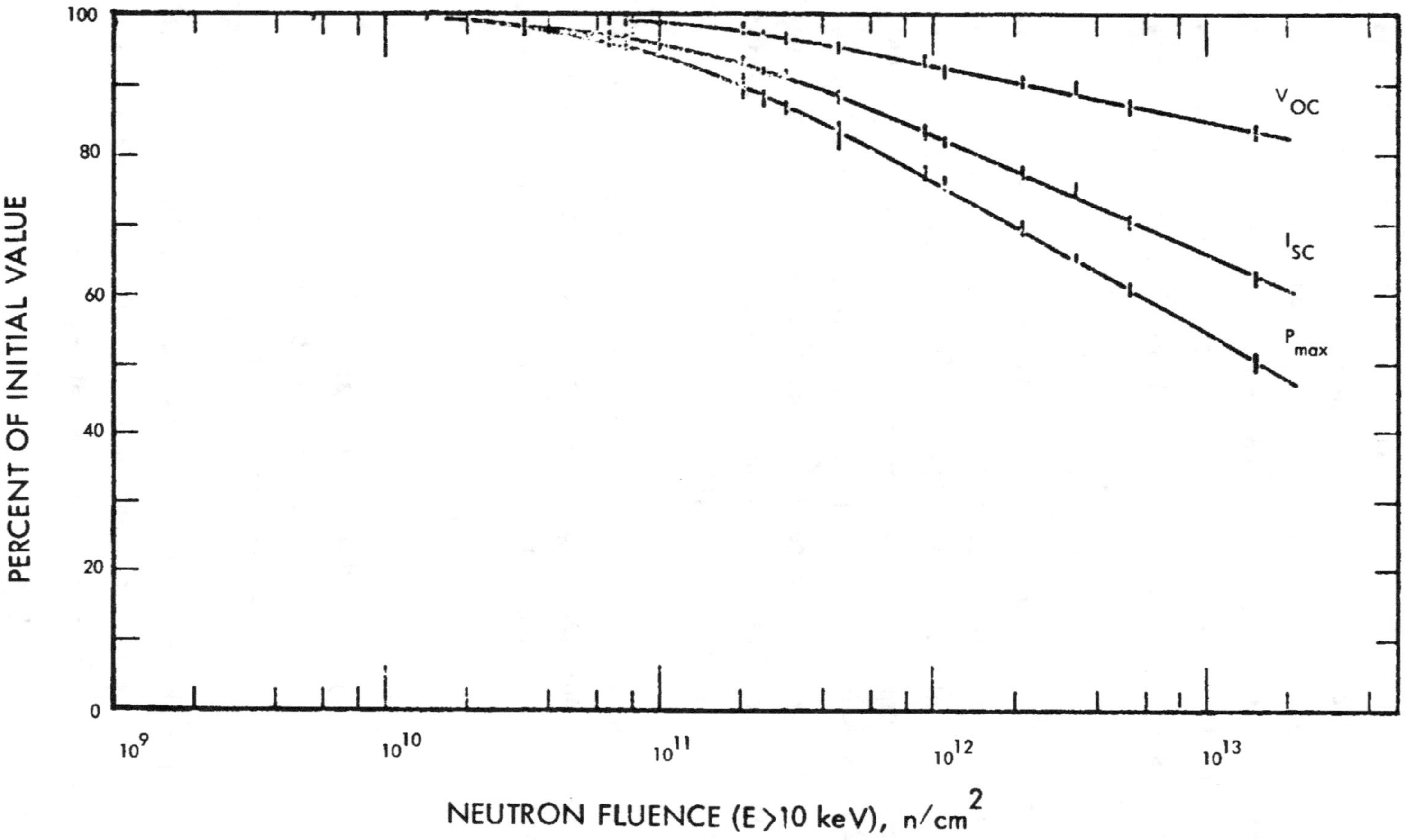

Figure 3.13 Neutron-Induced Change in n/p Silicon Solar Cells 3.81

When neutron damage is evaluated with a solar simulator and described by equation (3.2.1), the constant C is approximately equal to 6.5 mA/cm^2 per decade fluence. This value is significantly larger than that found for electron irradiation. Similar slope values are found in cells irradiated with high energy protons. Work by Gregory [3.33] and Stofel [3.81] has shown that diffusion lengths measured in neutron-irradiated solar cells depend on carrier injection level and increase with the excess minority carrier concentration (see Figure 3.5). This behavior is similar to that reported for proton-irradiated solar cells.

Gamma ray radiation interacts with silicon mainly by the production of Compton electrons. These secondary particles have energies high enough to cause displacement damage in silicon solar cells. The effect of gamma radiation on silicon solar cells has been reported by Fang [3.84] and Hicks. [3.82] The results of cobalt 60 gamma irradiation of n/p silcon solar cells are shown in Figure 3.6. The displacement cross section of prompt gammas is very small as compared with that of other radiation species and the damage can usually be neglected.

The most important aspect of gamma radiation from weapons is the transient photocurrent generated in the array during a nuclear event. The primary photocurrent can be estimated from the following expression:

$$I_{pp} = 6.4(\text{ A cm}^{-3}\text{ rad}^{-1}\text{ sec})\dot{\gamma} A\ L \qquad (3.8.2)$$

where
- $\dot{\gamma}$ = dose rate (rad/sec)
- L = diffusion length (cm)
- A = cell junction area (cm^2)

The transient rise and fall of the photocurrent has been treated by Wirth and Rogers. [3.85] The peak current values developed by solar cells under these conditions can be very large and may cause problems in circuits interfacing with the solar array. Current limiting by the external load and the internal cell series resistance may limit the observed photocurrents to values well below the generated current. Under very intense pulses of such ionizing radiation at room temperature the cell V_{oc} saturates at approximately 0.7 V.[3.86-3.88] This value appears to be related to the barrier potential (V_b) of the junction as determined by capacitance-voltage measurements.

3.9 Lithium Doped Solar Cells

Interest in this field began with Vavilov's report of a radiation resistant diode made with lithium-doped, crucible grown silicon.[3.89] Wysocki later reported lithium-doped solar cells which degraded under electron irradiation, but rapidly recovered at room temperature.[3.90] Float zone silicon, with a characteristic lower oxygen concentration, was used to achieve this result. Subsequent work indicated that recovery also occurred in lithium-doped, quartz-crucible silicon solar cells. Since this initial work, the general subject was studied in two ways. Empirical changes in the manufacturing techniques for lithium-doped solar cells were evaluated with the aim of optimizing the recovery effect.[3.91, 3.92] Other studies were directed at the development of a physical model of the degradation and recovery processes in lithium-doped silicon.

Some of the more pertinent facts gained during these studies are as follows. The lithium concentration in a solar cell is not uniform, but increases in a linear or near linear manner with distance from the solar cell junction. This characteristic can be used to advantage to produce cells with exceptionally high open-circuit voltages. Solar cells with low or insufficient lithium concentrations do not recover in a satisfactory manner. Float zone silicon solar cells with exceptionally high lithium concentrations lose efficiency during storage in the unirradiated condition. These same cells, when irradiated and recovered, also exhibit a time-dependent loss of efficiency. This loss has been related to the room temperature diffusion of lithium into the active area of the cell. It has also been observed that higher lithium concentrations cause faster recovery rates. Because of the recovery rate dependence of the radiation damage in lithium-doped solar cells, it was difficult to evaluate cell performance by accelerator irradiations. Real time irradiations of lithium-doped solar cells have been done with beta particle sources. The results of these beta irradiations indicated that some types of lithium-doped solar cells are slightly superior to n/p cells under some temperature conditions. The major potential advantages of lithium-doped solar cells over conventional n/p solar cells are in regard to proton [3.60, 3.90] and neutron damage. [3.79, 3.93] Figure 3.14 shows that lithium-doped solar cells are clearly superior to conventional cells under proton irradiation. However, the long recovery period following a neutron exposure would probably be a severe limitation in military spacecraft. The most advantageous uses of lithium-doped solar cells would be for spacecraft in proton-dominated orbits with high proton fluxes. Since such orbits are not commonly used and since substantial improvements in ordinary cell efficiencies began appearing in the early 1970's, research efforts on lithium cells were suspended. A summary of the lithium cell work was published by Berman in 1972. [3.94]

3.10 Annealing of Irradiated Solar Cells

Annealing and reverse annealing of irradiated solar cells as a function of temperature has received considerable study. The kinetics and energy levels involved are functions of type of radiation, type of solar cells, and other parameters as yet undetermined. Though the situation is quite complex, it can be generally stated that irradiated conventional silicon solar cells cannot be significantly annealed at temperatures below 200°C, which is considered a practical limit for space applications. Significant annealing of conventional silicon solar cells irradiated with electrons or protons typically occurs in the 200 to 400°C range. [3.61]

Of more practical importance is the fact that some ambient annealing of charged particle radiation damage exists. In the laboratory, the radiation exposure rate is usually many orders of magnitude greater than natural space radiation rates. In space, the damage and annealing processes occur simultanously, with the annealing rate much closer to the damage rate than in the laboratory. For laboratory electron irradiation,

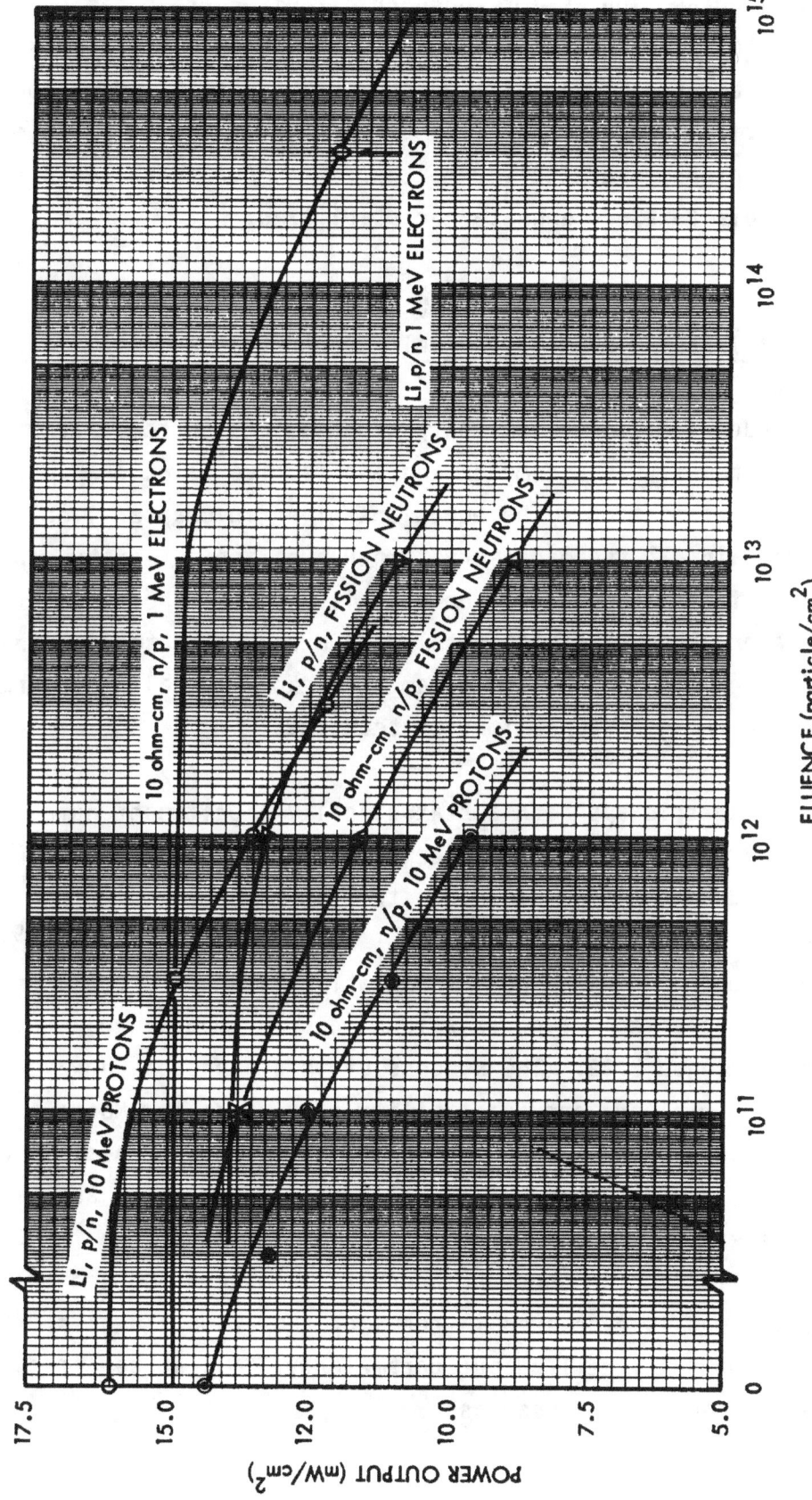

Figure 3.14 Recovered Power Output of Irradiated Conventional and Lithium-Doped Solar Cells 3.90,3.93,3.60

ambient annealing as high as 10% in short circuit current has been observed in a few days to a month, predominantly in 10 ohm-cm cells. This annealing can be induced and stabilized by a 24-hour soak at 60°C. For laboratory proton irradiation, ambient annealing of as high as 20% of short circuit current has been observed after 22 months.[3.61] For these reasons, all the data used in this text are annealed or stabilized.

3.11 Effect of Irradiation on Solar Cell Temperature and Illumination Intensity Dependences

The dependence of solar cell output parameters on temperature and illumination intensity is a complex and interactive relationship even in the absence of irradiation. The equations given in Chapter 1 do not explicitly contain all the necessary temperature and intensity dependent terms since in fact many are not functionally known. Therefore, two alternative approaches have historically been used to determine these relationships; namely, a specific parametric experimental determination or a more general linear approximation technique.

In the parametric approach a statistically significant number of a particular type of solar cell is experimentally measured in a matrix of temperatures and intensities of interest. The resulting data are then analyzed to yield the required functional relationships.[3.95-3.99] A typical example of a pre- and post-irradiation parametric characterization of a particular solar cell type as a function of temperature and intensity is given in Reference 3.99. The success of this approach, however, depends on the quantity, quality, and availability of test data. Moreover, the application of such data may be limited exclusively to a specific type or group of cells tested. Therefore, an alternate approach is adopted in this text: whenever a linear approximation is warranted, the first-order temperature coefficient at one sun intensity is determined and the variation of this coefficient is expressed in terms of radiation fluences. With this technique, a solar cell output parameter $y(T,\Phi)$, at temperature T and fluence level Φ can be expressed as:

$$y(T,\Phi) = y(T_0,\Phi) + b(\Phi)(T - T_0). \qquad (3.11.1)$$

where $y(T_0,\Phi)$ = the value of parameter y at temperature T_0 and fluence level Φ

$b(\Phi)$ = temperature coefficient of y at fluence level Φ

Φ = radiation fluence

The dependence of solar cell output on illumination intensity is somewhat predictable from the equations in Chapters 1 and 2. The spectral response $R(\lambda)$ in equation (2.1.1) is independent of spectral irradiance $E(\lambda)$, and the light-generated current, I_L, becomes proportional to illumination intensity. The short circuit current is therefore almost equal to I_L and hence is proportional to the illumination intensity.

$$I_{sc}(T,S) \cong I_L(T,S)$$

$$= S\, I_{sc}(T,1) \qquad (3.11.2)$$

where

S = intensity scale factor (unity = 1 solar constant)

and

$I_{sc}(T,1)$ = short circuit current at one sun intensity and temperature T°C

In general, equation (3.11.1) can be expanded in terms of the first order of illumination intensity while making use of temperature coefficient data.

For short circuit current,

$$I_{sc}(T,\Phi,S) = S\, I_{sc}(T_0,\Phi,1) + b(\Phi,S)(T-T_0) \qquad (3.11.3)$$

Similarly,

$$V_{oc}(T,\Phi,S) = V_{oc}(T_0,\Phi,S) + b(\Phi,S)(T-T_0) \qquad (3.11.4)$$

and

$$P_{max}(T,\Phi,S) = [P_{max}(T_0,\Phi,1) + b(\Phi,1)(T-T_0)] S \qquad (3.11.5)$$

For unirradiated n/p silicon solar cells at one sun illumination and ambient temperature, dI_{sc}/dT is approximately 0.02 mA/cm^2-°C, dV_{oc}/dT ranges from -2.0 mV/°C (2 ohm-cm) to -2.4 mV/°C (10 ohm-cm), and dP_{max}/dT is approximately -0.07 mW/cm^2-°C.

For practical applications, the maximum power point must be specified by either the current or the voltage at maximum power. The current at maximum power, I_{mp}, varies almost quadratically with respect to temperature, and the temperature coefficient becomes temperature dependent. The voltage at maximum power, V_{mp}, on the other hand shows a large and almost linear variation with respect to temperature and hence is a better candidate than I_{mp} for presenting simpler and more reliable data. In this context, V_{mp} is used and discussed in this text. The dV_{mp}/dT of unirradiated 10 ohm-cm n-p cells is approximately -2.2 mV/°C.

Anspaugh[3.100] made thorough measurements from -20 to 40°C for 2 and 10 ohm-cm n/p cells bombarded with 1 MeV electrons. Cells from the same production lots were used on a flight experiment aboard ATS-5. The effects of radiation on various temperature coefficients are shown in Figures 3.15 through 3.18. Other data are also plotted on these figures for comparison.[3.49, 3.100-3.103] Reference 3.138 may also be consulted for the temperature variation of damage coefficients.

The dV_{oc}/dT of 10 ohm-cm n/p silicon solar cells does not change significantly after 1 MeV electron bombardment but it does change for 2 ohm-cm cells (from -2.0 mV/°C to -2.3 mV/°C) as the fluence increases from 10^{12} to 10^{16} electrons/cm^2 (see Figure 3.16). Luft[3.103] has reported similar results, as did Haynes and Ellis,[3.49] who irradiated cells with 2.4 MeV electrons. The dI_{sc}/dT of 10 ohm-cm cells changes by a factor of 3 (from 0.018 to 0.06 mA/cm^2-°C) when irradiated with 1 MeV electron fluences to 10^{16} e/cm^2, but 2 ohm-cm cells change by a factor of 2 (see Figure 3.15).

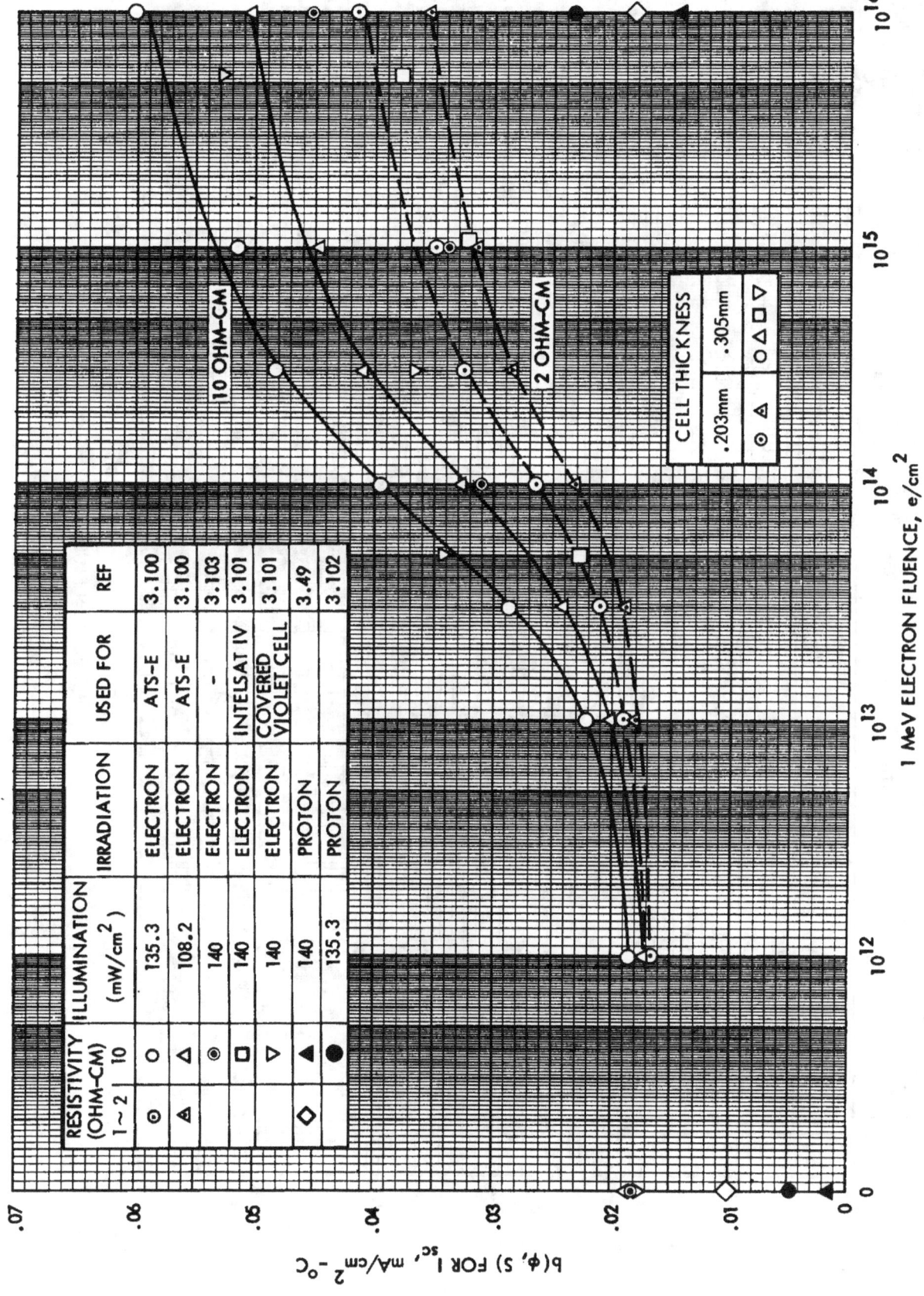

Figure 3.15 I_{sc} Temperature Coefficient vs 1 MeV Electron Fluence

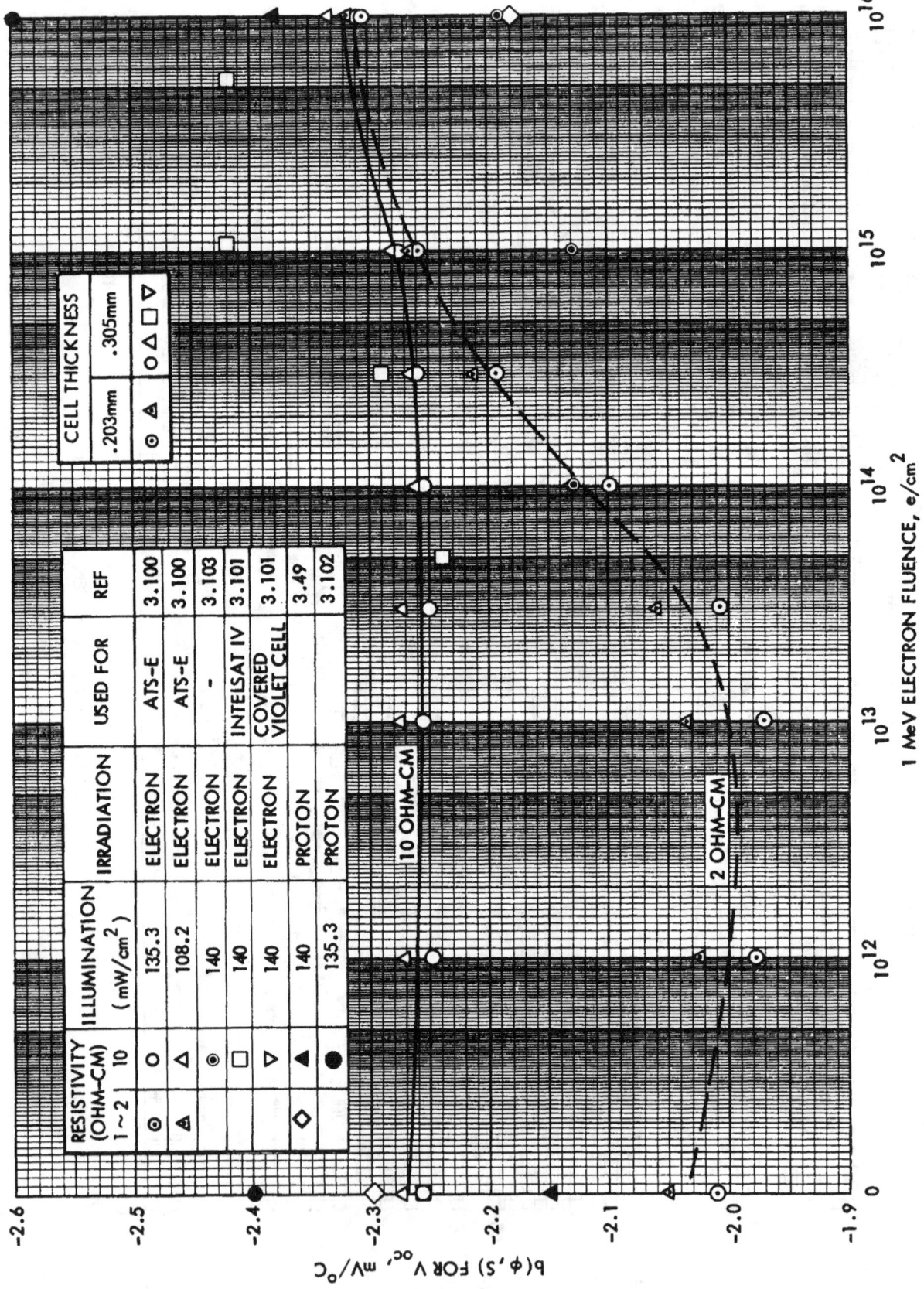

Figure 3.16 V_{OC} Temperature Coefficient vs 1 MeV Electron Fluence

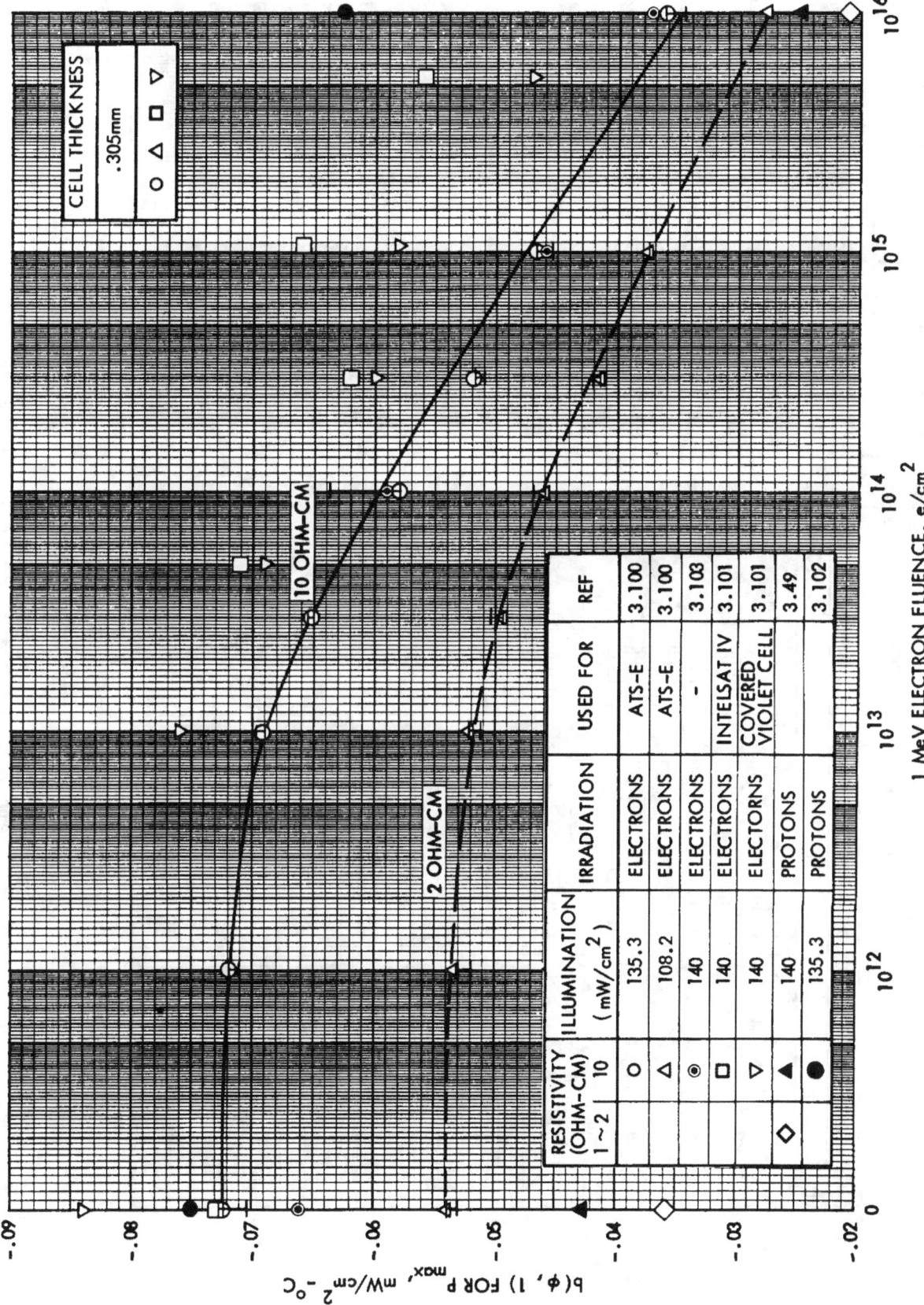

Figure 3.17 P_{max} Temperature Coefficient vs 1 MeV Electron Fluence

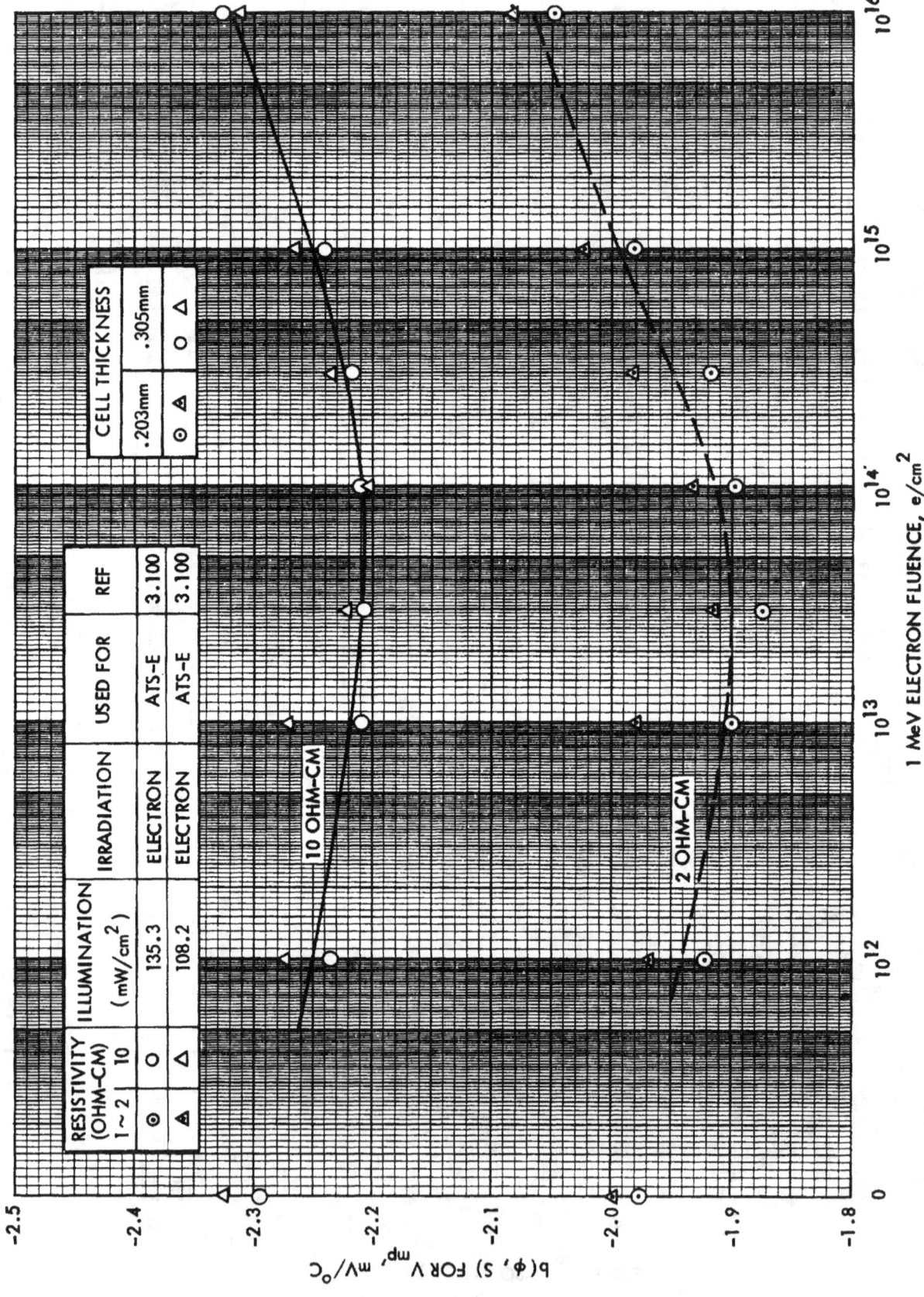

Figure 3.18 V_{mp} Temperature Coefficient vs 1 MeV Electron Fluence

Both 2 and 10 ohm-cm cells exhibit a monotonic decrease in dP_{max}/dT by about a factor of 2 after exposure to 10^{16} e/cm^2 (see Figure 3.17). However, the 2 ohm-cm cells have a much smaller temperature coefficient than 10 ohm-cm cells. The dV_{mp}/dT is also negative and the magnitude decreases to a minimum at a fluence level of approximatley 5 X 10^{13} electrons/cm^2, then starts increasing to its initial value. The variation of dV_{mp}/dT ranges from -2.2 to -2.3 mV/°C for 10 ohm-cm n-p cells and from -1.9 to -2.1 for 2 ohm-cm cells (see Figure 3.18).

Data are extremely limited on the variation of temperature coefficients with proton irradiation. The data for 22 MeV protons with a fluence of 2 X 10^{12} protons/cm^2 are shown in Figures 3.15 through 3.18 for comparison with electron data. The temperature coefficients after proton irradiation to the 25% power-degradation point for proton energies from 2 to 155 MeV [3.102] are shown in the same figures. These data, though sparse, indicate trends similar to the electron data.

Although illumination has been shown to affect the evaluation of radiation damage in silicon solar cells through injection level effects, it has been assumed that the production of displacement-type radiation damage in silicon solar cells is independent of illumination intensity during irradiation. Reynard has reported that during real-time beta ray irradiation, silicon solar cells, illuminated and electrically loaded, degraded more severely than similar cells irradiated dark without load. [3.104] The results of a similar study did not confirm the above result. [3.105]

Crabb [3.106] reported that 10 ohm-cm float zone silicon solar cells, which had been degraded with 1 MeV electrons, exhibited a further degradation when illuminated by a 10-sun source. Further investigations by many workers [3.107-3.113] revealed that photon degradation depends not only on crystal growth technique but also type and amount of dopant as well as radiation particle species as summarized below.

- ° Many investigators agreed that crucible (Czochralski) grown silicon cells did not exhibit photon degradation except a case reported by Crabb.[3.107] According to Crabb, the float zone, boron-doped cells exhibited no photon degradation, whereas the crucible-grown B-doped cells suffered a 6% power loss due to photon degradation.

- Gallium-doped float zone silicon cells did not exhibit photon degradation.

- Boron- and aluminum-doped float zone silicon cells suffered from photon degradation. The degradation was more pronounced for lower resistivity cells, practically no degradation for 85 ohm-cm cells, about 5% for 10 ohm-cm cells, and greater than 10% degradation for 0.2 ohm-cm cells.

- No photon degradation was observed following 2.5 and 10 MeV proton irradiation.[3.112, 3.113]

3.12 Radiation Effects on Shielding Materials

The degradation due to radiation effects on solar cell coverglass material in space is difficult to assess. The different radiation components of the environment act individually and synergistically on the elements of the shielding material and also cause changes in the interaction of shielding elements. The complexity is illustrated in Table 3.2, where the various effects reported for commonly used cover materials are summarized and referenced. In addition to the data in Table 3.2, a large volume of data has been presented in the literature regarding materials currently not in use for shielding solar cells. In this section, the emphasis will be on solar cell shielding material currently used in array construction.

The coverglass shielding currently in use in most spacecraft construction is usually fabricated from Corning 0211 Microsheet or Corning 7940 fused silica. Where thin covers are desired, the usage tends toward Microsheet, because it is relatively inexpensive in thin sections. Where thicker covers are desired, Corning 7940 fused silica is used to avoid the darkening due

Table 3.2
Radiation Effects on Shielding Materials

	Anti-Reflective Coating on Coverglass	COVERGLASS		Blue Filter On Coverglass	Silicone Adhesives
		Corning 0211 Microsheet	Corning 7940 Fused Quartz		
keV Protons	Degrades Transmission 3.121, 3.122, 3.123				
MeV Protons		Degrades Transmission 3.121, 3.116, 3.118	No Transmission Loss 3.121, 3.116	Degrades Transmission 3.121, 3.116	No Transmission Loss 3.124
MeV Electrons		Degrades Transmission 3.116, 3.125, 3.126, 3.128, 3.129, 3.130	No Transmission Loss 3.116, 3.126, 3.127, 3.128, 3.129, 3.130	Degrades Transmission 3.126, 3.130	No Transmission Loss 3.124
Ultraviolet Light		Bleaches Transmission Loss Due To Radiation 3.117	Degrades Transmission 3.131 No Absorptance Change 3.132	Degrades Transmission Reduced UV Rejection 3.133	Degrades Transmission 3.124, 3.126

to radiation. Coverglasses are usually used with a MgF_2 antireflecting front coating and an ultraviolet rejecting filter on the rear surface. Coverglasses are usually attached to solar cells with silicone elastomers (Dow Corning DC 93-500).

Most experimental assessments of radiation effects are based on accelerated testing in which a complete space environment is not simulated. This may account for some of the differences between darkening of coverglass material observed in laboratory radiation studies and space flight data for covered solar cells which indicated that radiation effects in cover materials were insignificant. [3.114]

The radiation effects observed in cover materials can be characterized as ionization damage rather than displacement damage. In general, ionization effects are usually dependent upon the absorbed dose and to that degree are independent of particle type or energy. Some exceptions to this rule occur in the case of highly charged massive particles. In such cases, the ionization effects may be concentrated along the particle track rather than uniformly distributed. [3.115] It is reasonable to assume that the ionization damage produced in cover materials by space electrons and protons is related to the total absorbed dose. This assumption allows the various radiation components of the space environment to be reduced to a total dose, without a laborious determination of degradation constants for each energy and particle. It also allows the use of experimental data from a single ionizing environment such as 1 MeV electrons.

The most significant radiation effects in cover materials involve changes in the transmission of light in the visible and near infrared region. These data are commonly reported as spectral transmission data. The use of coverglass spectral-transmission data in determining changes in solar cell output is rather cumbersome. This procedure was outlined by Campbell. [3.116] An alternate approach to the reporting of the data is the use of so-called "wide-band" transmission loss. In this method, solar cell short circuit currents are measured under sun simulated conditions, with coverglasses attached. The coverglasses are attached with a thin liquid film with an index of refraction (n = 1.4) similar to that of silicone adhesive. Cyclohexane and n-amyl alcohol have been used for this purpose. The "wide-band" transmittance is defined as the solar cell I_{sc} with an irradiated coverglass in place divided by the solar cell I_{sc} with the unirradiated coverglass in place. Such measurements are influenced by solar cell spectral response. Results determined with unirradiated solar cells will not be representative of those for irradiated solar cells. This error is probably negligible compared to the uncertainty of the available experimental data.

Since the "wide-band" transmission loss is a measure of the loss in light transmitted, it directly affects the light generated current (I_L) and likewise the short circuit current (I_{sc}). It is desirable to use the "wide-band" transmission data to estimate the change in solar cell P_{max}. Equation (3.2.9) indicates that cell P_{max} is proportional to the product of I_{sc} and V_{oc}. Because V_{oc} is proportional to $\ln I_{sc}$, the following relation can be developed to estimate the change in P_{max} due to coverglass darkening from transmission data:

$$\frac{P_{max}}{P_{maxo}} = T \left[\frac{\ln (T\, I_{sc})}{\ln (I_{sc})} \right] \quad (3.12.1)$$

where P_{max}/P_{maxo} = the fractional change in P_{max}
T = "wide-band" transmission of irradiated coverglass
I_{sc} = short circuit current with unirradiated coverglass

To aid in the estimation of solar array losses due to reduced transmission from radiation effects in coverglass materials, data relating transmittance to absorbed dose is required. In Figure 3.19, "wide-band" transmittance is shown for various absorbed doses. The absorbed doses were produced by 1 MeV electron irradiations in a room temperature, air environment which included no ultraviolet illumination. This electron radiation is sufficiently penetrating to produce a relatively uniform dose through the entire coverglass, coating, and filter. The P_{max}/P_{maxo} data shown in Figure 3.19 were calculated from the "wide-band" transmittance value by use of equation (3.12.1). The data in Figure 3.19 include 0.0152 cm

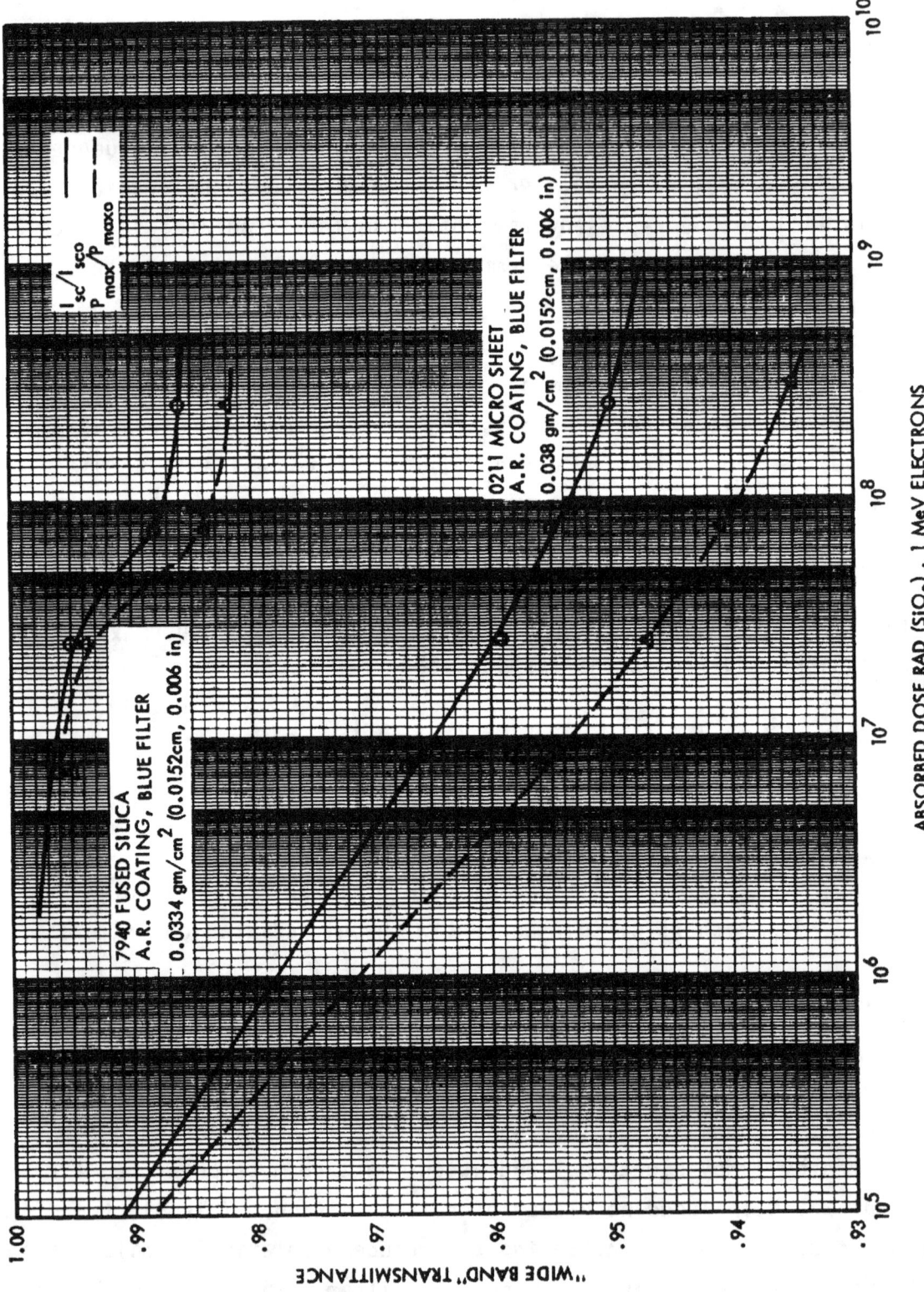

Figure 3.19 Variation of Coverglass Transmittance with Absorbed Dose

(0.006 in.) 7940 fused silica and 0211 Microsheet coverglass with antireflecting coating and blue filter. It has been established that Corning 7940 fused silica exhibits little or no radiation darkening in the visible region. Since the transmission loss for 7940 coverglass must be assumed to be due to changes in the filter, the data can also be used for thicker coverglasses. For thicker 0211 Microsheet coverglass, the data in Figure 3.19 cannot be used.

The dose-depth profiles experienced by coverglass shielding in space are highly non-uniform due to the low energy protons stopped in the front surface. An accurate estimation of the transmission through a coverglass with such a dose-depth profile would require the integration of absorption coefficients (as a function of dose) through the coverglass and its thin film layer. The lack of absorption coefficient data for these materials for various doses in a total space environment does not allow such evaluations at this time.

The diversity of technical opinions on transmission loss in coverglass due to space radiation also includes those who do not include this factor in array power estimates and those who simply allow for a 2 to 4% initial loss due to coverglass and adhesive darkening due to radiation and ultraviolet effects. Studies by Luedke at TRW indicated that nearly all darkening produced in 0211 Microsheet by a dose of 10^7 rad(SiO_2) was bleached by a relatively short ultraviolet light exposure. [3.117] Such results indicate that the use of data such as that in Figure 3.19 is probably an overly conservative practice and emphasizes the importance of performing coverglass darkening studies in a realistic environment. Some investigations have reported results which indicate that cerium doping of glass reduces or eliminates darkening due to irradiation. [3.118, 3.119] Other studies indicated that hydrogen impregnation of glasses reduces transmission loses due to irradiation effects. [3.120]

In addition to the glass/adhesive system discussed above, numerous attempts have been made to develop more practical and economical protective systems. Included in these attempts were direct deposition of glass, electrostatic bonding, spray-on FEP teflon, [3.135] laminated FEP teflon and spray-on polyimide. Details of the various systems investigated and their merits and shortcomings are contained in the Solar Cell Array Design Handbook [3.134] for activities through 1976. More recently, Russell [3.136] has reported the results of testing several of these newer systems with electron, proton and UV exposure followed by thermal cycling. He found that FEP-A teflon, PFA hardcoat, DC 93-500, GE 615/UV-24, and GR 650 Glass Resin when used in place of the usual coverglass all either became brittle and cracked after particle irradiation and thermal cycling or suffered large transmission losses after UV exposure. One system consisting of 7070 glass electrostatically bonded (ESB) to the solar cells did not degrade in radiation but may have difficulty withstanding thermal cycling. Preliminary evaluation of a recently developed polyimide film [3.137] has not yet revealed any major difficulties and may prove to be a useful cover material. Research in this area is continuing; however, no new system has yet been able to achieve space qualification.

3.13 Solar Cell Output vs. 1 MeV Electron Irradiation

In this section, solar cell output parameter degradation data are presented as a function of 1 MeV electron fluence. Five basic output parameters (I_{sc}, V_{oc}, P_{max}, I_{mp}, and V_{mp}) of various cell types (see Appendix C for the definition of cell types) were measured at JPL and are shown in Figures 3-20 through 3-111 for base resistivities of 2 and 10 ohm-cm. The cells chosen for inclusion in this section are tabulated in Table 3.3. They represent a cross section of cells which are estimated to be available from the production lines of space qualified solar cell manufacturers for the next three to five years. The production cells were purchased from both ASEC and Spectrolab for use as test samples. Although Table 3.3 includes a cell description column which utilizes the K designations assigned by Spectrolab, we have applied the k designations to ASEC cell as well. Vertical junction cells and LPE gallium arsenide cells are included for comparison, even though they are less likely to be mass produced within that time period. Also included are some of the cell types previously published in this Handbook which, although now obsolete, are still producing power in operational spacecraft.

Cell selection and measurement were performed as follows. Lots of 50 solar cells were purchased from the solar cell manufacturers to a specification designed to yield representative production cells. From each lot, 14 cells were selected for mounting on a test plate and subsequent irradiation. They were selected by requiring their I_{sc}, V_{oc}, and P_{max} to all be near the center of the lot distribution. These 14 cells were then mounted to a test plate using RTV 560. Individual current and voltage leads were soldered to each cell. They were measured in situ after each radiation fluence using an Aerospace Controls Model 302 Simulator as the illumination source. A balloon flight standard cell carefully matched to the spectral response of the unirradiated test cell was used to set illumination intensity. After the cumulated fluence reached 10^{14} e/cm^2, the cells were annealed for approximately 16 hours at 60°C prior to measurement. During the irradiation and during the I-V curve measurement, the cell temperature was held at 28°C by monitoring a thermocouple soldered to the busbar of one of the test cells. After the cell measurement and irradiation was complete, averages of the 5 output parameters were computed for each fluence level. These averages were then plotted against the log of electron fluence with the help of a computer which was programmed to calculate cubic spline fits to the data and then to produce the plots. Current and power curves are given in terms of output/cm^2 where the total area of the cell was used in the computation. For each set of curves, standard deviations are plotted for one set of cells in the form of error bars at the pre-irradiation and 10^{15} e/cm^2 fluence levels. The standard deviations plotted are the largest of all the cell types on that paticular figure.

Table 3.3

Test Cell Descriptions

CELL TYPE OR K-DESIGNATION	(Ω-cm)	t (mils)	BSF	AR	BSR	CELL MFR YEAR	SAMPLE SIZE	FIGURE NOS.
K4 3/4	2	4	NO	DAR	YES	1980	14	50-59
K4 3/4	2	4	NO	DAR	YES	1980	13	50-59
K6 3/4	2	4	PASTE	DAR	YES	1981	14	60-69
CONVENTIONAL	2	8	NO	SiO	NO	1976	7	20-29
K4	2	8	NO	Ta_2O_5	NO	1976	7	20-29
K4 3/4	2	8	NO	DAR	YES	1980	14	50-59
K4 1/2	2	8	NO	Ta_2O_5	YES	1980	14	50-59
K6 1/2	2	8	PASTE	Ta_2O_5	YES	1980	14	60-69
K6 1/2	2	8	PASTE	DAR	YES	1980	14	60-69
CONVENTIONAL	2	12	NO	SiO	NO	1976	7	20-29
K4	2	12	NO	Ta_2O_5	NO	1976	7	20-29
VERTICAL JCN.	2	12	PASTE	Ta_2O_5	YES	1980	14	100-111
K6 3/4	10	2	BORON	DAR	YES	1980	13	80-89
K6 3/4	10	2	PASTE	DAR	YES	1981	12	80-89
k4 3/4	10	4	NO	DAR	YES	1980	14	70-79
K4 3/4	10	4	NO	DAR	YES	1980	13	70-79
K6 3/4	10	4	BORON	DAR	YES	1980	12	80-89
K6 3/4	10	4	PASTE	DAR	YES	1981	14	80-89
CONVENTIONAL	10	8	NO	SiO	NO	1976	6	30-39
K4	10	8	NO	Ta_2O_5	NO	1976	7	30-39
K6	10	8	EVAP Al	Ta_2O_5	NO	1976	7	40-49
K7 TEXTURED	10	8	EVAP Al	Ta_2O_5	NO	1976	5	40-49 100-111
K4 3/4	10	8	NO	DAR	YES	1980	14	70-79
K4 1/2	10	8	NO	Ta_2O_5	YES	1980	14	70-79
K6 3/4	10	8	PASTE	DAR	YES	1980	13	90-99
K6 3/4	10	8	PASTE	DAR	YES	1981	14	90-99
K6 1/2	10	8	PASTE	Ta_2O_5	YES	1980	14	90-99
K7 TEXTURED	10	8	PASTE	DAR	YES	1982	14	100-111
CONVENTIONAL	10	12	NO	SiO	NO	1976	7	30-39
K4	10	12	NO	Ta_2O_5	NO	1976	6	30-39
K6	10	12	EVAP Al	Ta_2O_5	NO	1976	7	40-49
GALLIUM ARSENIDE		12	GaAlAs	Ta_2O_5	NO	1979	7	100-111

NOTE: ALL CELLS ARE 2 x 2 cm

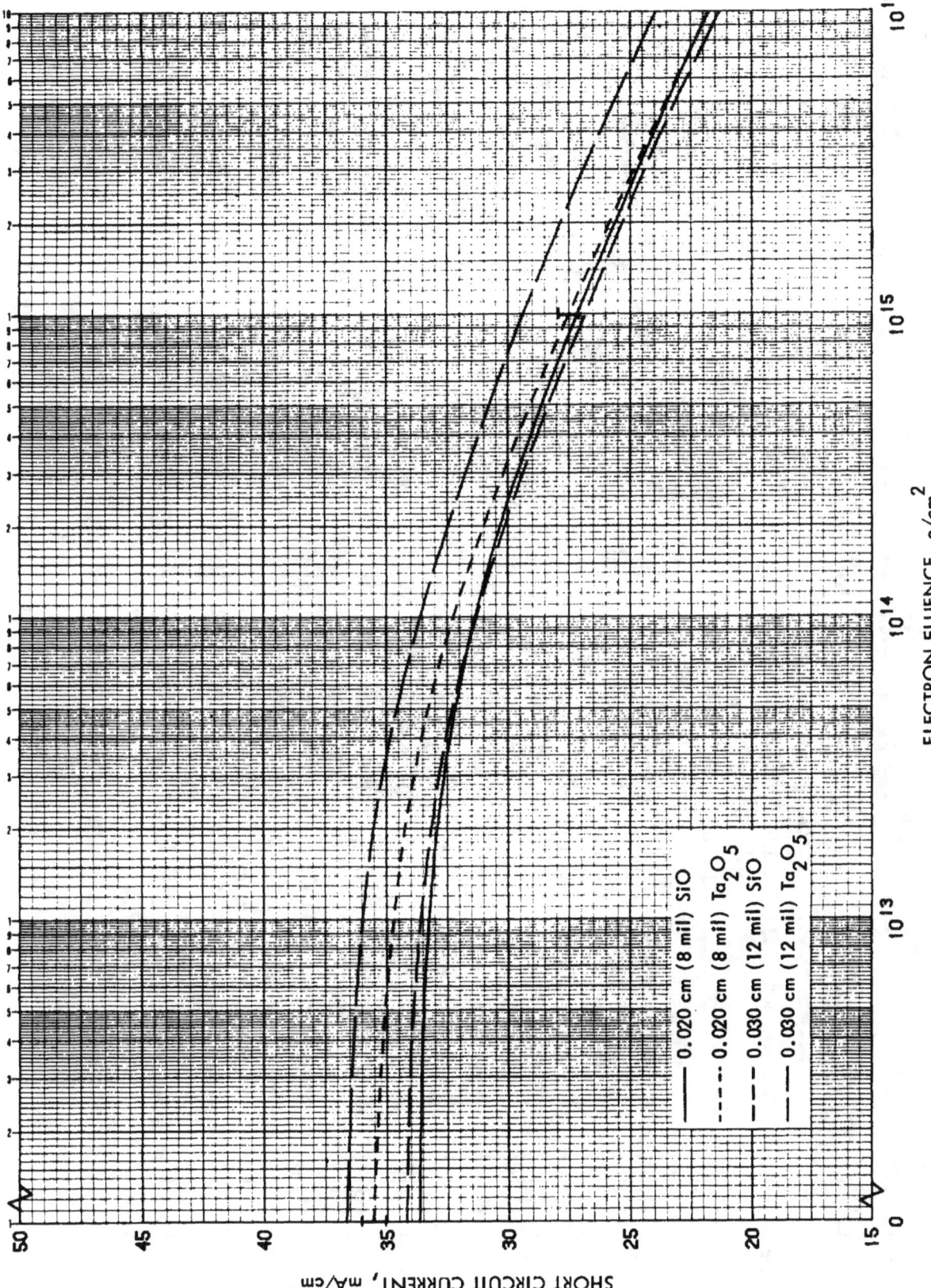

Figure 3.20 I_{sc} vs 1 MeV Electron Fluence for 2 Ohm-cm n/p Conventional and Shallow Junction Silicon Cells

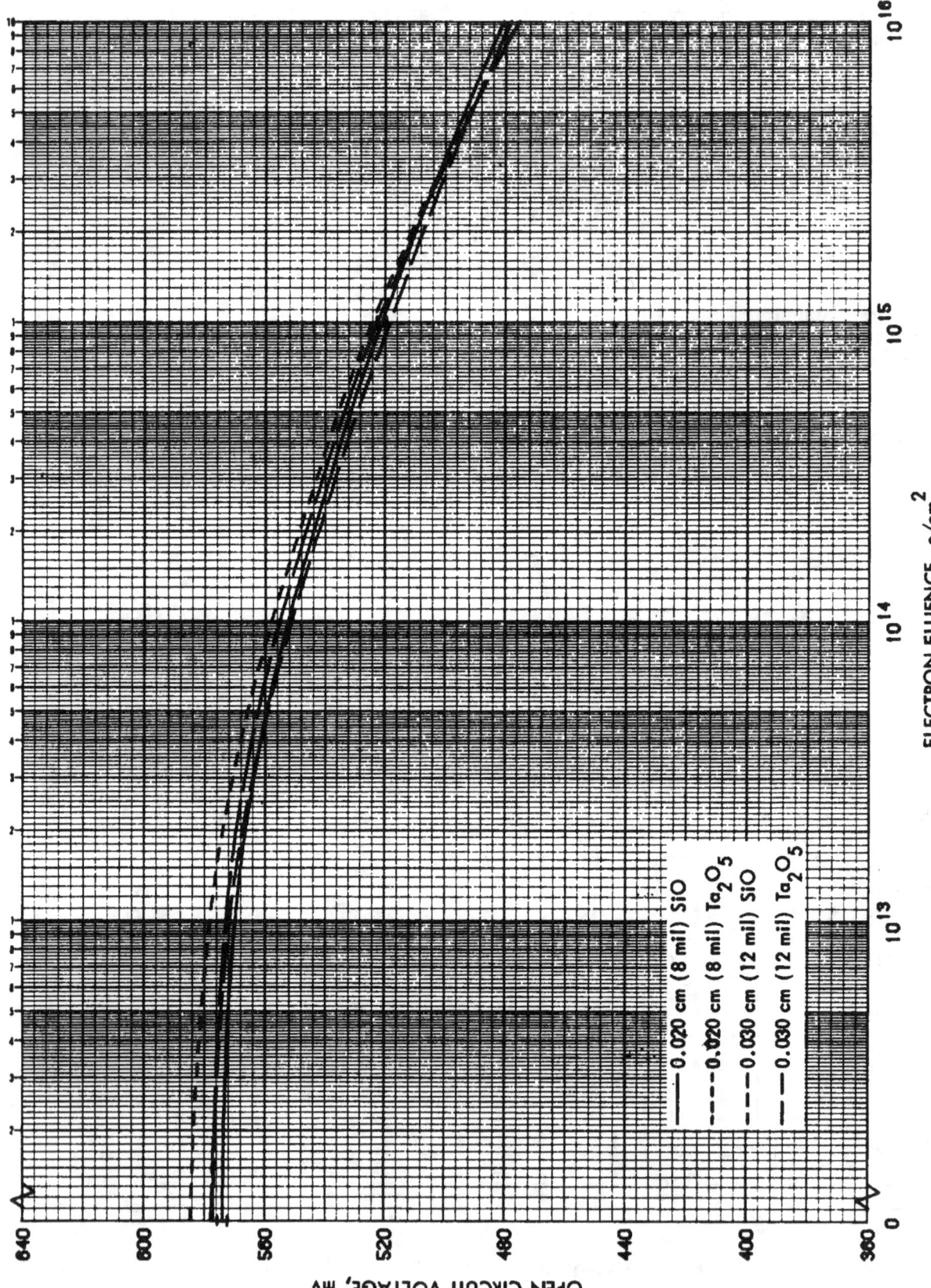

Figure 3.21 V_{oc} vs 1 MeV Electron Fluence for 2 Ohm-cm n/p Conventional and Shallow Junction Silicon Cells

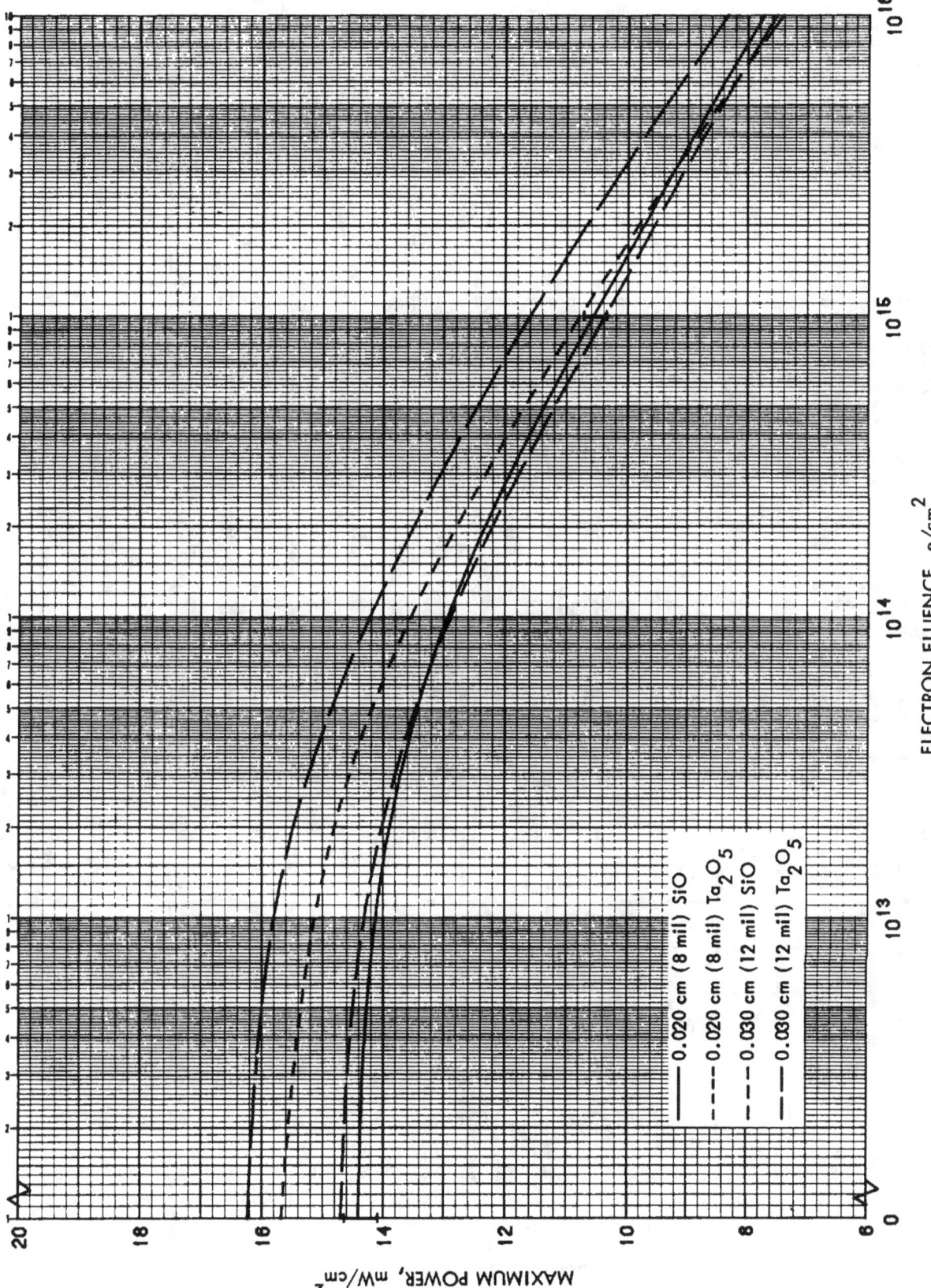

Figure 3.22 P_{max} vs 1 MeV Electron Fluence for 2 Ohm-cm n/p Conventional and Shallow Junction Silicon Cells

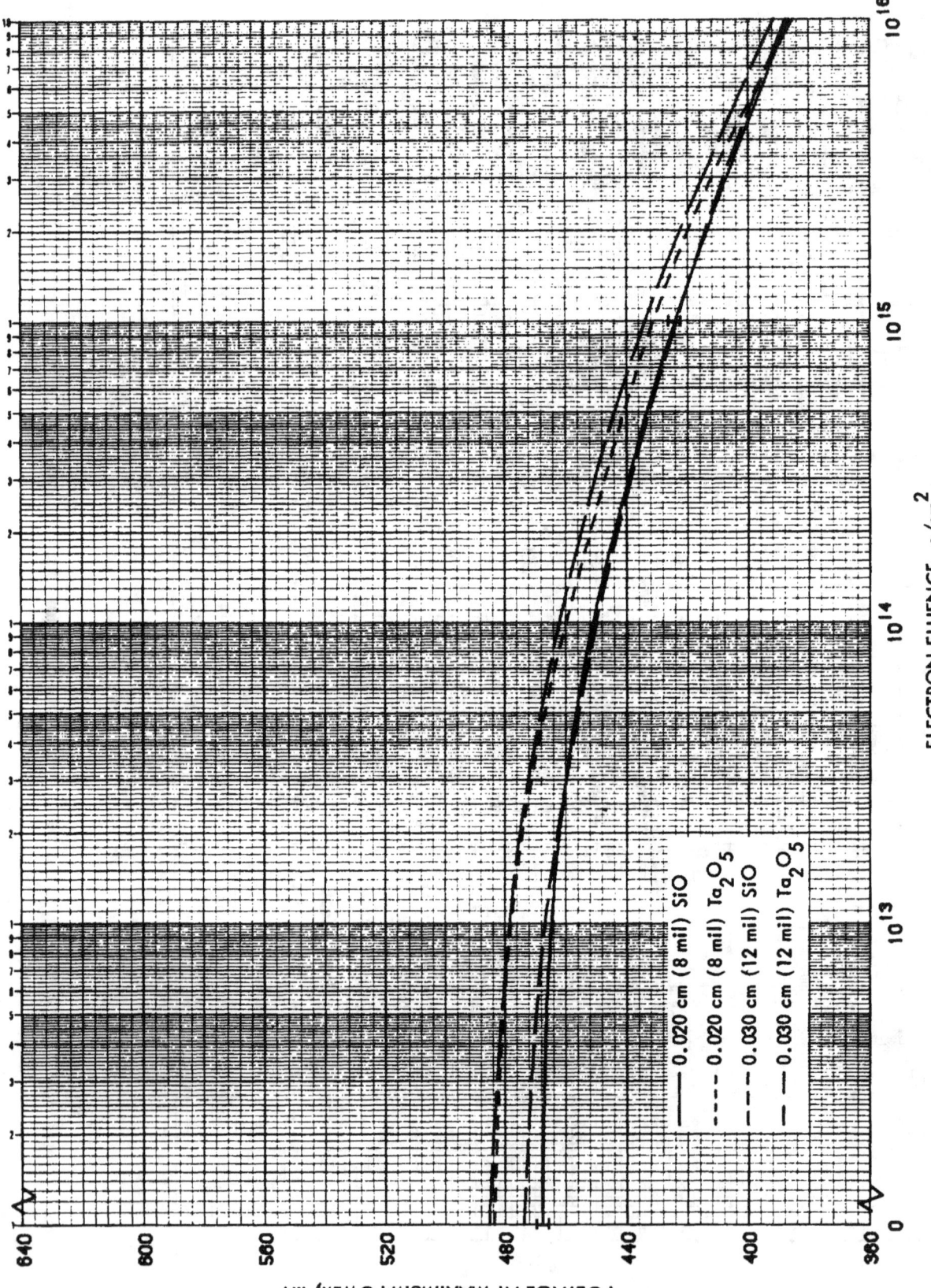

Figure 3.23 V_{mp} vs 1 MeV Electron Fluence for 2 Ohm-cm n/p Conventional and Shallow Junction Silicon Cells

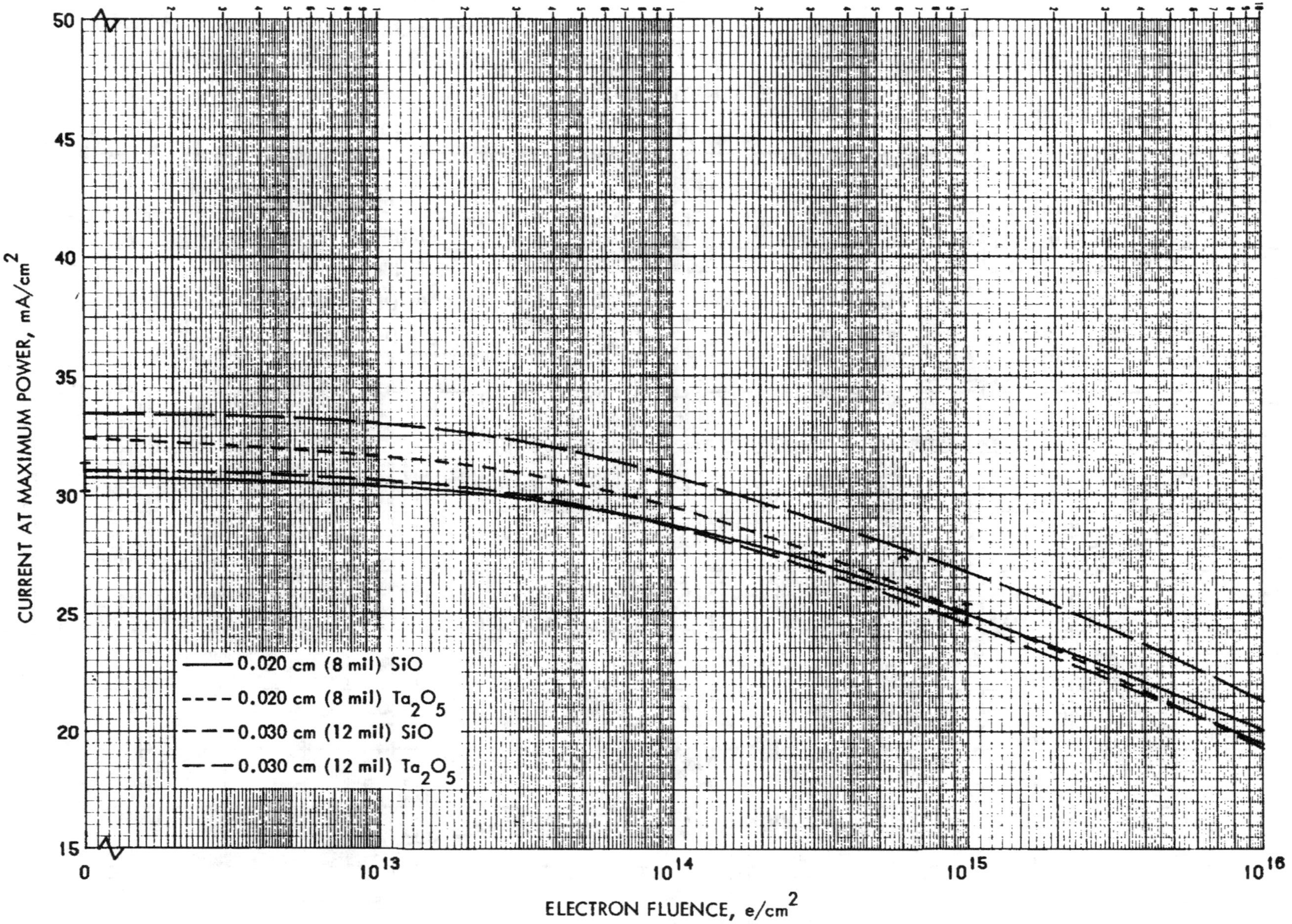

Figure 3.24 I_{mp} vs 1 MeV Electron Fluence for 2 Ohm-cm n/p Conventional and Shallow Junction Silicon Cells

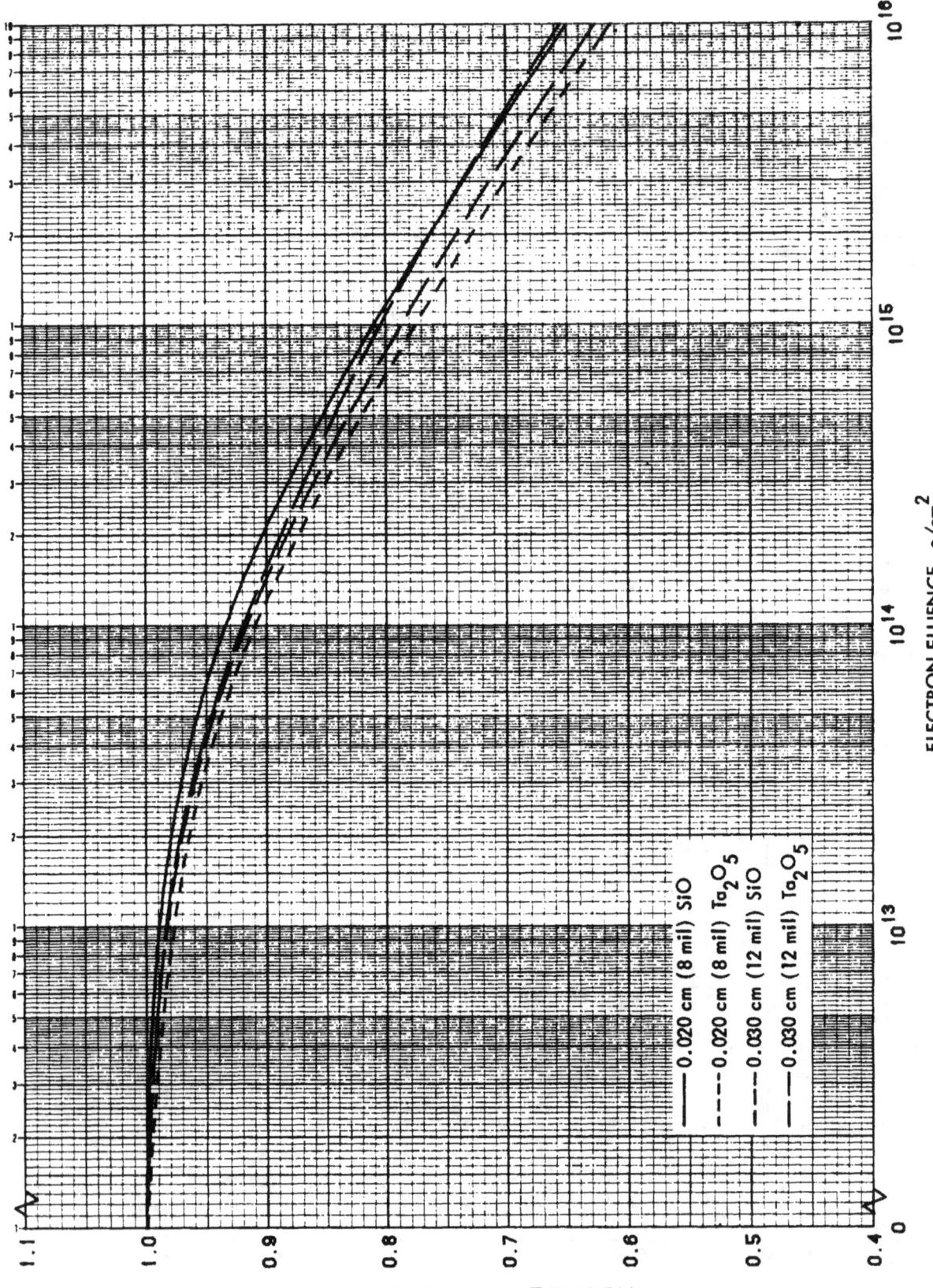

Figure 3.25 Normalized I_{sc} vs 1 MeV Electron Fluence for 2 Ohm-cm n/p Conventional and Shallow Junction Silicon Cells

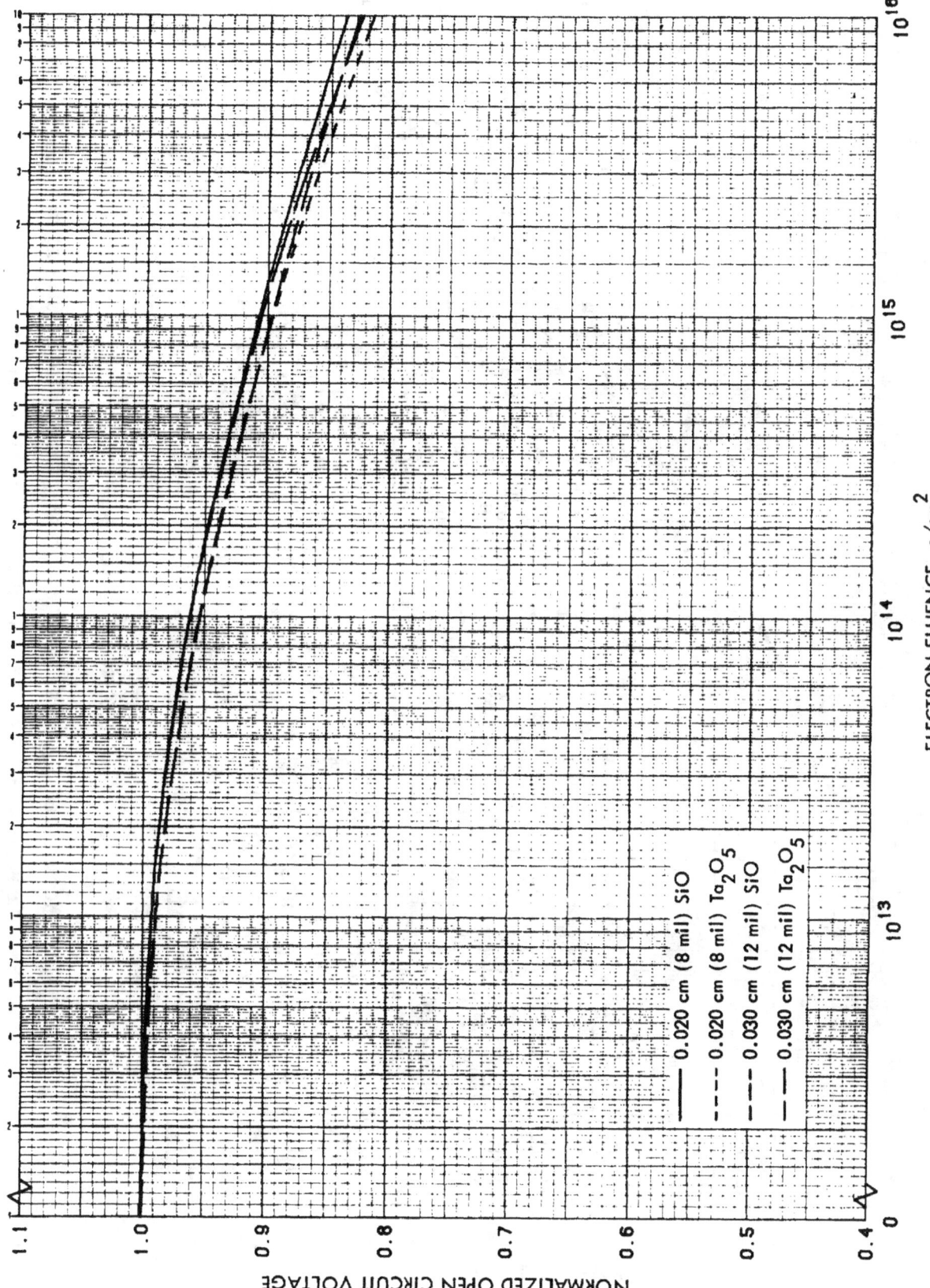

Figure 3.26 Normalized V_{oc} vs 1 MeV Electron Fluence for 2 Ohm-cm n/p Conventional and Shallow Junction Silicon Cells

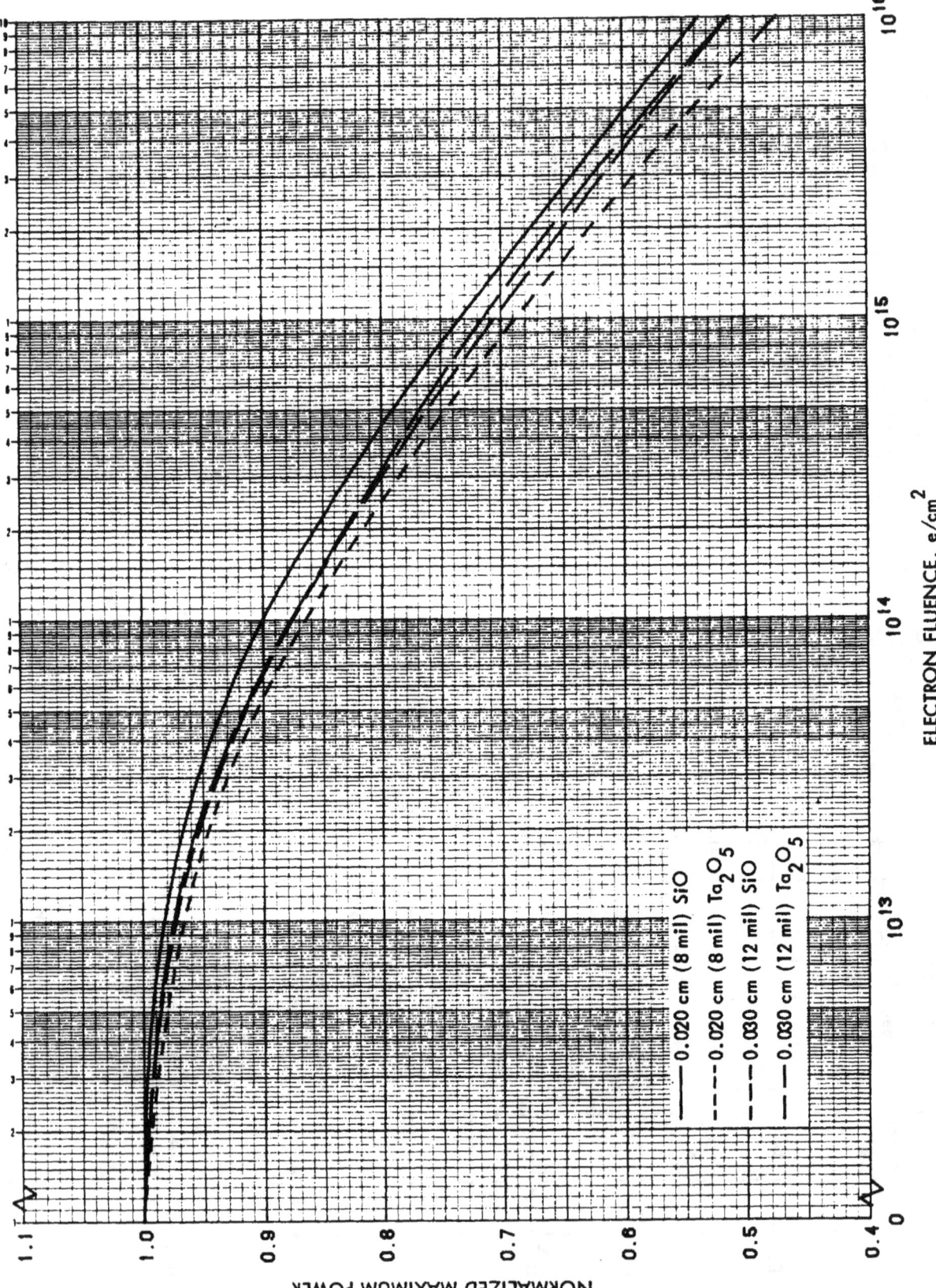

Figure 3.27 Normalized P_{max} vs 1 MeV Electron Fluence for 2 Ohm-cm n/p Conventional and Shallow Junction Silicon Cells

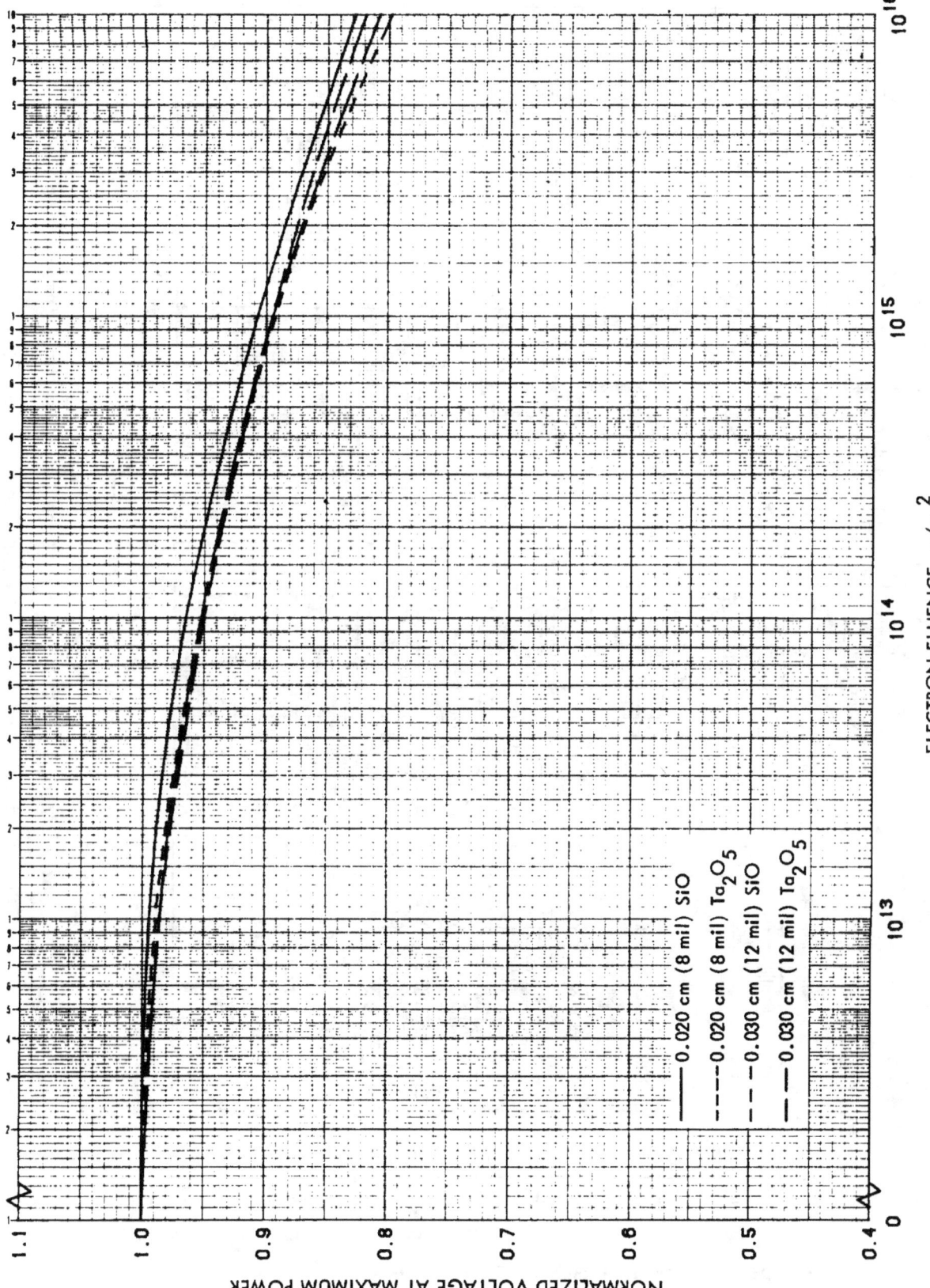

Figure 3.28 Normalized V_{mp} vs 1 MeV Electron Fluence for 2 Ohm-cm n/p Conventional and Shallow Junction Silicon Cells

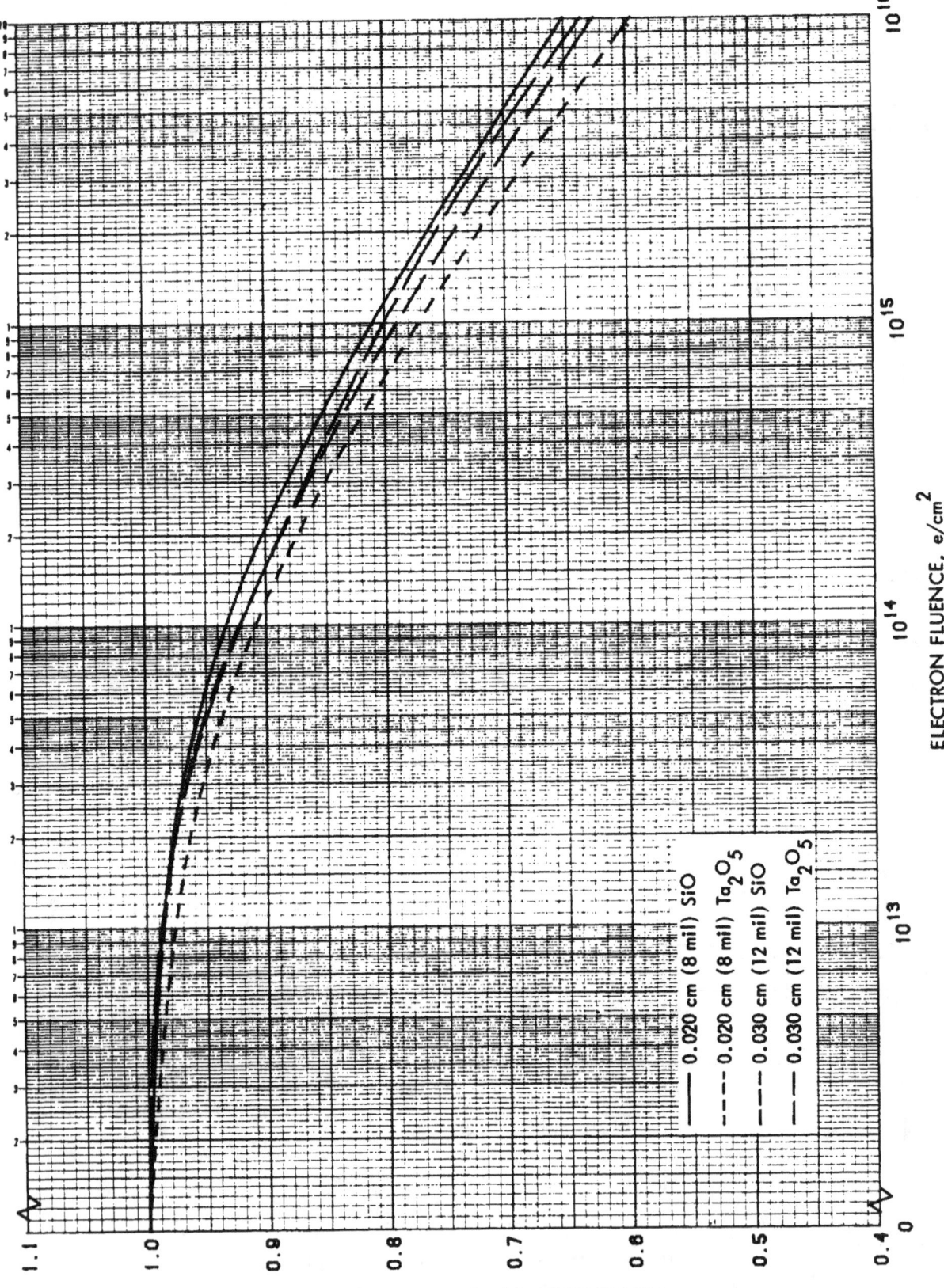

Figure 3.29 Normalized I_{mp} vs 1 MeV Electron Fluence for 2 Ohm-cm n/p Conventional and Shallow Junction Silicon Cells

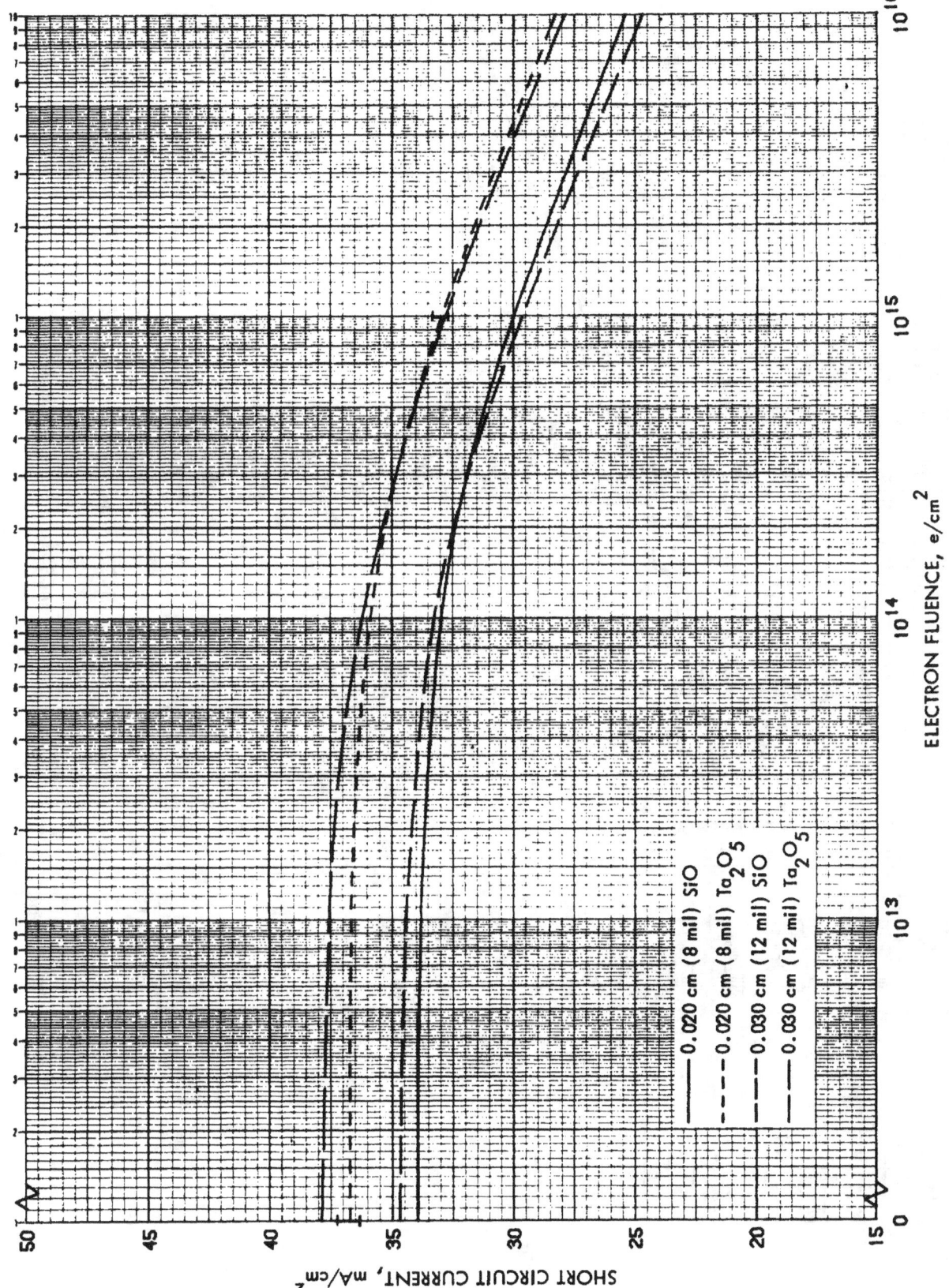

Figure 3.30 I_{SC} vs 1 MeV Electron Fluence for 10 Ohm-cm n/p Conventional and Shallow Junction Silicon Cells

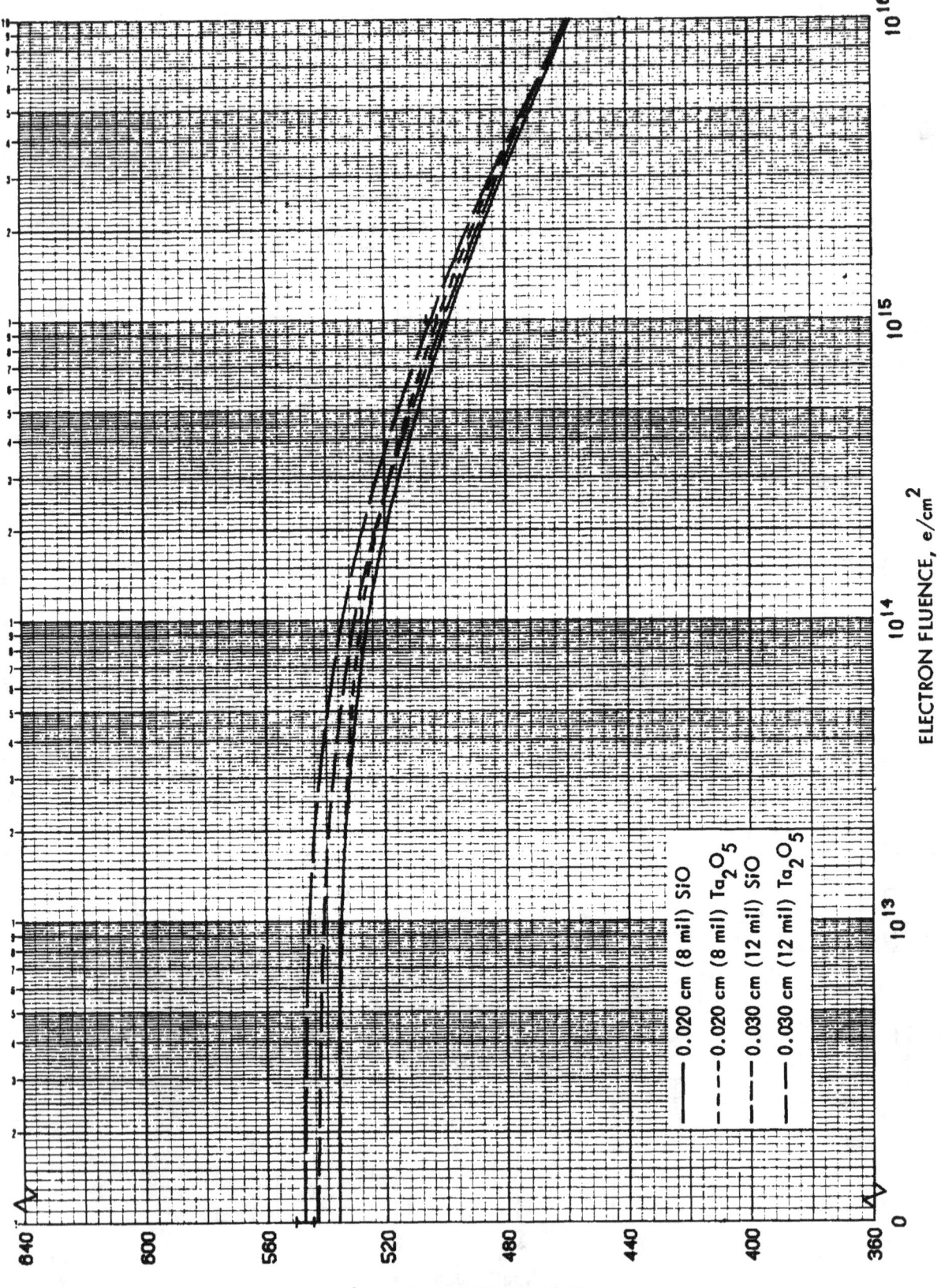

Figure 3.31 V_{oc} vs 1 MeV Electron Fluence for 10 Ohm-cm n/p Conventional and Shallow Junction Silicon Cells

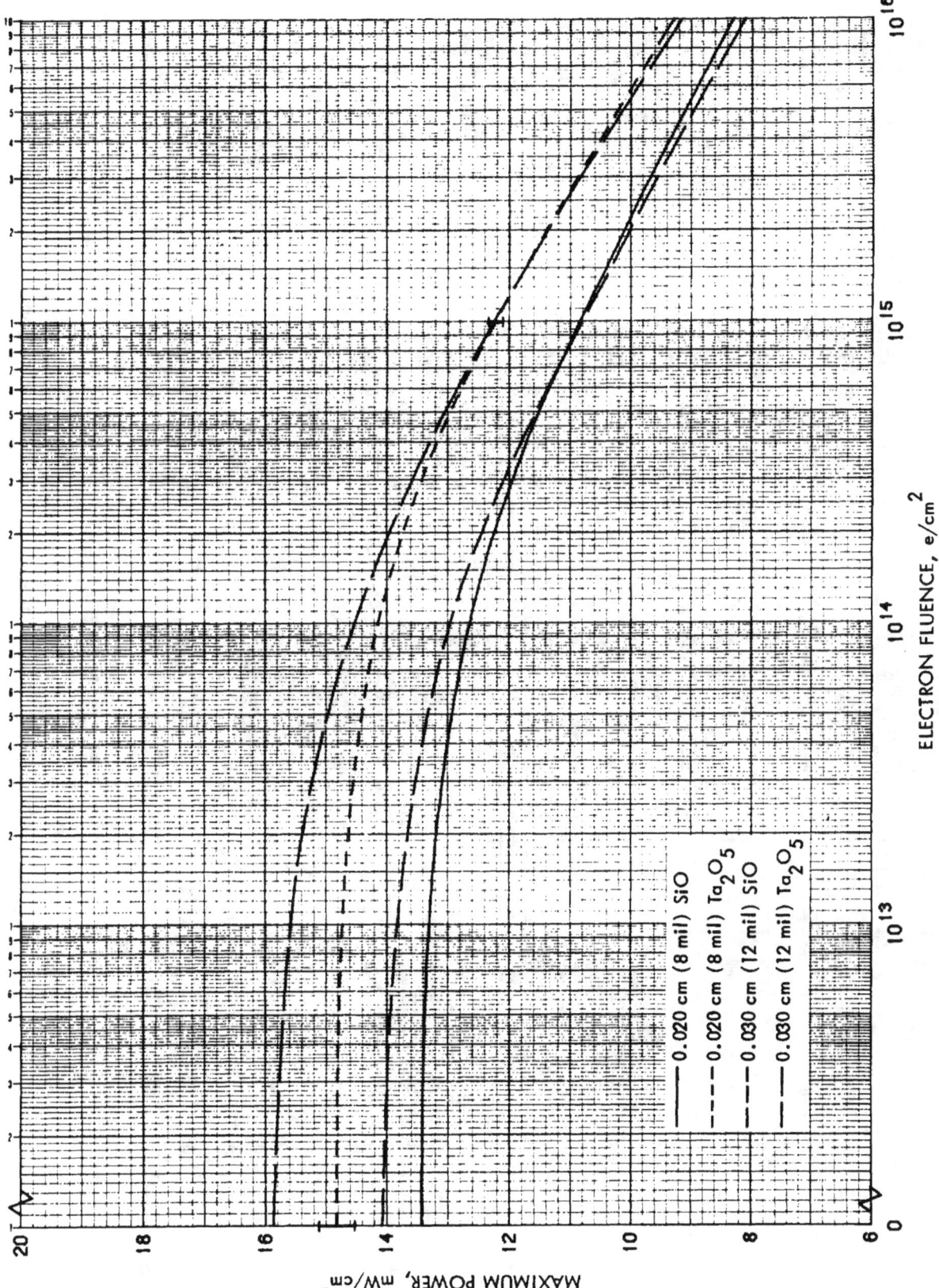

Figure 3.32 P_{max} vs 1 MeV Electron Fluence for 10 Ohm-cm n/p Conventional and Shallow Junction Silicon Cells

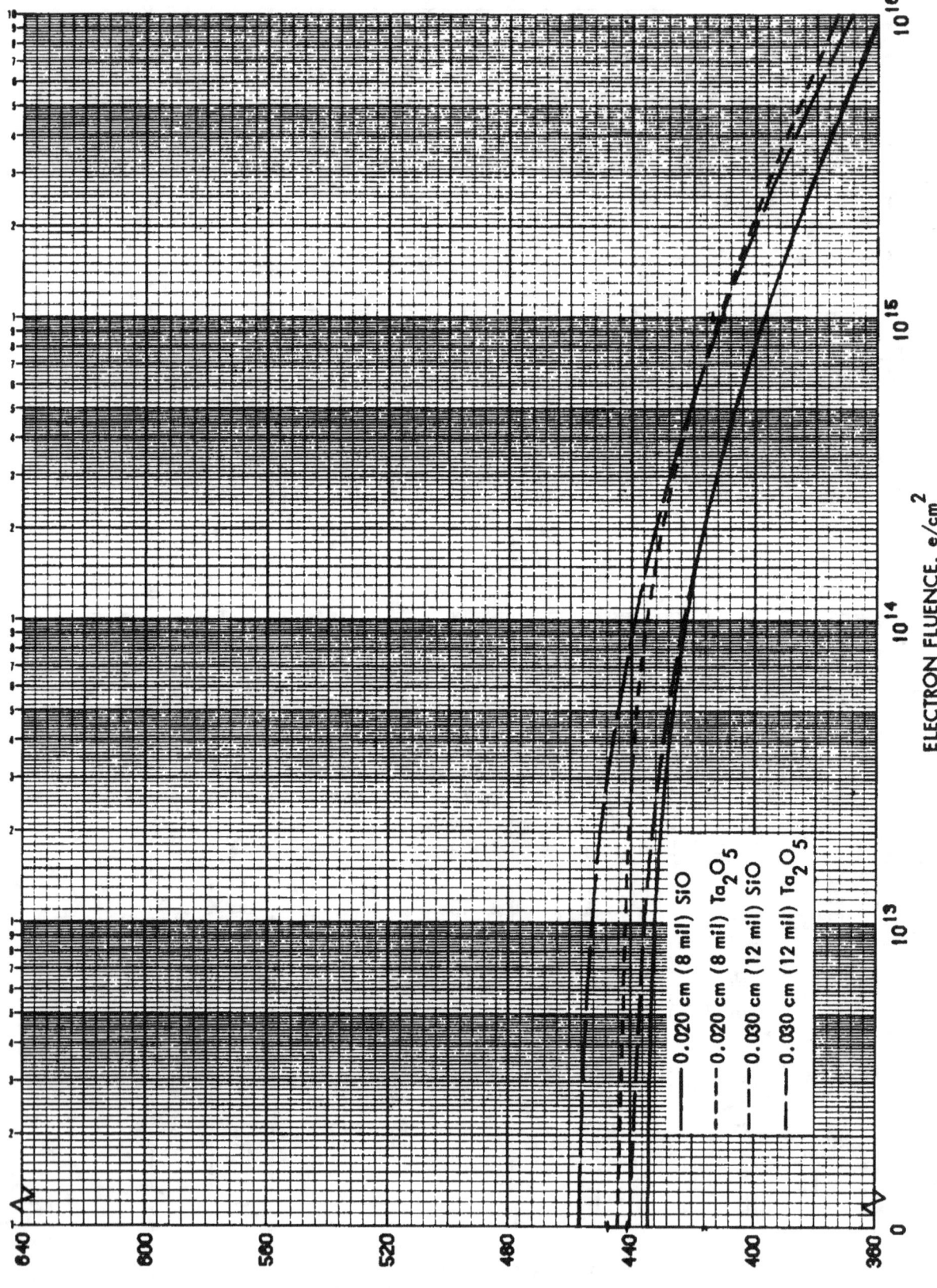

Figure 3.33 V_{mp} vs 1 MeV Electron Fluence for 10 Ohm-cm n/p Conventional and Shallow Junction Silicon Cells

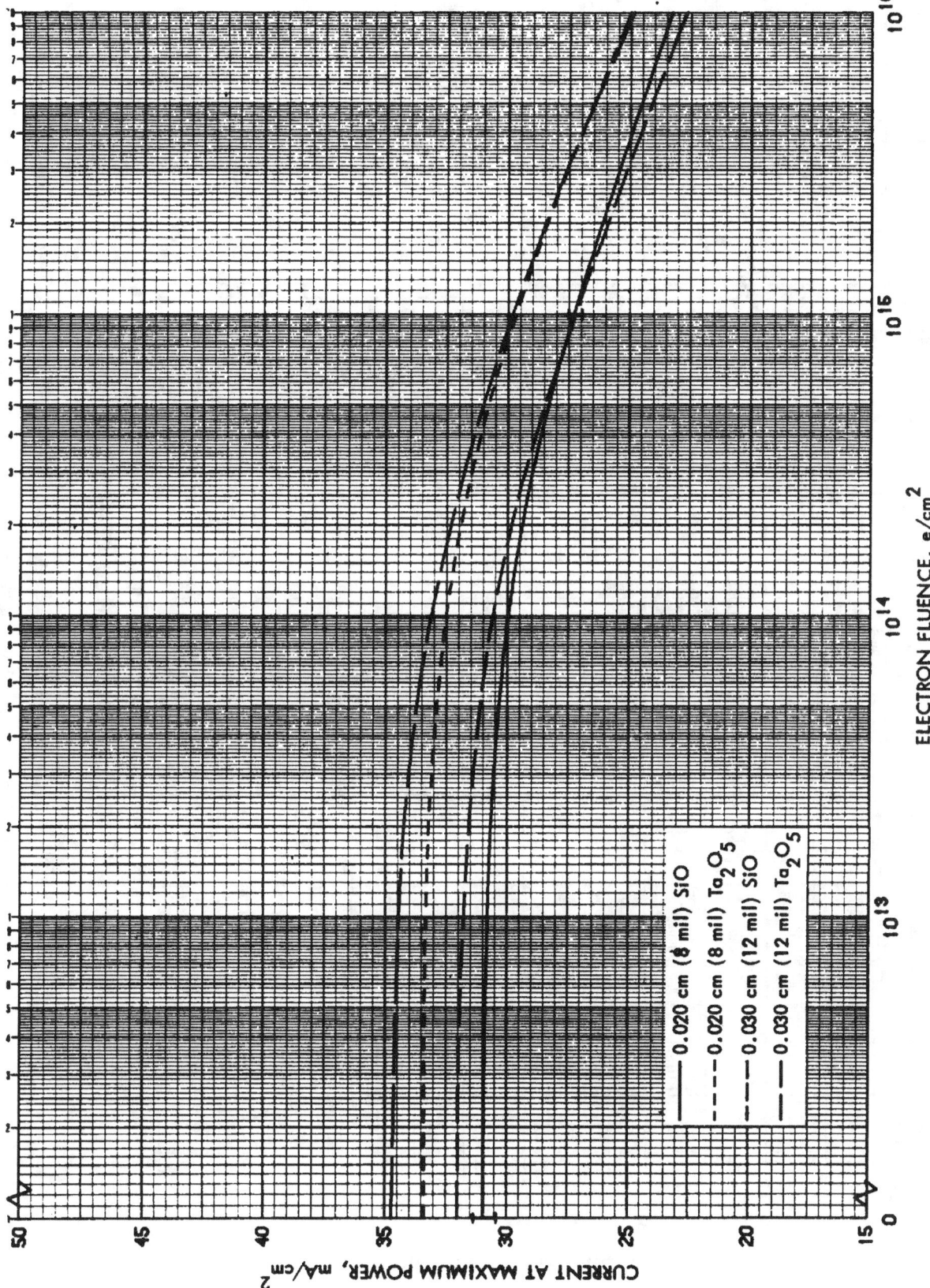

Figure 3.34 I_{mp} vs 1 MeV Electron Fluence for 10 Ohm-cm n/p Conventional and Shallow Junction Silicon Cells

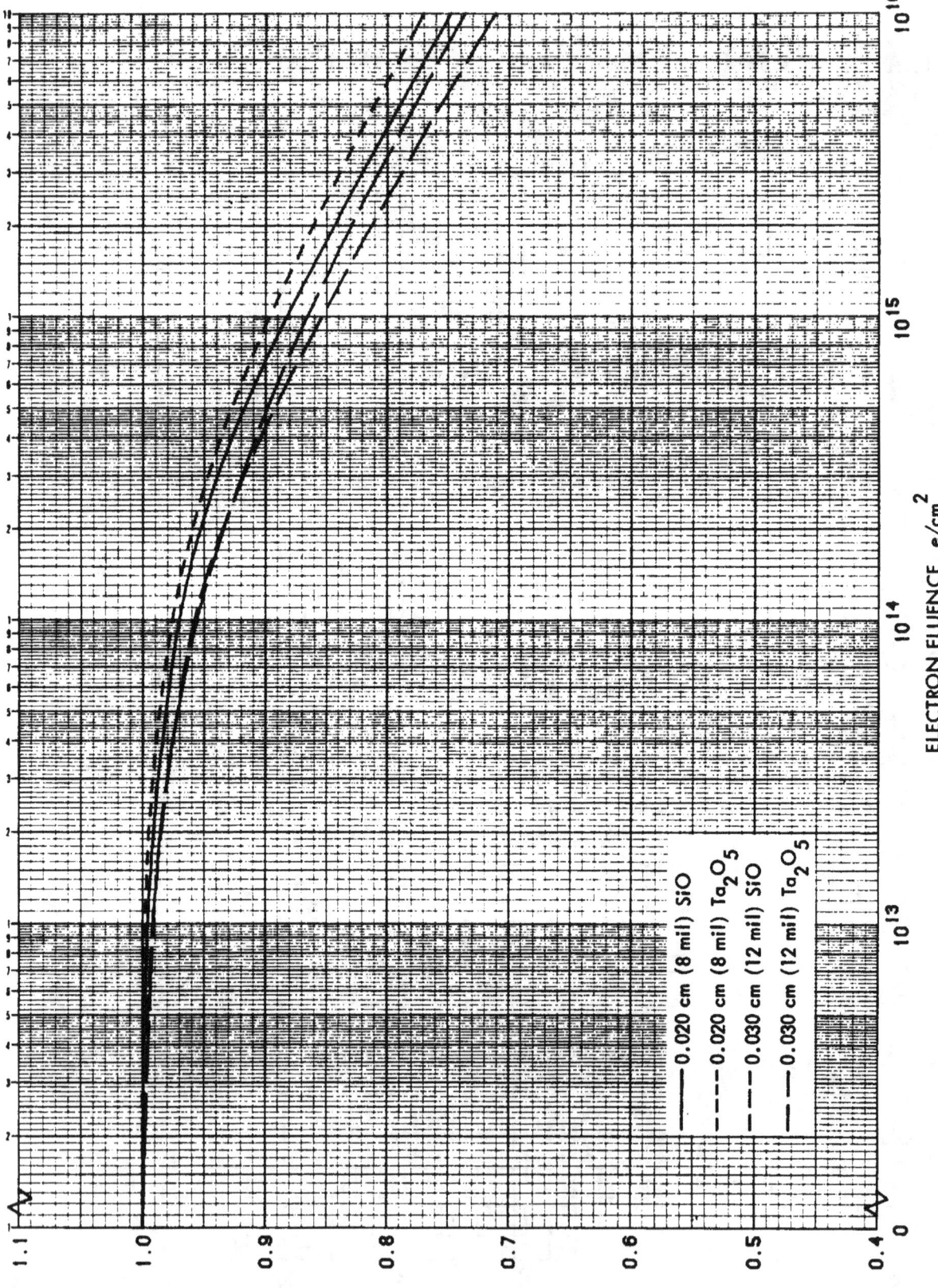

Figure 3.35 Normalized I_{sc} vs 1 MeV Electron Fluence for 10 Ohm-cm n/p Conventional and Shallow Junction Silicon Cells

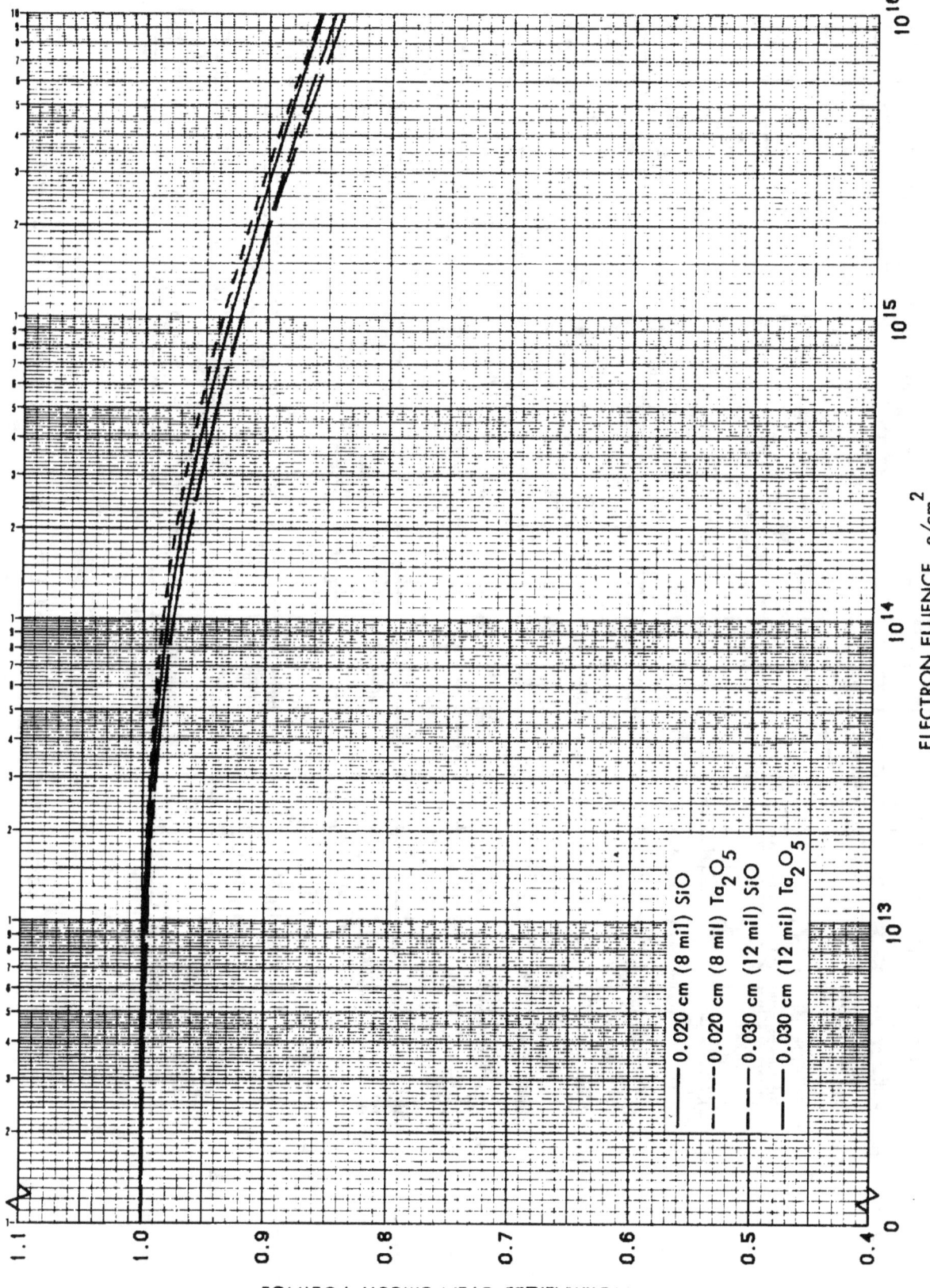

Figure 3.36 Normalized V_{oc} vs 1 MeV Electron Fluence for 10 Ohm-cm n/p Conventional and Shallow Junction Silicon Cells

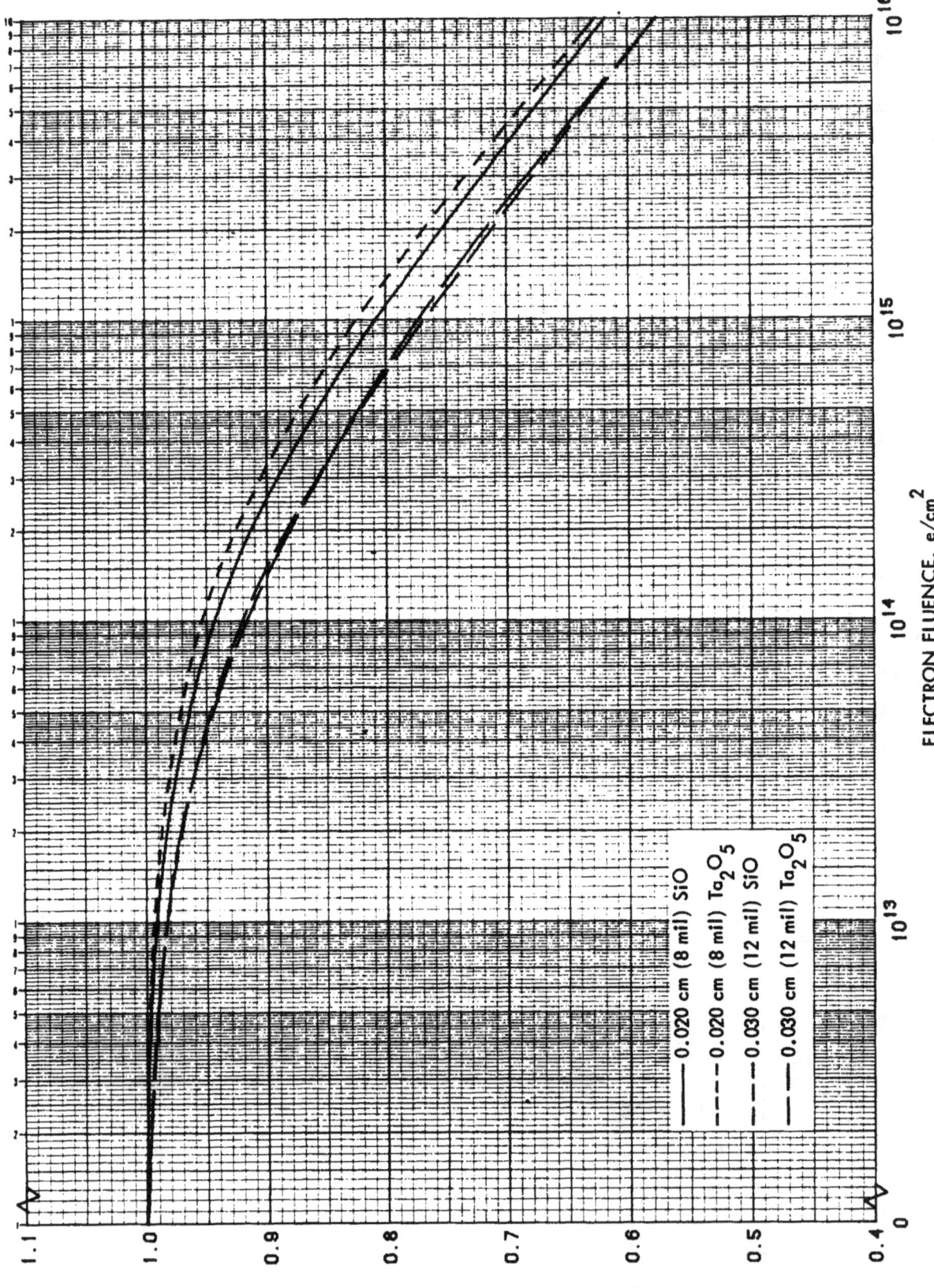

Figure 3.37 Normalized P_{max} vs 1 MeV Electron Fluence for 10 Ohm-cm n/p Conventional and Shallow Junction Silicon Cells

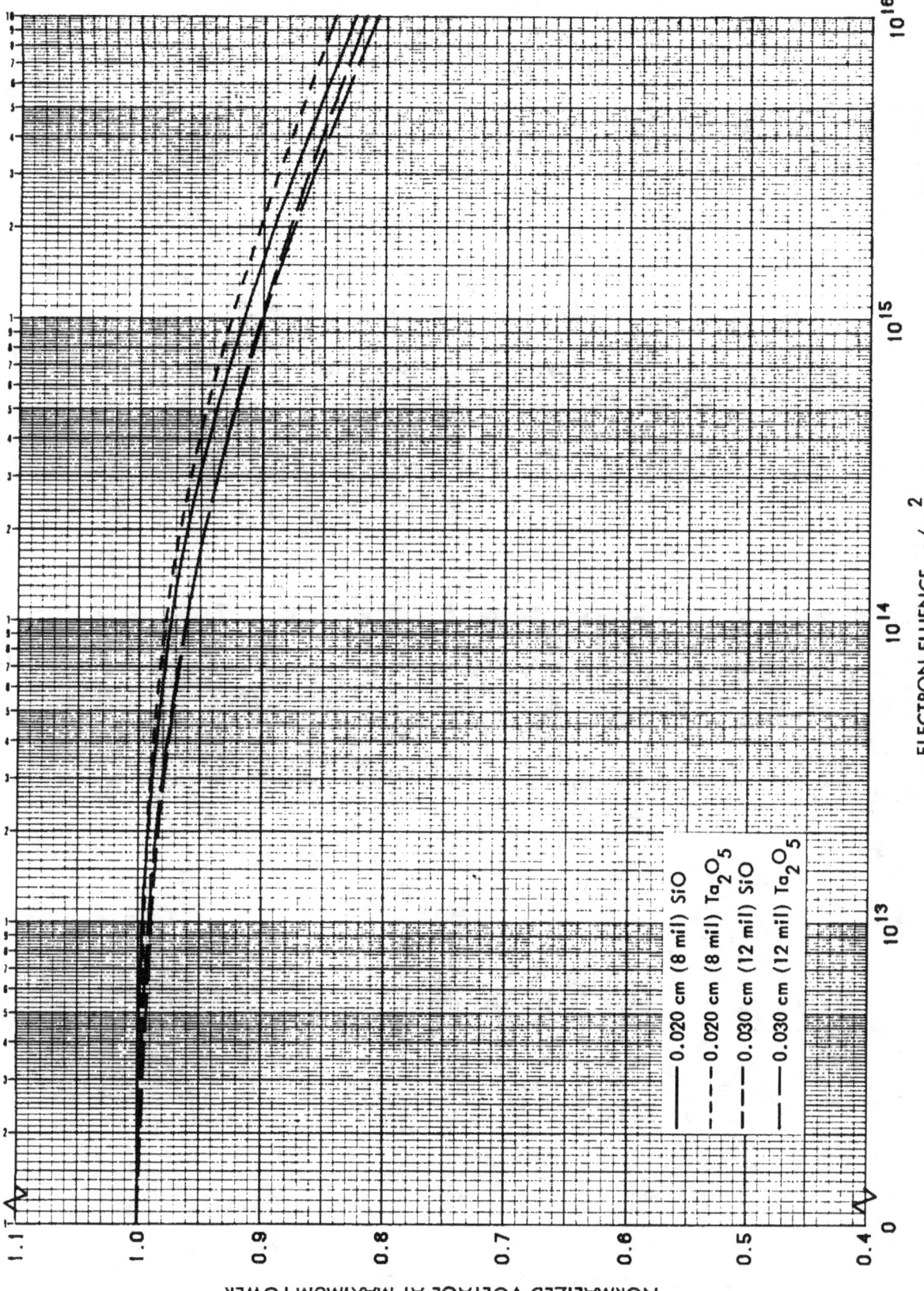

Figure 3.38 Normalized V_{mp} vs 1 MeV Electron Fluence for 10 Ohm-cm n/p Conventional and Shallow Junction Silicon Cells

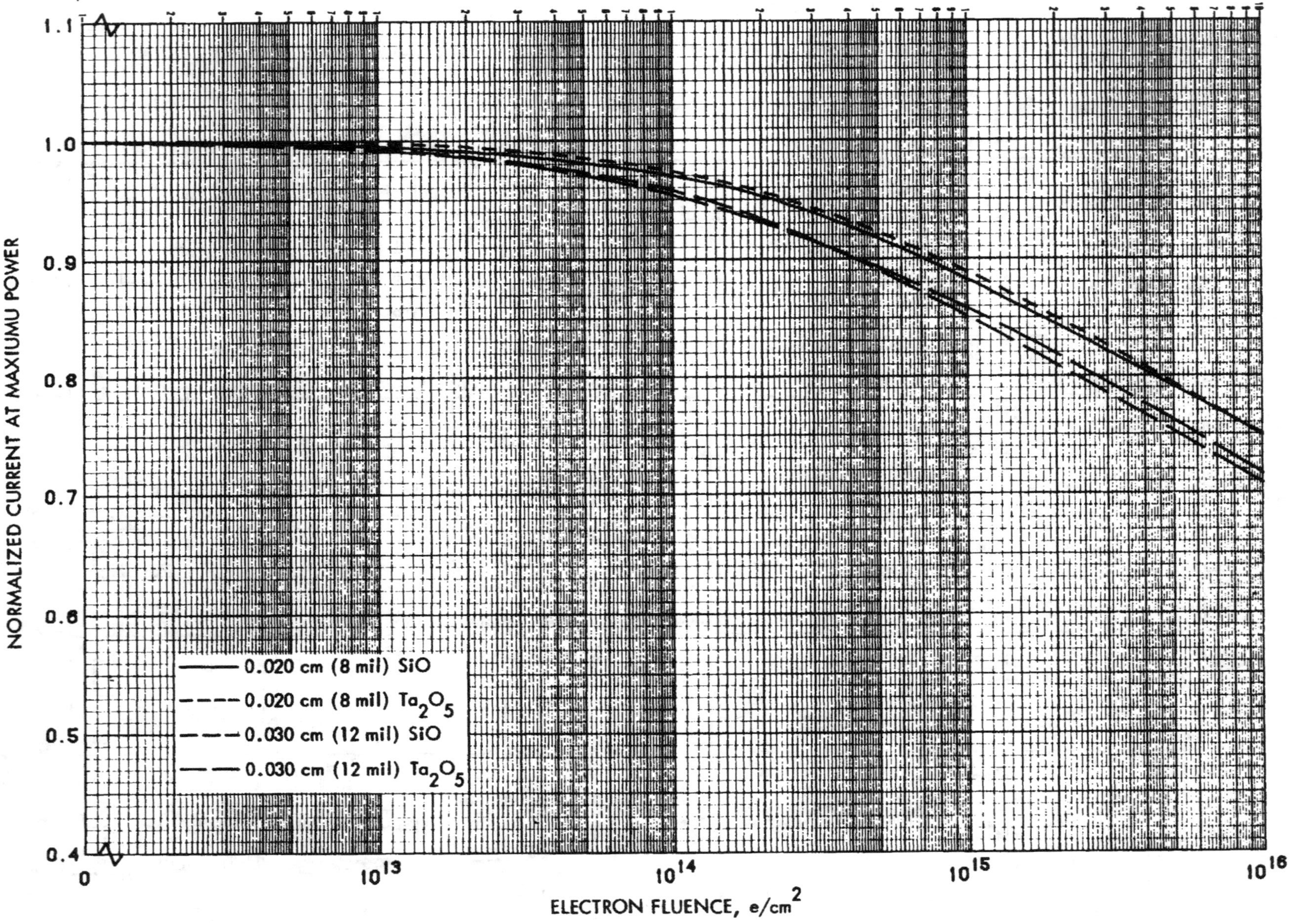

Figure 3.39 Normalized I_{mp} vs 1 MeV Electron Fluence for 10 Ohm-cm n/p Conventional and Shallow Junction Silicon Cells

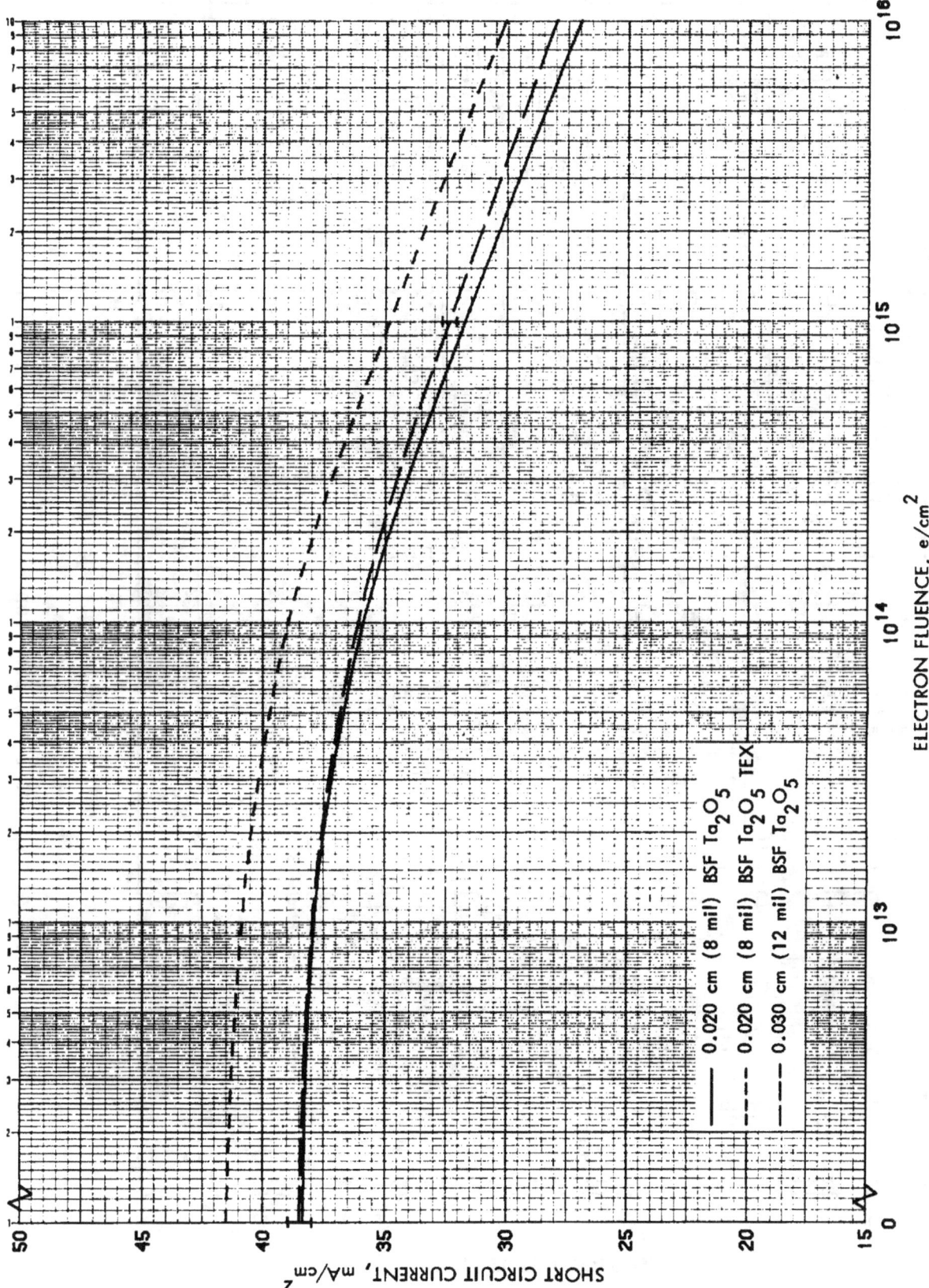

Figure 3.40 I_{SC} vs 1 MeV Electron Fluence for 10 Ohm-cm n/p Back Surface Field Silicon Cells (+TEX)

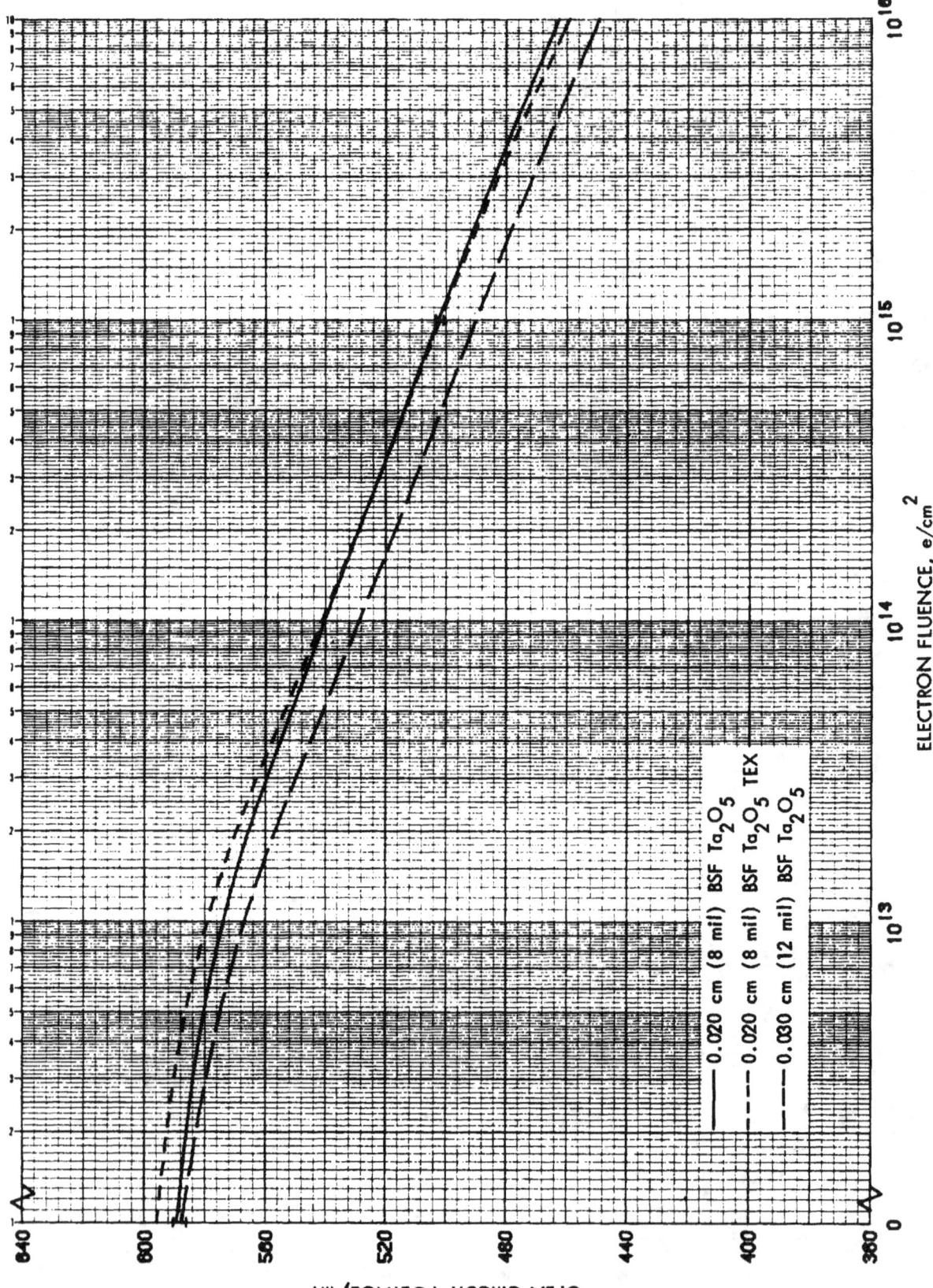

Figure 3.41 V_{OC} vs 1 MeV Electron Fluence for 10 Ohm-cm n/p Back Surface Field Silicon Cells (+TEX)

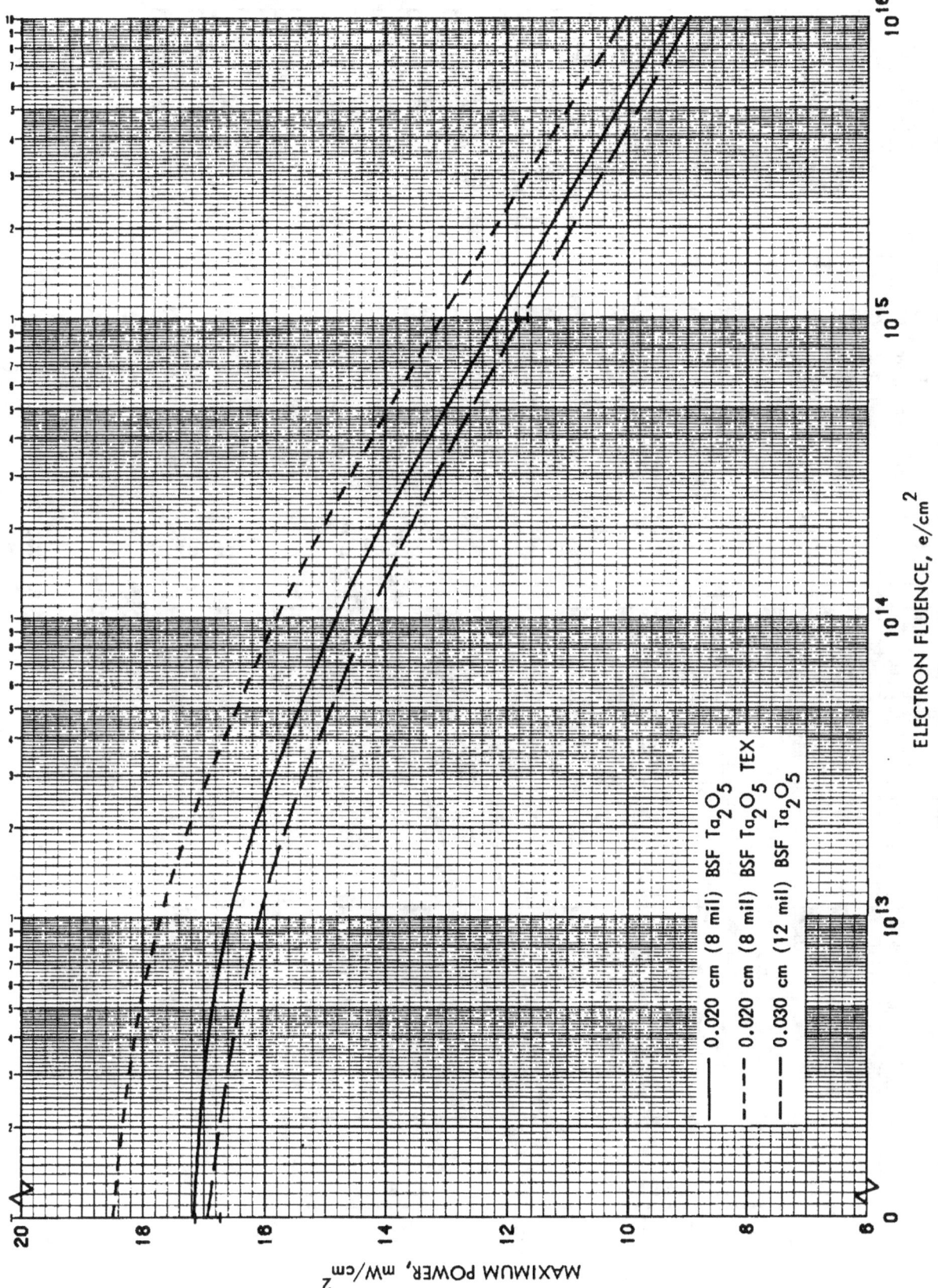

Figure 3.42 P_{max} vs 1 MeV Electron Fluence for 10 Ohm-cm n/p Back Surface Field Silicon Cells (+TEX)

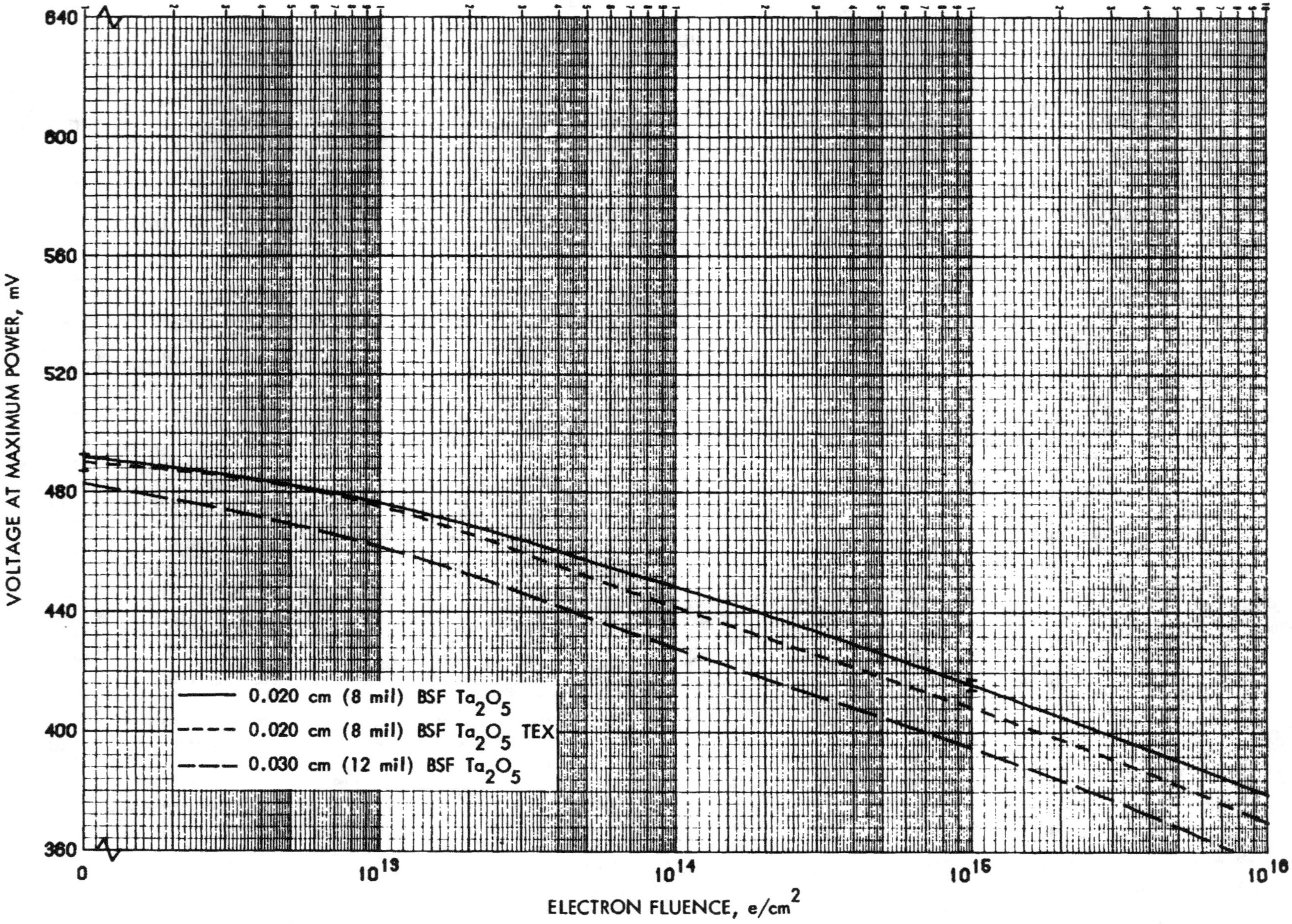

Figure 3.43 V_{mp} vs 1 MeV Electron Fluence for 10 Ohm-cm n/p Back Surface Field Silicon Cells (+TEX)

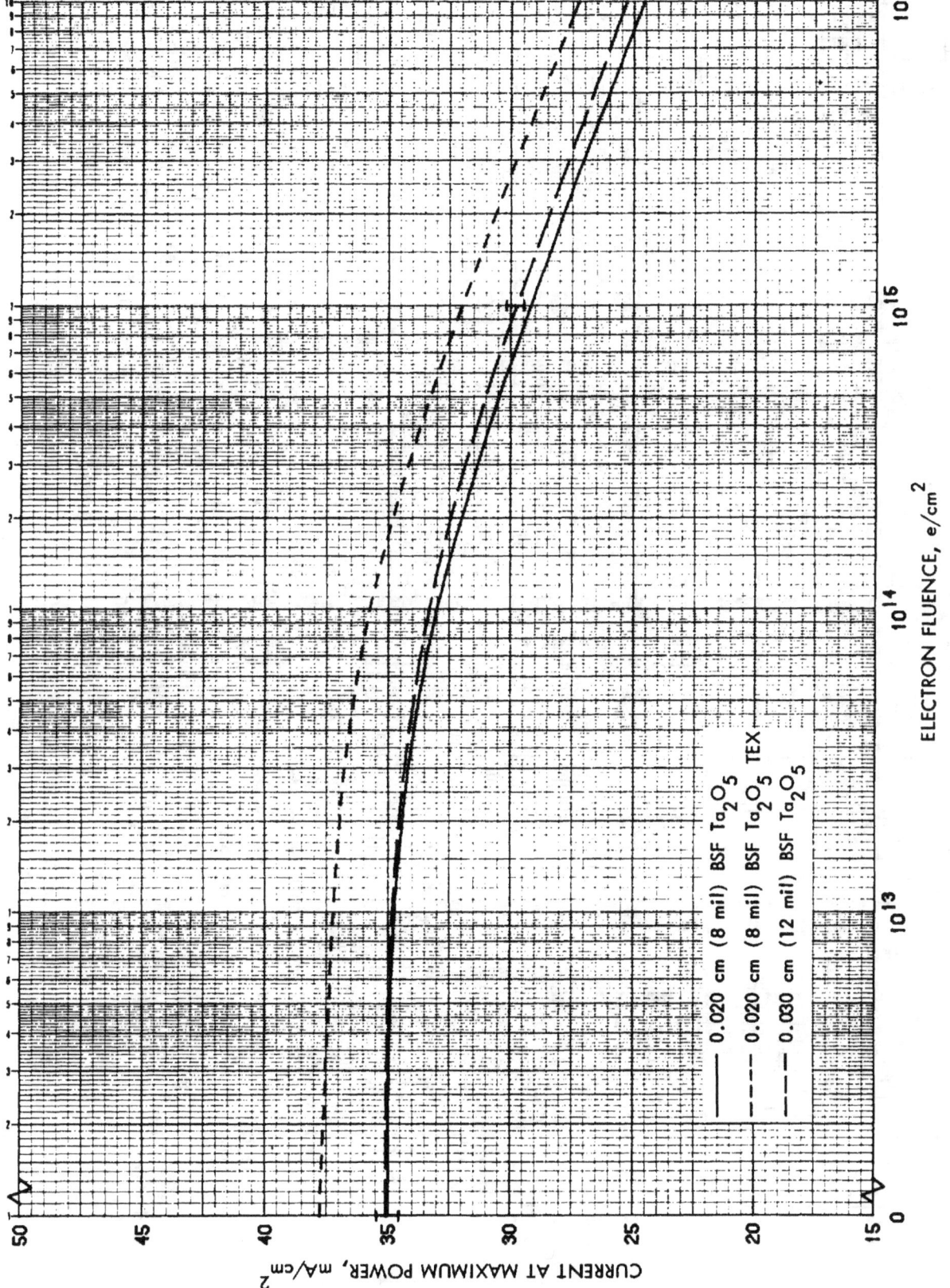

Figure 3.44 I_{mp} vs 1 MeV Electron Fluence for 10 Ohm-cm n/p Back Surface Field Silicon Cells (+TEX)

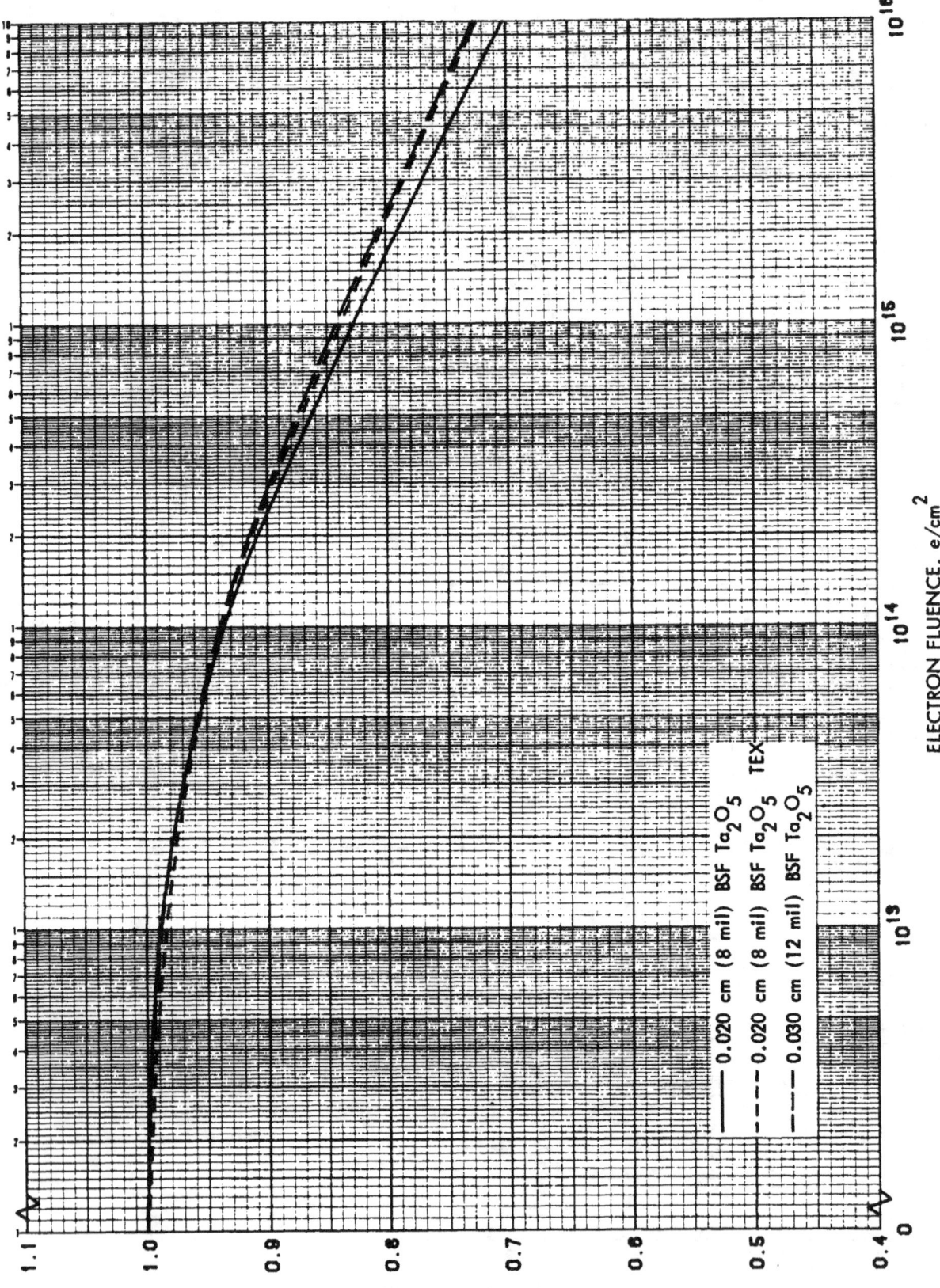

Figure 3.45 Normalized I_{sc} vs 1 MeV Electron Fluence for 10 Ohm-cm n/p Back Surface Field Silicon Cells (+TEX)

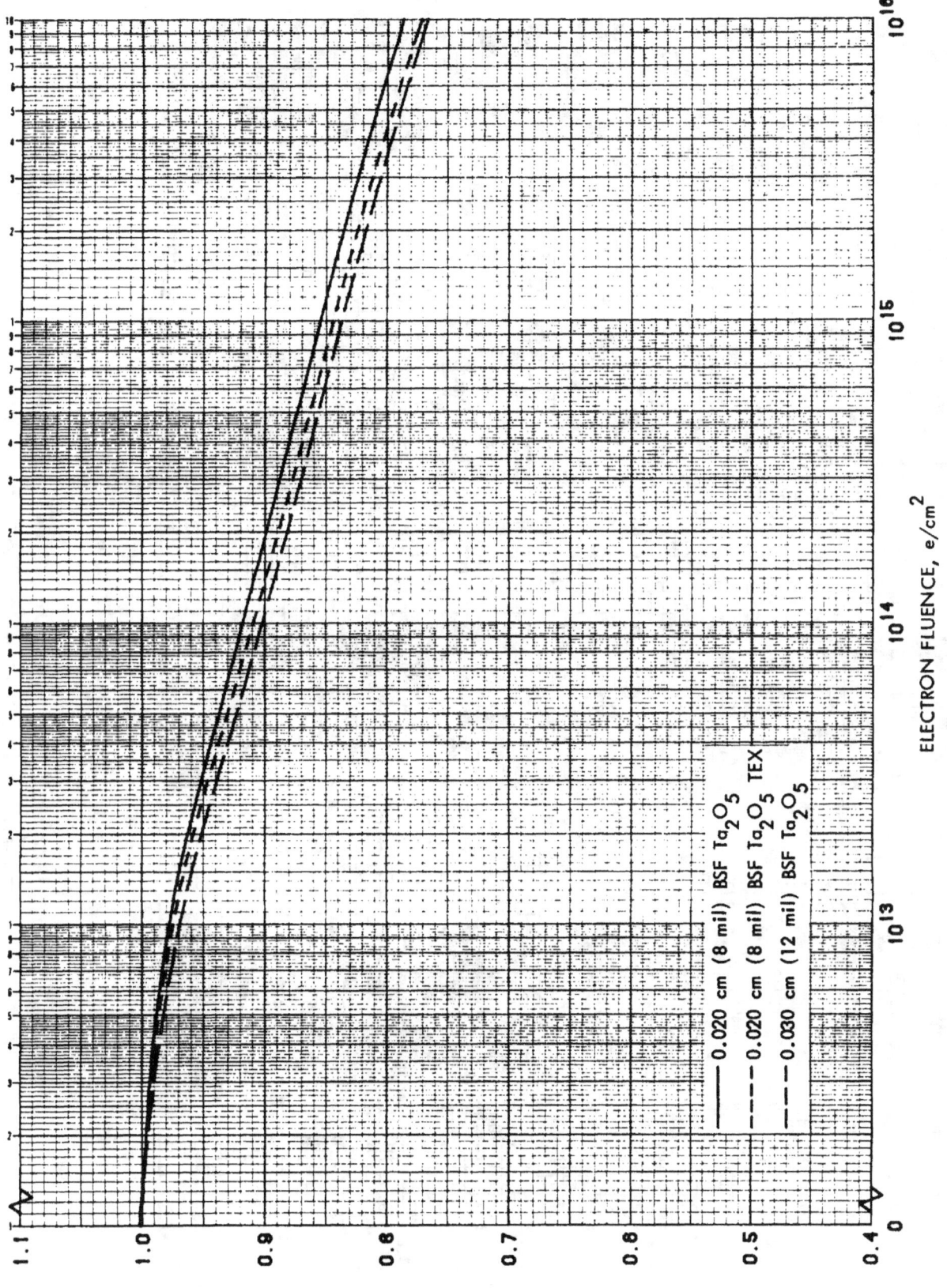

Figure 3.46 Normalized V_{OC} vs 1 MeV Electron Fluence for 10 Ohm-cm n/p Back Surface Field Silicon Cells (+TEX)

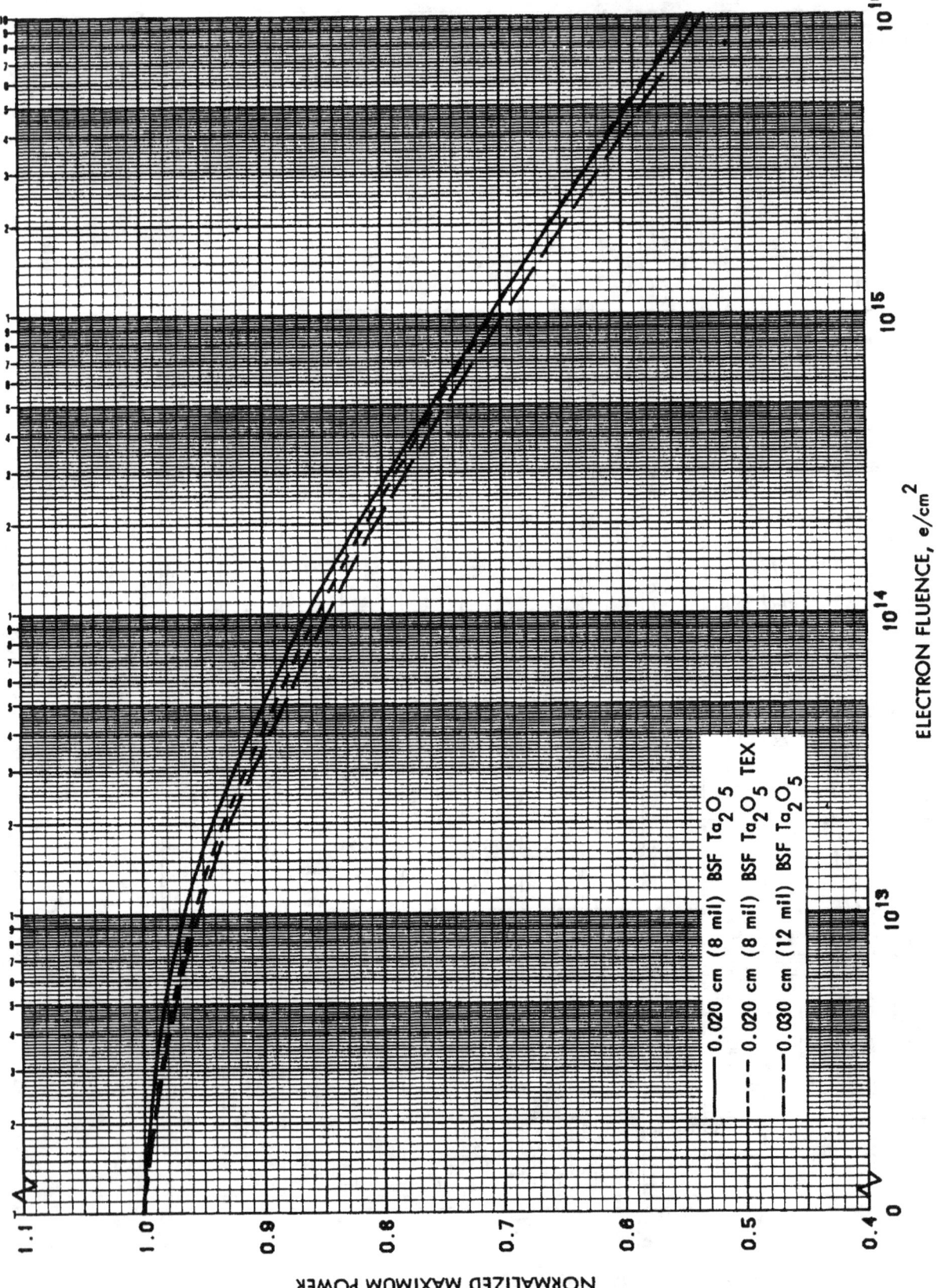

Figure 3.47 Normalized P_{max} vs 1 MeV Electron Fluence for 10 Ohm-cm n/p Back Surface Field Silicon Cells (+TEX)

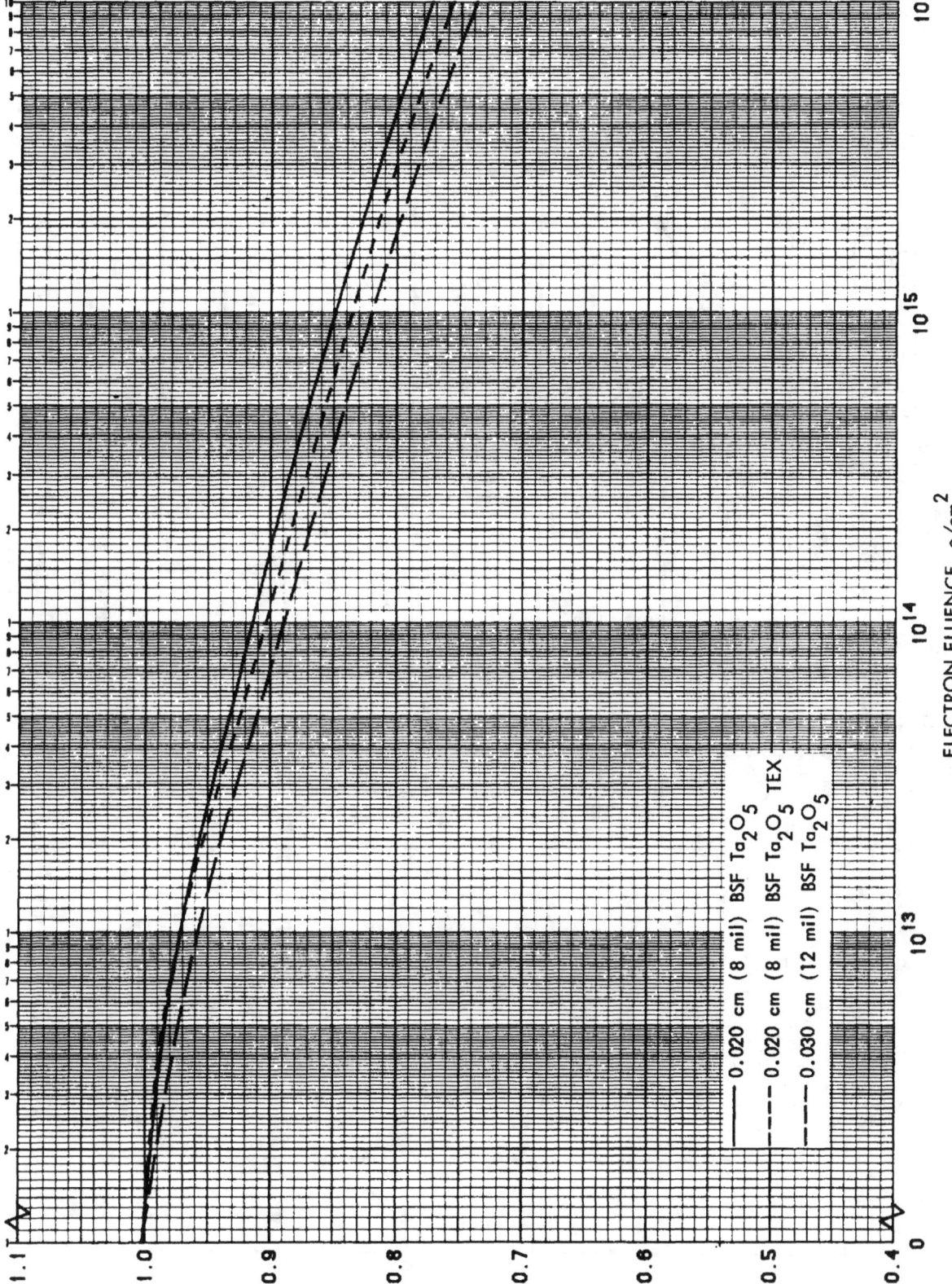

Figure 3.48 Normalized V_{mp} vs 1 MeV Electron Fluence for 10 Ohm-cm n/p Back Surface Field Silicon Cells (+TEX)

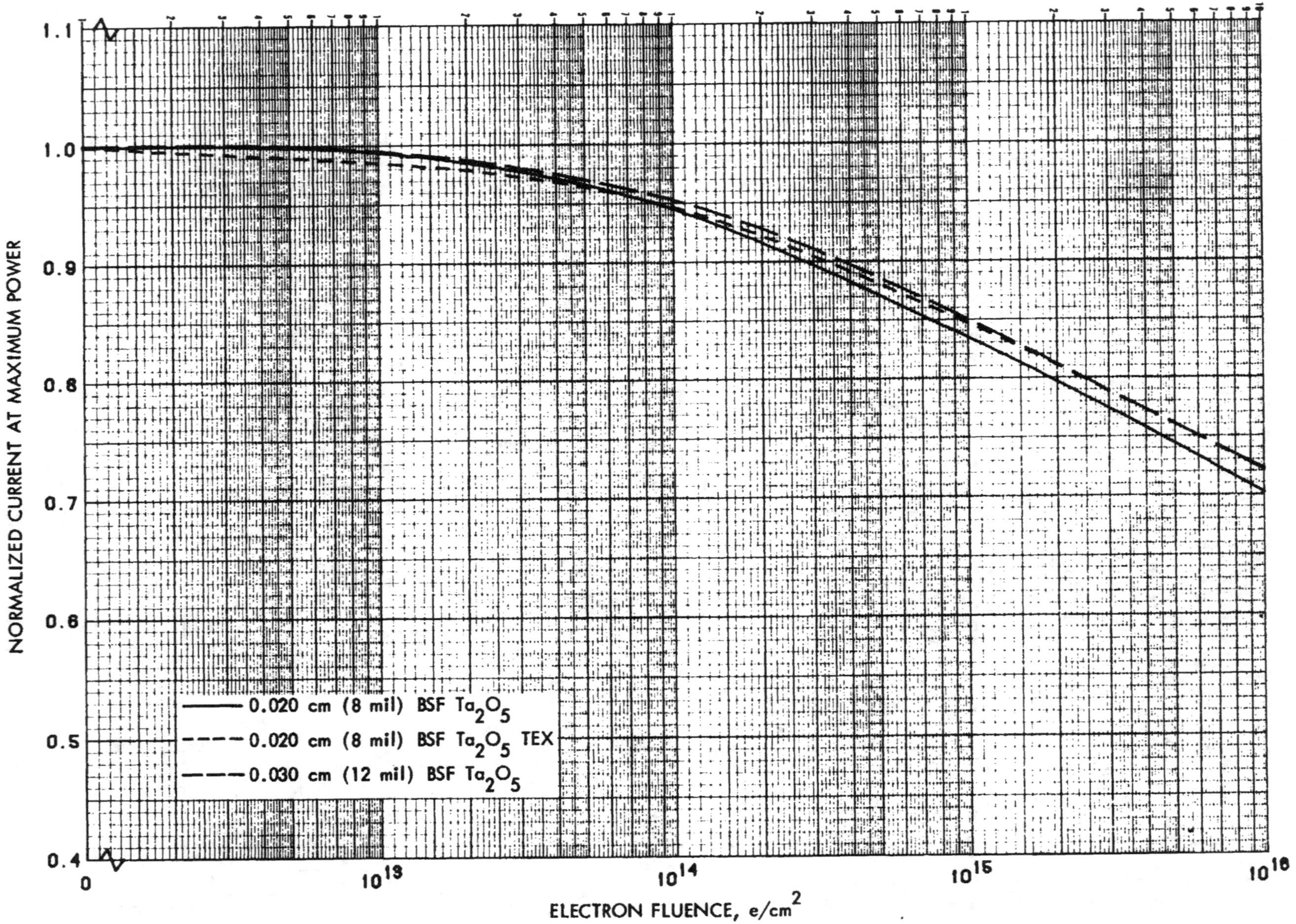

Figure 3.49 Normalized I_{mp} vs 1 MeV Electron Fluence for 10 Ohm-cm n/p Back Surface Field Silicon Cells (+TEX)

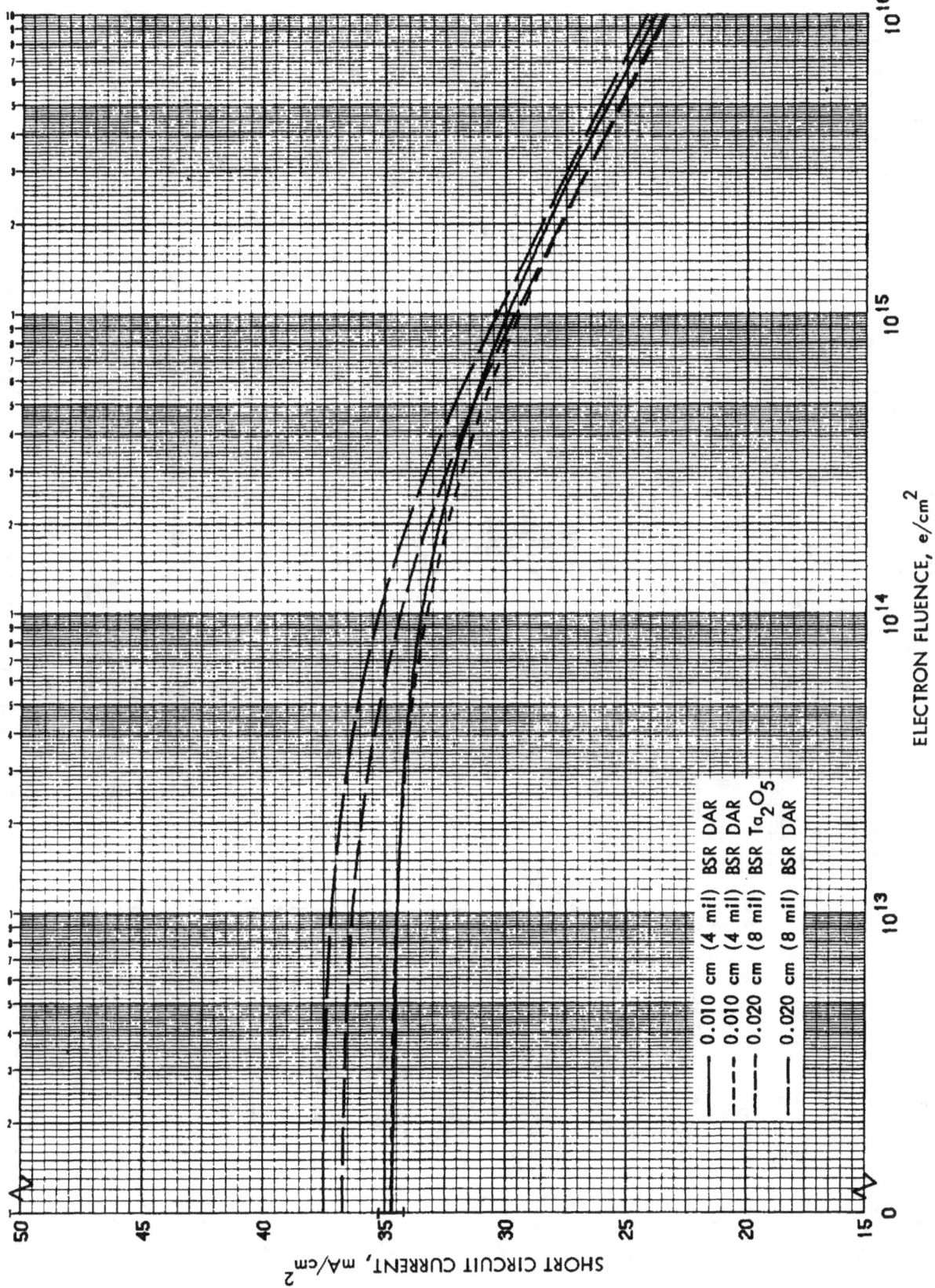

Figure 3.50 I_{sc} vs 1 MeV Electron Fluence for 2 Ohm-cm n/p Silicon Cells with BSR

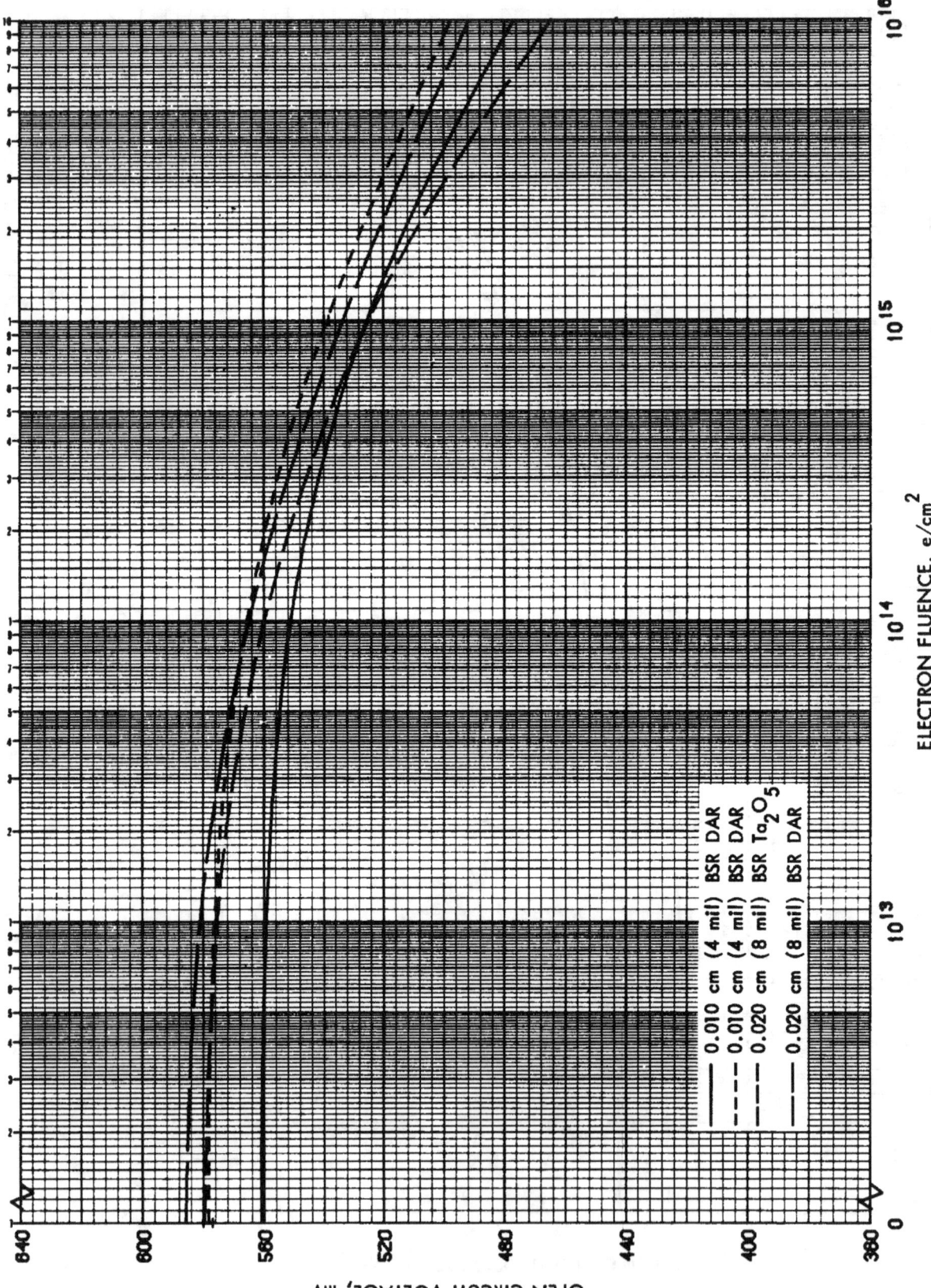

Figure 3.51 V_{oc} vs 1 MeV Electron Fluence for 2 Ohm-cm n/p Silicon Cells with BSR

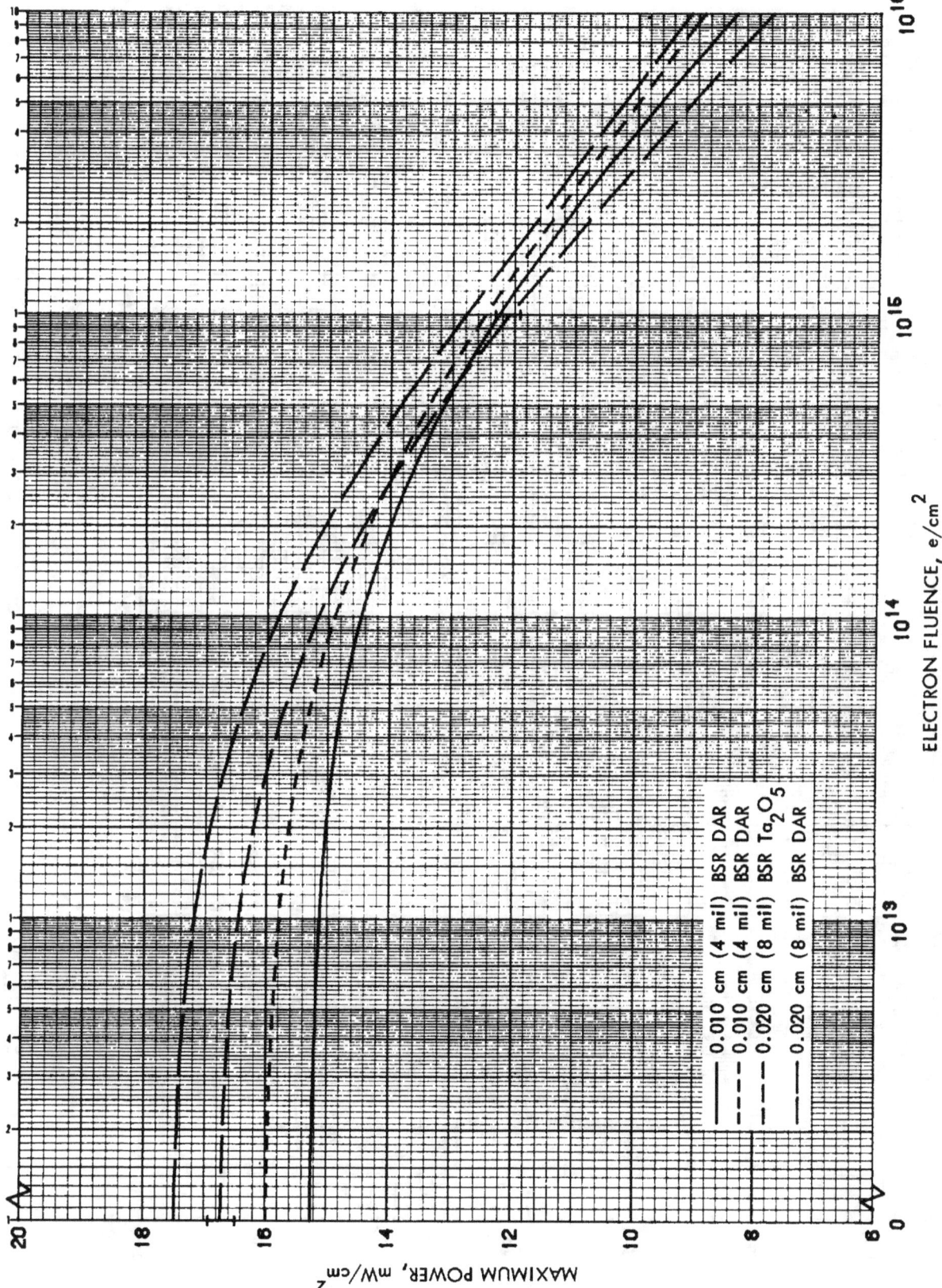

Figure 3.52 P_{max} vs 1 MeV Electron Fluence for 2 Ohm-cm n/p Silicon Cells with BSR

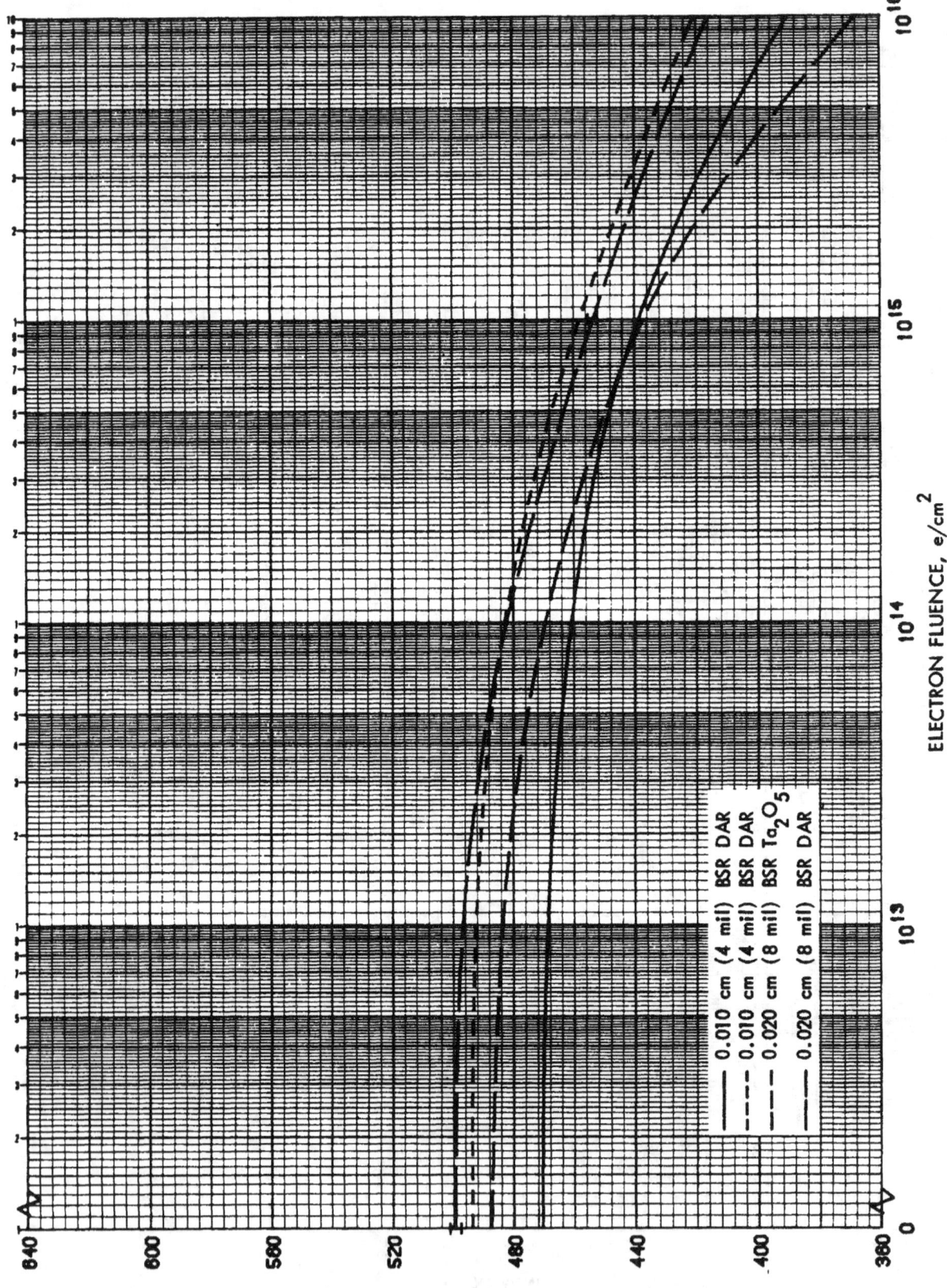

Figure 3.53 V_{mp} vs 1 MeV Electron Fluence for 2 Ohm-cm n/p Silicon Cells with BSR

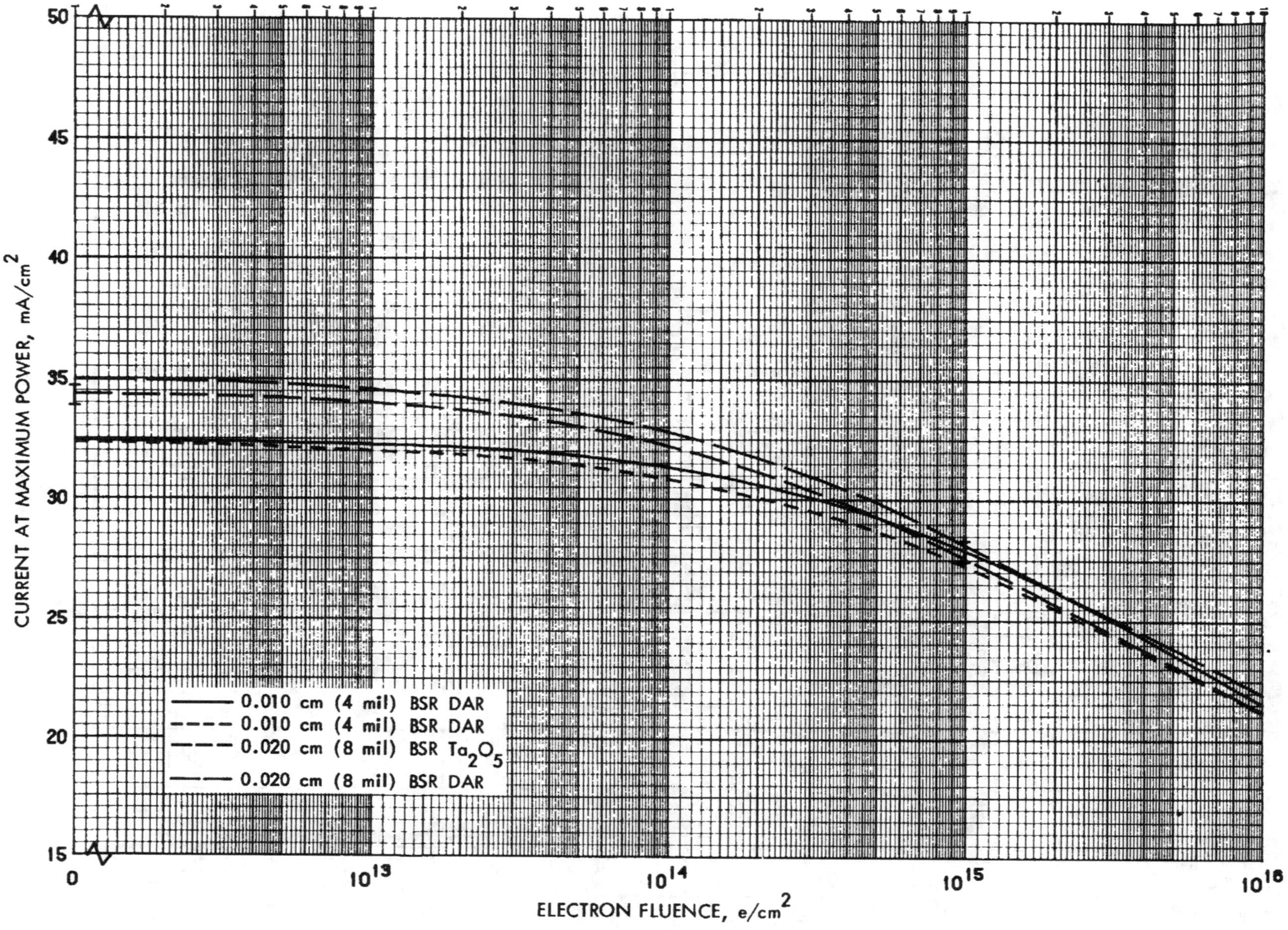

Figure 3.54 I_{mp} vs 1 MeV Electron Fluence for 2 Ohm-cm n/p Silicon Cells with BSR

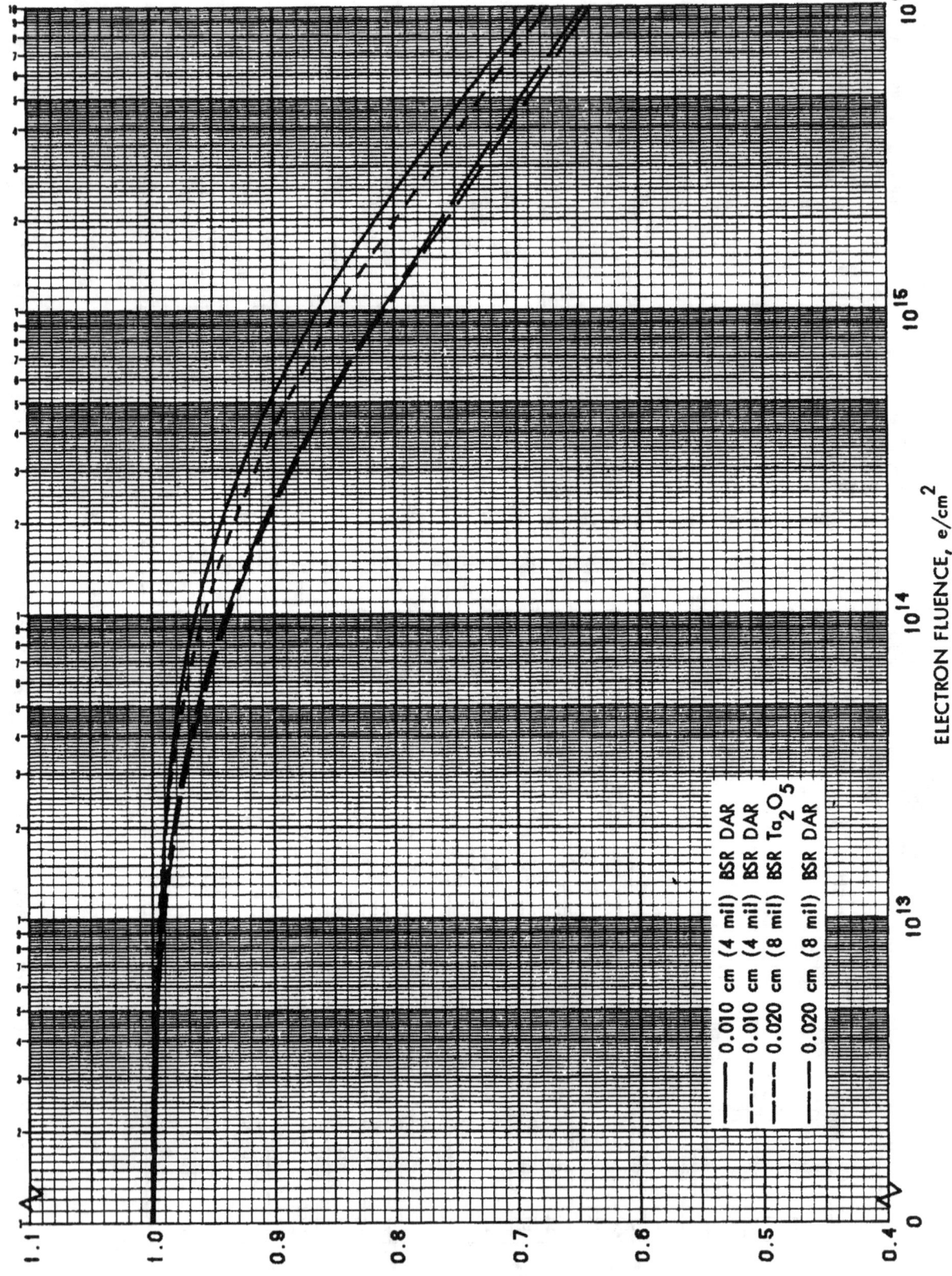

Figure 3.55 Normalized I_{SC} vs 1 MeV Electron Fluence for 2 Ohm-cm n/p Silicon Cells with BSR

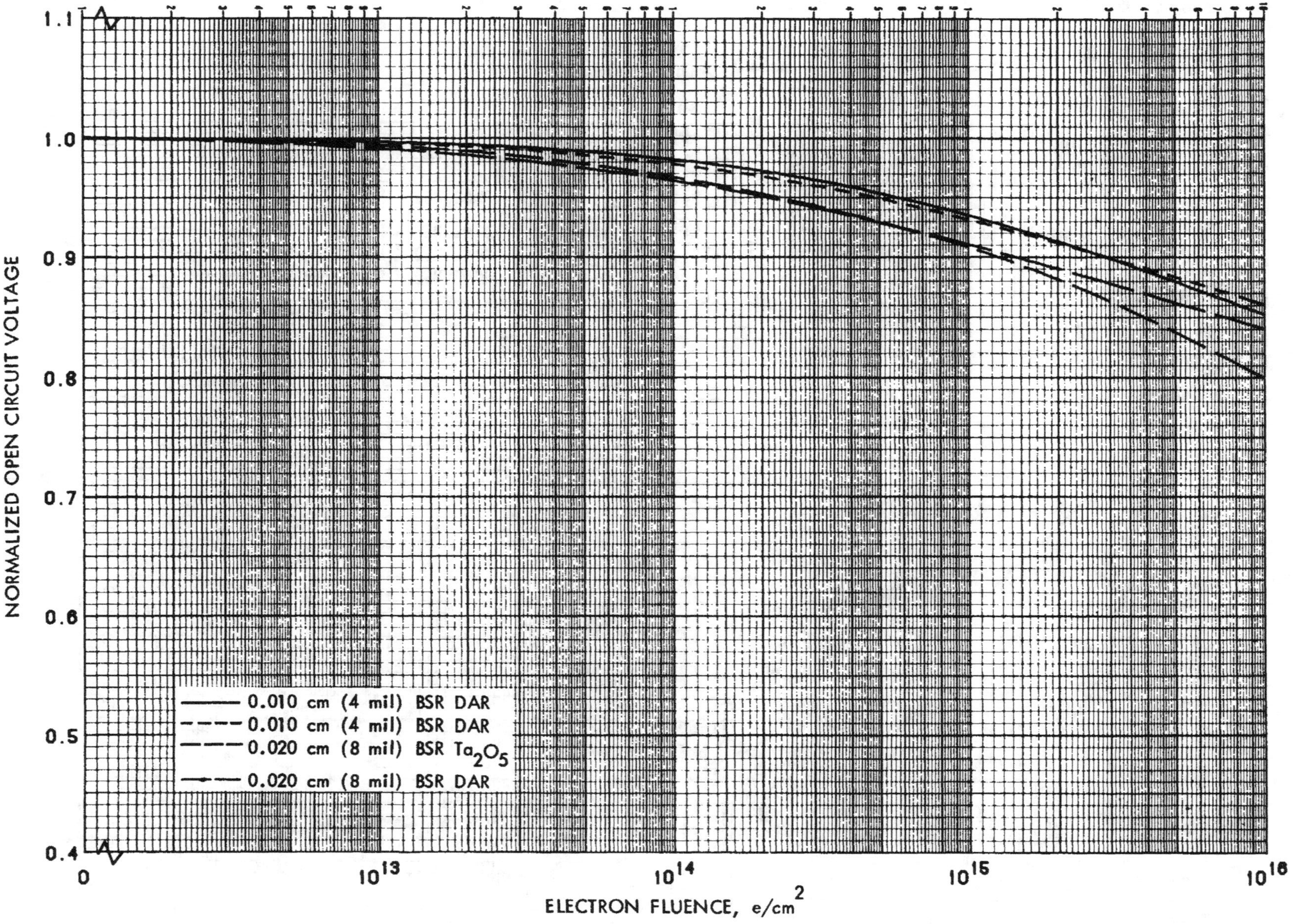

Figure 3.56 Normalized V_{OC} vs 1 MeV Electron Fluence for 2 Ohm-cm n/p Silicon Cells with BSR

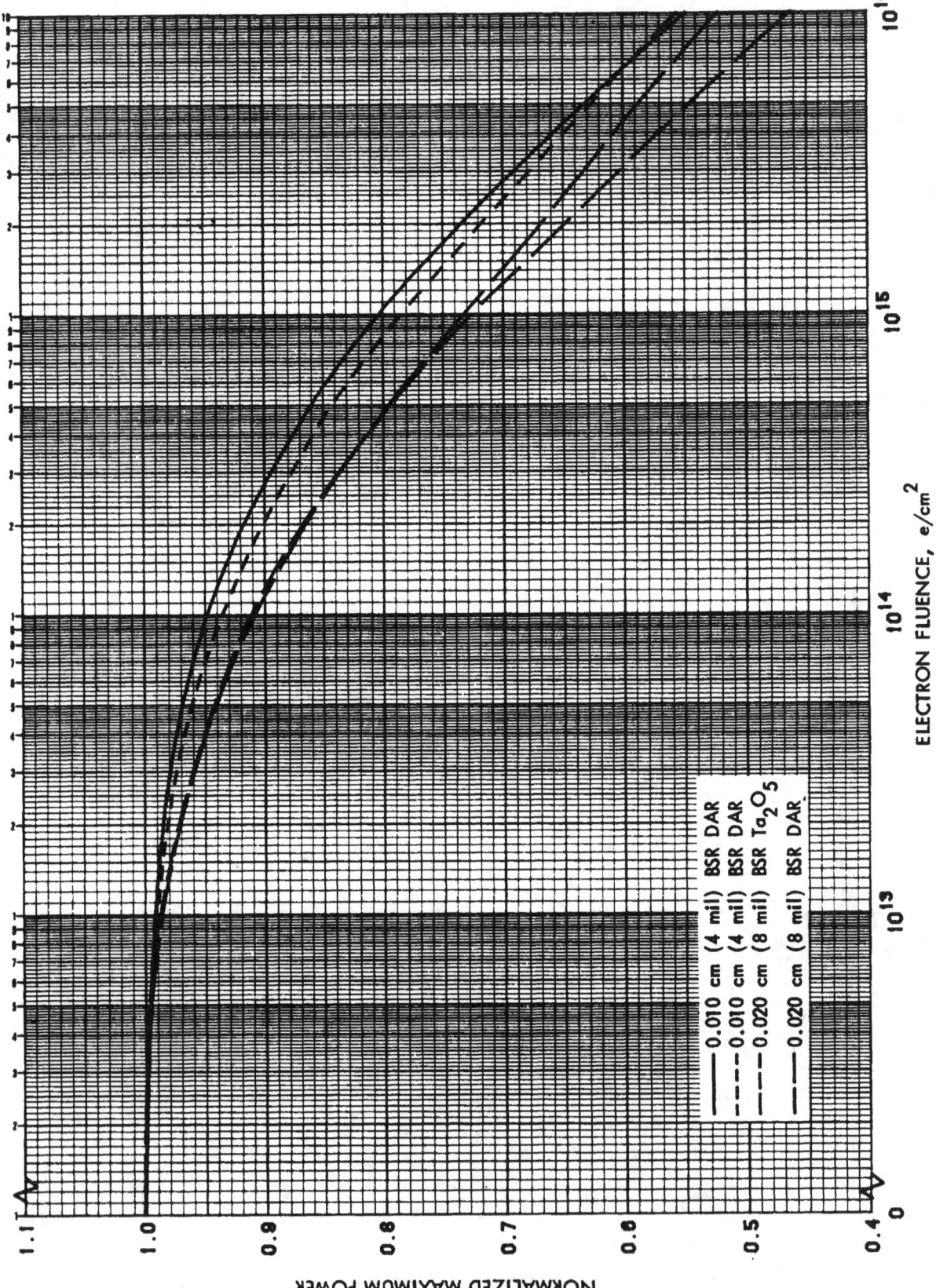

Figure 3.57 Normalized P_{max} vs 1 MeV Electron Fluence for 2 Ohm-cm n/p Silicon Cells with BSR

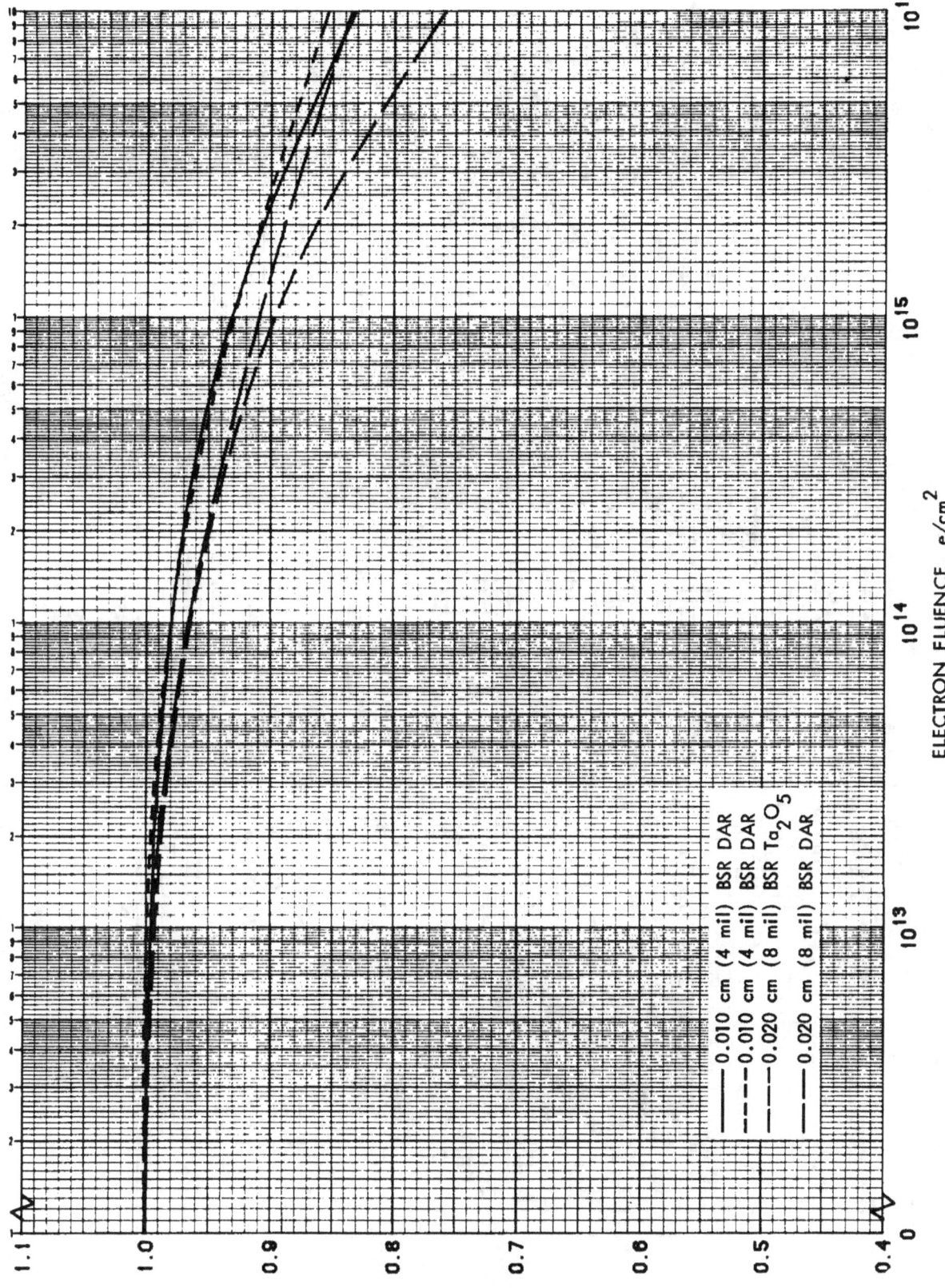

Figure 3.58 Normalized V_{mp} vs 1 MeV Electron Fluence for 2 Ohm-cm n/p Silicon Cells with BSR

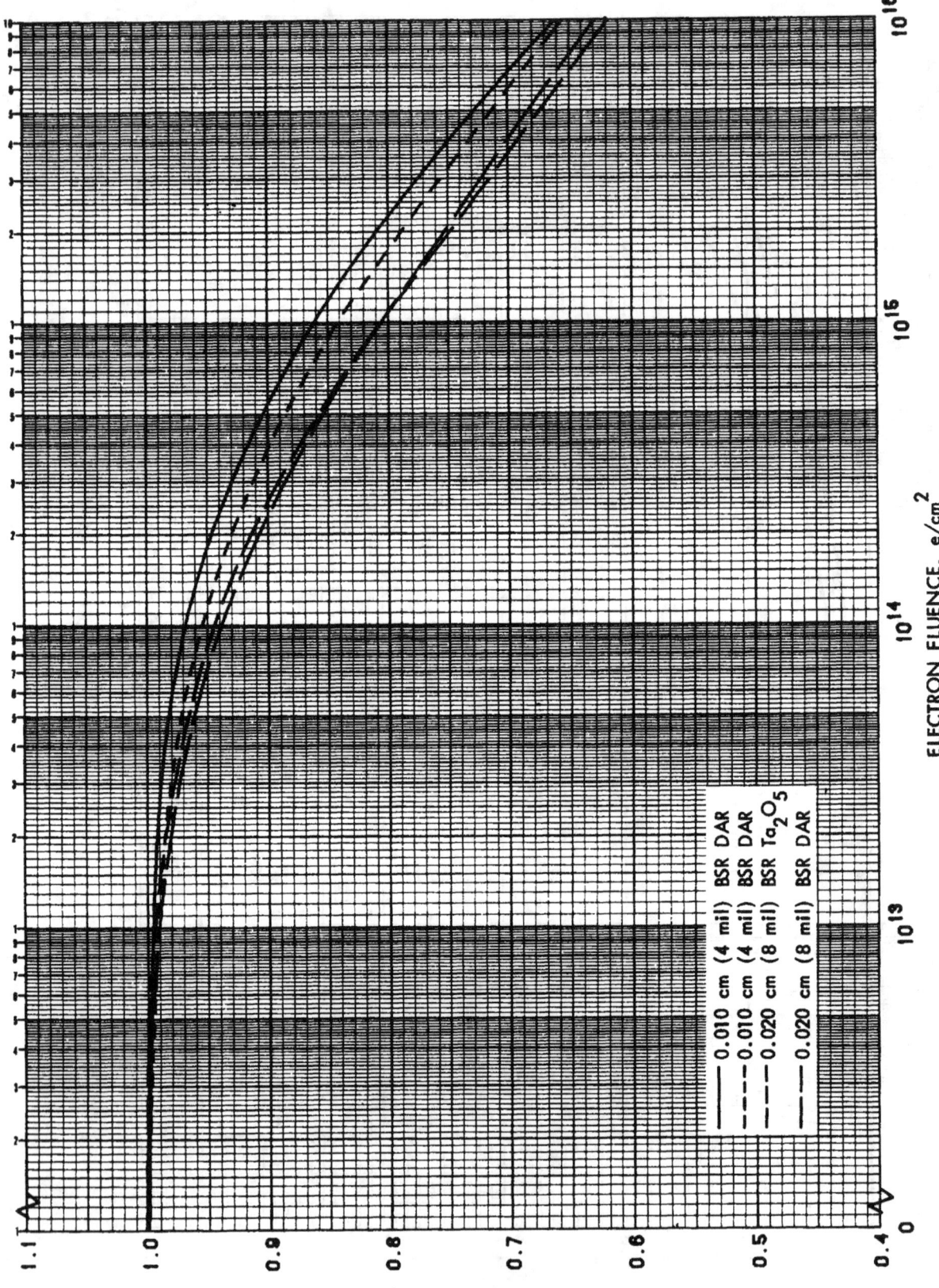

Figure 3.59 Normalized I_{mp} vs 1 MeV Electron Fluence for 2 Ohm-cm n/p Silicon Cells with BSR

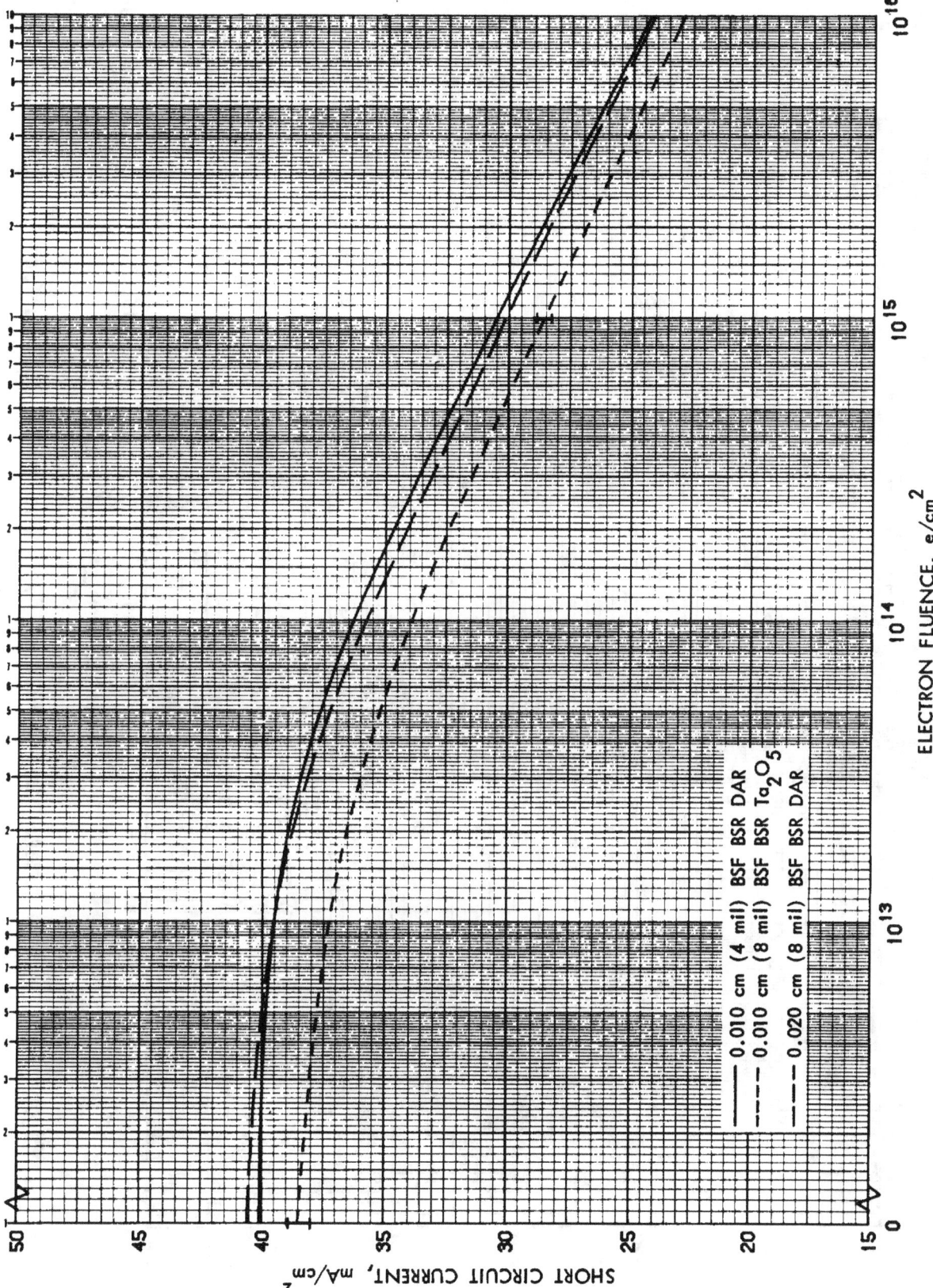

Figure 3.60 I_{sc} vs 1 MeV Electron Fluence for 2 Ohm-cm n/p Back Surface Field Silicon Cells with BSR

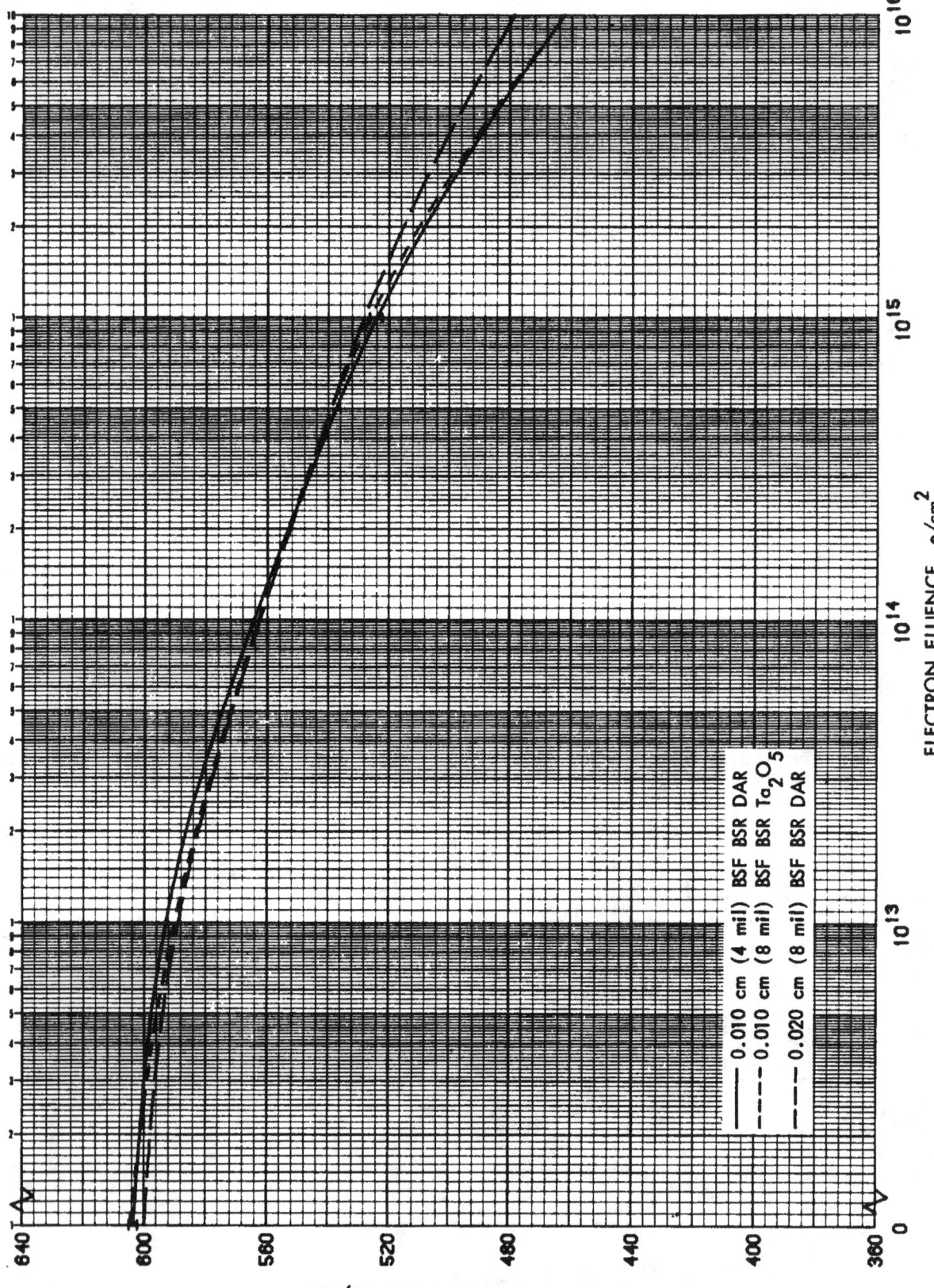

Figure 3.61 V_{oc} vs 1 MeV Electron Fluence for 2 Ohm-cm n/p Back Surface Field Silicon Cells with BSR

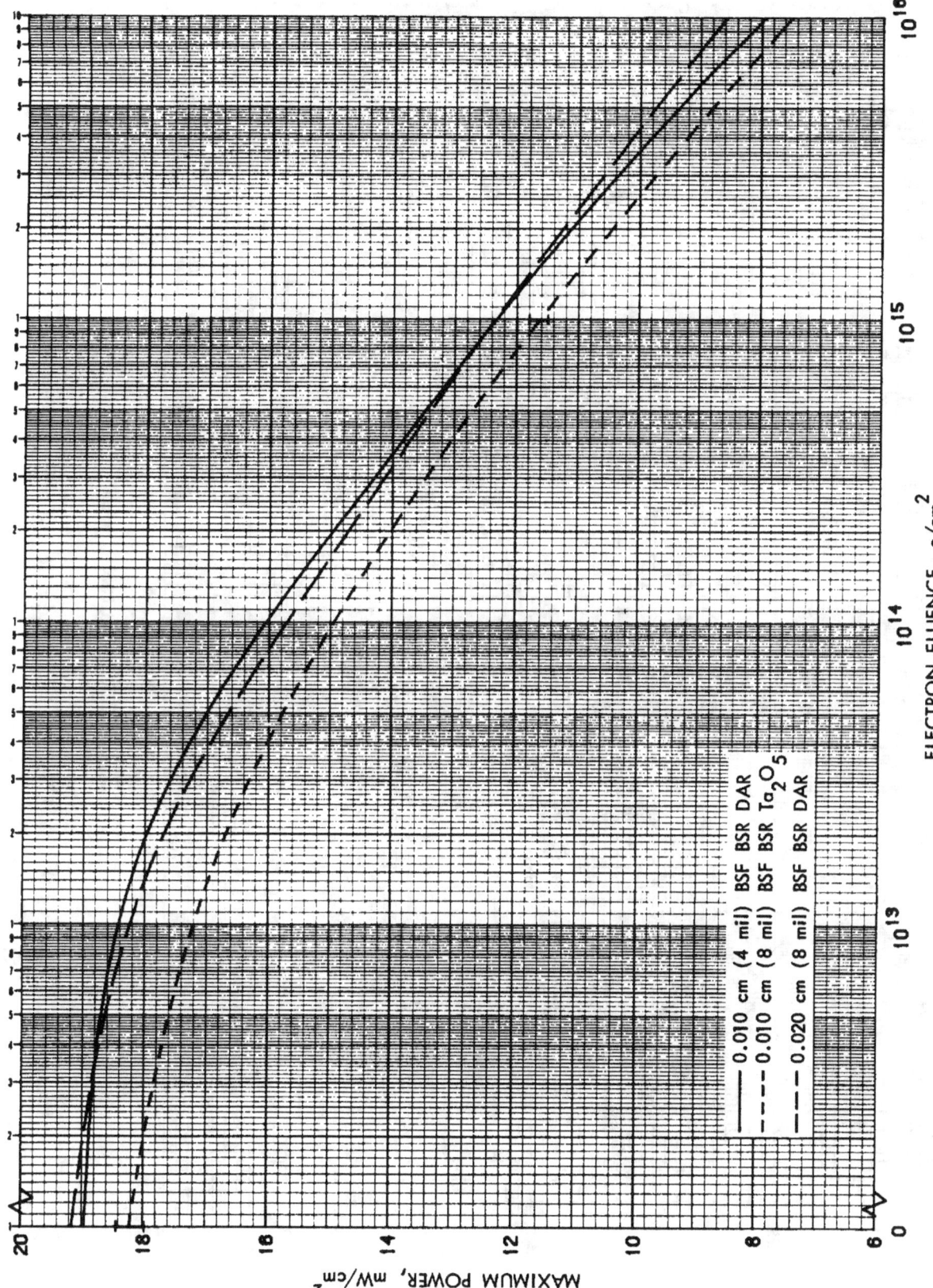

Figure 3.62 P_{max} vs 1 MeV Electron Fluence for 2 Ohm-cm n/p Back Surface Field Silicon Cells with BSR

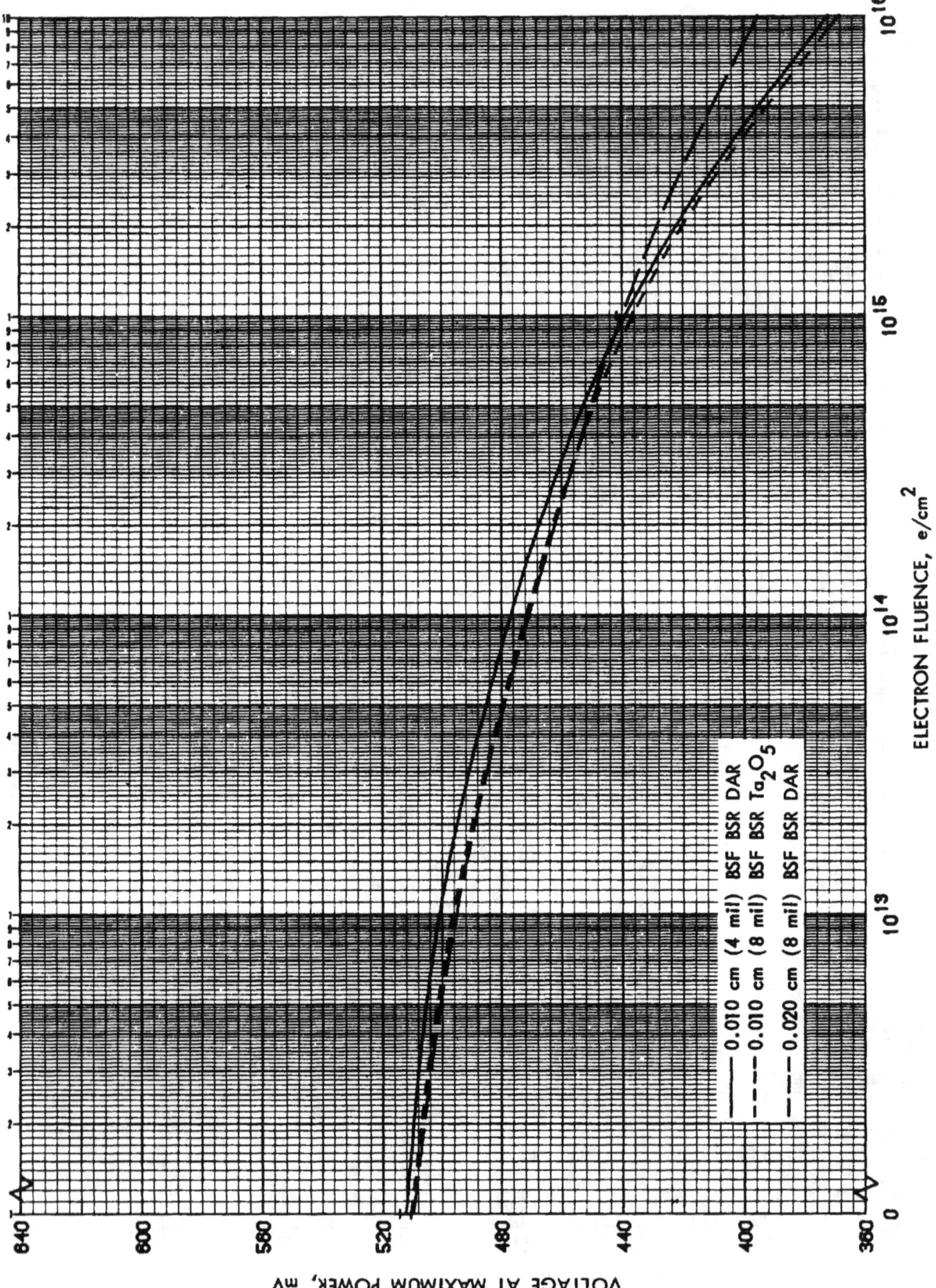

Figure 3.63 V_{mp} vs 1 MeV Electron Fluence for 2 Ohm-cm n/p Back Surface Field Silicon Cells with BSR

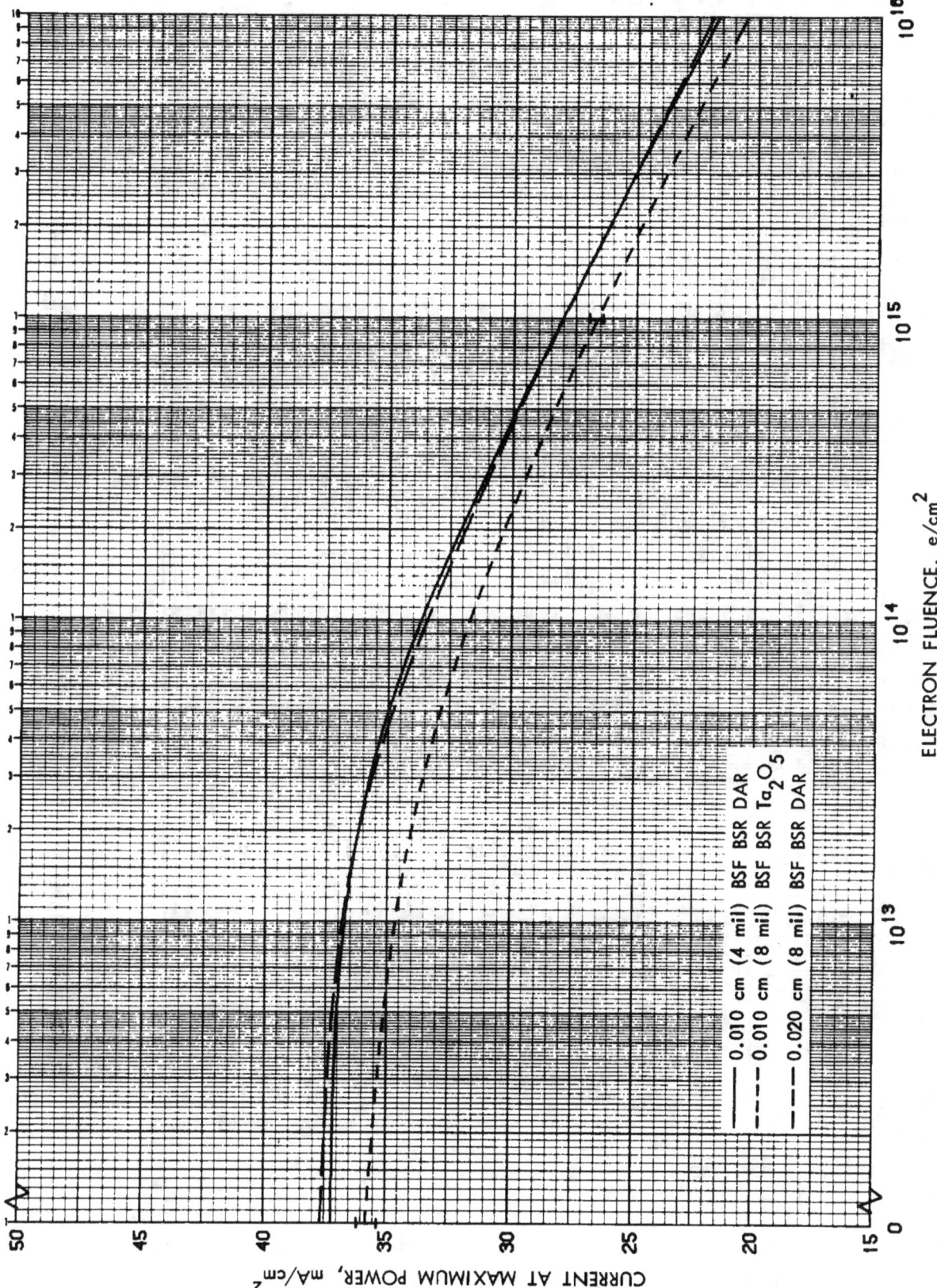

Figure 3.64 I_{mp} vs 1 MeV Electron Fluence for 2 Ohm-cm n/p Back Surface Field Silicon Cells with BSR

3-105

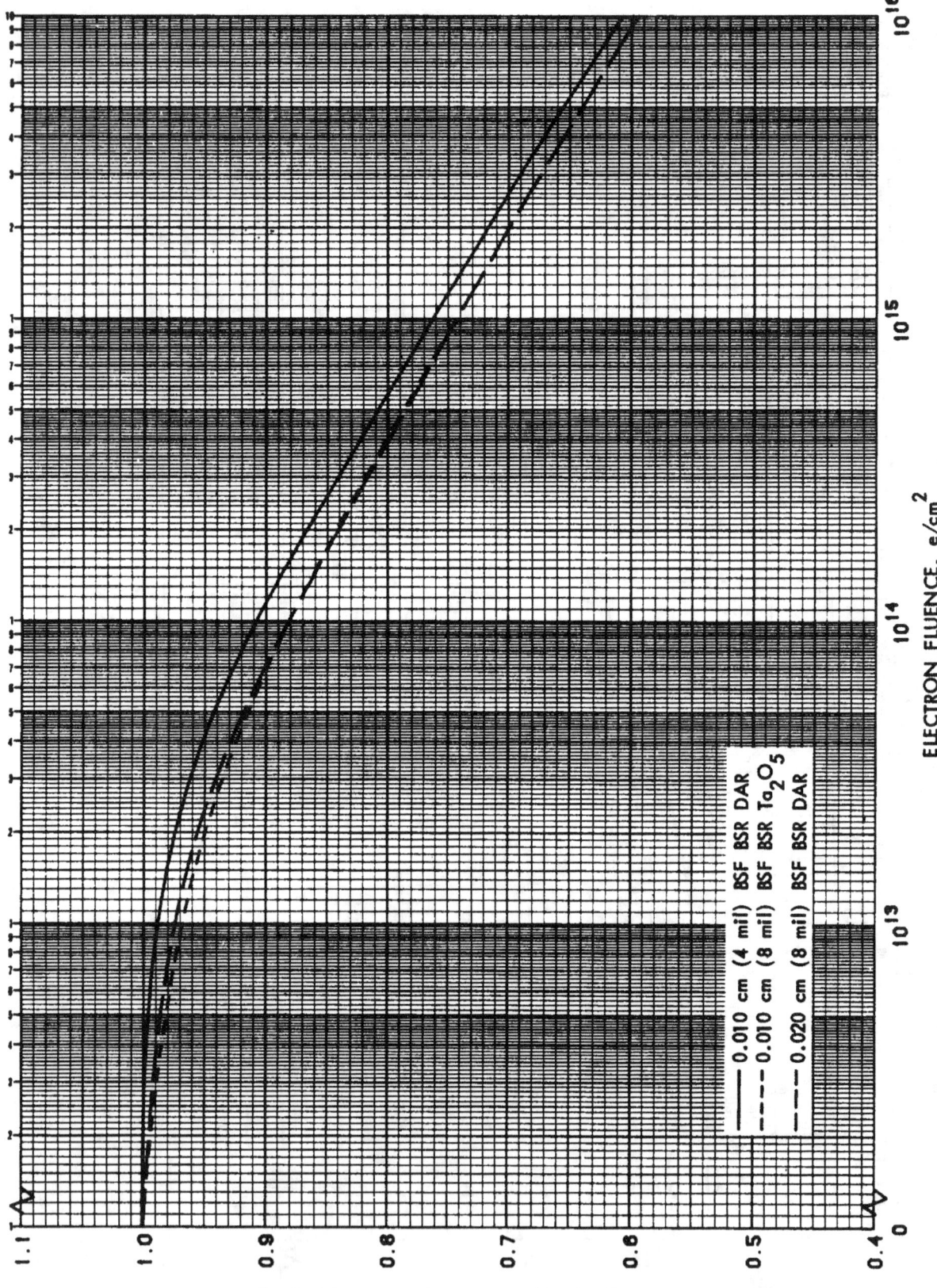

Figure 3.65 Normalized I_{sc} vs 1 MeV Electron Fluence for 2 Ohm-cm n/p Back Surface Field Silicon Cells with BSR

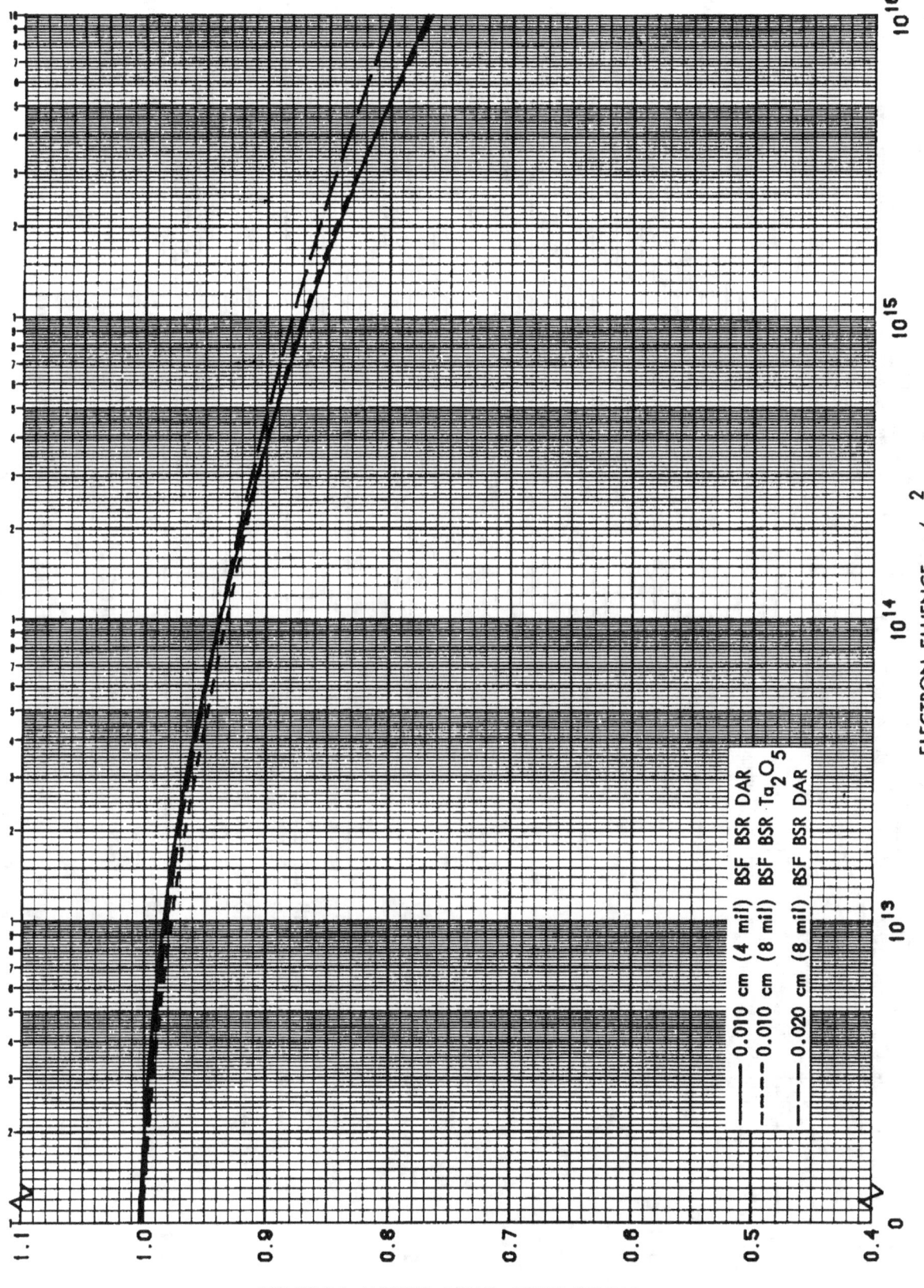

Figure 3.66 Normalized V_{oc} vs 1 MeV Electron Fluence for 2 Ohm-cm n/p Back Surface Field Silicon Cells with BSR

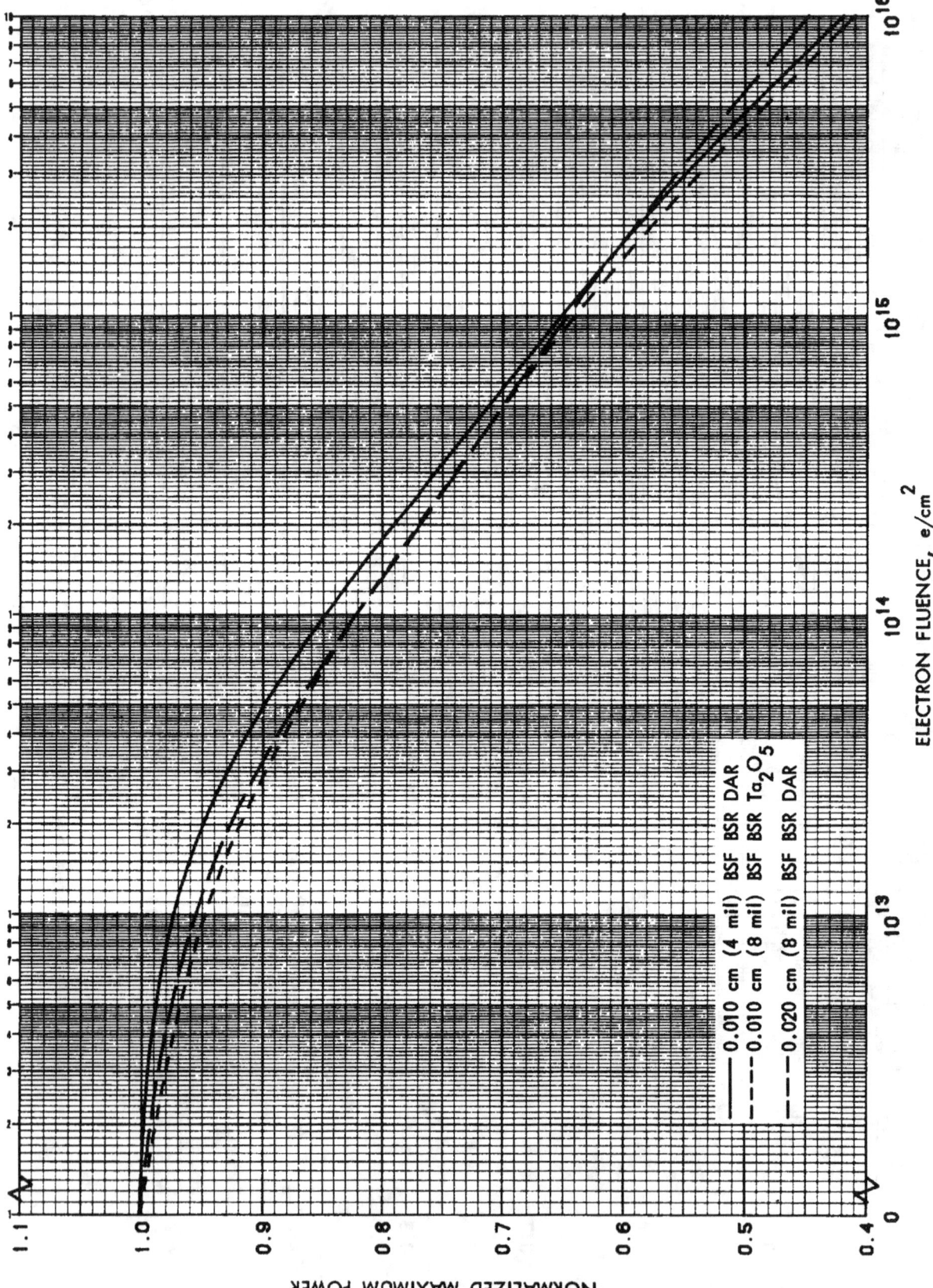

Figure 3.67 Normalized P_{max} vs 1 MeV Electron Fluence for 2 Ohm-cm n/p Back Surface Field Silicon Cells with BSR

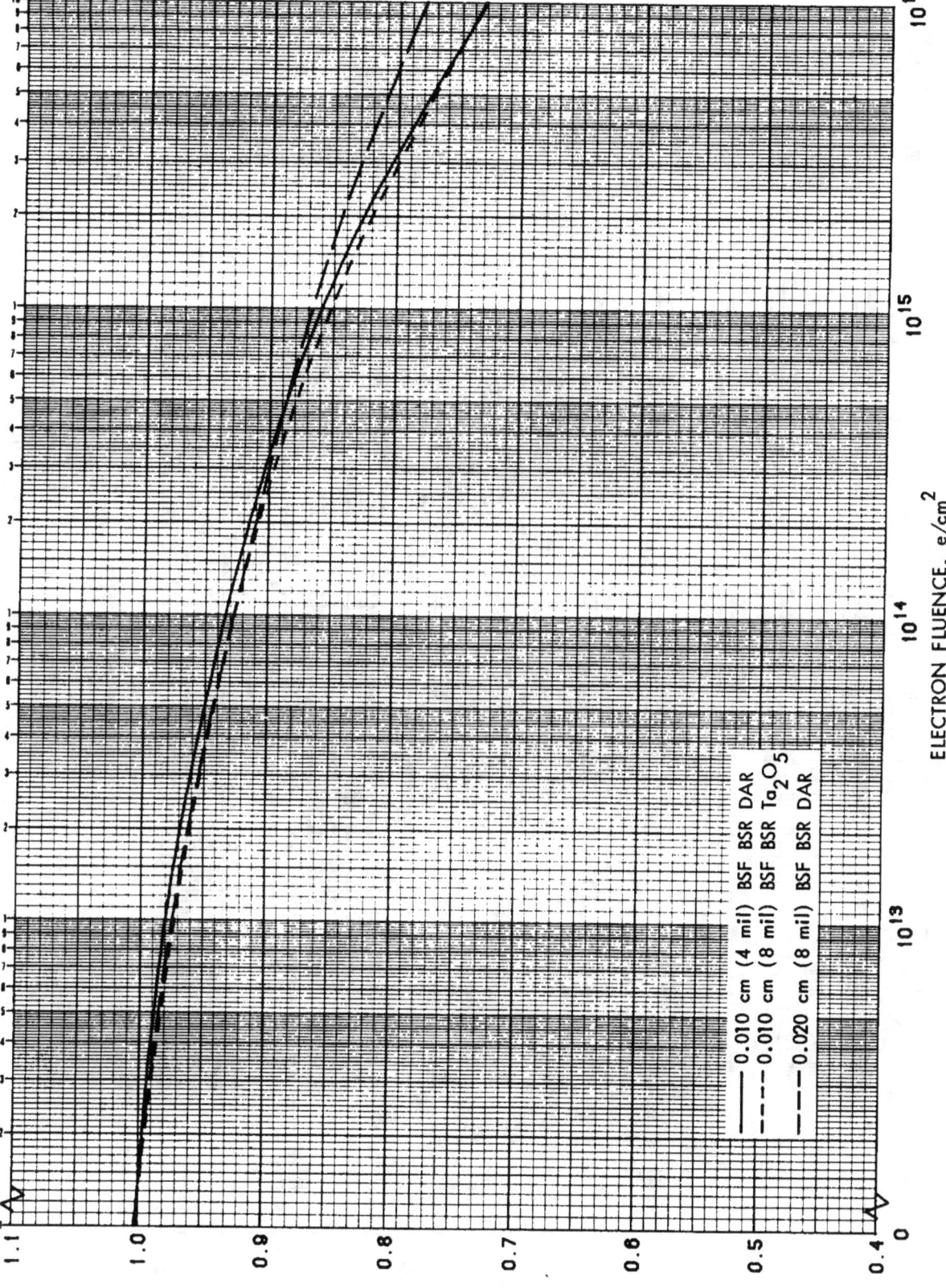

Figure 3.68 Normalized V_{mp} vs 1 MeV Electron Fluence for 2 Ohm-cm n/p Back Surface Field Silicon Cells with BSR

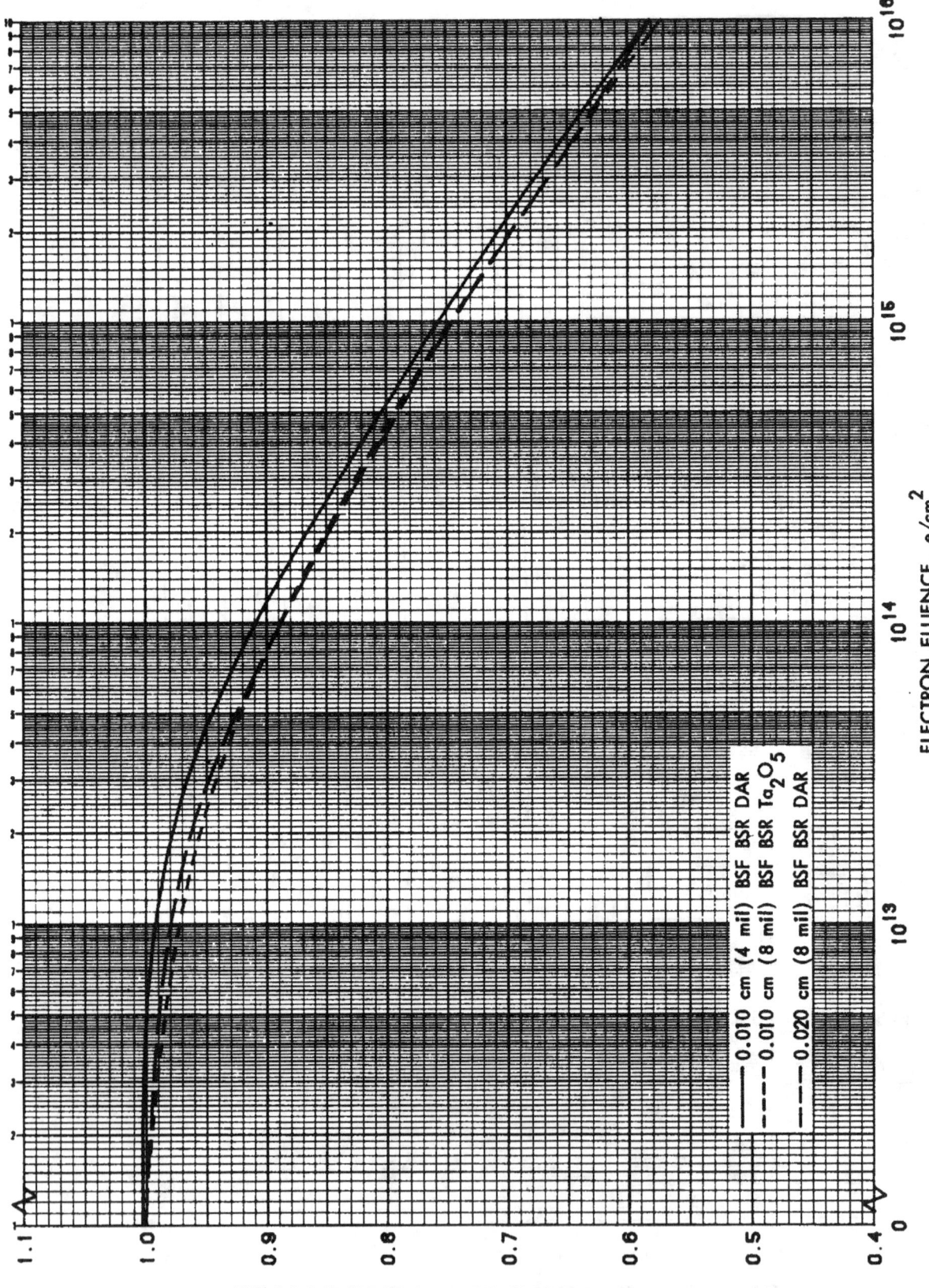

Figure 3.69 Normalized I_{mp} vs 1 MeV Electron Fluence for 2 Ohm-cm n/p Back Surface Field Silicon Cells with BSR

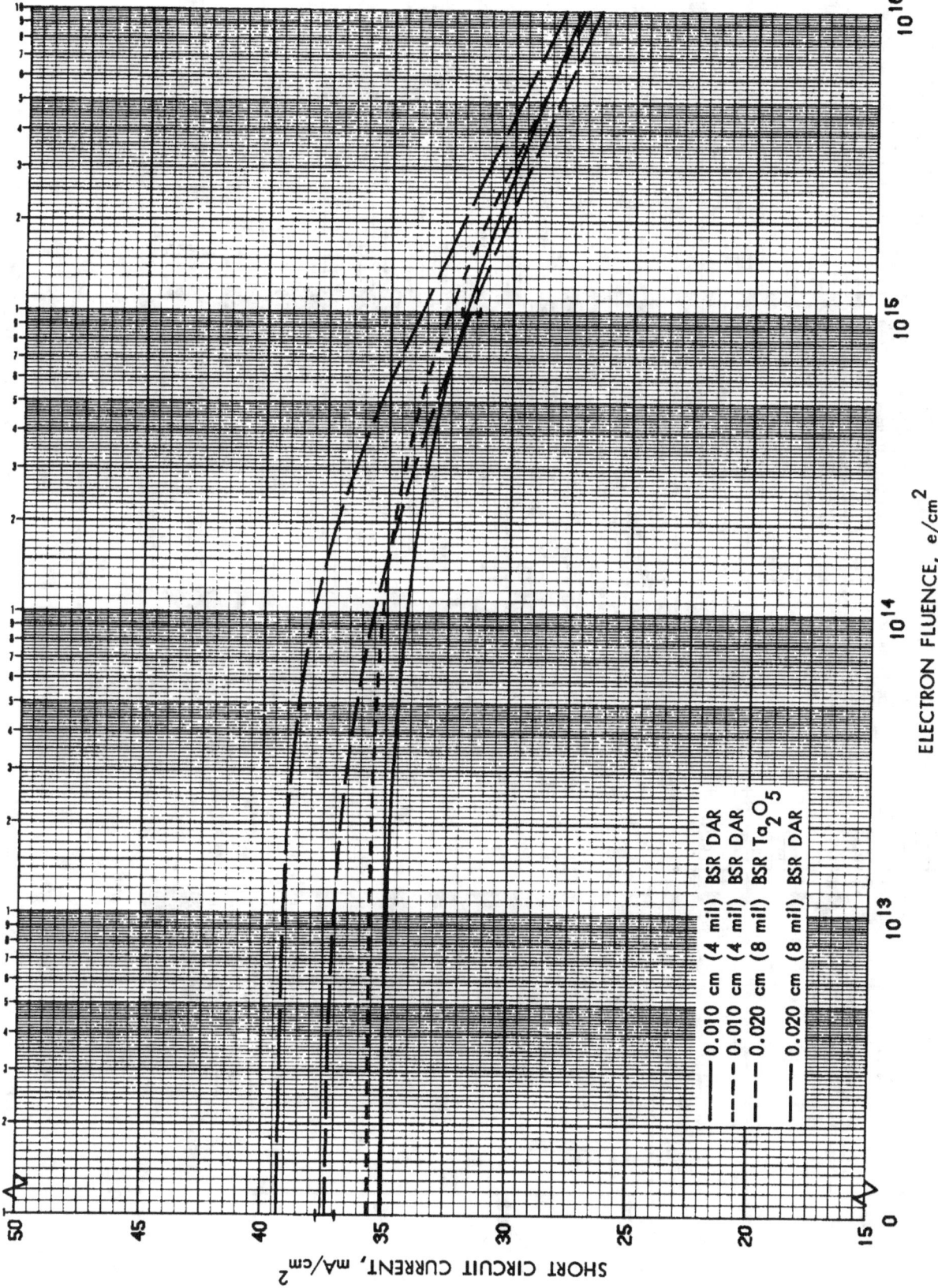

Figure 3.70 I_{sc} vs 1 MeV Electron Fluence for 10 Ohm-cm n/p Silicon Cells with BSR

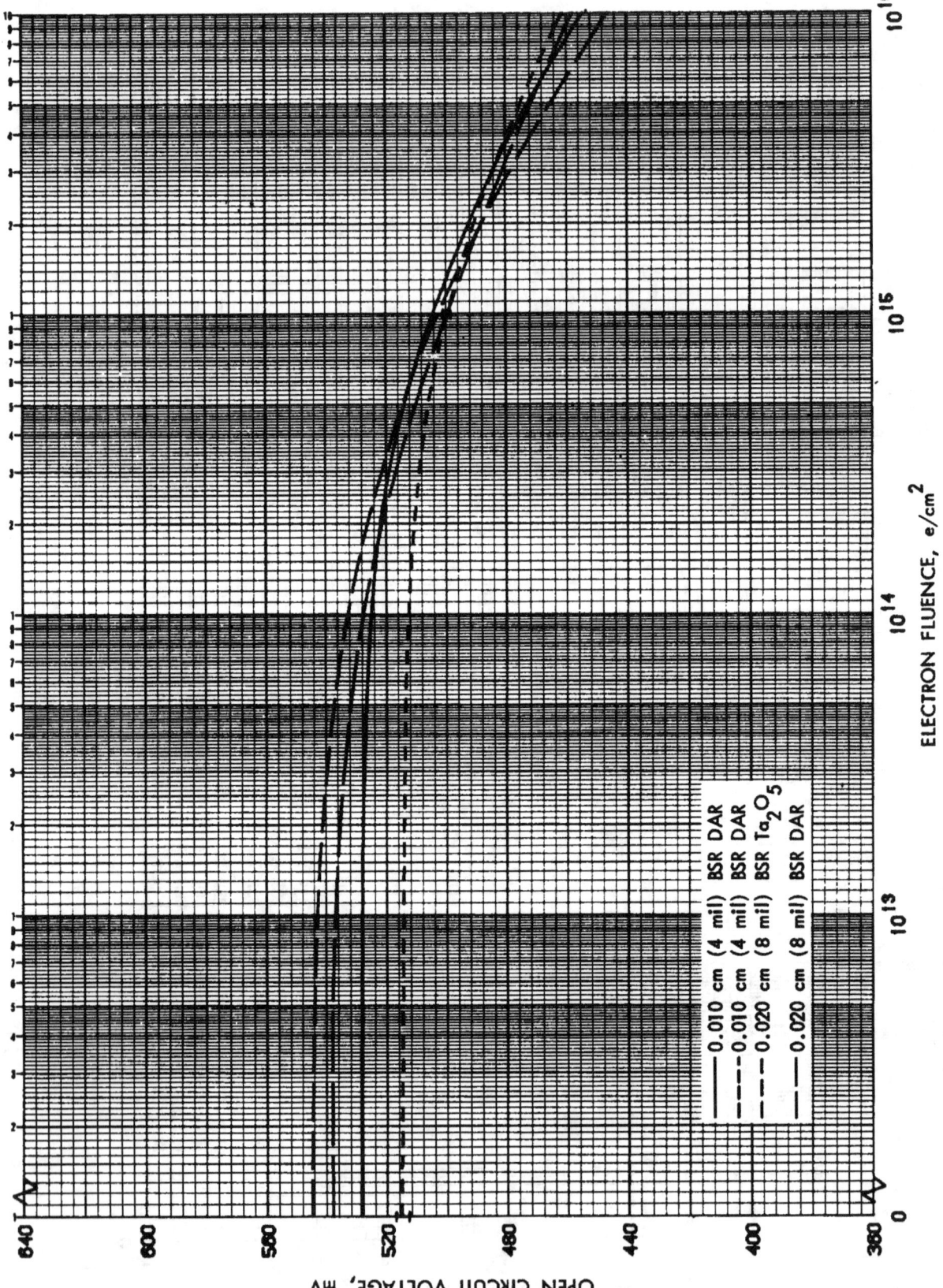

Figure 3.71 V_{OC} vs 1 MeV Electron Fluence for 10 Ohm-cm n/p Silicon Cells with BSR

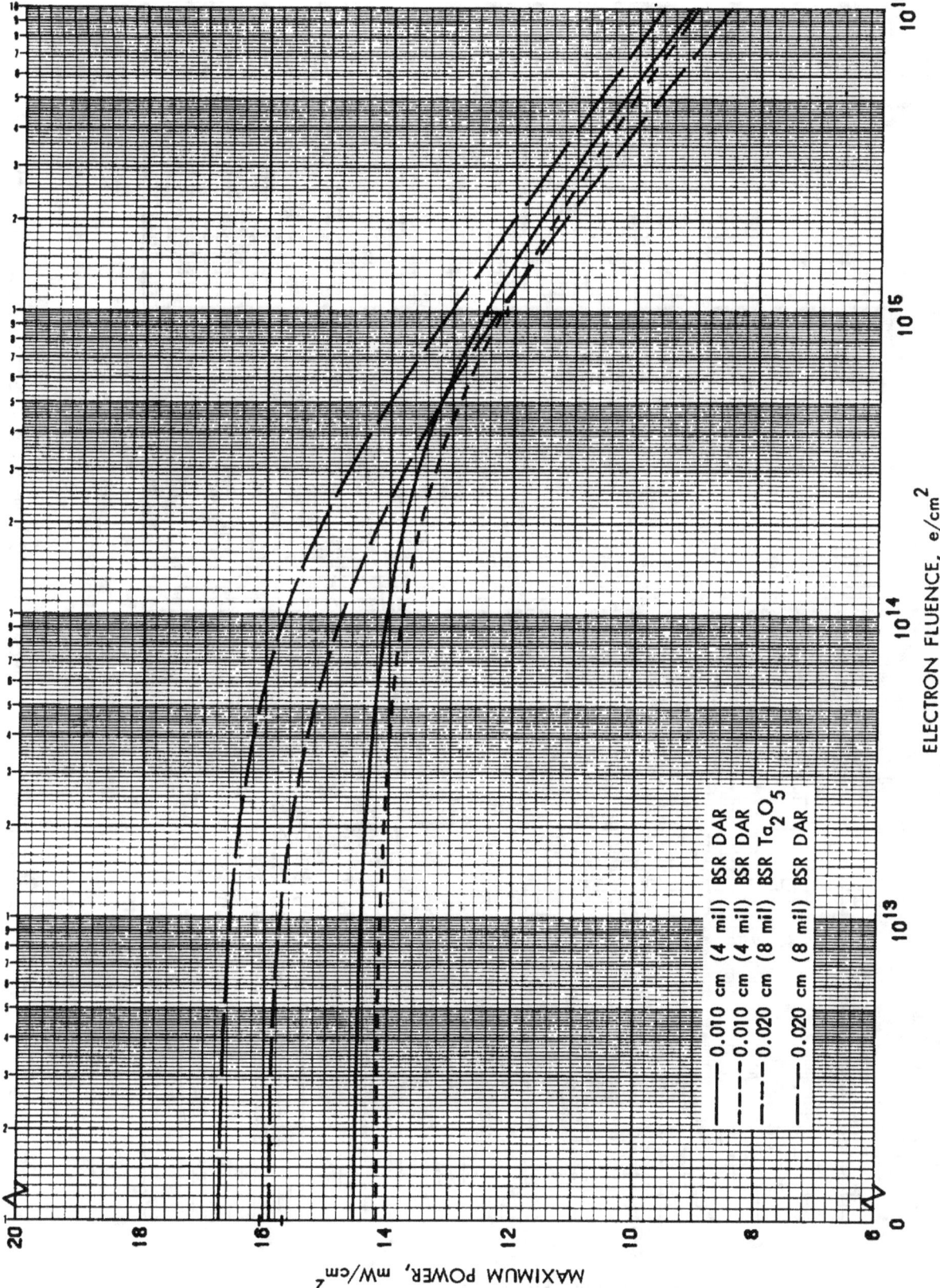

Figure 3.72 P_{max} vs 1 MeV Electron Fluence for 10 Ohm-cm n/p Silicon Cells with BSR

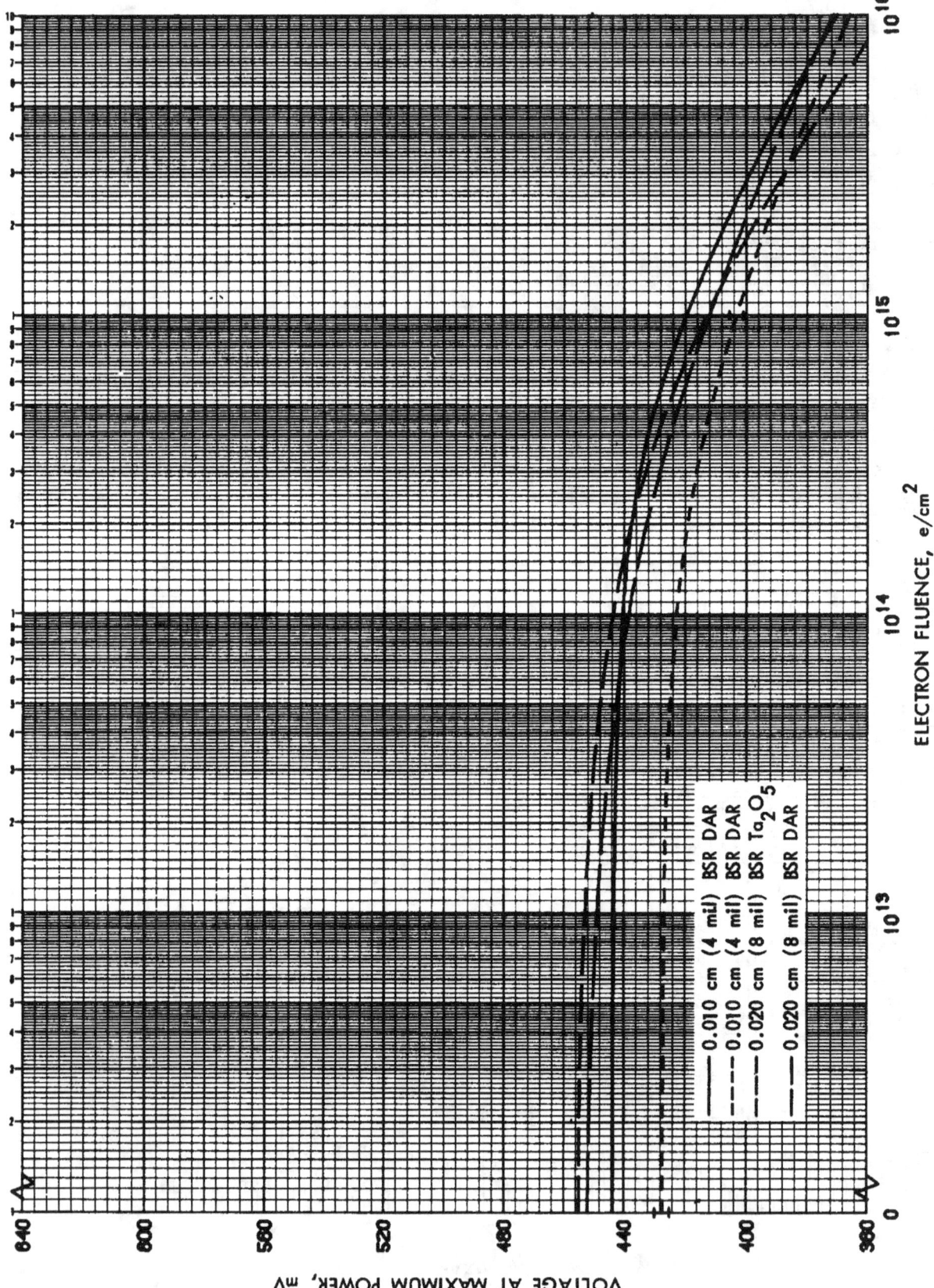

Figure 3.73 V_{mp} vs 1 MeV Electron Fluence for 10 Ohm-cm n/p Silicon Cells with BSR

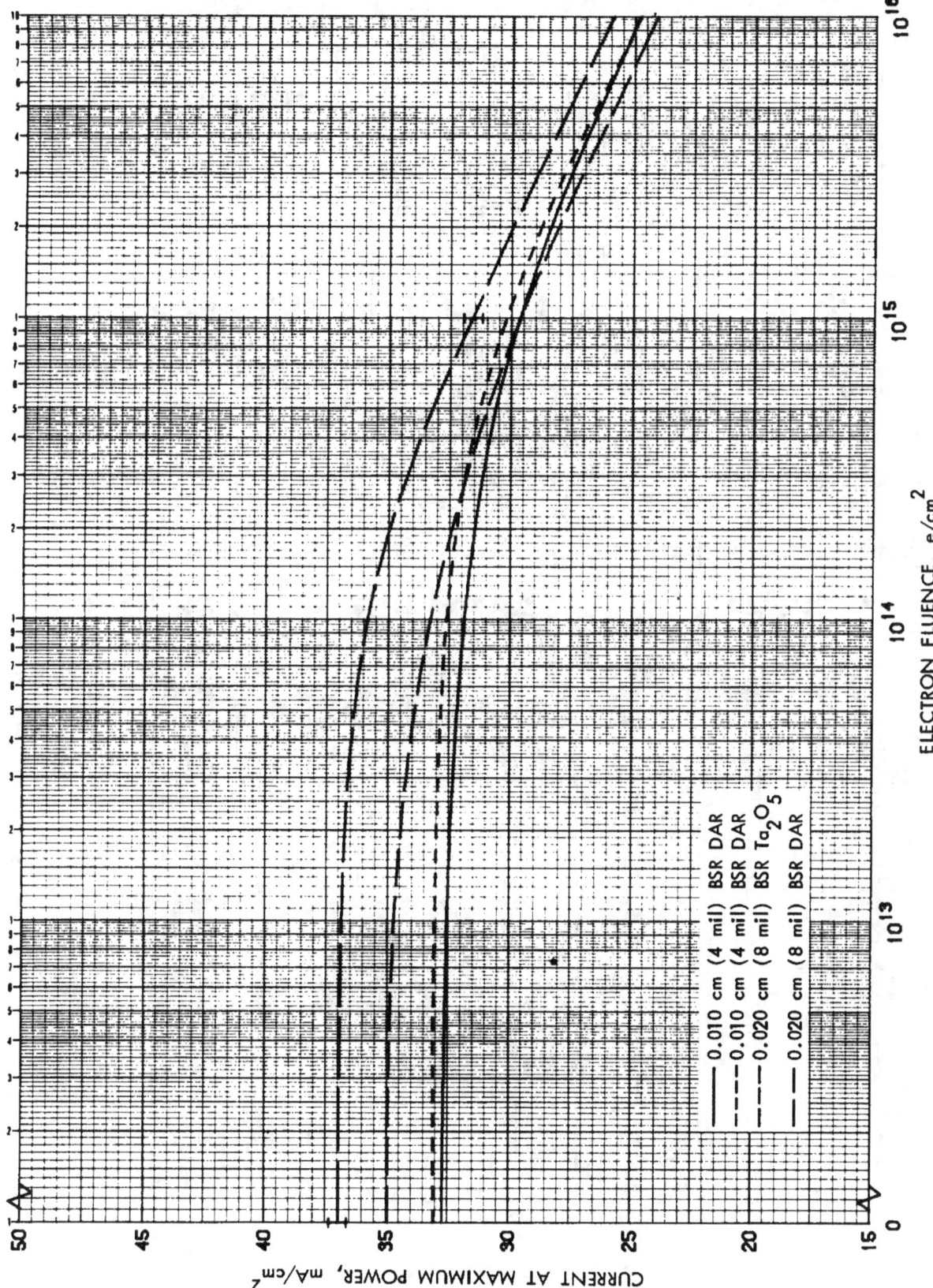

Figure 3.74 I_{mp} vs 1 MeV Electron Fluence for 10 Ohm-cm n/p Silicon Cells with BSR

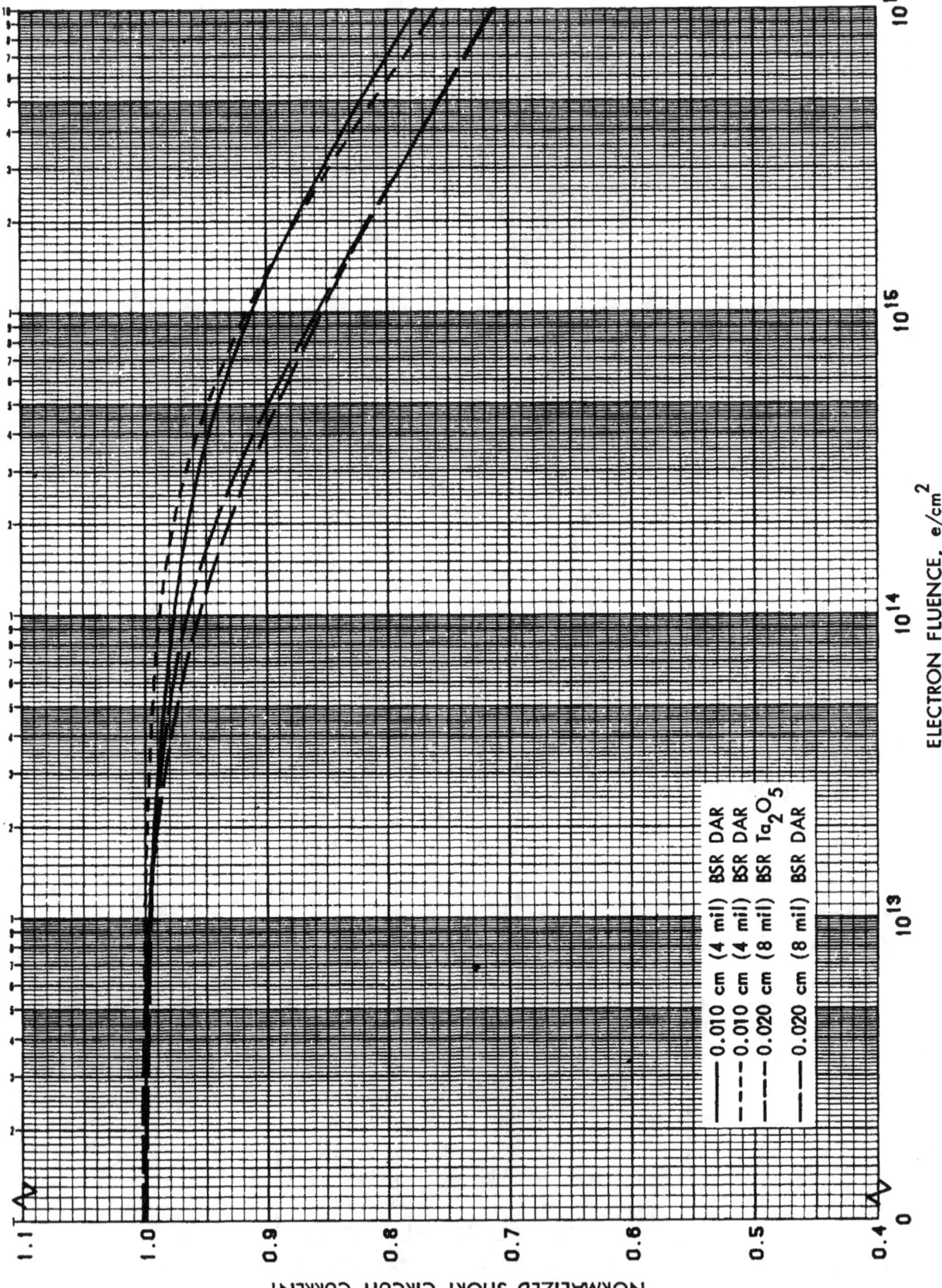

Figure 3.75 Normalized I_{SC} vs 1 MeV Electron Fluence for 10 Ohm-cm n/p Silicon Cells with BSR

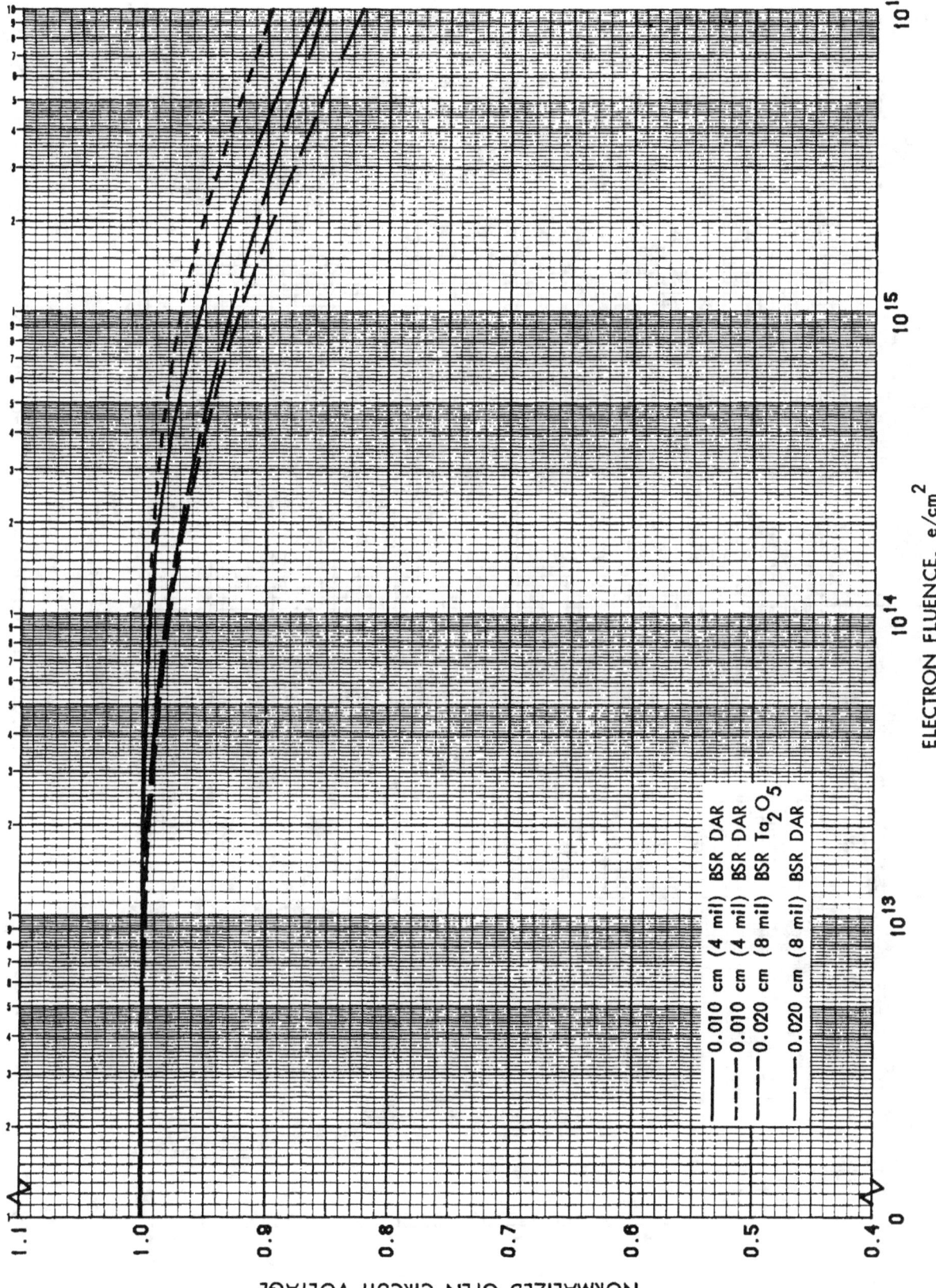

Figure 3.76 Normalized V_{OC} vs 1 MeV Electron Fluence for 10 Ohm-cm n/p Silicon Cells with BSR

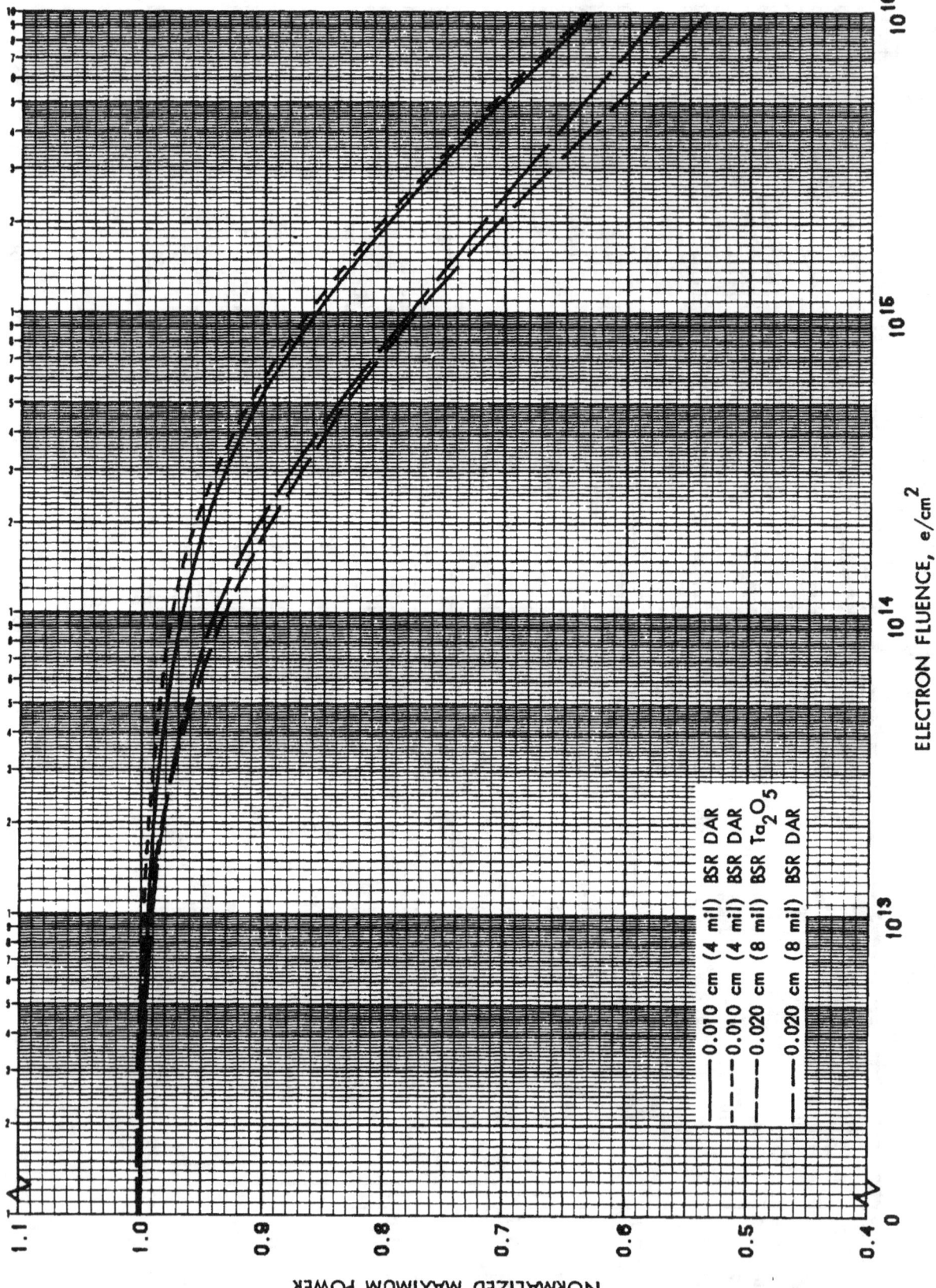

Figure 3.77 Normalized P_{max} vs 1 MeV Electron Fluence for 10 Ohm-cm n/p Silicon Cells with BSR

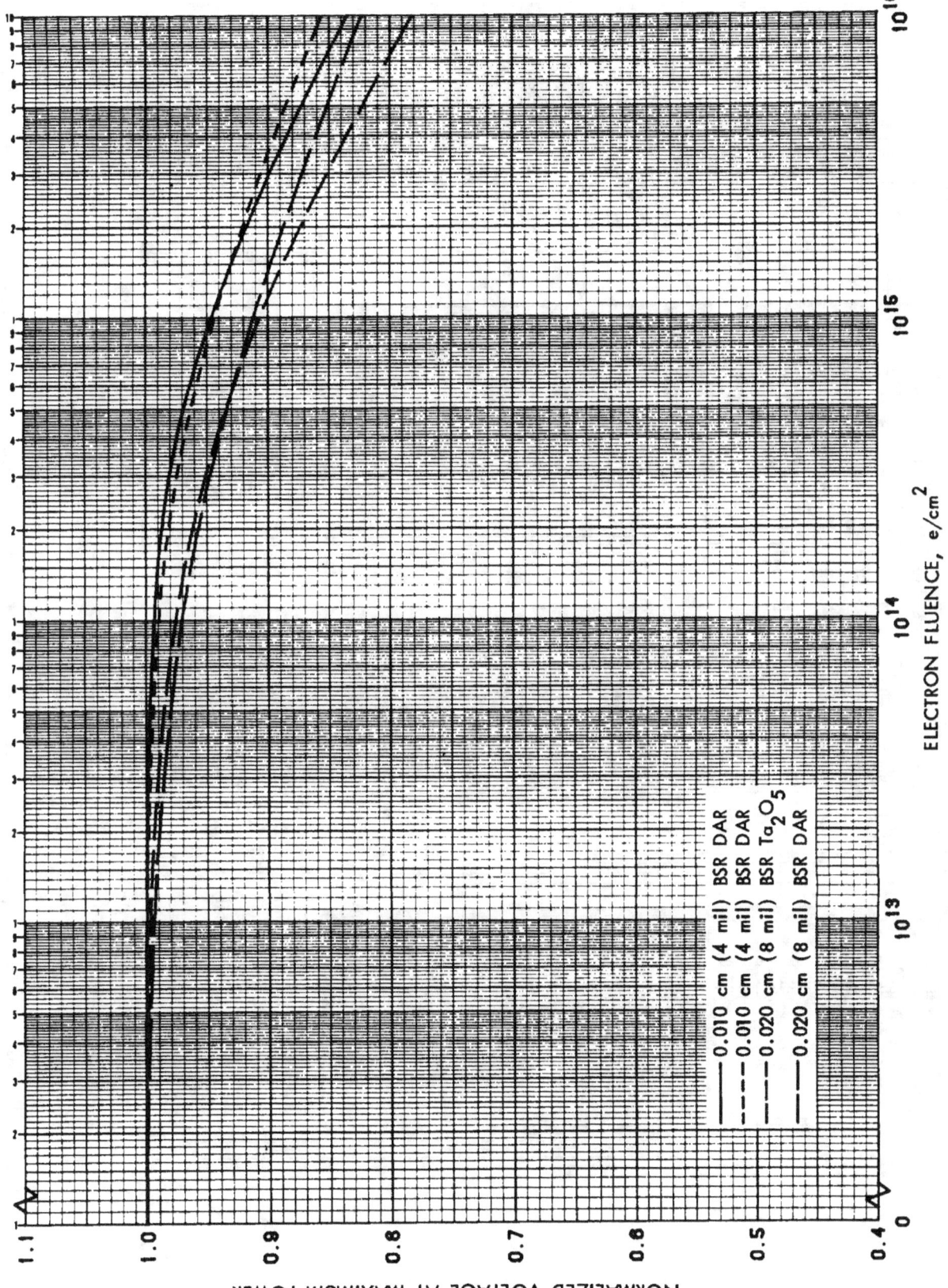

Figure 3.78 Normalized V_{mp} vs 1 MeV Electron Fluence for 10 Ohm-cm n/p Silicon Cells with BSR

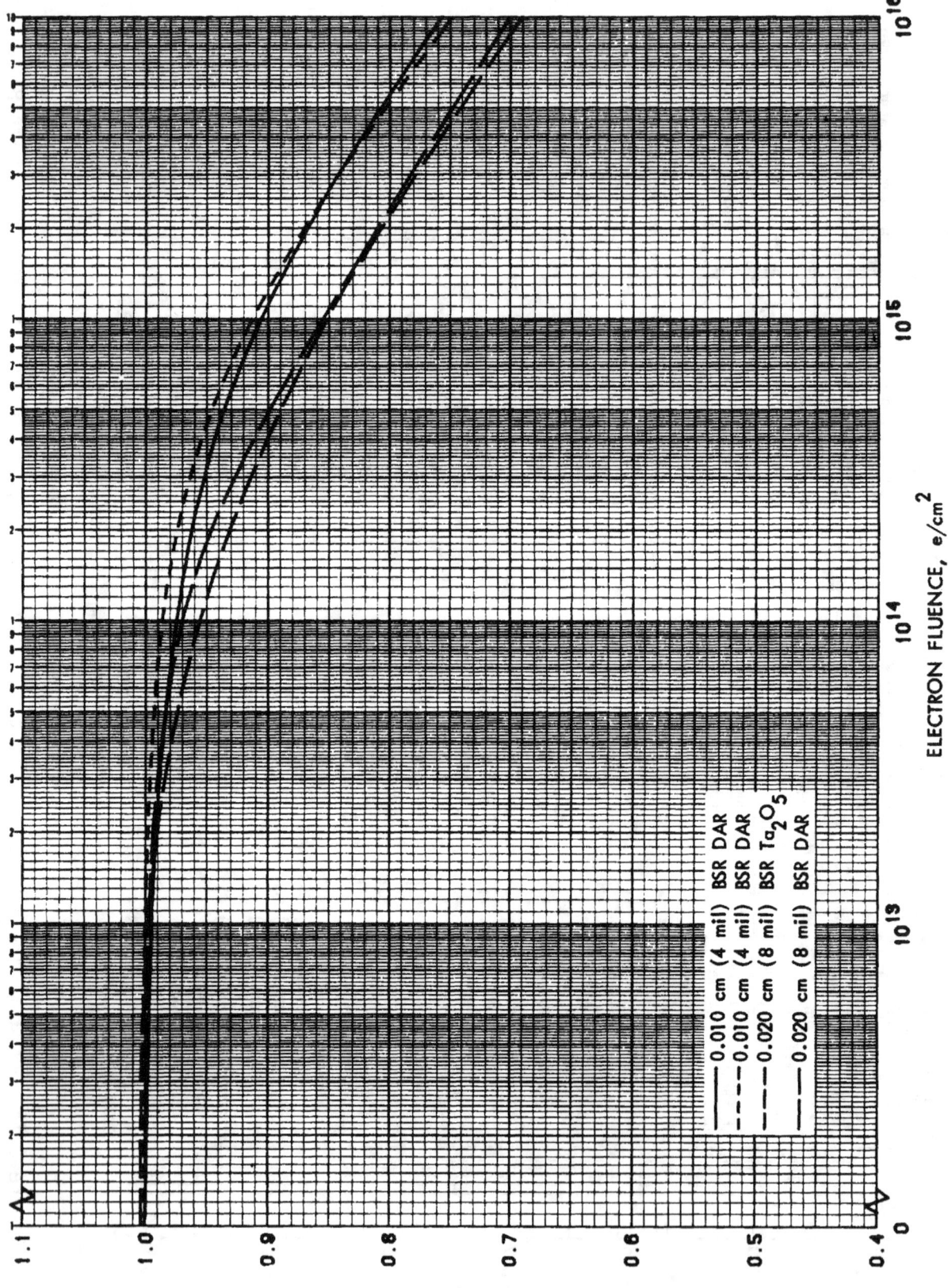

Figure 3.79 Normalized I_{mp} vs 1 MeV Electron Fluence for 10 Ohm-cm n/p Silicon Cells with BSR

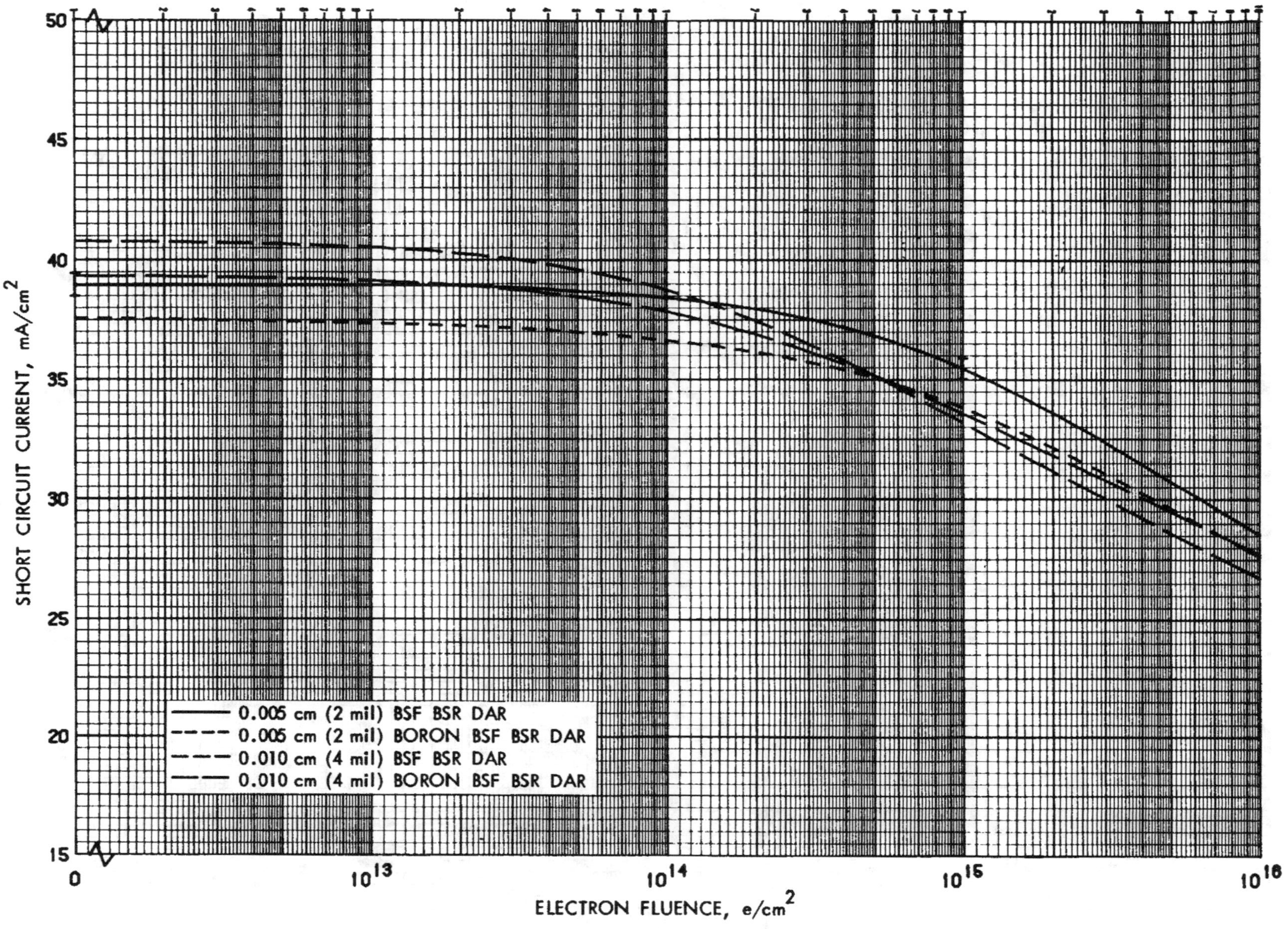

Figure 3.80 I_{sc} vs 1 MeV Electron Fluence for 10 Ohm-cm n/p Back Surface Field Thin Silicon Cells with BSR

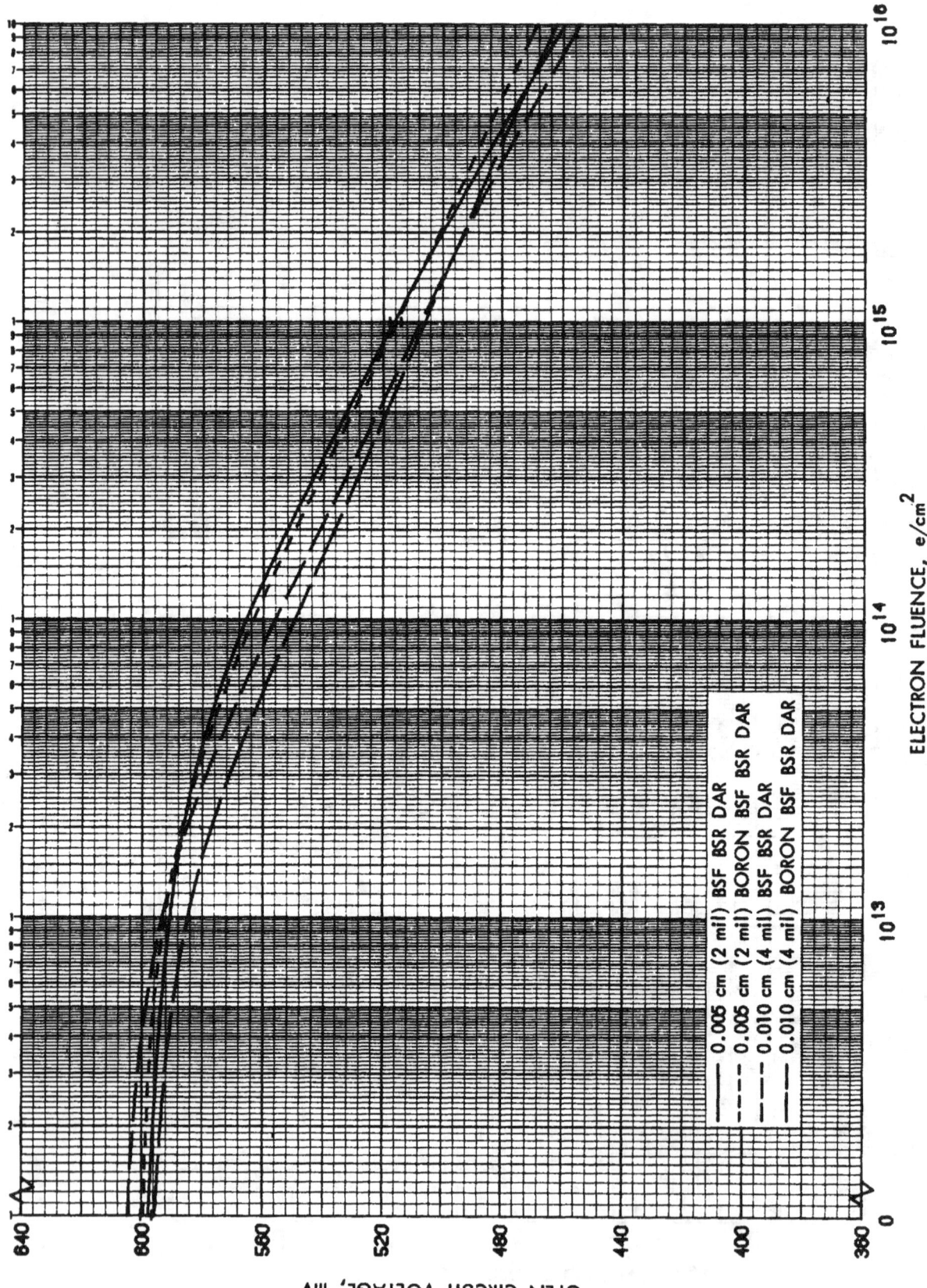

Figure 3.81 V_{oc} vs 1 MeV Electron Fluence for 10 Ohm-cm n/p Back Surface Field Thin Silicon Cells with BSR

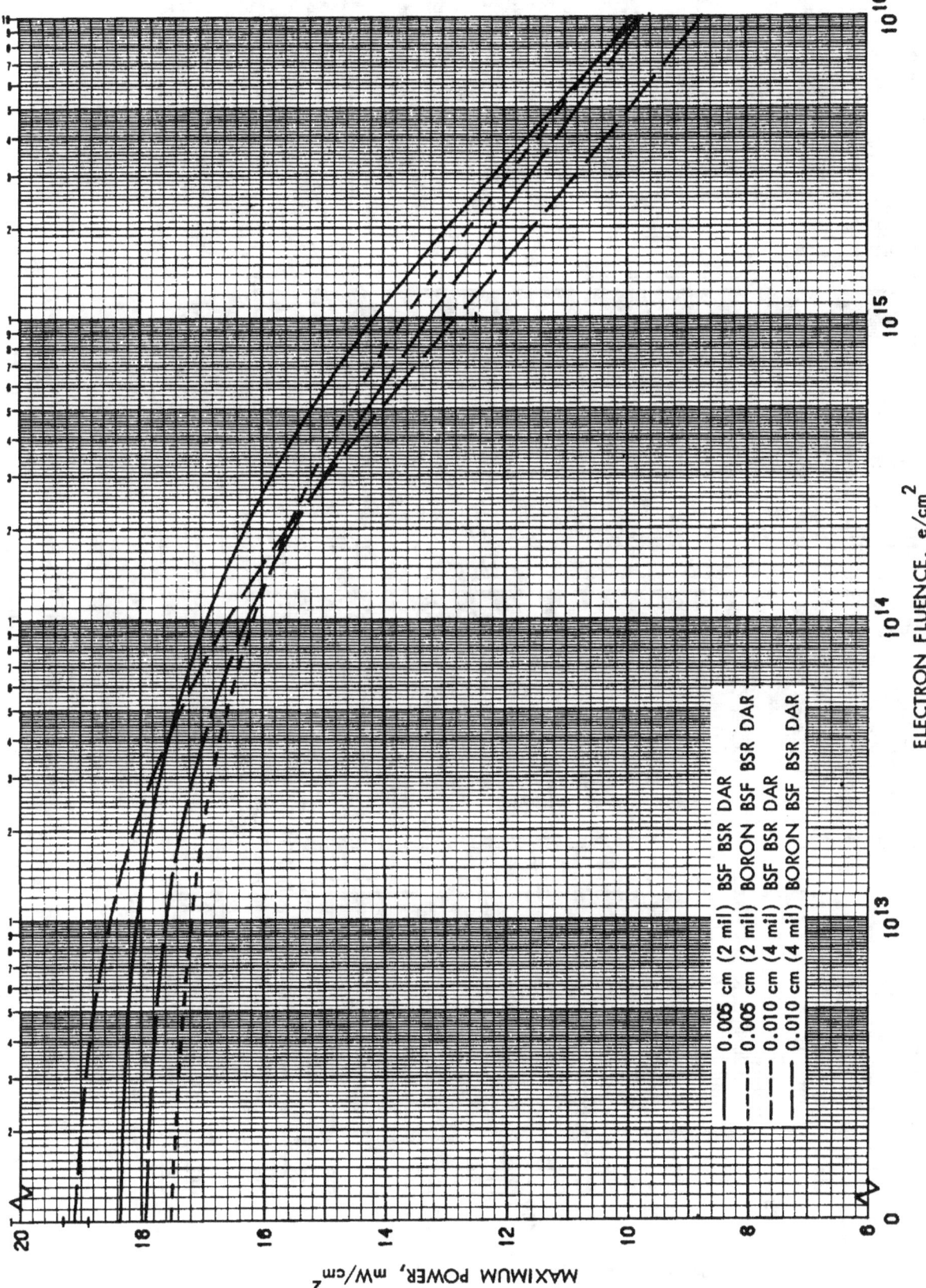

Figure 3.82 P_{max} vs 1 MeV Electron Fluence for 10 Ohm-cm n/p Back Surface Field Thin Silicon Cells with BSR

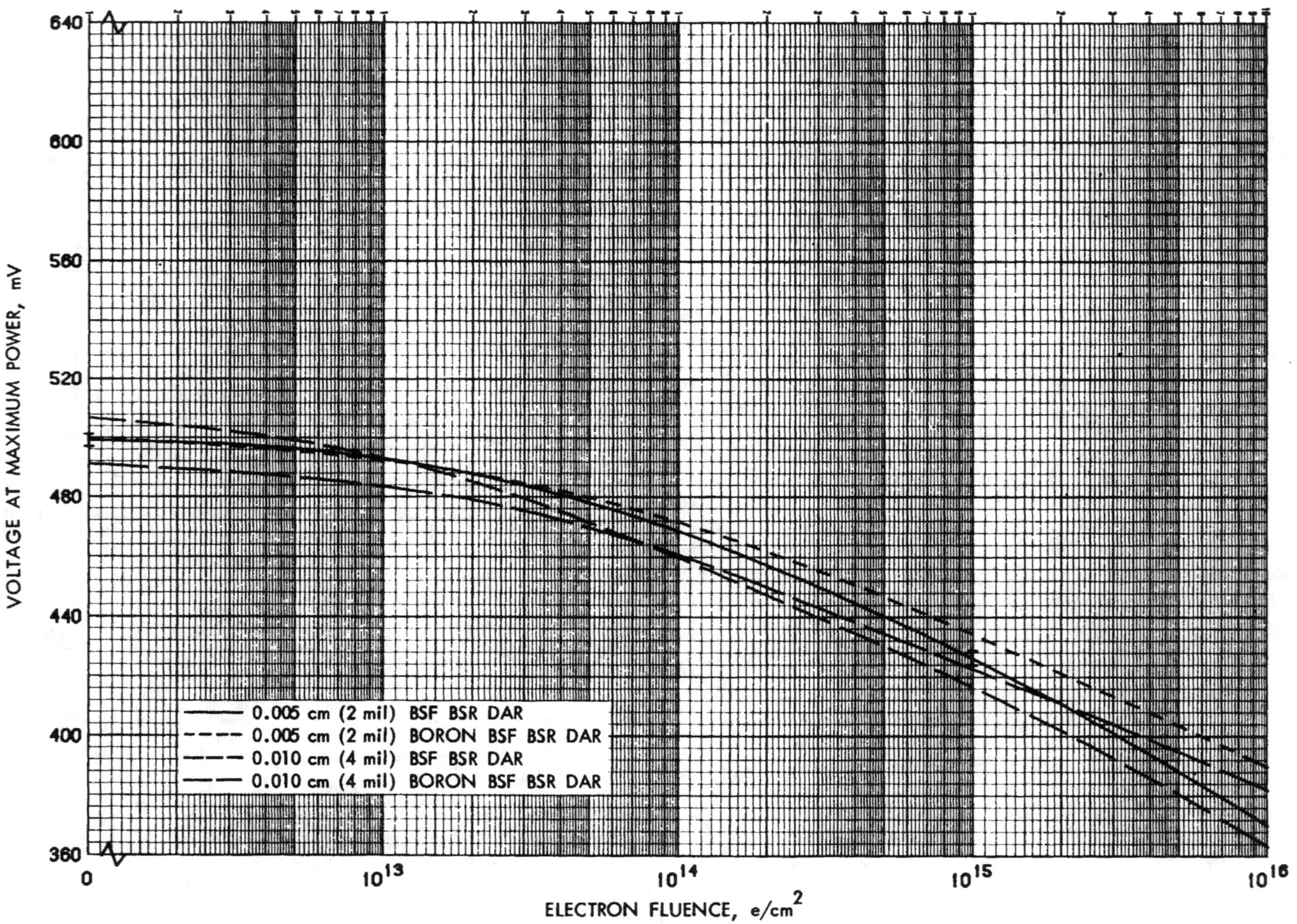

Figure 3.83 V_{mp} vs 1 MeV Electron Fluence for 10 Ohm-cm n/p Back Surface Field Thin Silicon Cells with BSR

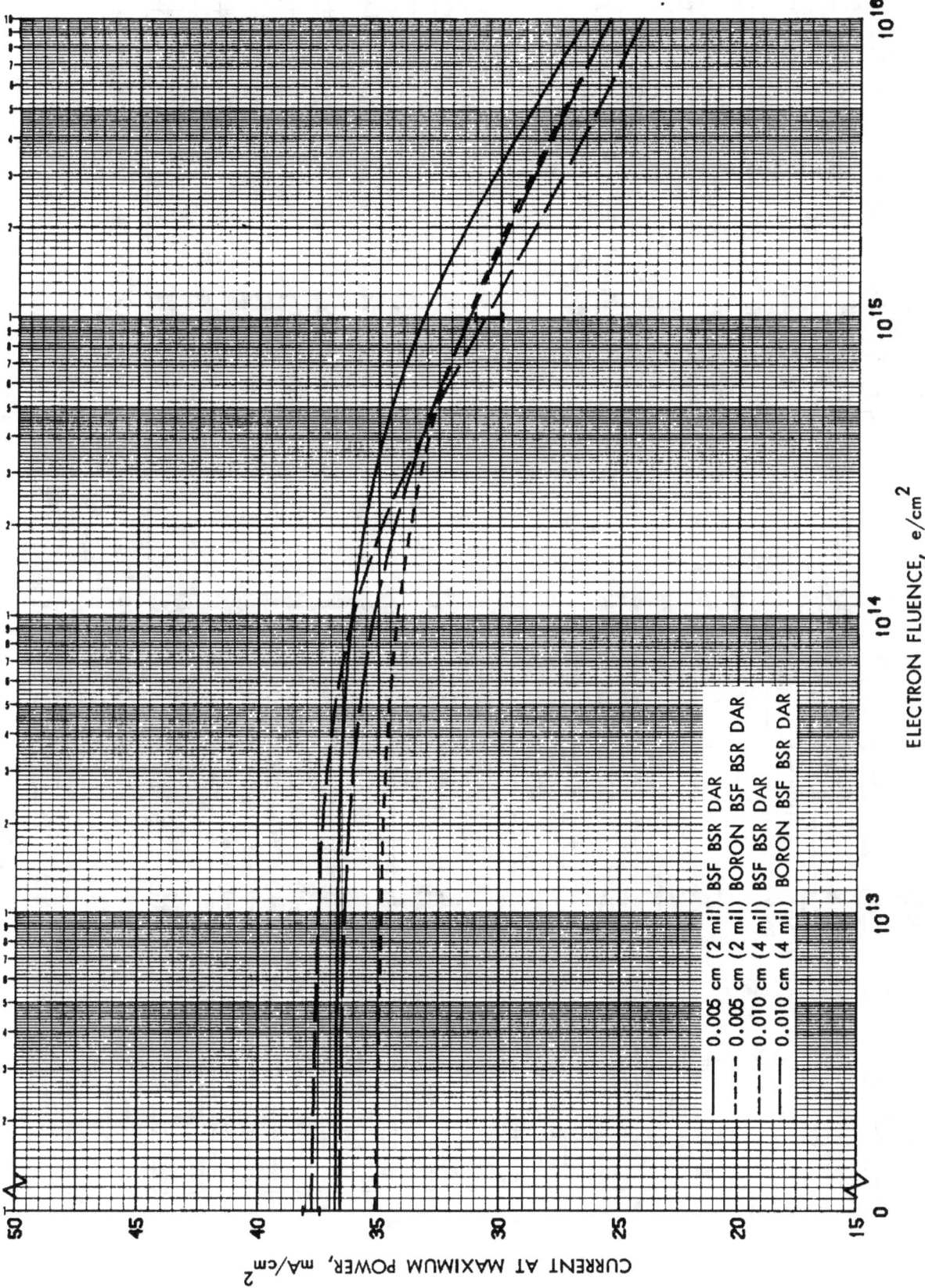

Figure 3.84 I_{mp} vs 1 MeV Electron Fluence for 10 Ohm-cm n/p Back Surface Field Thin Silicon Cells with BSR

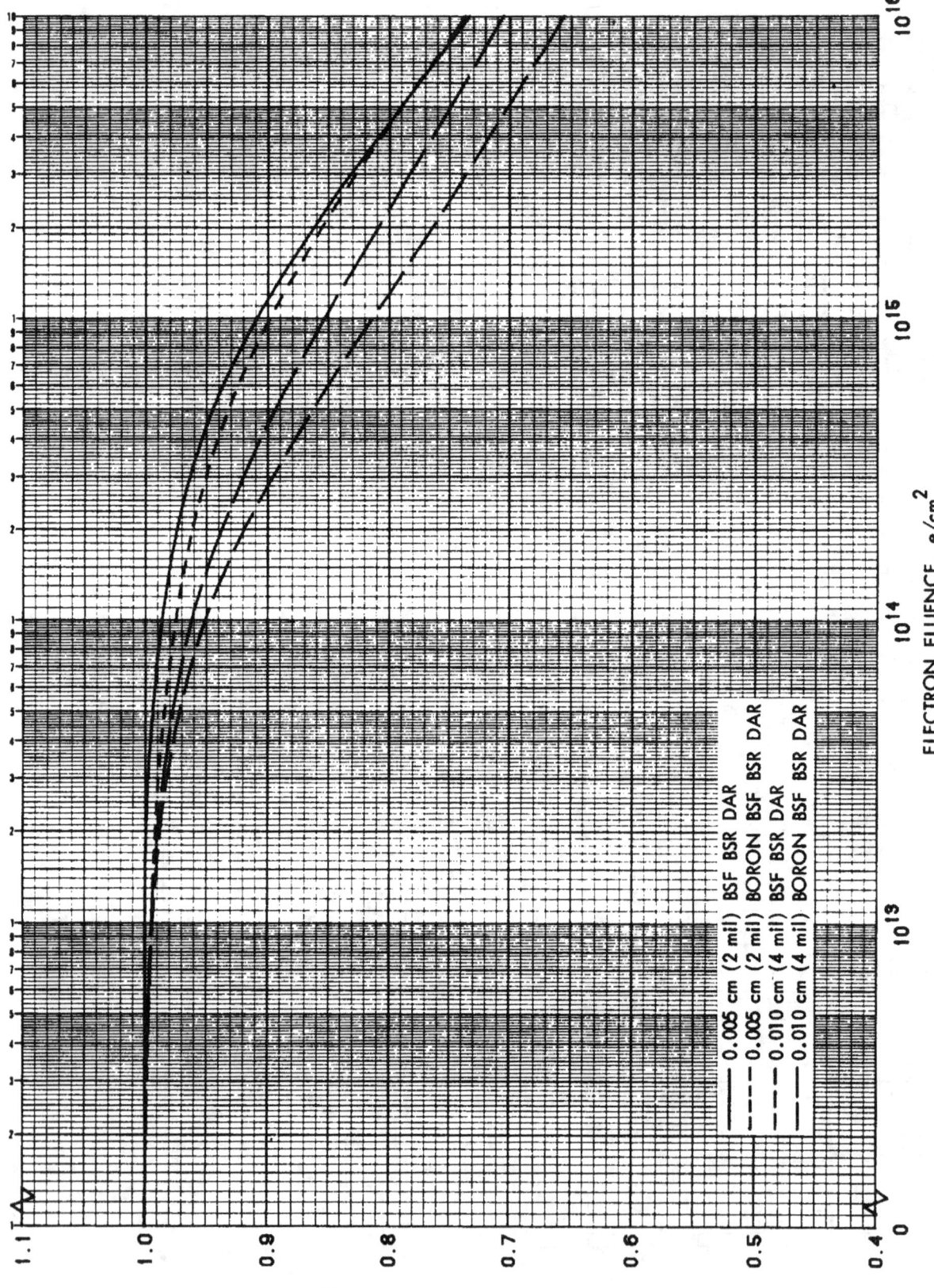

Figure 3.85 Normalized I_{sc} vs 1 MeV Electron Fluence for 10 Ohm-cm n/p Back Surface Field Thin Silicon Cells with BSR

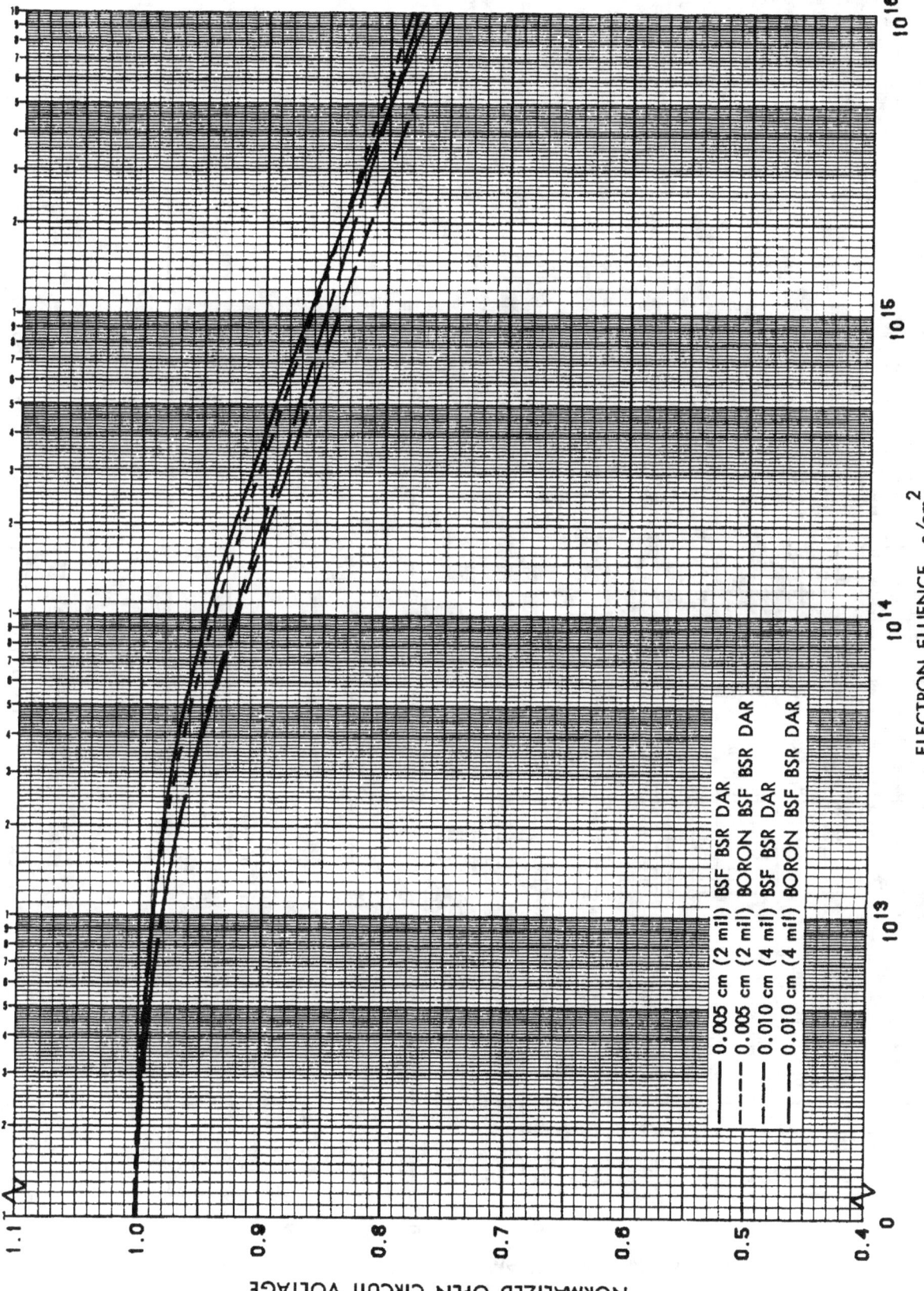

Figure 3.86 Normalized V_{OC} vs 1 MeV Electron Fluence for 10 Ohm-cm n/p Back Surface Field Thin Silicon Cells with BSR

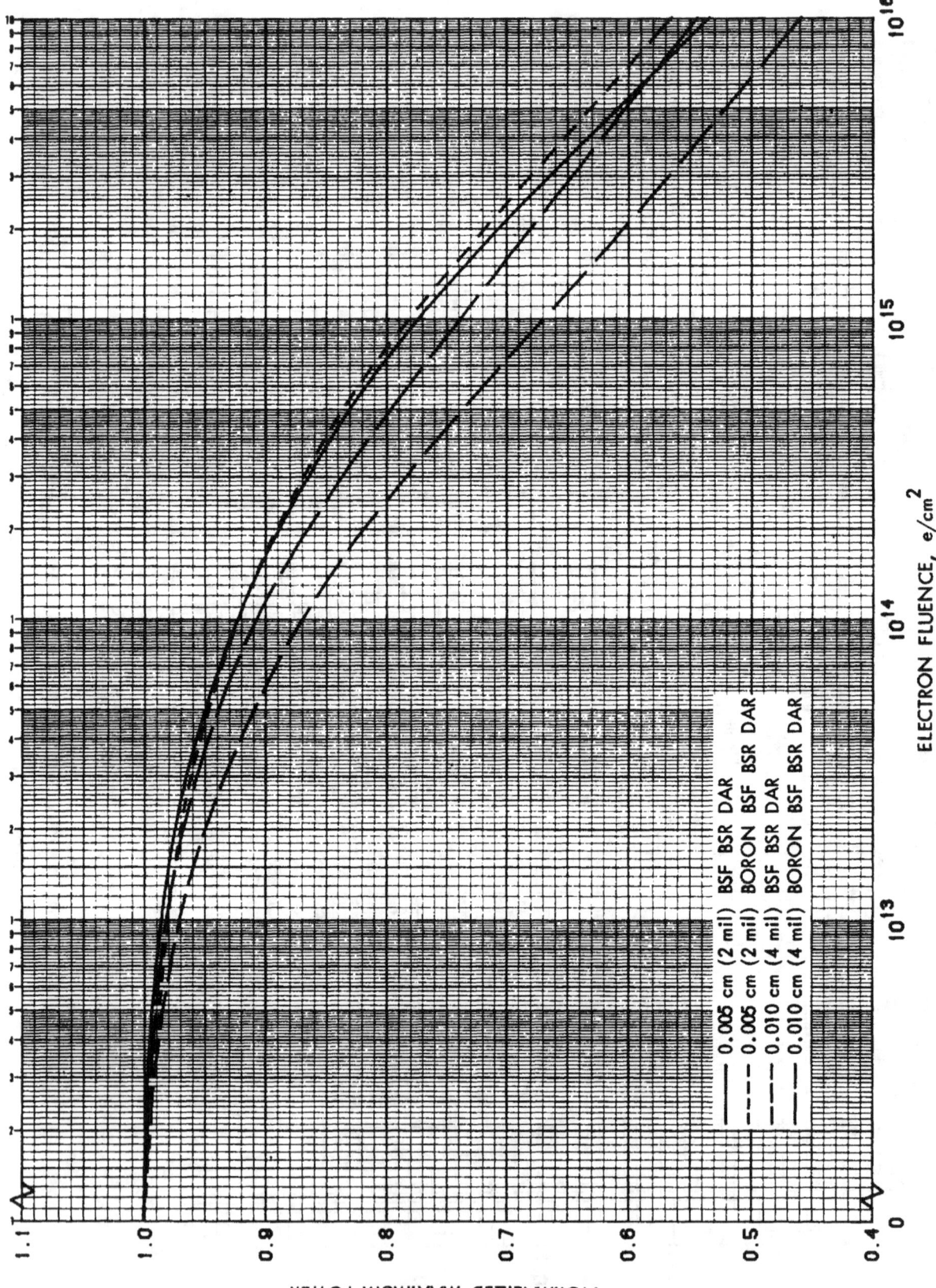

Figure 3.87 Normalized P_{max} vs 1 MeV Electron Fluence for 10 Ohm-cm n/p Back Surface Field Thin Silicon Cells with BSR

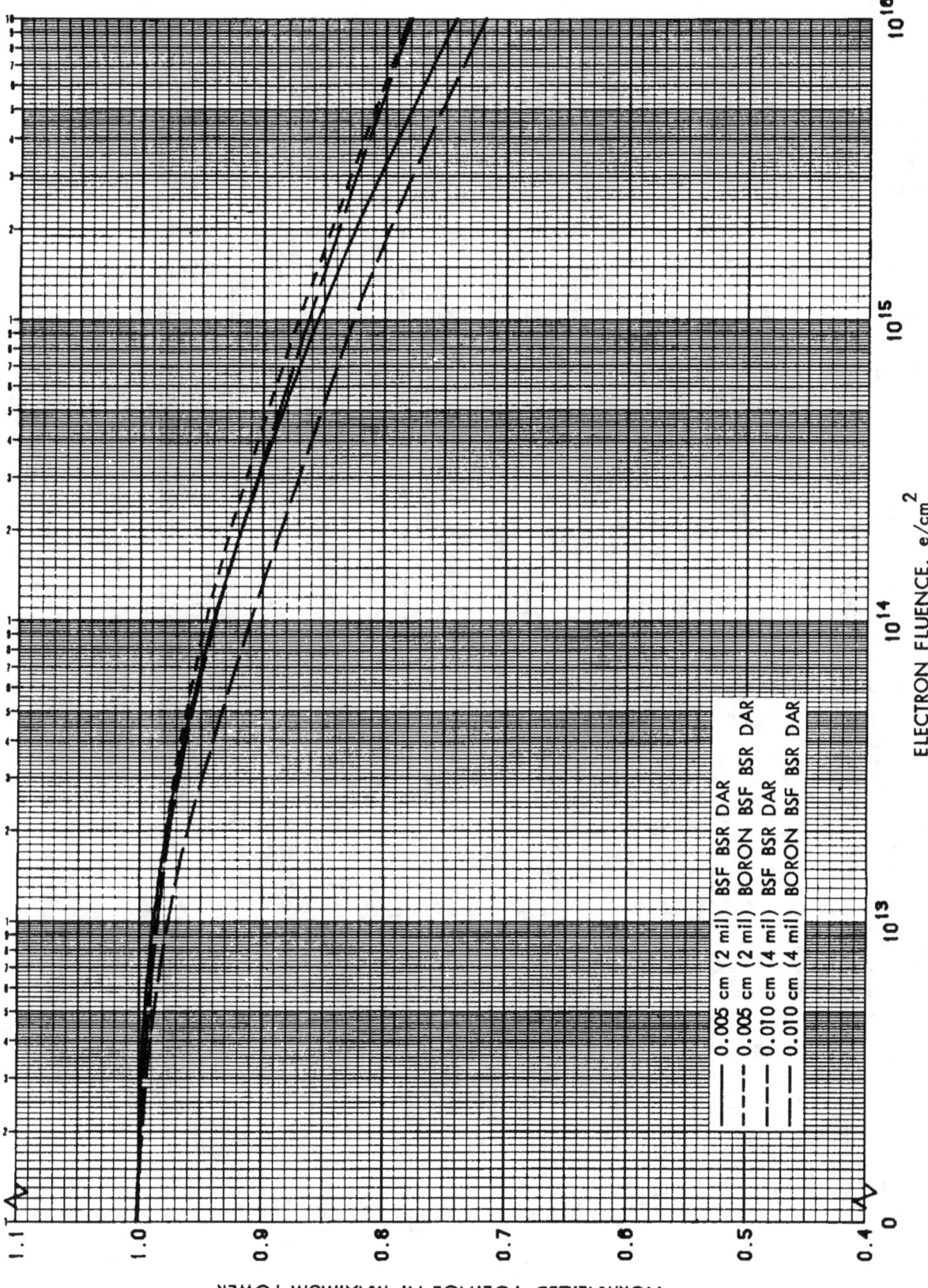

Figure 3.88 Normalized V_{mp} vs 1 MeV Electron Fluence for 10 Ohm-cm n/p Back Surface Field Thin Silicon Cells with BSR

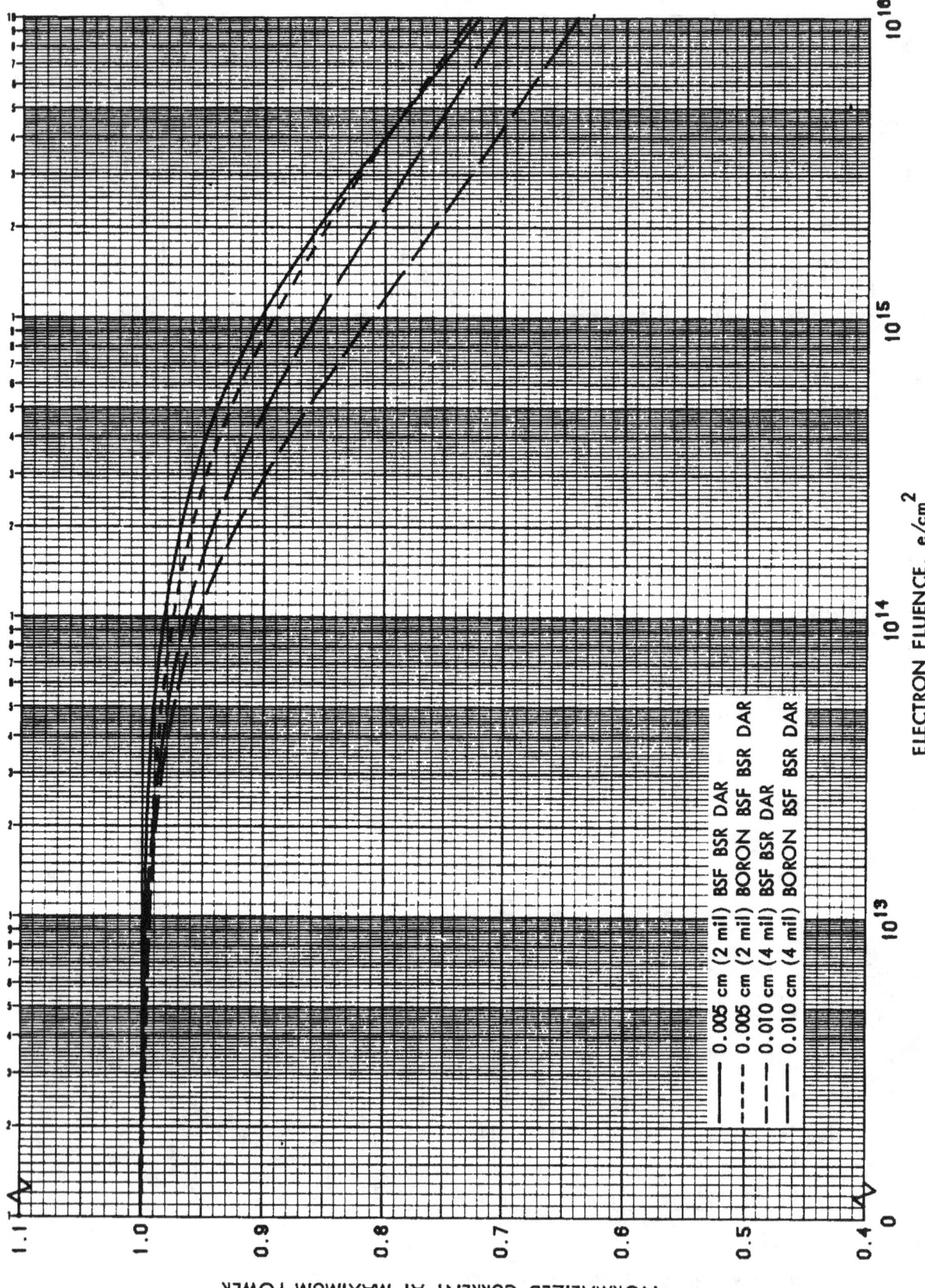

Figure 3.89 Normalized I_{mp} vs 1 MeV Electron Fluence for 10 Ohm-cm n/p Back Surface Field Thin Silicon Cells with BSR

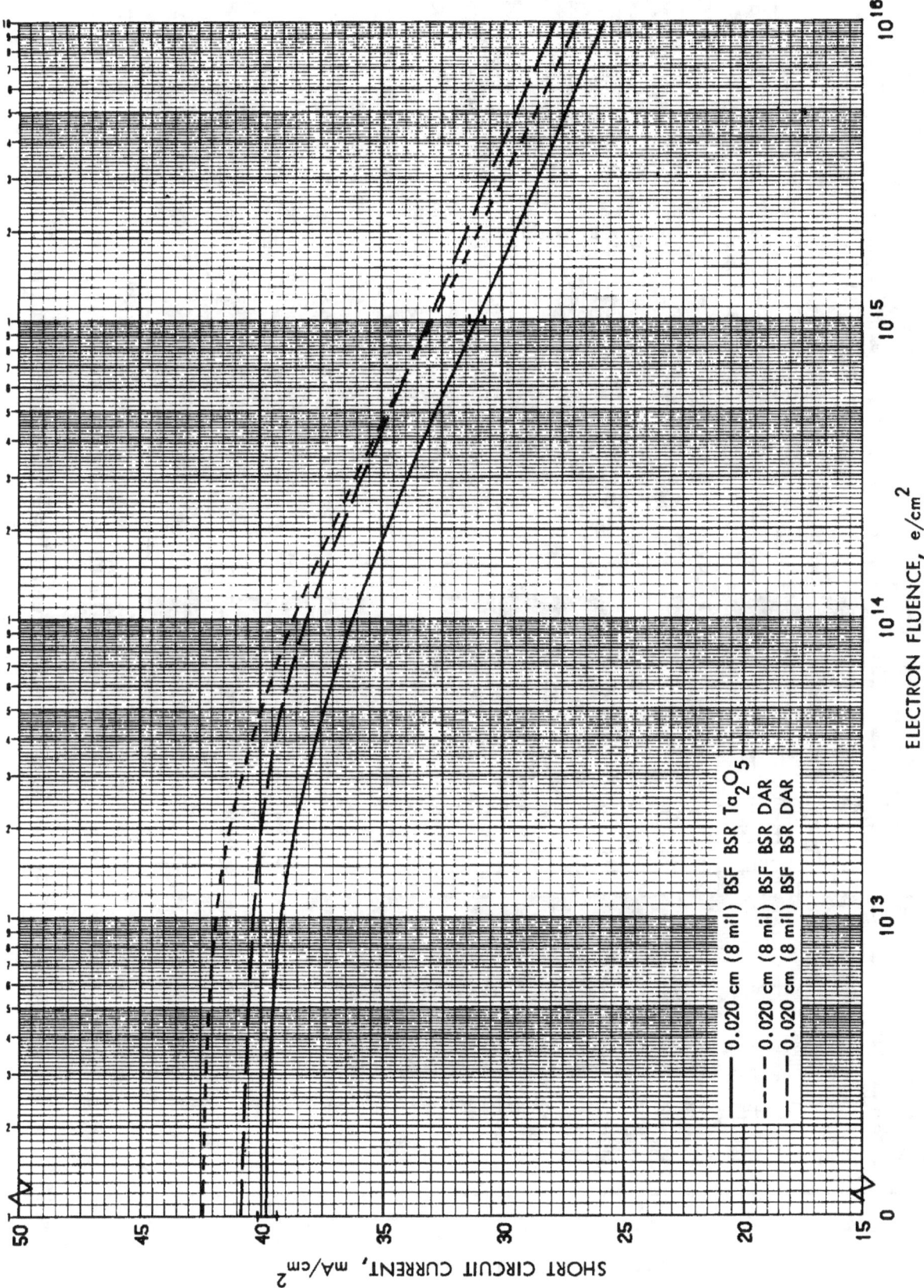

Figure 3.90 I_{sc} vs 1 MeV Electron Fluence for 10 Ohm-cm n/p Back Surface Field Silicon Cells with BSR

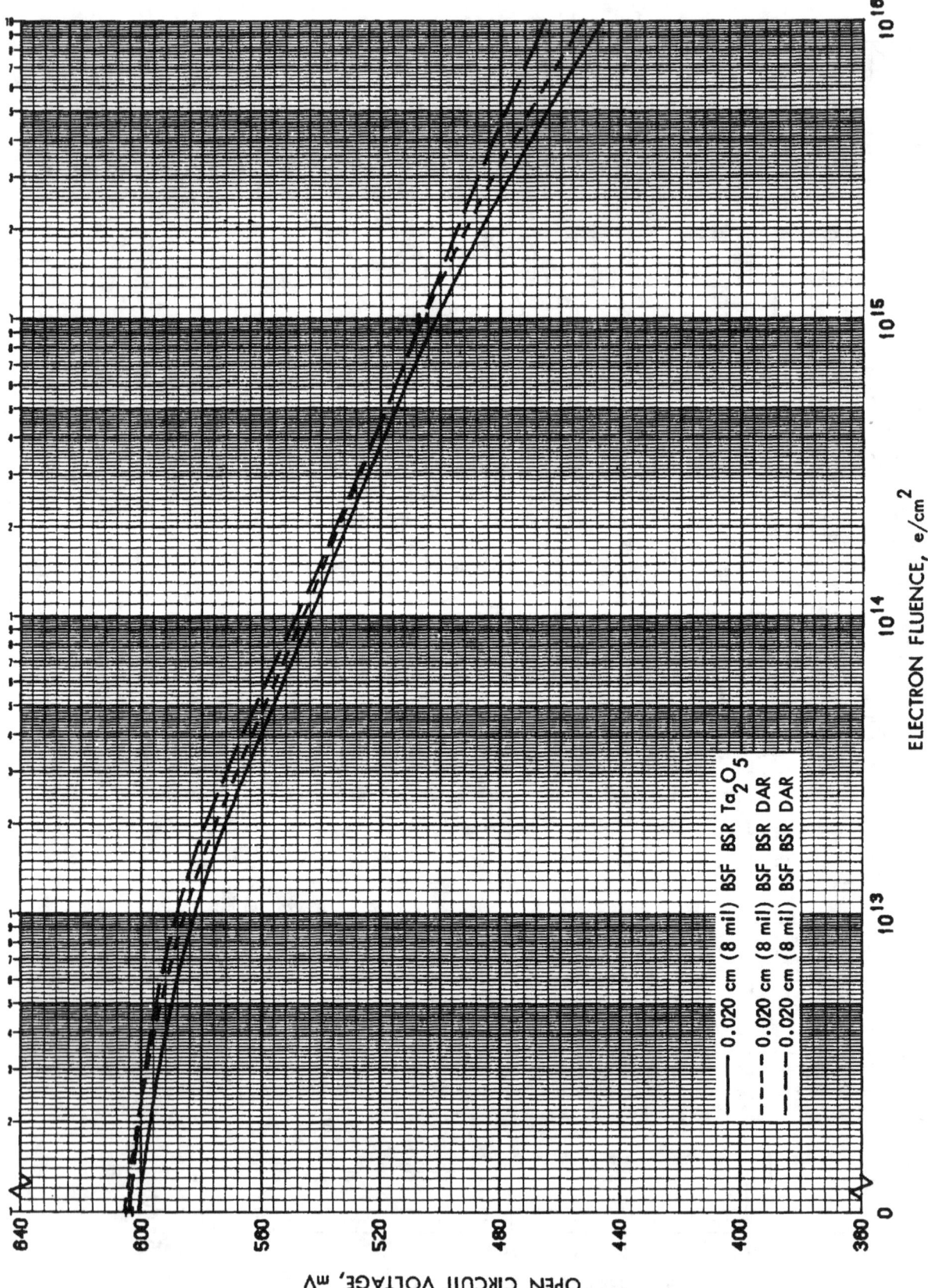

Figure 3.91 V_{oc} vs 1 MeV Electron Fluence for 10 Ohm-cm n/p Back Surface Field Silicon Cells with BSR

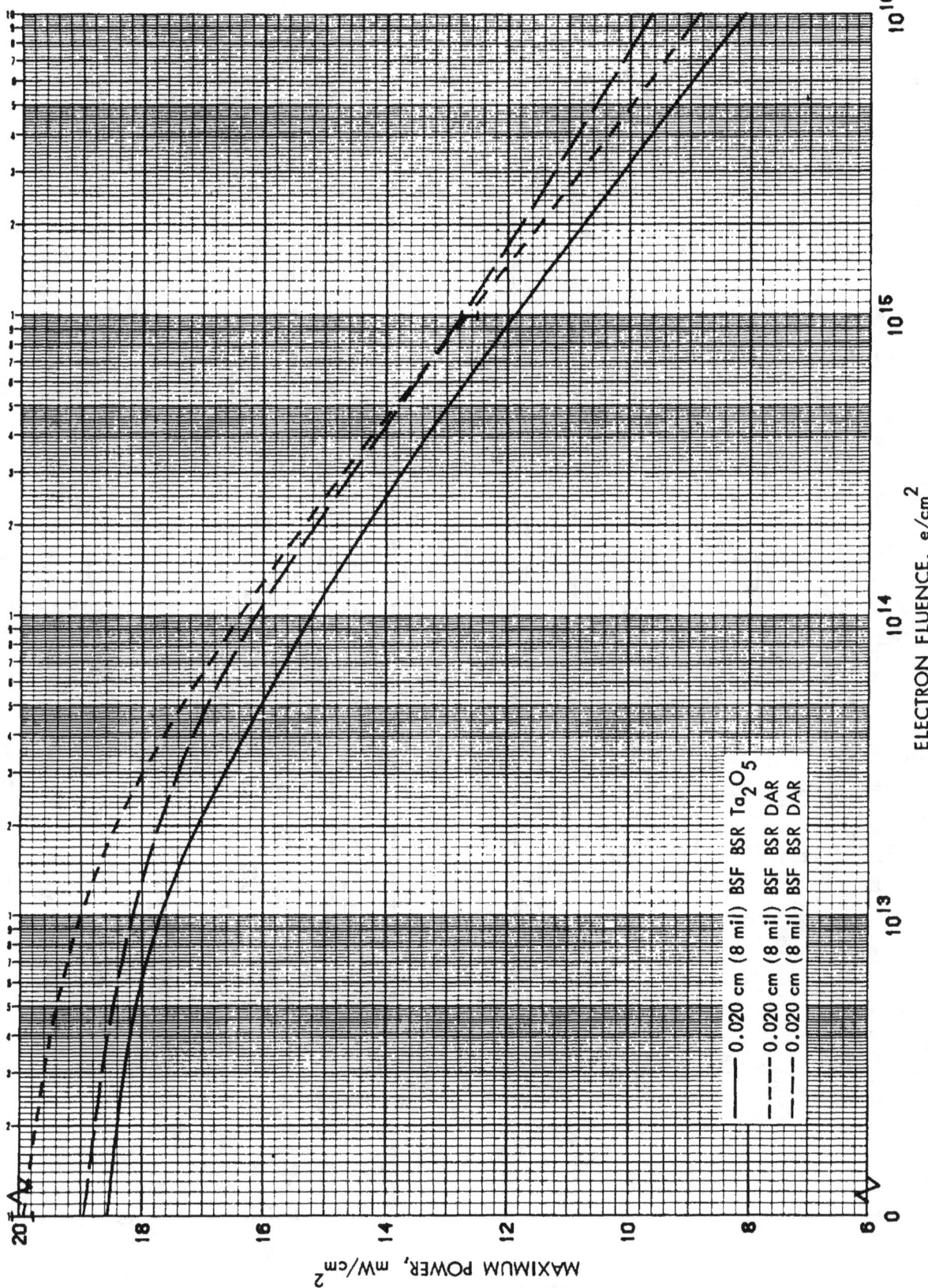

Figure 3.92 P_{max} vs 1 MeV Electron Fluence for 10 Ohm-cm n/p Back Surface Field Silicon Cells with BSR

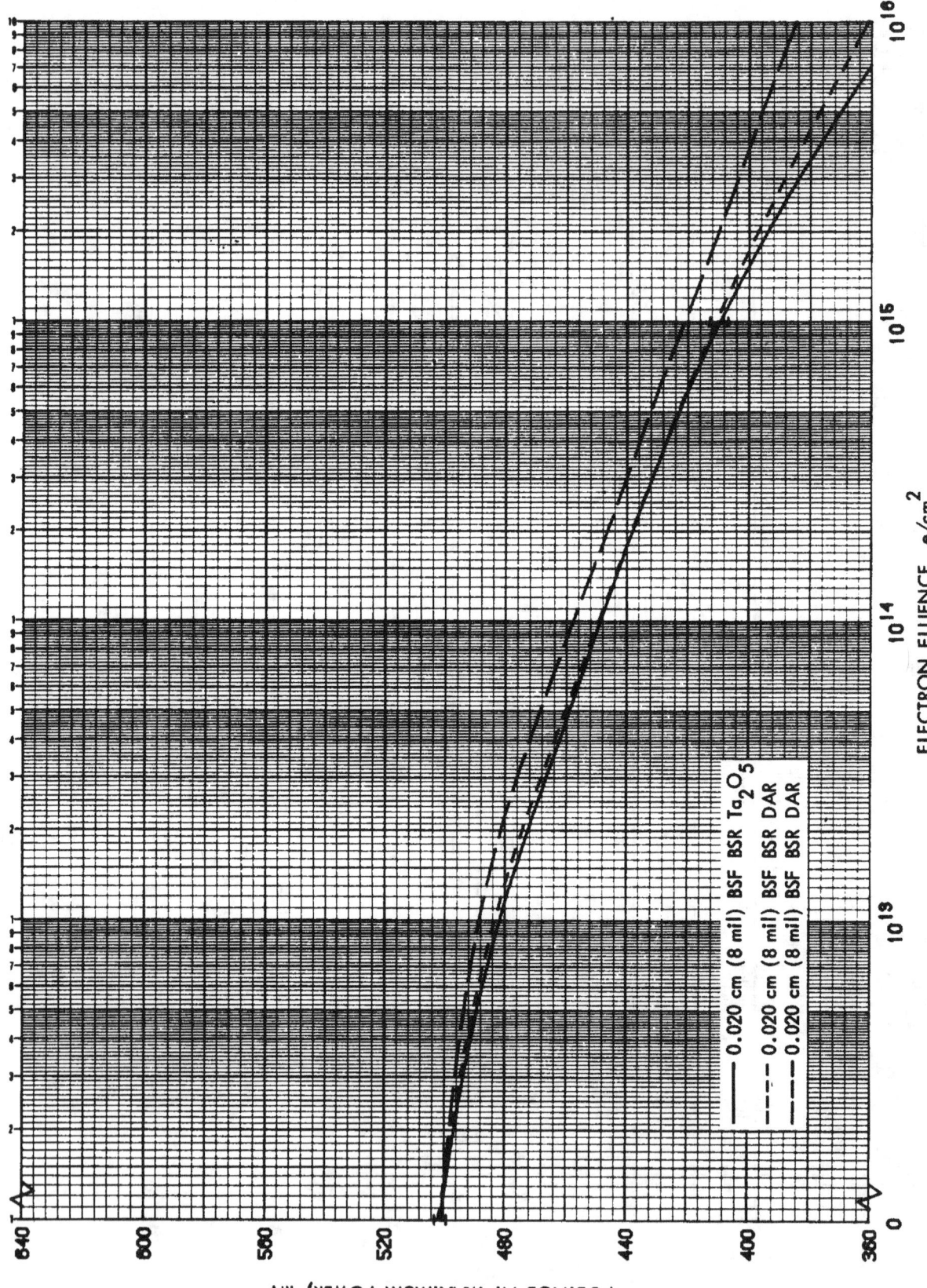

Figure 3.93 V_{mp} vs 1 MeV Electron Fluence for 10 Ohm-cm n/p Back Surface Field Silicon Cells with BSR

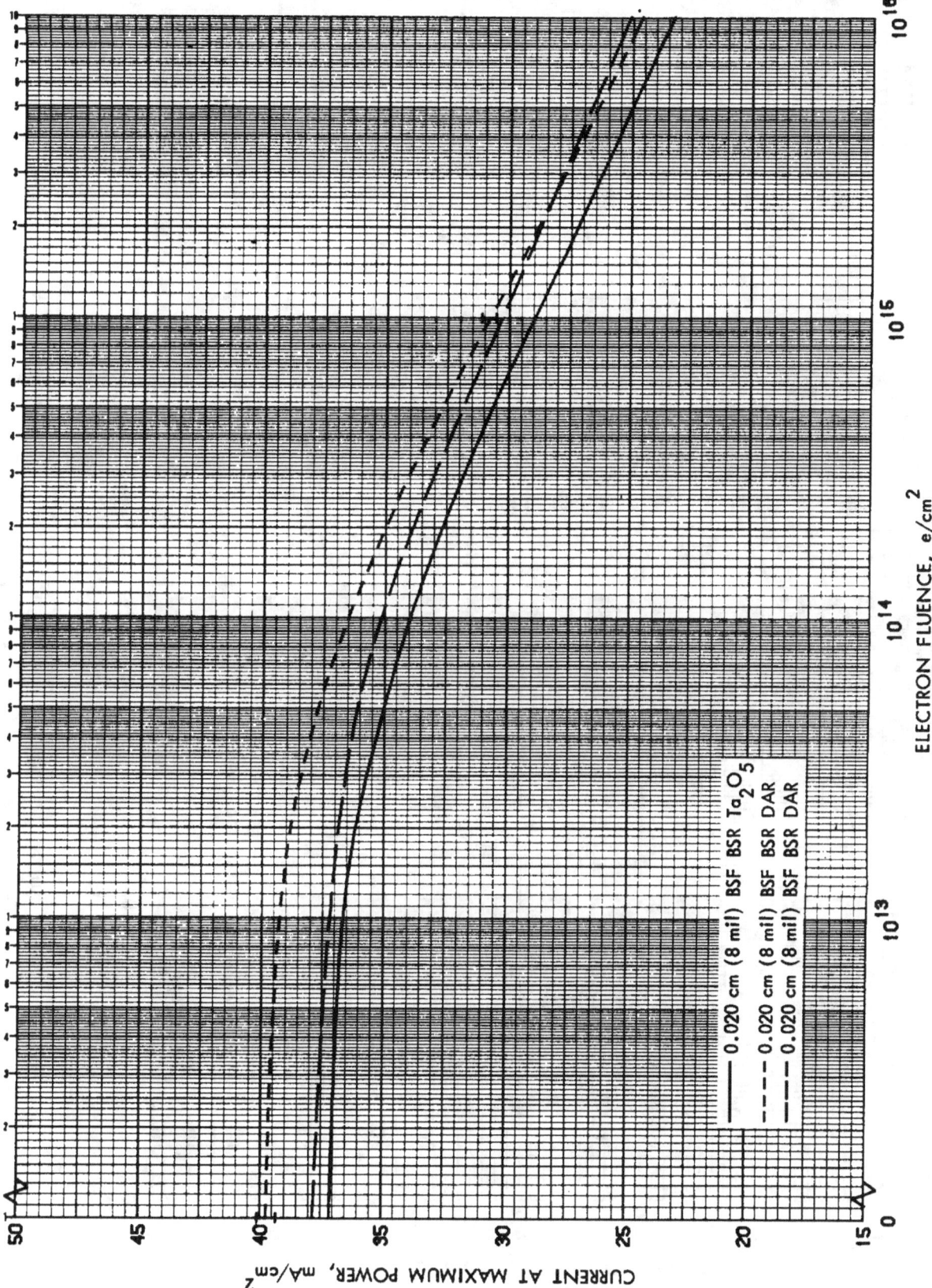

Figure 3.94 I_{mp} vs 1 MeV Electron Fluence for 10 Ohm-cm n/p Back Surface Field Silicon Cells with BSR

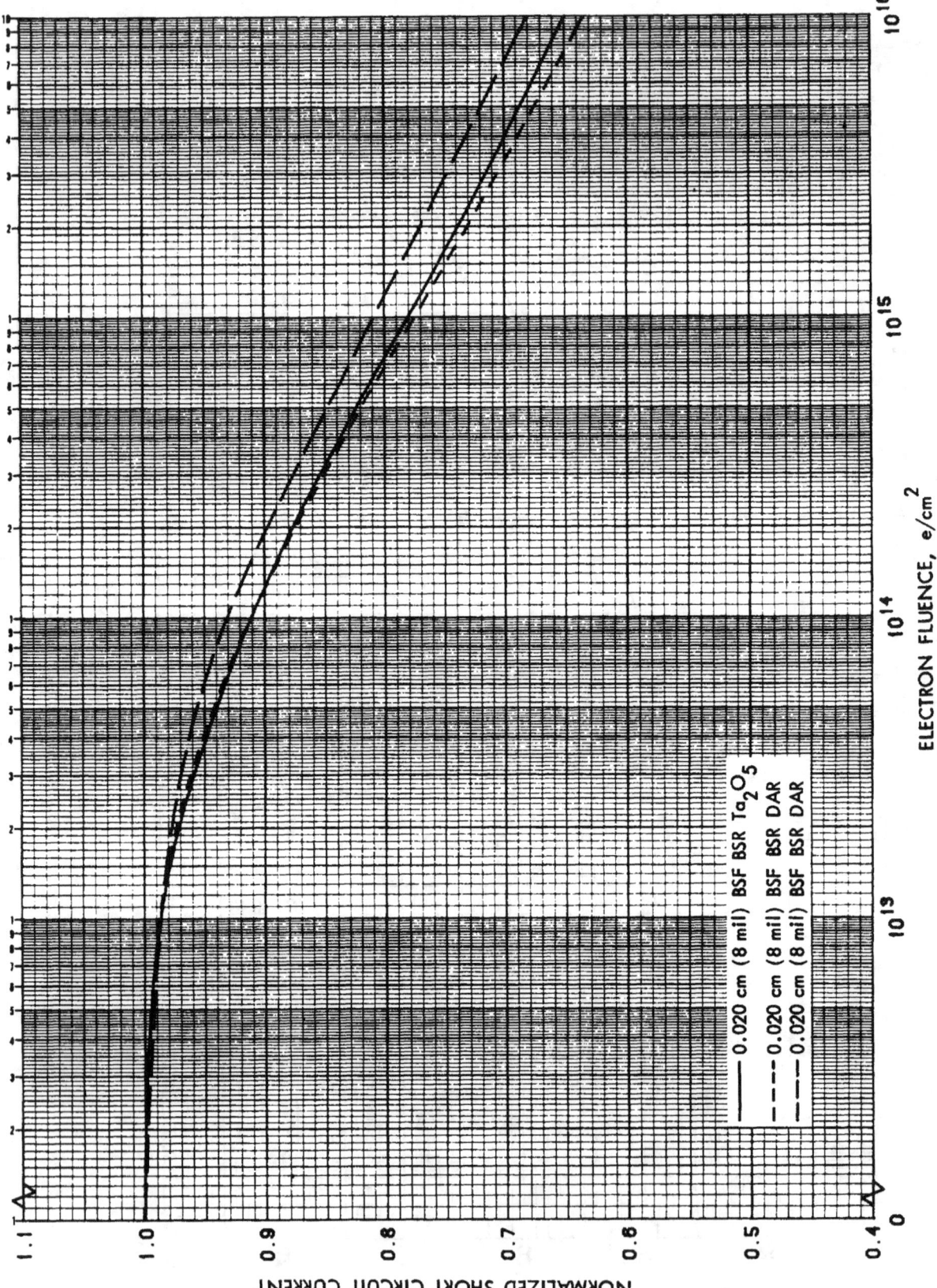

Figure 3.95 Normalized I_{sc} vs 1 MeV Electron Fluence for 10 Ohm-cm n/p Back Surface Field Silicon Cells with BSR

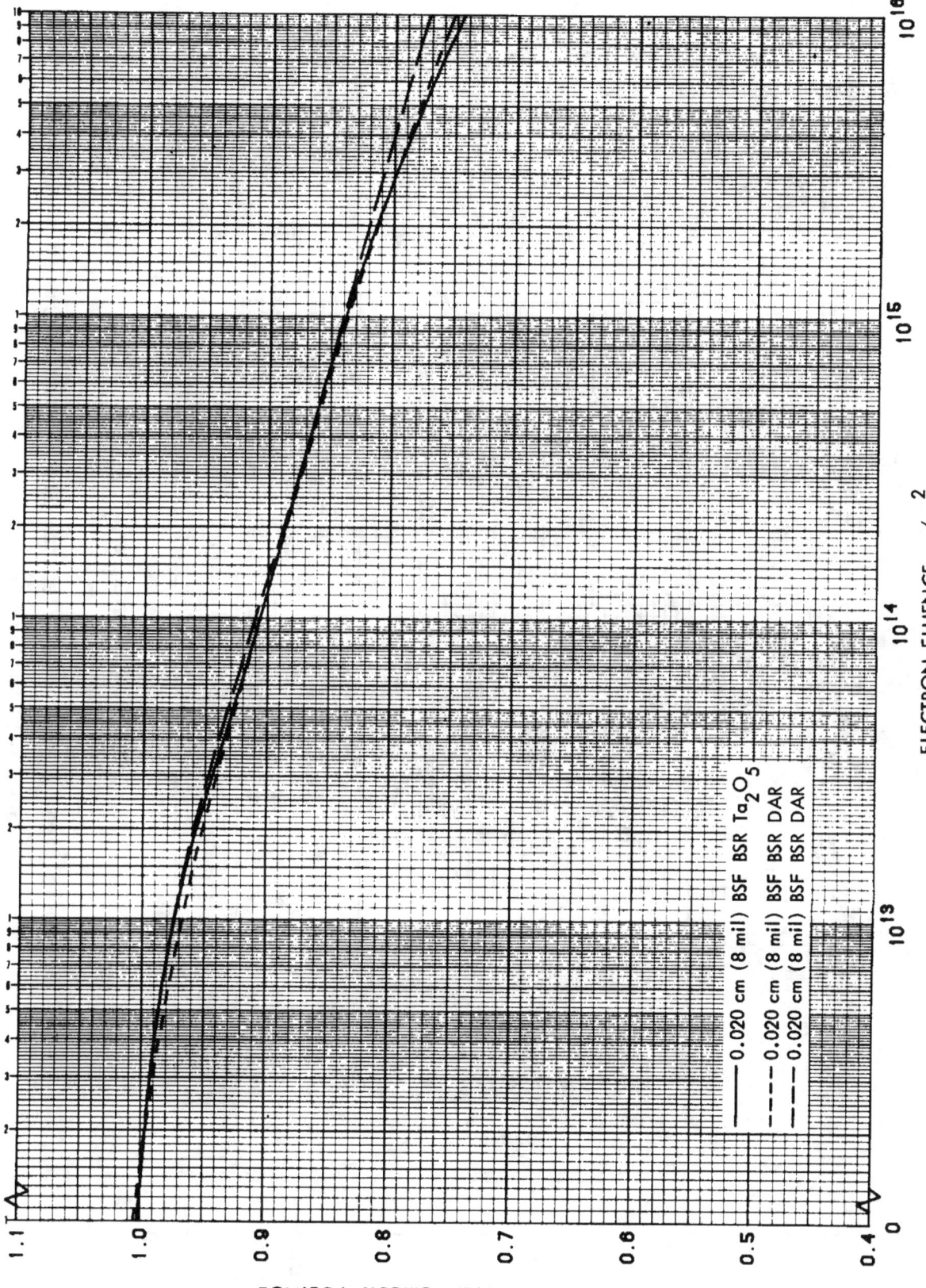

Figure 3.96 Normalized V_{oc} vs 1 MeV Electron Fluence for 10 Ohm-cm n/p Back Surface Field Silicon Cells with BSR

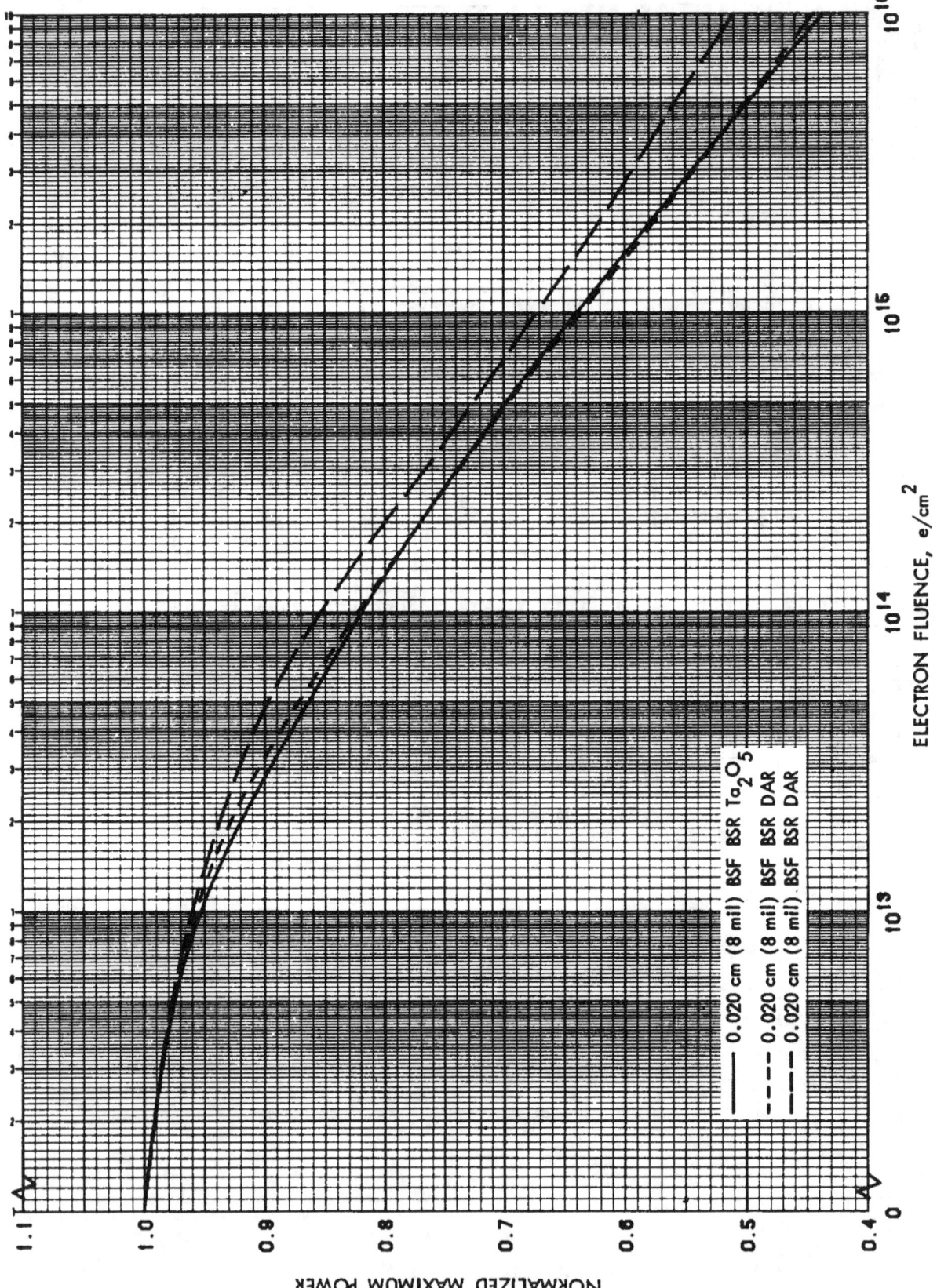

Figure 3.97 Normalized P_{max} vs 1 MeV Electron Fluence for 10 Ohm-cm n/p Back Surface Field Silicon Cells with BSR

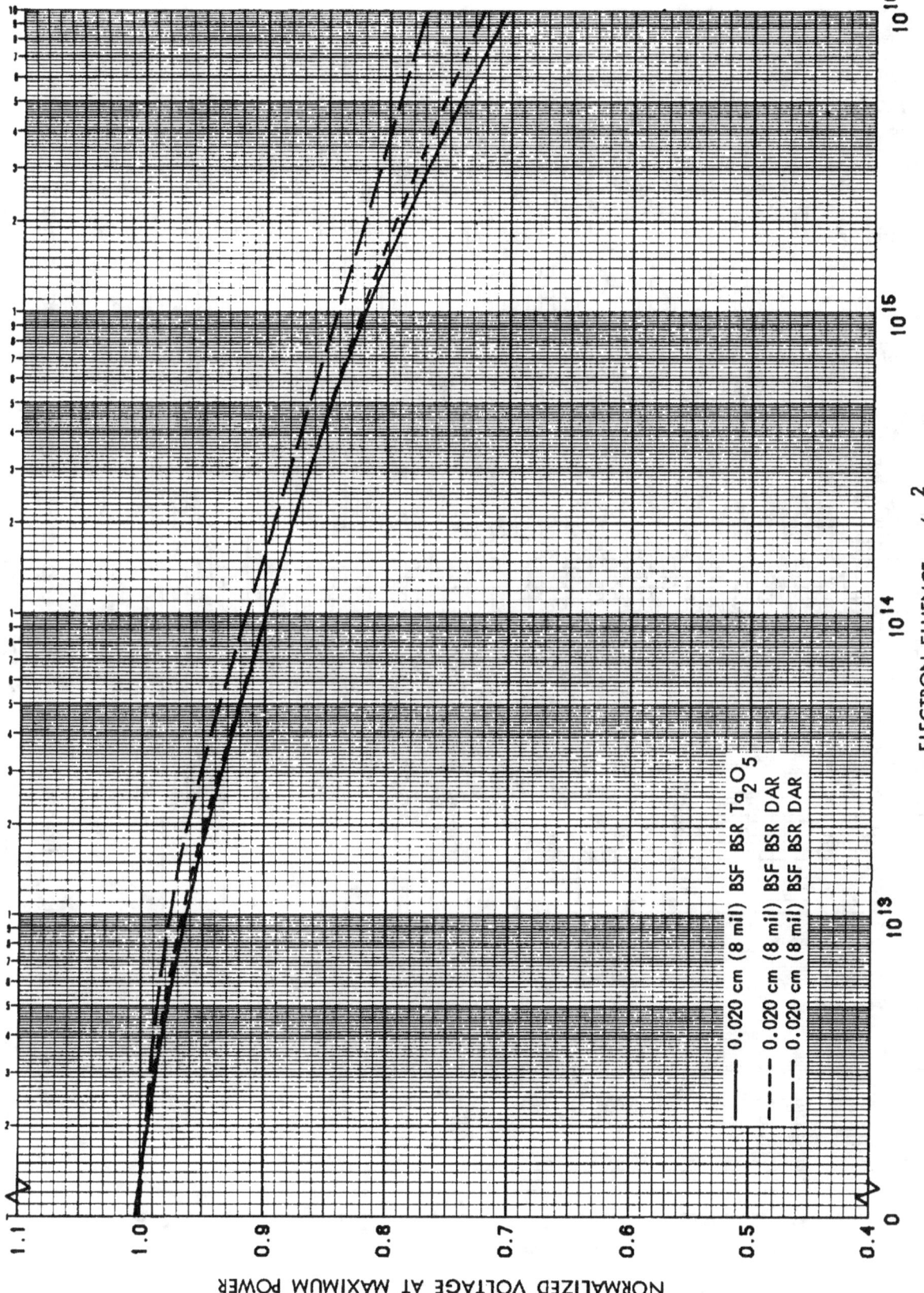

Figure 3.98 Normalized V_{mp} vs 1 MeV Electron Fluence for 10 Ohm-cm n/p Back Surface Field Silicon Cells with BSR

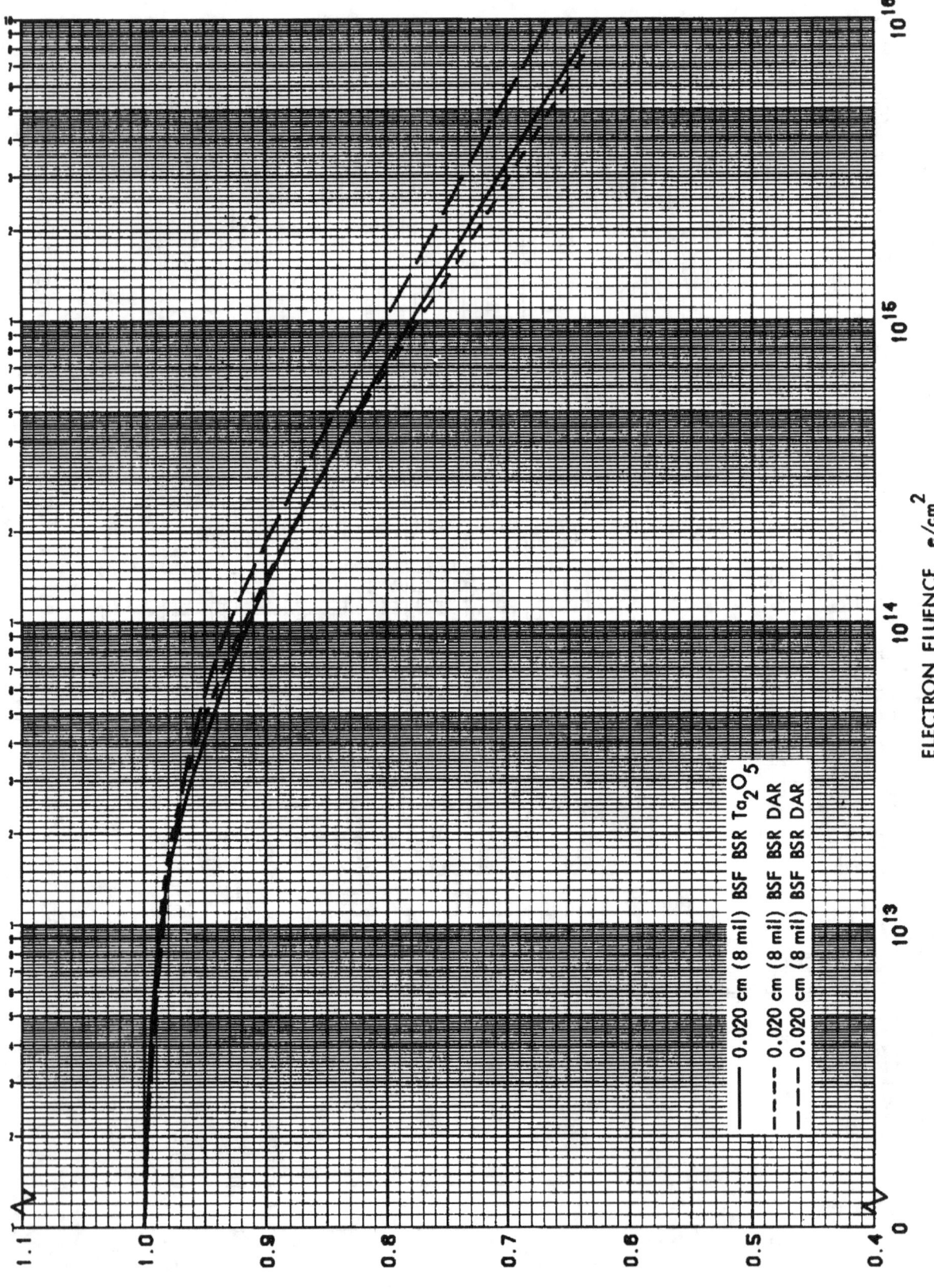

Figure 3.99 Normalized I_{mp} vs 1 MeV Electron Fluence for 10 Ohm-cm n/p Back Surface Field Silicon Cells with BSR

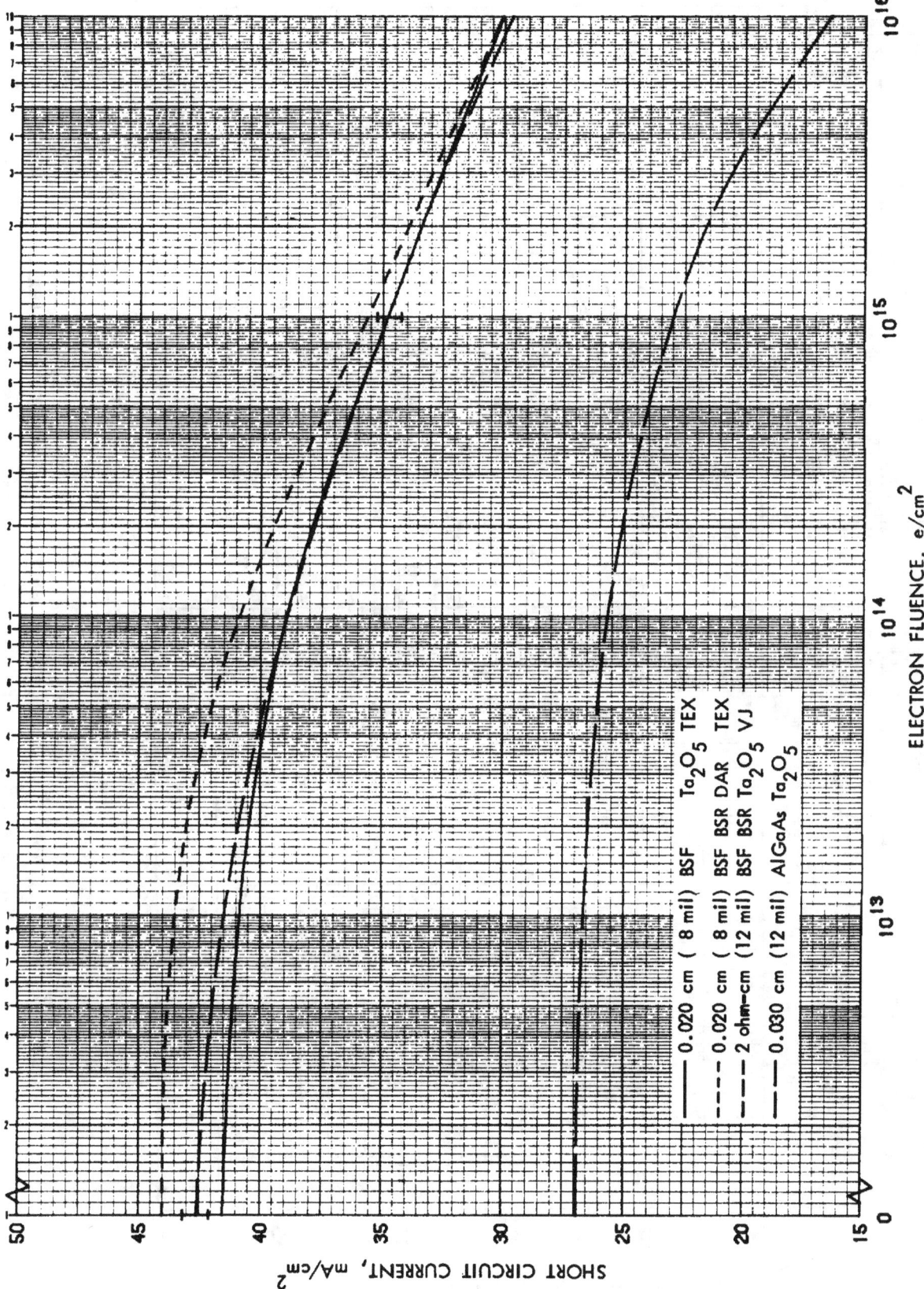

Figure 3.100 I_{sc} vs 1 MeV Electron Fluence for 10 Ohm-cm n/p Textured, 2 Ohm-cm Vertical Junction, and p/n AlGaAs Cells

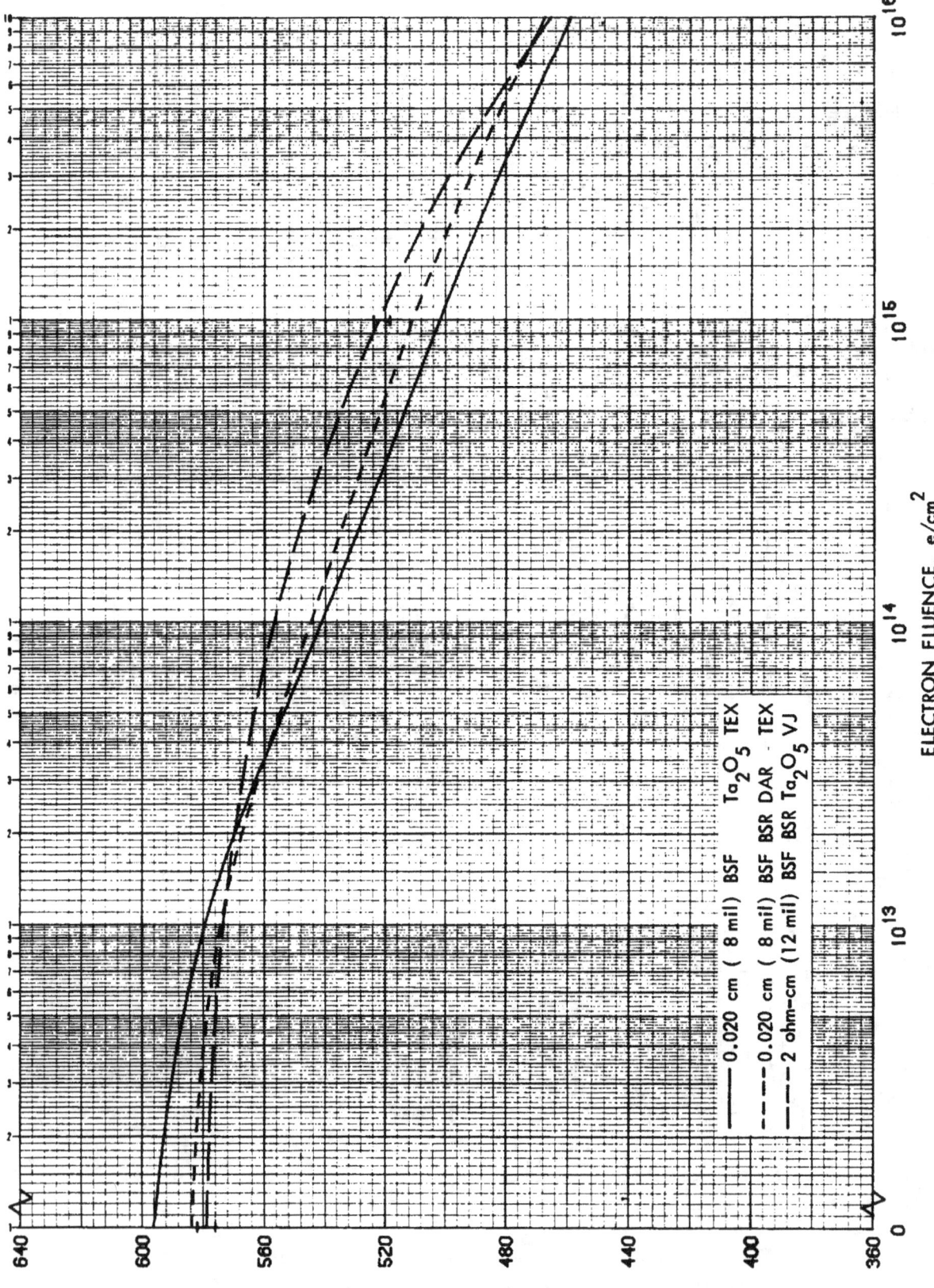

Figure 3.101 V_{oc} vs 1 MeV Electron Fluence for 10 Ohm-cm n/p Textured and 2 Ohm-cm Vertical Junction Cells

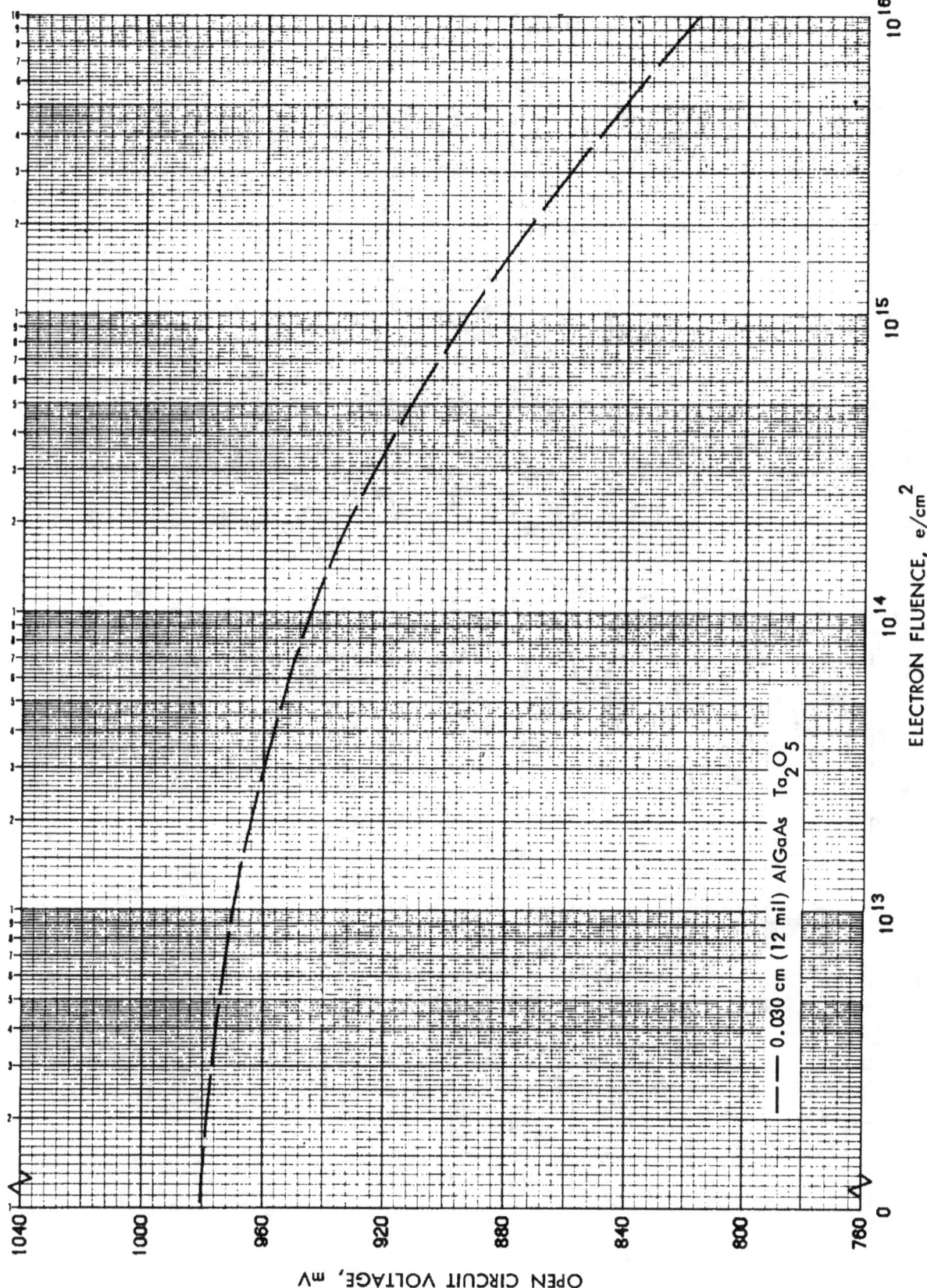

Figure 3.102 V_{oc} vs 1 MeV Electron Fluence for p/n AlGaAs Cells

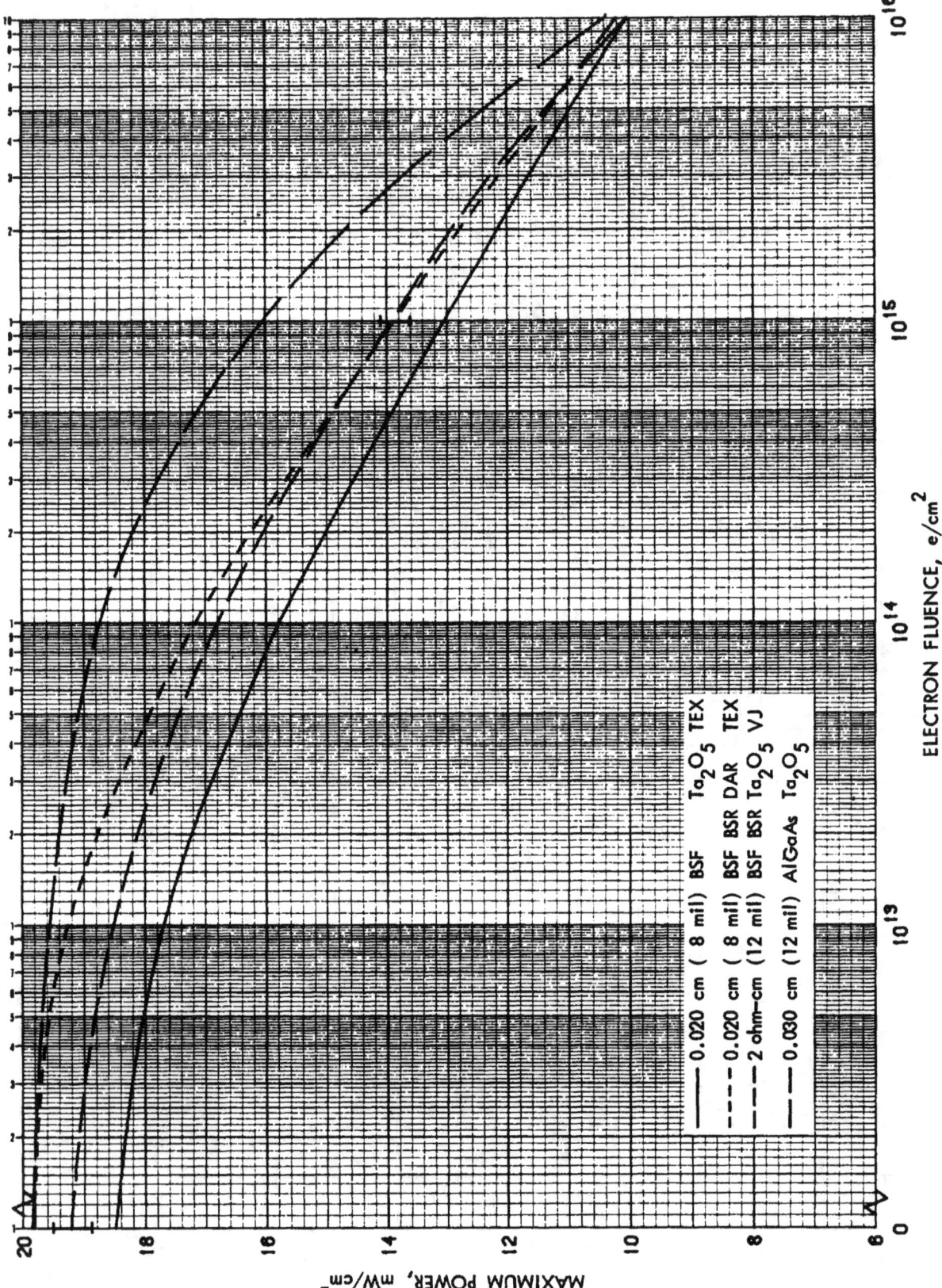

Figure 3.103 P_{max} vs 1 MeV Electron Fluence for 10 Ohm-cm n/p Textured, 2 Ohm-cm Vertical Junction, and p/n AlGaAs Cells

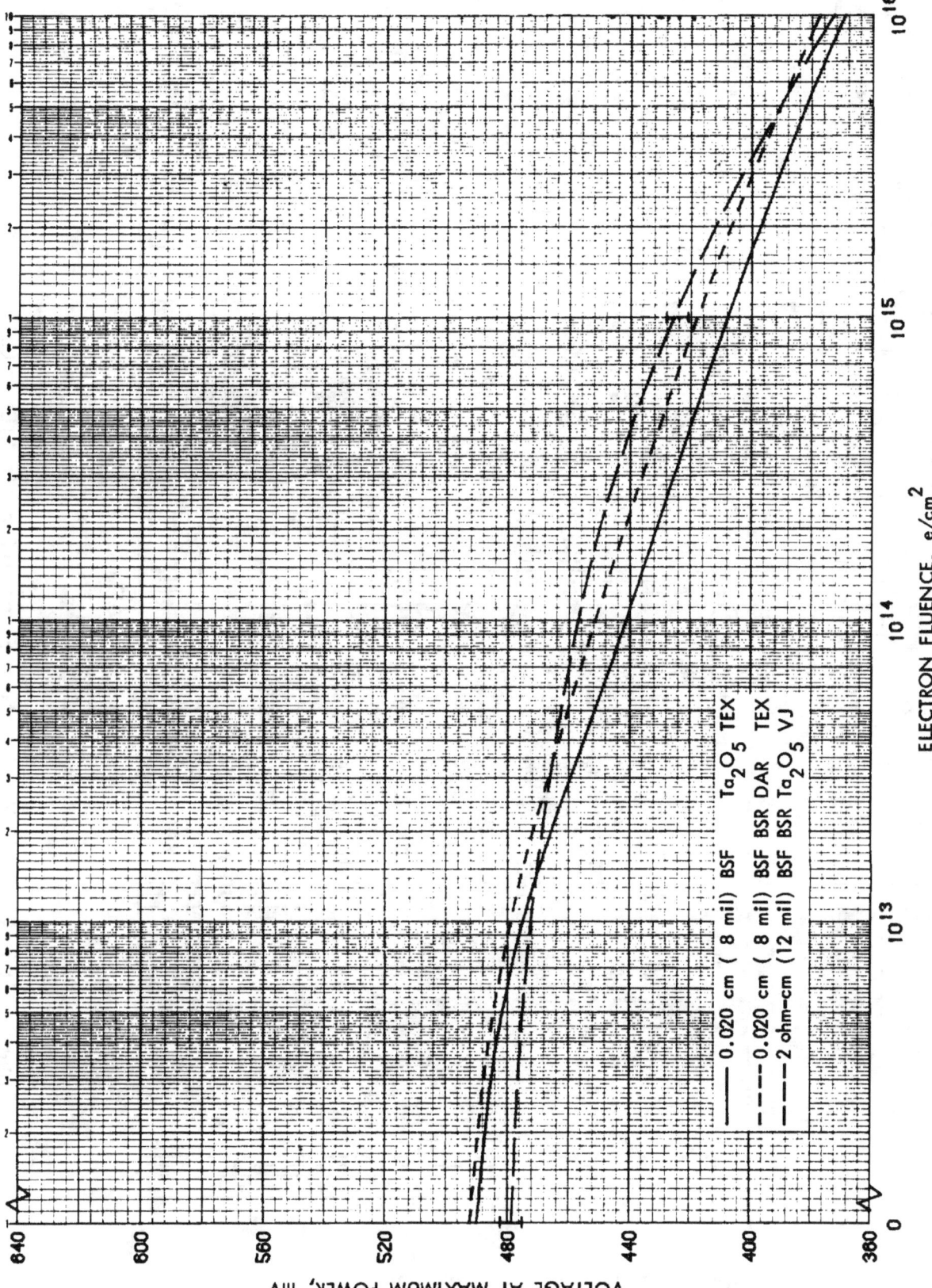

Figure 3.104 V_{mp} vs 1 MeV Electron Fluence for 10 Ohm-cm n/p Textured and 2 Ohm-cm Vertical Junction Cells

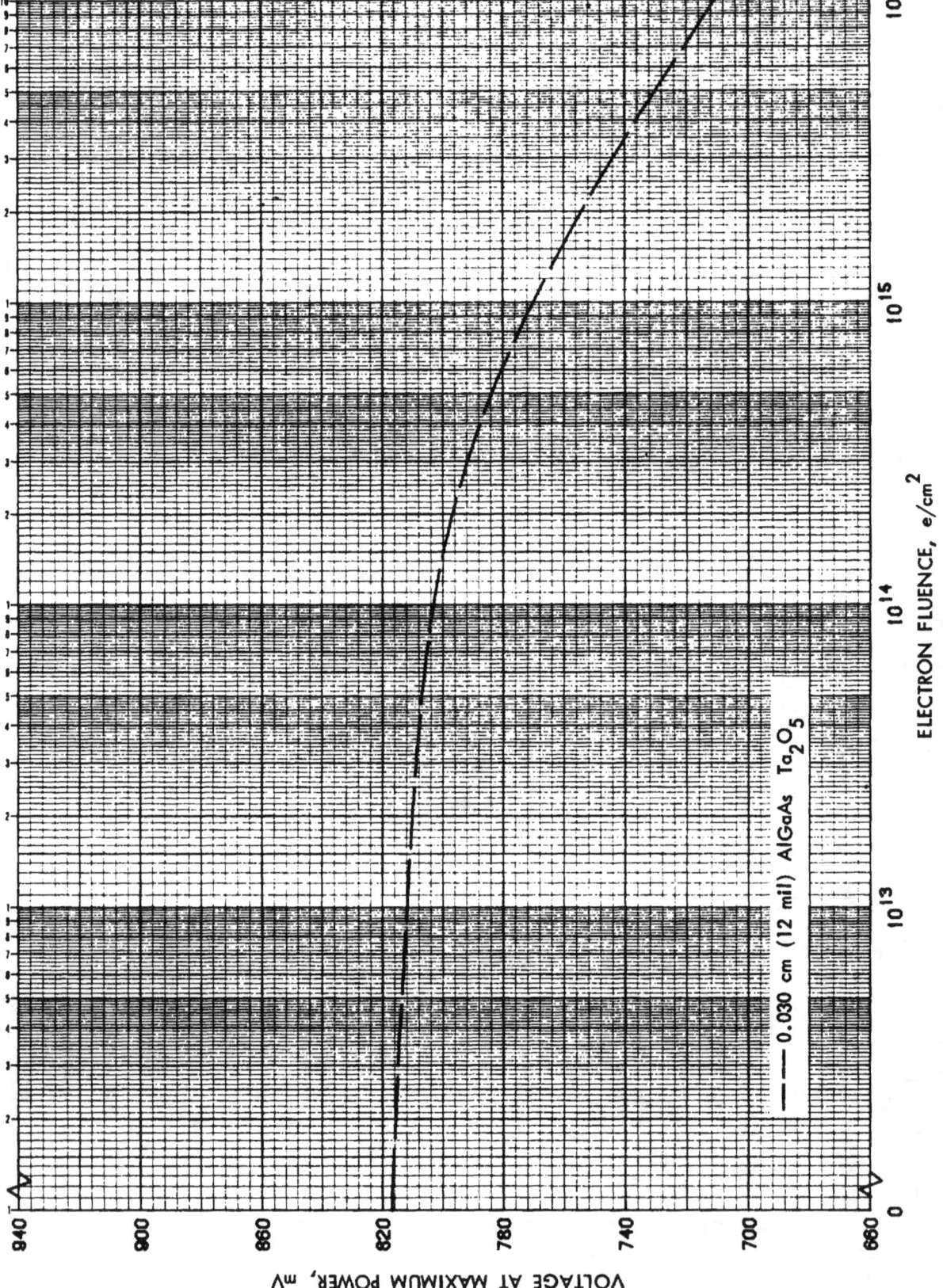

Figure 3.105 V_{mp} vs 1 MeV Electron Fluence for p/n AlGaAs Cells

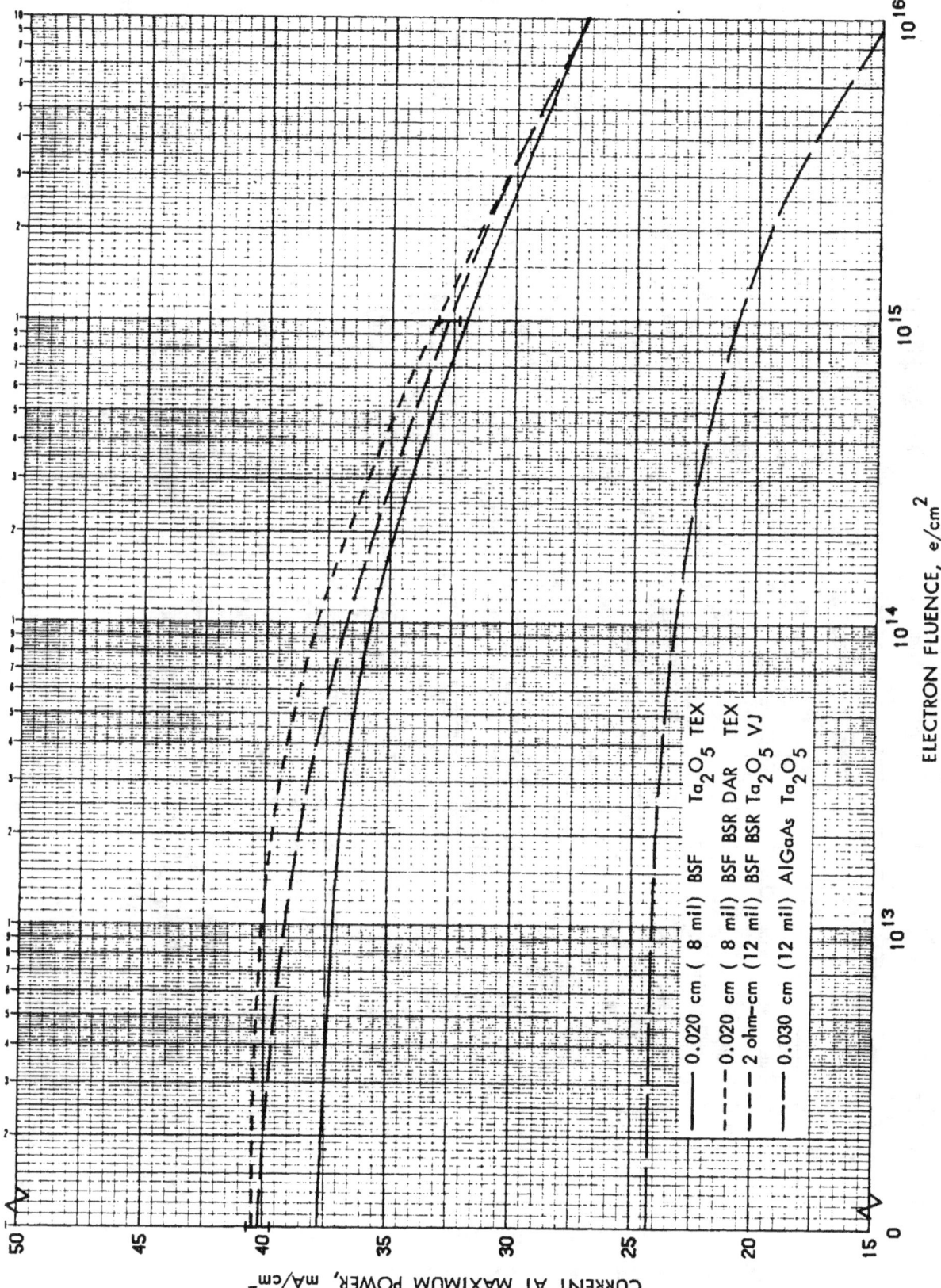

Figure 3.106 I_{mp} vs 1 MeV Electron Fluence for 10 Ohm-cm n/p Textured, 2 Ohm-cm Vertical Junction, and p/n AlGaAs Cells

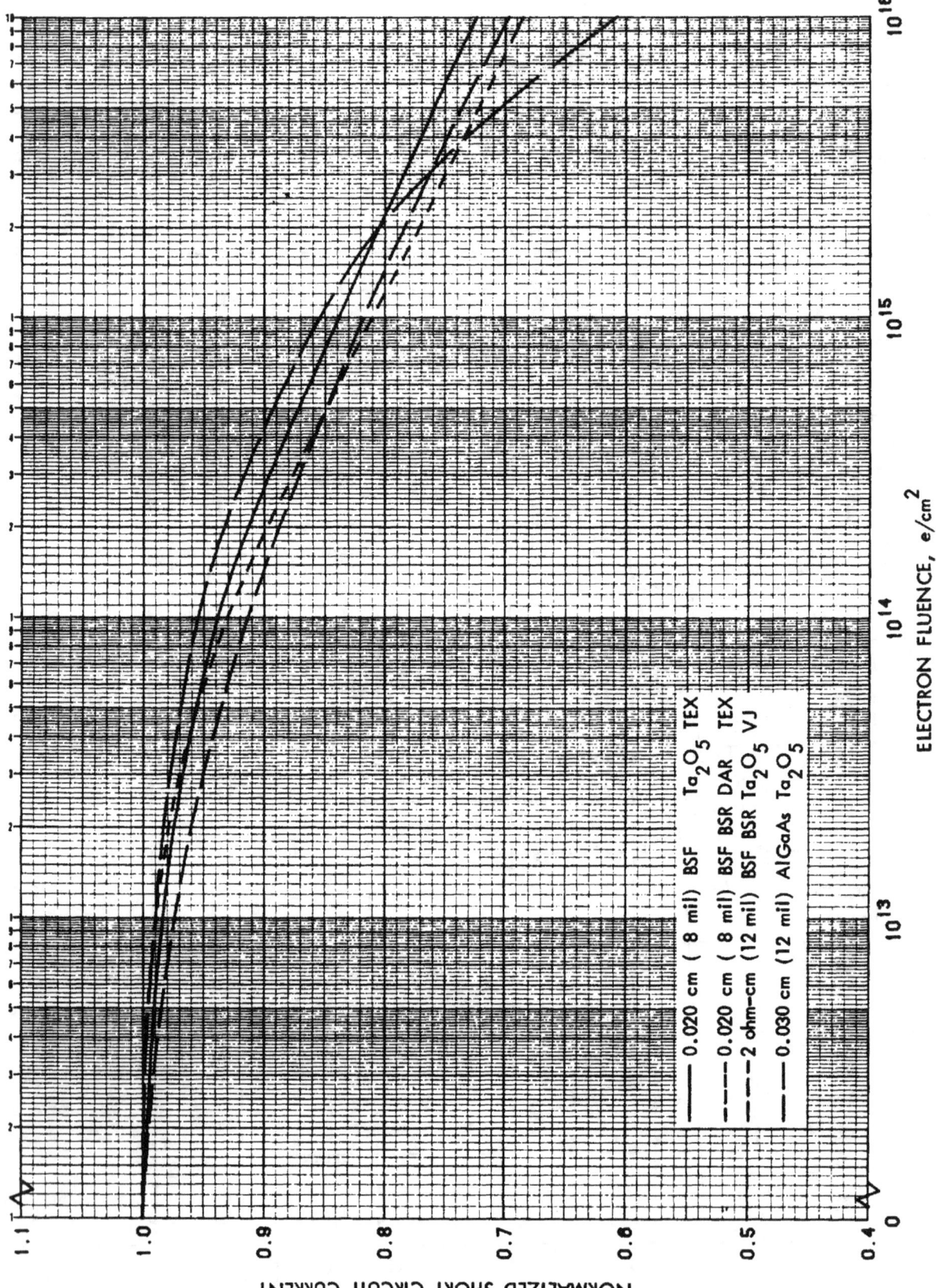

Figure 3.107 Normalized I_{SC} vs 1 MeV Electron Fluence for 10 Ohm-cm n/p Textured, 2 Ohm-cm Vertical Junction, and p/n AlGaAs Cells

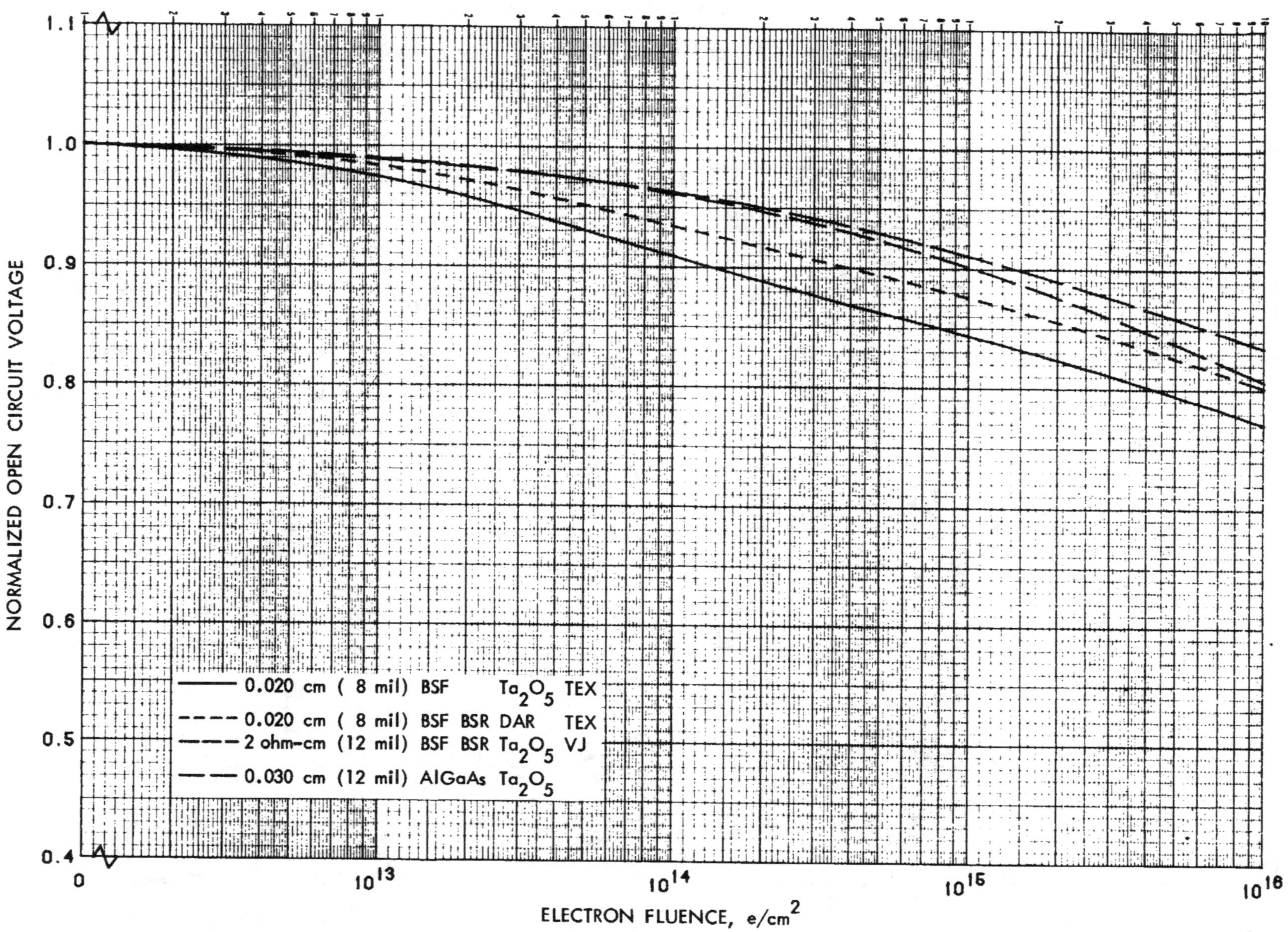

Figure 3.108 Normalized V_{oc} vs 1 MeV Electron Fluence for 10 Ohm-cm n/p Textured, 2 Ohm-cm Vertical Junction, and p/n AlGaAs Cells

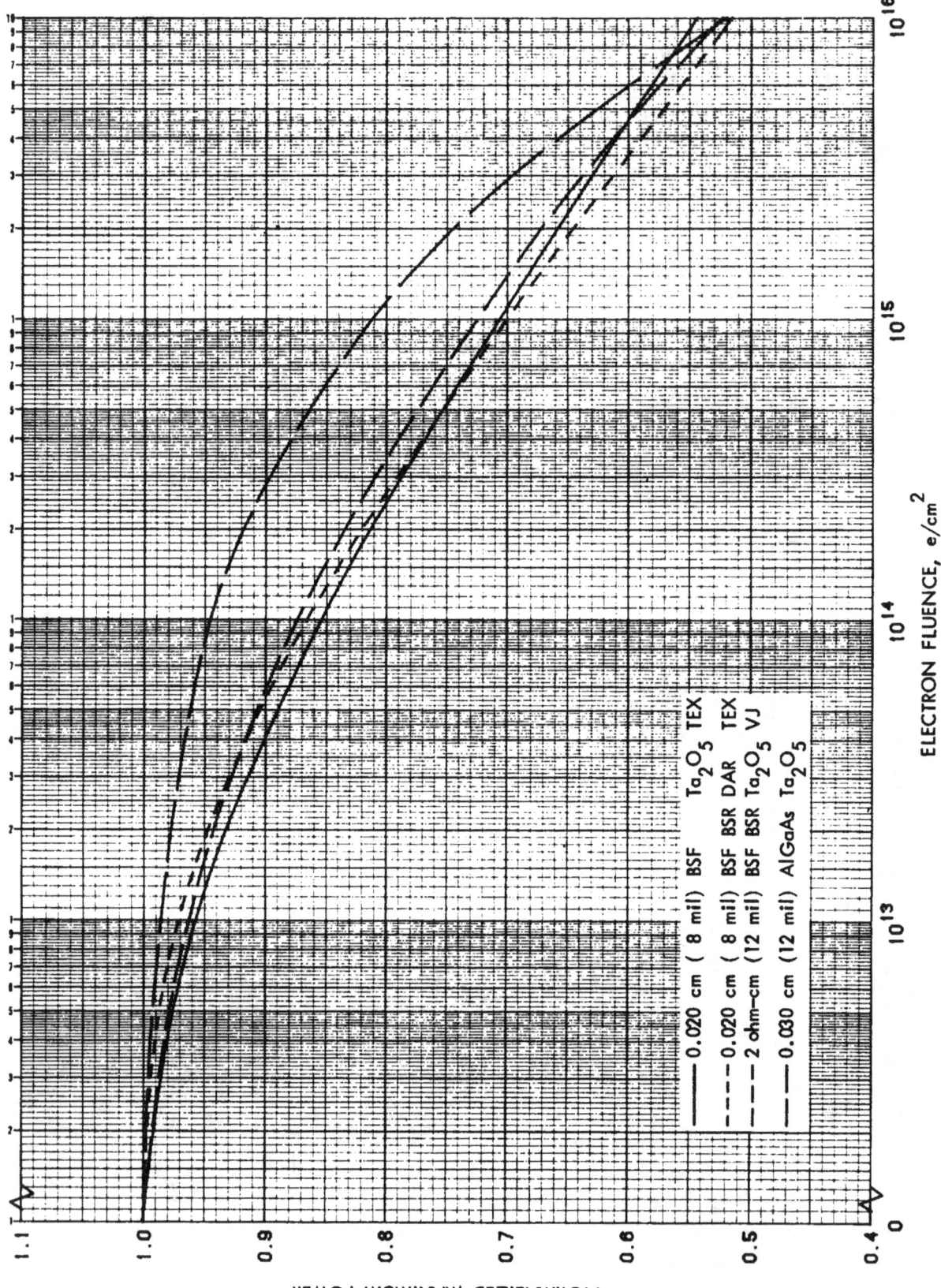

Figure 3.109 Normalized P_{max} vs 1 MeV Electron Fluence for 10 Ohm-cm n/p Textured, 2 Ohm-cm Vertical Junction, and p/n AlGaAs Cells

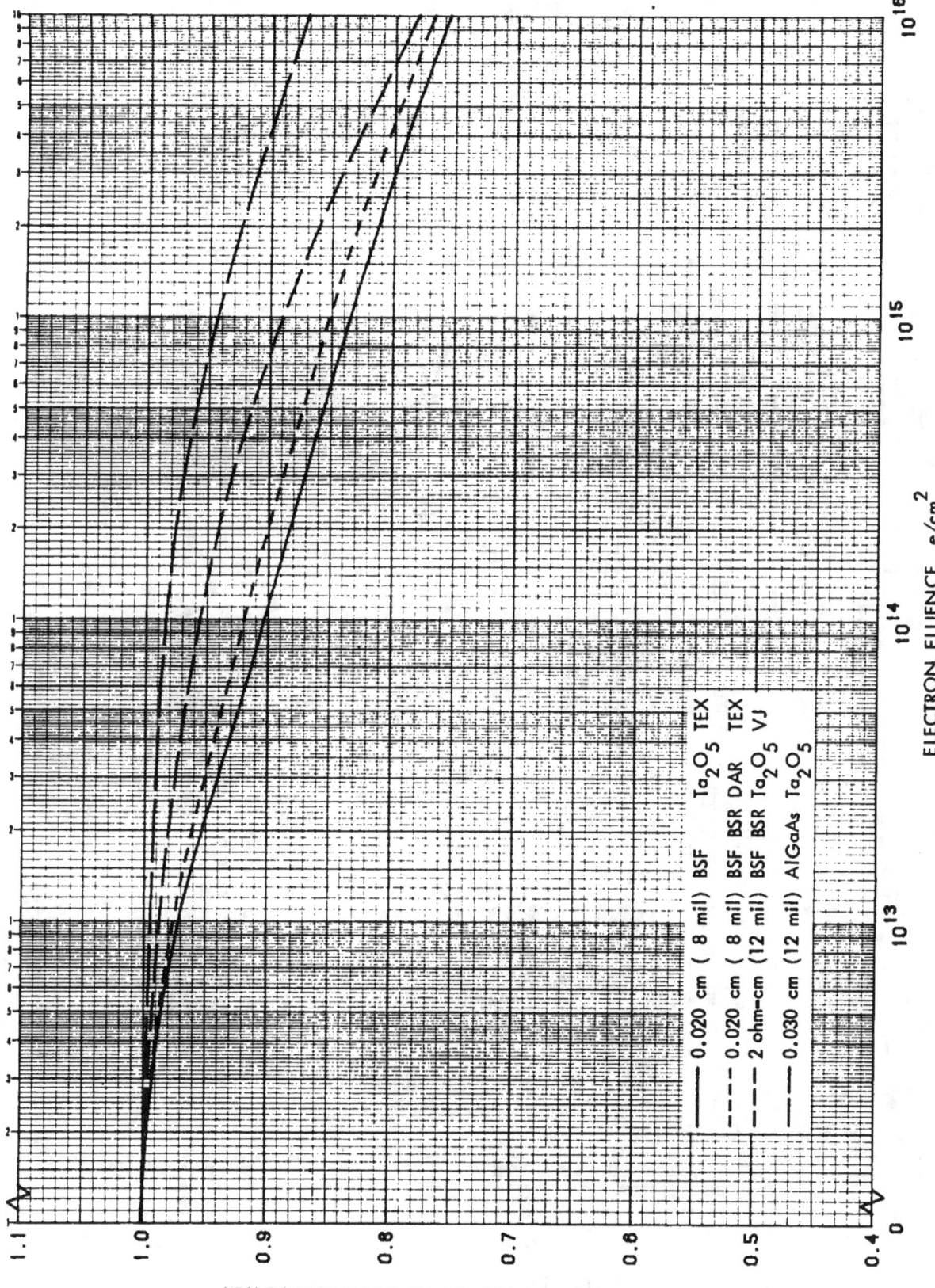

Figure 3.110 Normalized V_{mp} vs 1 MeV Electron Fluence for 10 Ohm-cm n/p Textured, 2 Ohm-cm Vertical Junction, and p/n AlGaAs Cells

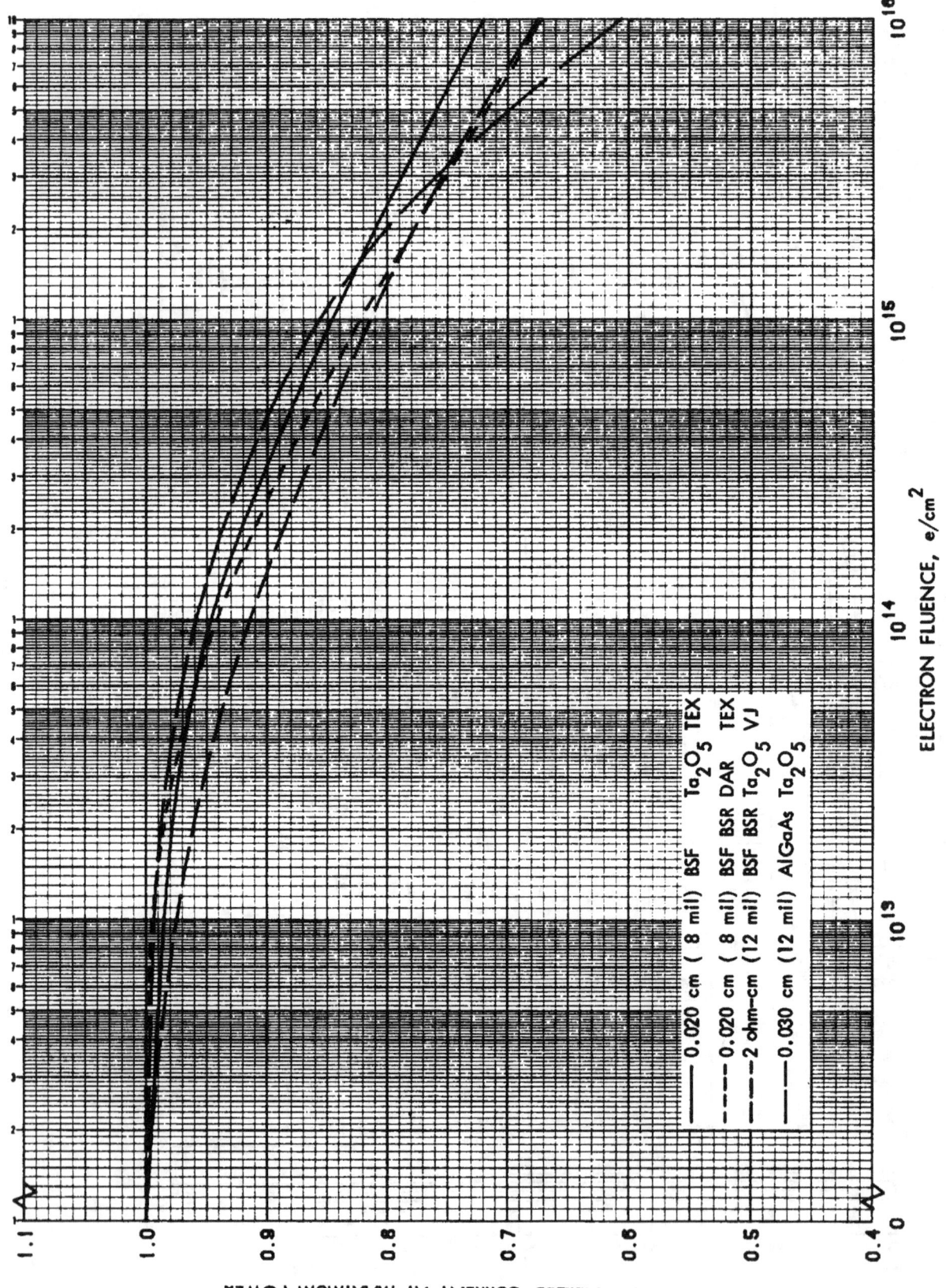

Figure 3.111 Normalized I_{mp} vs 1 MeV Electron Fluence for 10 Ohm-cm n/p Textured, 2 Ohm-cm Vertical Junction, and p/n AlGaAs Cells

REFERENCES

3.1 F. Seitz and J. S. Koehler, "Displacement of Atoms During Irradiation," Solid State Physics, 2, p.305, Academic Press, 1956.

3.2 J. J. Loferski and P. Rappaport, "Radiation Damage in Ge and Si Detected by Carrier Lifetime Changes - Damage Thresholds," Phys. Rev., 111, 2, 432, 1958.

3.3 H. Flicker, J. J. Loferski and J. Scott-Monck, "Radiation Defect Introduction Rates in n- and p- Type Silicon in the Vicinity of the Radiation Damage Threshold," Phys. Rev., 128, 6, 2557, 1962.

3.4 N. N. Gersimenko, et al., "Threshold Energy for the Formation of Radiation Defects in Semiconductors," Soviet Physics - Semiconductors, 5, 8, 1439, 1972.

3.5 G. W. Kinchin and R. S. Pease, "The Displacement of Atoms in Solids by Radiation," Report Prog. Phys. 18, 1, 1955.

3.6 G. Benski, "Paramagnetic Resonance in Electron Irradiated Silicon," J. Appl. Phys., 30, 8, 1195, 1959.

3.7 G. D. Watkins, J. W. Corbett, and R. M. Walker, "Spin Resonance in Electron Irradiated Silicon," J. Appl. Phys., 30, 8, 1198, 1959.

3.8 G. D. Watkins and J. W. Corbett, "Defects in Irradiated Silicon. I. Electron Spin Resonance of the Si-A Center," Phys. Rev., 121, 4, 1001, 1961.

3.9 J. W. Corbett, et al., "Defects in Irradiated Silicon. II. Infrared Absorption of the Si-A Center," Phys. Rev., 121, 4, 1015, 1961.

3.10 G. D. Watkins and J. W. Corbett, "Defects in Irradiated Silicon: Electron Paramagnetic Resonance and Electron-Nuclear Double Resonance of the Si-E Center," Phys. Rev., 134, 5A, A1359, 1964.

3.11 G. K. Wertheim, "Energy Levels in Electron-Bombarded Silicon," Phys. Rev., 105, 1730, 1957.

3.12 G. K. Wertheim, "Electron-Bombardment Damage in Silicon," Phys. Rev., 110, 1272, 1958.

3.13 H. Saito and M. Hirata, "Nature of Radiation Defects in Silicon Single Crystals," Japanese J. of Appl. Phys., 2, 11, 678, 1963.

3.14 J. R. Carter, Jr., "Effect of Electron Energy on Defect Introduction in Silicon," J. Phys. Chem. Solids, 27, 913, 1966.

3.15　M. Hirata, et al., "Effects of Impurities on the Annealing Behavior of Irradiated Silicon," J. Appl. Phys., 30, 6, 2433, 1967.

3.16　H. Flicker and W. R. Patterson, III, "Theoretical Calculation of the Direct Production of Divacancies in Silicon," J. Appl. Phys., 37, 13, 4998, 1966.

3.17　S. M. Gorodetskii and L. B. Kreinin, "A Possible Mechanism for the Formation of Complex Radiation Defects in Silicon Caused by Electron Bombardment," Soviet Physics-Doklady, 13, 7, 660, 1969.

3.18　G. Benski, "A New Paramagnetic Center in Electron Irradiated Silicon," J. Phys. Chem. Solids, 24, 1, 1963.

3.19　G. D. Watkins and J. W. Corbett, "Defects in Irradiated Silicon: Electron Paramagnetic Resonance of the Divacancy," Phys. Rev., 138, 2A, A543, 1965.

3.20　N. Almeleh and B. Goldstein, "Electron Paramagnetic Resonance and Electrical Properties in Electron-Irradiated p-Type Silicon," Phys. Rev., 149, 687, 1966.

3.21　J. A. Baicker, "Recombination and Trapping in Normal and Electron-Irradiated Silicon," Phys. Rev., 129, 6, 2454, 1963.

3.22　S. M. Gorodetskii, et al., "Recombination in p-Type Silicon Irradiated with Fast Electrons," Soviet Physics-Semiconductors, 5, 7, 1280, 1972.

3.23　R. G. Downing, J. R. Carter, Jr. and J. M. Denney, "The Energy Dependence of Electron Damage in Silicon," Proc. of the 4th Photovoltaic Specialists Conf., Vol. I, A-5-1, 1964.

3.24　J. A. Baicker, H. Flicker and J. Vilms, "Proton Induced Lattice Displacement in Silicon," Appl. Phys. Letters, 2, 5, 104, 1963.

3.25　G. W. Simon, J. M. Denney, and R. G. Downing, "Energy Dependence of Proton Damage in Silicon," Phys. Rev., 129, 6, 2454, 1963.

3.26　Yu. V. Bulgakov and M. A. Kumakhov, "Spatial Distribution of Radiation Defects in Materials Irradiated with Beams of Mono-energetic Particles," Soviet Physics-Semiconductors, 2, 11, 1334, 1969.

3.27　J. R. Bilinski, et al., "Proton-Neutron Damage Equivalence in Si and Ge Semiconductors,". IEEE Trans. on Nuc. Sci., NS-10, 5, 71, 1963.

3.28　Y. Gervais de Lafond, "Interactions Proton-Silicium Et Proton-Germanium Entre 1 Et 3000 MeV," Thesis, University of Toulouse, 12 May 1969.

3.29 J. R. Carter, Jr., "Study of Energy Levels in High Energy Proton Damaged Silicon," IEEE Trans. on Nuc. Sci., NS-11, 1, 290, 1964.

3.30 R. Breckenridge, "Proton-Produced Defects in n-Type Silicon," NASA TN D-4830, 1968.

3.31 D. Bielle-Daspet, "Effects De Protons De Grande Energie Sur Les Proprietes Electriques Du Silicium Et Du Germanium," Thesis, University of Toulouse, 2 March 1970.

3.32 B. R. Gossick, "Disordered Regions in Semiconductors Bombarded by Fast Neutrons," J. Appl. Phys., 30, 8, 1214, 1959.

3.33 B. L. Gregory, "Minority Carrier Recombination in Neutron Irradiated Silicon," IEEE Trans. on Nuc. Sci., NS-16, 6, 53, 1969.

3.34 R. E. Leadon, "Model for Short-Term Annealing of Neutron Damage in p-Type Silicon," IEEE Trans. on Nuc. Sci., NS-17, 6, 110, 1970.

3.35 R. R. Holmes, "Carrier Removal in Neutron Irradiated Silicon," IEEE Trans. on Nuc. Sci., NS-17, 6, 137, 1970.

3.36 L. C. Kimerling, "Defect States in Electron-Bombarded Silicon: Capacitance Transient Analysis," International Conf. on Radiation Effects in Semiconductors, Dubrovnik, Yugoslavia, 1976, N. B. Urli and J. W. Corbett, Eds., Institute of Physics (London) Conf. Series, 31, 221, 1977.

3.37 I. Weinnberg and C. K. Swartz, "Annealing of Radiation Damage in Low Resistivity Silicon Solar Cells," Proc. Space Photovoltaic Research and Technology 1980 Conf., NASA Conf. Pub. 2169, 181, 1980.

3.38 P. J. Drevinsky and H. M. DeAngelis, "Defect Behavior in Electron-Irradiated Boron- and Gallium-Doped Silicon," Proc. Space Photovoltaic and Technology Conf., 1982, To be published.

3.39 P. M. Mooney, L. J. Cheng, M. Suli, J. D. Gerson and J. W. Corbett, "Defect Energy Levels in Boron-Doped Silicon Irradiated with 1-Mev Electrons," Phys. Rev. B, 15, 8, 3836, 1977.

3.40 J. W. Corbett, L. J. Cheng, A. Jaworowski, J. P. Karins, Y. H. Lee, L. Lindstron, P. M. Mooney, G. Oehrlein, and K. L. Wang, "High-Energy Electron-Induced Damage Production at Room Temperature in Aluminum-Doped Silicon," Proc. Solar Cell High Efficiency and Radiation Damage 1979 Conf., NASA Conf. Pub. 2097, 185, 1979.

3.41 K. L. Brower, "EPR of a Jahn-Teller Distorted <111> Carbon Interstitialcy in Irradiated Silicon," Phys. Rev. B, 9, 2607, 1974.

3.42 K. L. Brower, "EPR of a <001> Si Interstitial Complex in Irradiated Silicon," Phys. Rev. B, 14, 872, 1976.

3.43 J. M. Denney, et al., "Effect of 1 MeV Electron Bombardment on Solar Cells," TRW Systems Report 8653-6018-KU-000, 11 Feb. 1963.

3.44 J. Bernard, et al., "Effects of Proton and Electron Radiation on French Silicon Solar Cells," Solar Cells, Gordon and Breach, London, p. 623, 1971.

3.45 H. Y. Tada, "A Theoretical Model for Low-Energy Proton Irradiated Silicon Solar Cells," Proc. of the 5th Photovoltaic Specialists Conf., Vol. II, D-8-1, 1966.

3.46 J. H. Martin, R. L. Statler and E. L. Ralph, "Radiation Damage to Thin Silicon Solar Cells," Advances in Energy Conversion Engineering, ASME 1967.

3.47 J. R. Carter, Jr., R. G. Downing and H. Flicker, "Effects of High Energy Electrons in Silicon and Silicon Solar Cells," TRW Report 4161-6023-R000, 25 May 1966.

3.48 J. R. Carter, Jr. and R. G. Downing, "Effects of Low Energy Protons and High Energy Electrons on Silicon," NASA CR-404, March 1966.

3.49 G. H. Haynes and W. E. Ellis, "Effects of 22 MeV Proton and 2.4 MeV Electron Radiation on Boron and Aluminum-Doped Silicon Solar Cells," NASA TND-4407, April 1968.

3.50 D. Lesbre, "Amelioration De La Resistance De Cellules Solaires Aux Rayonnements," Thesis, University of Toulouse, 12 May 1969.

3.51 W. R. Cherry and R. L. Statler, "Photovoltaic Properties of U.S. and European Silicon Cells Under 1-MeV Electron Irradiation," Goddard Space Flight Center, X-716-68-204, April 1968.

3.52 W. Luft and H. S. Rauschenbach, "Effects of Base Resistivity on the Characteristics of N-ON-P Silicon Solar Cells," Conf. Rec. of the 6th Photovoltaic Specialists Conf., Vol. III, 75, 1967.

3.53 E. L. Ralph, "Performance of Very Thin Silicon Solar Cells," Conf. Rec. of the 6th Photovoltaic Specialists Conf., Vol. III, 98, 1967.

3.54 R. L. Crabb, "Status Report on Thin Silicon Solar Cells For Large Flexible Arrays," Solar Cells, Gordon and Breach, p. 35, 1971.

3.55 W. Brown, I. Gabbe and W. Rosenzweig, "Results of the Telstar Radiation Experiments," Bell System Tech. J. XLII, 4, Part 2, 1505, July 1963.

3.56 J. J. Wysocki, "Radiation Studies on GaAs and Si Devices," IEEE Trans. on Nuc. Sci., NS-10, 5, 60 1963.

3.57 S. M. Gorodetskii, et al., "Influence of Electron Bombardment on Some Parameters of Silicon Photocells," Soviet Physics-Semiconductors, 2, 1, 90, 1968.

3.58 W. Rosenzweig, F. M. Smits and W. L. Brown, "Energy Dependence of Proton Irradiation Damage in Silicon," J. Appl. Phys., 35, 9, 2707, 1964.

3.59 J. M. Denney and R. G. Downing, "Proton Radiation Damage in Silicon Solar Cells," TRW Systems Report 8653-6026-KV-000, 16 July 1963.

3.60 B. E. Anspaugh and J. R. Carter, Jr., "Proton Irradiation of Conventional and Lithium Solar Cells: 11-37 MeV," Conf. Rec. of the 10th IEEE Photovoltaic Specialists Conf., 366, 1973.

3.61 B. E. Anspaugh and R. G. Downing, "Damage Coefficients and Thermal Annealing of Irradiated Silicon and GaAs Solar Cells," Conf. Rec. of the 15th IEEE Photovoltaic Specialists Conf., 499, 1981.

3.62 A. Meulenberg, Jr., and F. C. Treble, "Damage in Silicon Solar Cells from 2 to 155 MeV Protons," Conf. Rec. of the 10th IEEE Photovoltaic Specialists Conf., 359, 1973.

3.63 D. Reynard, "Proton and Electron Irradiation of N/P Silicon Solar Cells," Lockheed Missile and Space Co. Report 3-56-65-4, 12 April 1965.

3.64 J. J. Wysocki, et al., "Low-Energy Proton Bombardment of GaAs and Si Solar Cells," IEEE Trans. on Electronic Devices, ED-13, 4, 420, 1966.

3.65 E. A. Lodi, et al., "Low Energy Proton Degradation in Silicon Solar Cells," Proc. of the 4th Photovoltaic Specialists Conf. Vol. I, A-4-1, 1964.

3.66 A. G. Stanley, "Comparison of Low-Energy Proton Damage in Ion-Implanted and Diffused Silicon Solar Cells," Proc. of IEEE, 59, 2, 271, 1971.

3.67 I. Nashiyama, et al., "Proton and Deuteron Irradiation Damage in Silicon Solar Cells," Japanese J. of Appl. Phys. 10, 11, 1564, 1971.

3.68 R. R. Brown, "Surface Effects in Silicon Solar Cells," IEEE Trans. on Nuc. Sci., NS-14, 6, 260, 1967.

3.69 E. Stofel and D. Joslin, "Low-Energy Proton Damage to Silicon Solar Cells," IEEE Trans. on Nuc. Sci., NS-17, 6, 250, 1970.

3.70 E. Stofel and D. Joslin, "Low-Energy Proton Irradiation of Solar Cell Back Contacts," Conf. Rec. of the 8th IEEE Photovoltaic Specialists Conf., 209, 1970.

3.71 G. J. Brucker, et al., "Low Energy Proton Damage in Partially Shielded Solar Cells," Proc. of IEEE, 54, 798, 1966.

3.72 W. D. Brown, "ATS Power Subsystem Radiation Effects Study," Hughes Aircraft Company, NAS 5-3823, SSD-80089R, 1968.

3.73 L. J. Goldhammer, "Solar Cell Radiation Flight Report," Hughes Aircraft Co., Report SSD-90329R, 30 May 1969.

3.74 R. L. Statler and D. J. Curtin, "Low Energy Proton Damage in Silicon Solar Cells," Conf. Rec. of the 7th Photovoltaic Specialists Conf., 177, 1968.

3.75 R. L. Statler and D. J. Curtin, "Radiation Damage in Silicon Solar Cells from Low Energy Protons," IEEE Trans. Electron Devices ED-18, 7, 412, 1971.

3.76 D. J. Curtin, "Testing of Solar Cells for Communication Satellites," Solar Cells, Gordon and Breach, London, 605, 1971.

3.77 H. H. Sander and B. L. Gregory, "Transient Annealing in Semiconductor Devices Following Pulsed Neutron Irradiation," IEEE Trans. on Nuc. Sci., NS-13, 6, 63, 1966.

3.78 G. J. Brucker and B. Markow, "Neutron Damage in Silicon Solar Cells," Conf. Rec. of the 6th Photovoltaic Specialists Conf., Vol. III, 53, 1967.

3.79 J. R. Carter, R. G. Downing, C M. McDonnel, J. D. O'Keefe and E. L. Ralph, "Lithium Doped Hardened Solar Cell Optimization," AFAPL-TR-70-35, May 1970.

3.80 W. M. Morris, "Effect of 14 MeV Neutrons on Silicon Solar Cells," Thesis, University of Utah, June 1970.

3.81 E. J. Stofel, T. B. Stewart and J. R. Ornelas, "Neutron Damage to Silicon Solar Cells," IEEE Trans. on Nuc. Sci., NS-16, 5, 97, 1969.

3.82 J. M. Hicks, "Solar Cell Neutron Damage Investigations," AFAPL-TR-70-24, May 1970.

3.83 E. C. Smith, "Theoretical and Experimental Determination of Neutron Energy Deposition in Silicon," IEEE Trans. on Nuc. Sci., NS-13, 6, 11, 1966.

3.84 P. H. Fang, "Gamma Irradiation of Lithium Silicon Solar Cells," Conf., Rec. of the 7th Photovoltaic Specialists Conf., 113, 1968.

3.85 J. L. Wirth and S. C. Rogers, "The Transient Response of Transistors and Diodes to Ionizing Radiation," IEEE Trans. on Nuc. Sci., NS-11, 5, 24, 1964.

3.86 J. M. Blinov, et al., "Photo-EMF of PN Junctions in a Strongly Excited Semiconductor," Soviet Physics-JETP Letters, 3, 234, 1966.

3.87 N. Holoyak, et al., "An Experimental Investigation of the Maximum Photo-EMF of a PN Junction," J. Appl. Phys., 38, 5422, 1967.

3.88 D. Girton, "P-N Diode Saturation Using Lasers," Proc. of IEEE, 51, 938, 1963.

3.89 V. S. Vavilov, *Radiation Damage in Semiconductors*, p 115, Academic Press, N.Y., 1964.

3.90 J. J. Wysocki, "Lithium-Doped Radiation-Resistant Silicon Solar Cells," IEEE Trans. on Nuc. Sci., NS-13, 6, 168, 1966.

3.91 T. J. Faith, "Damage and Recovery Characteristics of Lithium-Containing Solar Cells," IEEE Trans. on Nuc. Sci., NS-18, 6, 371, 1971.

3.92 T. J. Faith, "Radiation Characteristics and Mission Consideration in Lithium-Containing Solar Cells," IEEE Trans. on Nuc. Sci., NS-19, 6, 371, 1972.

3.93 J. R. Carter, Jr. and R. G. Downing, "Behavior of Lithium in Irradiated Solar Cells," Conf. Rec. of the 8th IEEE Photovoltaic Specialists Conf., 240, 1970.

3.94 P. A. Berman, "Summary of Results of JPL Lithium-Doped Solar Cell Development Program," Conf. Rec. of the 9th IEEE Photovoltaic Specialists Conf., 281, 1972.

3.95 J. Bruno, "Sunlight Checkout Test for SAS," Martin-Marietta Corp. Report No. MCR-71-320, March 1972.

3.96 J. D. Sandstrom, "A Method for Predicting Solar Cell Current-Voltage Curve Characteristics as a Function of Incident Solar Intensity and Cell Temperature," Conf. Rec. of the 6th Photovoltaic Specialists Conf., Vol. II, 199, 1967.

3.97 J. H. Martin, R. L. Statler and E. L. Ralph, "Radiation Damage to Thin Silicon Solar Cells," Intersociety Energy Conversion Engineering Conf., Miami Beach, Florida, August 13-17, 1967.

3.98 R. E. Patterson, R. K. Yasui and B. Anspaugh, "The Determination and Treatment of Temperature Coefficients of Silicon Solar Cells for Interplanetary Spacecraft Application," Conf. Proc., 7th Intersociety Energy Conversion Engineering Conf. 1972, San Diego, Calif., September 1972.

3.99 B. E. Anspaugh, T. F. Miyahira and R. S. Weiss, "Characterization of Solar Cells for Space Applications: Volume V. Electrical Characteristics of OCLI 225-Micron MLAR Wraparound Cells as a Function of Intensity, Temperature, and Irradiation," JPL Publication 78-15, Vol. V., 1979.

3.100 B. E. Anspaugh, "The ATS-5 Solar Cell Experiment after 6-1/2 Years in Synchronous Orbit," Conf. Rec. of the 12th IEEE Photovoltaic Specialists Conf., 191, 1976.

3.101 D. J. Curtin and R. W. Cool, "Qualification Testing of Laboratory Produced Violet Solar Cells," Conf. Rec. of the 10th IEEE Photovoltaic Specialists Conf., 139, 1973.

3.102 A. Meulenberg and F. C. Treble, "Damage in Silicon Solar Cells from 2 to 155 MeV Protons," Conf. Rec. of the 10th IEEE Photovoltaic Specialists Conf., 359, 1973.

3.103 W. Luft, "Effects of Electron Irradiation on N on P Silicon Solar Cells," Advanced Energy Conversion, $\underline{5}$, 21, 1965.

3.104 D. L. Reynard and D. G. Peterson, "Results of Real-Time Irradiation of Lithium P/N and Conventional N/P Silicon Solar Cells," Conf. Rec. of the 9th IEEE Photovoltaic Specialists Conf., 303, 1972.

3.105 M. C. Whiffen and E. B. Trent, "The Effects of Radiation on Lithium Doped Solar Cells," Lockheed-Georgia Co. Report ER-11150, July 1971.

3.106 R. L. Crabb, "Photon Induced Degradation of Electron Irradiated Silicon Solar Cells," Conf. Rec. of the 9th IEEE Photovoltaic Specialists Conf., 329, 1972.

3.107 R. L. Crabb, "Photon Induced Degradation of Electron and Proton Irradiated Silicon Solar Cells," Conf. Rec. of the 10th IEEE Photovoltaic Specialists Conf., 396, 1973.

3.108 H. Fischer and W. Pschunder, "Investigation of Photon and Thermal Induced Changes in Silicon Solar Cells," Conf. Rec. of the 10th IEEE Photovoltaic Specialists Conf., 404, 1973.

3.109 J. Bernard and R. L. Crabb, "Simultaneous Electron-Photon Irradiation of Crucible Grown and Float-Zone Silicon Solar Cells," Conf. Rec. of the 7th Intersociety Energy Conversion Engineering Conf., 1972, 639, San Diego, California, 1972.

3.110 R. L. Crabb, "Photon Degradation in Silicon Solar Cells - Status Report, 1974," Conf. Rec. of International Photovoltaic Power Generation Conf., 337, Hamburg, 1974.

3.111   W. P. Rahilly, J. Scott-Monck, B. Anspaugh and D. Locker, "Electron and Photon Degradation in Aluminum, Gallium and Boron Doped Float Zone Silicon Solar Cells," Conf. Rec. of the 12th IEEE Photovoltaic Specialists Conf., 276, 1976.

3.112   J. Bernard, S. Mottet and R. L. Crabb, "Photon Degradation of Electron and Proton Irradiated Silicon Solar Cells," Conf. Rec. of the 12th IEEE Photolvoltaic Specialists Conf., 262, 1976.

3.113   A. A. Dollery, M. W. Walkden and R. L. Crabb, "The Effect of Protons, Electrons, and Photons on the Performance of Some New Types of High Efficiency Solar Cells," Conf. Rec. of the 13th IEEE Photovoltaic Specialists Conf., 116, 1978.

3.114   J. M. Denney, et al., "Final Flight Report Tetrahedral Research Satellites 1963-14B, 1963-14C, 1963-30B, Vol. II, Experiments and Results," TRW Systems Report 8685-6006-RU-000, 15 Feb 1964.

3.115   R. L. Fleischer, P. B. Price and R. M. Walker, "Nuclear Tracks in Solids," Scientific American, 220, 6, 30, 1969.

3.116   F. J. Campbell, "Effects of Space Radiation on Solar Cell Cover Materials," Transcript of the Photovoltaic Specialists Conf., Vol. II, D-2, 1963.

3.117   E. E. Luedke, "Charged Particle Exposure of Microsheet Second Surface Mirrors," TRW Systems IOC 8526.16-73-37, 1973.

3.118   G. A. Haynes, "Effect of Radiation on Cerium-Doped Solar-Cell Cover Glass," NASA TN D-6024, 1970.

3.119   R. L. Crabb, "Evaluation of Cerium Stabilized Microsheet Coverslips for Higher Solar Cell Outputs," Conf. Rec of the 9th IEEE Photovoltaic Specialists Conf., 185, 1972.

3.120   S. P. Faile, W. R. Harding and A. E. Wallis, "Hydrogen Impregnated Glass Covers for Hardened Solar Cells," Conf. Rec. of the 8th IEEE Photovoltaic Specialists Conf., 88, 1970.

3.121   A. C. Wilbur and D. L. Anderson, "An Exploratory Study of the Interplanetary Environmental Effects on Solar Cell Cover Glasses," Conf. Rec. of the 7th Photovoltaic Specialists Conf., 184, 1968.

3.122   E. E. Luedke, "Low Energy Proton Effects on MgF_2," TRW Systems IOC 68-3346.11-43, 26 September 1968.

3.123   E. E. Luedke, "Degradation of Antireflective Coating (MgF_2)," TRW Systems IOC 68-3346.11-47, 8 October 1968.

3.124   J. G. Haynos, "Investigation of Resinous Materials for Use as Solar Cell Adhesives," NASA TMX 55333, 1965.

3.125　F. J. Campbell and R. J. Lambert, "Effects of Shielding on Electron Damage to Solar Cells," Proc. of the 4th Photovoltaic Specialists Conf., Vol. I, A-9, 1964.

3.126　F. J. Campbell, "Status of Solar Cell Cover Material Radiation Damage," Proc. of the 5th Photovoltaic Specialists Conf., Vol. II, D-2.1, 1966.

3.127　J. L. Patterson and G. A. Haynes, "Effects of High-Energy Electron Radiation on Solar Cell Shields," Transcript of the Photovoltaic Specialists Conf., Vol. II, D-2, 1963.

3.128　G. A. Haynes, "High-Energy Radiation and Solar Cell Shields," Proc. of the 4th Photovoltaic Specialists Conf., Vol. I, A-10, 1964.

3.129　G. A. Haynes and W. E. Miller, "Effects of 1.2 and 0.3 MeV Electrons on the Optical Transmission Properties of Several Transparent Materials," NASA TN D-2620, 1965.

3.130　D. L. Reynard, "Irradiation of Solar Cell Cover Slides and Adhesives with 1.5 MeV Electrons," Lockheed Aircraft Company Report LMSC 3-56-64-5, August 1964.

3.131　L. B. Fogdall and S. S. Cannaday, "Space Radiation Effect of a Simulator Venus-Mercury Flyby on Solar Absorbance and Transmittance Properties of Solar Cells, Cover Glasses, Adhesives and Kapton Film," AIAA Paper No. 71-452, April 1971.

3.132　J. P. Millard, "Results from the Thermal Control Coating Experiment on OSO-III," *Thermal Design Principle of Spacecraft and Entry Bodies*, Academic Press, p 769, 1969.

3.133　E. E. Luedke, private communication.

3.134　H. S. Rauschenbach, *Solar Cell Array Design Handbook*, JPL SP 43-38, October 1976.

3.135　Anon, "Spraylon Fluorocarbon Encapsulation for Silicon Solar Cell Arrays," Lockheed Report No. LMSC-D558143 (JPL Contract 954410 Final Report), 1977.

3.136　D. A. Russell, "Testing of Solar Cell Covers and Encapsulants Conducted in a Simulated Space Environment," NASA CR-165475, D180-26590-1, 1981.

3.137.　J. A. Scott-Monck, "A Preliminary Evaluation of a Potential Space Worthy Encapsulant," To be published.

3.138.　T. J. Faith, "Temperature Dependence of Damage Coefficients in Electron Irradiated Solar Cells," IEEE Trans. on Nuc. Sci., NS-20, 6, 238, 1973.

CHAPTER 4

4.0 RELATIVE DAMAGE COEFFICIENTS FOR SPACE RADIATION

A large volume of experimental data is available for normal incidence irradiation of unshielded solar cells. These data are not directly applicable in the prediction of space radiation effects because of the omnidirectional nature of the space radiation and because of the energy degrading effects of coverglass shielding. In this section, the analytical methods of calculating the damage effectiveness of each component of the space radiation will be detailed. The damage effectiveness of space radiation is calculated relative to normal incidence 1 MeV electrons and 10 MeV protons on unshielded solar cells. This concept of the damage effectiveness or relative damage constant D is an extension of the previously discussed concept of equivalent fluence. It will allow the reduction of all components of the space radiation to an equivalent laboratory (normal incidence, monoenergetic) irradiation. In this way, laboratory data can be used to predict the behavior of shielded solar arrays in space. In addition, the similar problem of calculating energy deposition at various depths in shielding will be discussed.

4.1 Geometrical Aspects of Radiation Fluences

An omnidirectional flux is defined as the number of radiation particles of a particular type and energy which isotropically traverse a test sphere of unit cross-sectional area (radius = $1/\sqrt{\pi}$) per unit time. The commonly used sources of space radiation literature tabulate the environment in terms of omnidirectional fluxes with units of particles cm^{-2} day^{-1}. A commonly repeated derivation in the literature regarding the conversion of omnidirectional fluxes to unidirectional fluxes is as follows.[4.1] Assume a unit of plane area in space with an incident omnidirectional flux of particles.

Φ_n = the component of the omnidirectional flux which is normal to a surface

Φ_0 = the omnidirectional flux

4π = solid angle of test sphere (steradians)

θ = angle of radiation incidence (from normal)

$d\Omega$ = an increment of solid angle
     = $2\pi \sin\theta\, d\theta$ (for rotational symmetry)

$\cos\theta$ = projected area of unit plane area

$$\Phi_n = \frac{\Phi_0}{4\pi} \int_0^\pi \cos\theta\, d\Omega$$

$$= \frac{\Phi_0}{4\pi} \int_0^\pi 2\pi \sin\theta \cos\theta\, d\theta$$

$$= \frac{\Phi_0}{2}$$

(4.1.1)

The above derivation implies that the unidirectional fluence is equal in intensity or "equivalent" to the omnidirectional flux divided by 2. Likewise, if the unit plane area has infinite back shielding (i.e., integrate θ from 0 to $\pi/2$ only), one-fourth of the omnidirectional fluence is equal to the intensity of the unidirectional normally incident fluence. The above expression determines the normal component of an omnidirectional fluence, that is, the fluence which would pass through unit plane area. The conversion of an omnidirectional flux to an equivalent unidirectional flux must properly weight the damage effectiveness of all angular components.

The expression for the effectiveness or relative damage constant, weighted for all angular components of an omnidirectional monoenergetic flux and assuming infinite back shielding, is as follows:

$$D(E,t) = \frac{1}{4\pi} \int_0^{\pi/2} D(E_0,\theta)\, 2\pi \sin\theta \cos\theta\, d\theta \qquad (4.1.2)$$

where $D(E,t)$ = relative damage coefficient of omnidirectional radiation particles with energy E, relative to unidirectional 1 MeV electrons or 10 MeV protons for a cell protected by a coverglass of thickness t.

$D(E_0, \theta)$ = damage coefficient of unidirectional radiation particles with angle of incidence (θ) and energy (E_0) relative to unidirectional 1 MeV electrons or 10 MeV protons

t = shielding thickness; when $t = 0$, $E = E_0$

E_0 = proton energy as it enters the solar cell

The quantity $2\pi \sin\theta \, d\theta$ is an increment of solid angle as in equation (4.1.1). Equation (4.1.2) must be further modified to reflect the energy degradation in the coverglass shields used on silicon solar cells ($t \neq 0$).

4.2 Effect of Shielding on Radiation

A common solar cell configuration involves infinite back shielding and an optically transparent finite shield covering the front surface of the cell. The assumption of infinite back shielding is not always valid, and the differences in both shield thickness and material require separate treatments for front and back radiation. If an omnidirectional flux of radiation particles with energy E is incident on a solar cell shield of thickness t, the particles not stopped in the shielding will exit the shielding (i.e., enter the silicon) with an energy of E_0. The energy E_0 will be a strong function of the angle of incidence because of varying path length in the shield. The particle track length in the shield is equal to $t/\cos\theta$. By subtracting the particle track length in the shield ($t/\cos\theta$) from the range of the particle $R(E)$ in the shield material, one can determine the residual range $R(E_0)$ of a particle with energy E_0. Thus:

$$E_0(E, \theta, t) = R^{-1}\left[R(E) - \frac{t}{\cos\theta}\right] \qquad (4.2.1)$$

where R^{-1} is a convenient form used to represent an inverse function of the range-energy relation R. Proton and electron range-energy data suitable for this calculation have been tabulated by Janni [4.2] and Berger and Seltzer. [4.3, 4.4]

4.3 Electron Space Radiation Effects

The evaluation of $D(E,\theta)$ is necessary to complete the integration of equation (4.1.2). The data regarding the experimental evaluation of the relative damage coefficient for n/p silicon solar cells $D(E)$ for various electron energies at normal incidence is presented in Figure 4.1. Electrons in the MeV energy range penetrate silicon solar cells thoroughly enough that the damage produced by an electron can be considered uniform along its track. For this reason, the amount of displacement damage produced by a high energy electron is proportional to the total track length produced in the cell. The length of an individual electron track in a solar cell is proportional to $\sec\theta$ or $1/\cos\theta$, hence:

$$D(E_0,\theta) = \frac{D(E_0,0)}{\cos\theta} \quad (4.3.1)$$

The number of electrons intercepted by the cell is proportional to its projected area normal to the direction of the radiation (the $\cos\theta$ term in equation (4.1.2)). The net result of these two factors is a cancellation of the cos terms so that the damage induced in the solar cell is independent of θ. The fact that fast electron damage of unshielded silicon solar cells is independent of the angle of incidence was experimentally confirmed by Barrett.[4.5]

Equation (4.1.2) for the case of electron space radiation can be modified to the following expression:

$$D(E,t) = \frac{1}{4\pi} \int_0^{\pi/2} D(E_0,0) \, 2\pi \sin\theta \, d\theta \quad (4.3.2)$$

Equation (4.3.2) can be evaluated with the aid of equation (4.2.1) to evaluate E_0 and the data in Figure 4.1 to evaluate $D(E_0,0)$. The integration of equation (4.3.2) has been performed by machine and the results plotted in Figure 4.1. The results are also tabulated in Table 4.1. Because of electron straggling, there might be some question regarding the suitability of equation (4.2.1) to determine E_0; however, use of alternate Monte Carlo methods yielded results identical to those in Figure 4.1. Rosenzweig has

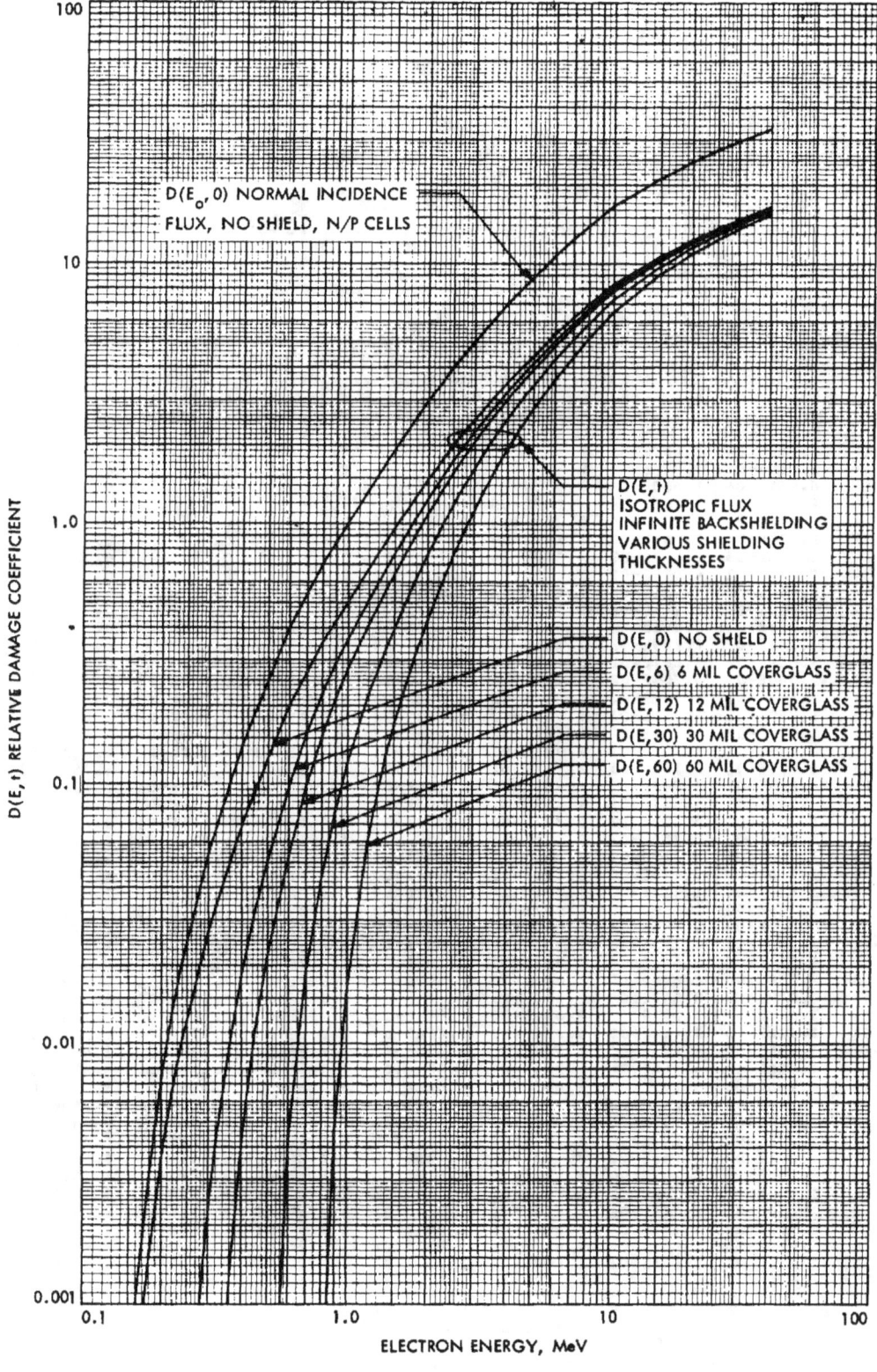

Figure 4.1 Relative Damage Coefficients for Space Electron Irradiation of Shielded n/p Silicon Solar Cells

Table 4.1. Electron Damage Coefficients

ELECTRON DAMAGE COEFFICIENT FOR JSC
OMNIDIRECTIONAL TO EQUIV. 1-MEV UNIDIRECTIONAL NORMAL ELECTRONS.

ENERGY (MEV)	(J)	0. (0.)	5.59E-03 (2.54E-3)	1.68E-02 (7.64E-3)	SHIELD THICKNESS, GM/CM2(CM) 3.35E-02 (1.52E-2)	6.71E-02 (3.05E-2)	1.12E-01 (5.09E-2)	1.68E-01 (7.64E-2)	3.35E-01 (1.52E-1)
.150	2.40E-14	2.690E-04	3.687E-05	0.	0.	0.	0.	0.	0.
.160	2.56E-14	5.000E-04	7.951E-05	0.	0.	0.	0.	0.	0.
.170	2.72E-14	8.951E-04	1.620E-04	0.	0.	0.	0.	0.	0.
.180	2.88E-14	1.550E-03	3.168E-04	2.227E-05	0.	0.	0.	0.	0.
.190	3.04E-14	2.406E-03	5.938E-04	5.228E-05	0.	0.	0.	0.	0.
.200	3.20E-14	3.650E-03	1.045E-03	1.143E-04	0.	0.	0.	0.	0.
.220	3.52E-14	6.750E-03	2.533E-03	4.375E-04	1.551E-05	0.	0.	0.	0.
.240	3.84E-14	1.035E-02	4.924E-03	1.263E-03	8.667E-05	0.	0.	0.	0.
.260	4.16E-14	1.450E-02	7.981E-03	2.814E-03	3.609E-04	0.	0.	0.	0.
.280	4.48E-14	2.010E-02	1.174E-02	5.052E-03	1.073E-03	0.	0.	0.	0.
.300	4.80E-14	2.725E-02	1.668E-02	7.941E-03	2.400E-03	2.828E-05	0.	0.	0.
.320	5.12E-14	3.385E-02	2.249E-02	1.156E-02	4.220E-03	1.481E-04	0.	0.	0.
.360	5.76E-14	5.004E-02	3.581E-02	2.142E-02	9.856E-03	1.314E-03	0.	0.	0.
.400	6.40E-14	7.000E-02	5.255E-02	3.423E-02	1.855E-02	4.311E-03	9.075E-05	0.	0.
.450	7.20E-14	9.506E-02	7.562E-02	5.344E-02	3.258E-02	1.106E-02	1.295E-03	0.	0.
.500	8.00E-14	1.250E-01	1.023E-01	7.595E-02	5.059E-02	2.146E-02	4.824E-03	7.759E-05	0.
.600	9.60E-14	2.000E-01	1.703E-01	1.343E-01	9.816E-02	5.347E-02	2.158E-02	4.315E-03	0.
.700	1.12E-13	2.700E-01	2.400E-01	2.004E-01	1.574E-01	9.769E-02	4.962E-02	1.802E-02	0.
.800	1.28E-13	3.500E-01	3.166E-01	2.718E-01	2.225E-01	1.527E-01	9.074E-02	4.262E-02	3.097E-04
.900	1.44E-13	4.225E-01	3.898E-01	3.438E-01	2.910E-01	2.121E-01	1.385E-01	7.726E-02	4.452E-03
1.000	1.60E-13	5.000E-01	4.657E-01	4.169E-01	3.607E-01	2.759E-01	1.934E-01	1.199E-01	1.566E-02
1.200	1.92E-13	6.700E-01	6.303E-01	5.733E-01	5.072E-01	4.068E-01	3.081E-01	2.172E-01	5.937E-02
1.400	2.24E-13	8.600E-01	8.160E-01	7.515E-01	6.759E-01	5.593E-01	4.419E-01	3.312E-01	1.281E-01
1.600	2.56E-13	1.060E+00	1.012E+00	9.405E-01	8.564E-01	7.256E-01	5.916E-01	4.614E-01	2.120E-01
1.800	2.88E-13	1.260E+00	1.210E+00	1.136E+00	1.045E+00	9.022E-01	7.521E-01	6.040E-01	3.099E-01
2.000	3.20E-13	1.470E+00	1.418E+00	1.339E+00	1.242E+00	1.088E+00	9.245E-01	7.611E-01	4.236E-01
2.250	3.60E-13	1.729E+00	1.676E+00	1.592E+00	1.489E+00	1.323E+00	1.145E+00	9.639E-01	5.793E-01
2.500	4.00E-13	2.000E+00	1.943E+00	1.854E+00	1.744E+00	1.566E+00	1.374E+00	1.178E+00	7.499E-01
2.750	4.40E-13	2.252E+00	2.197E+00	2.108E+00	1.997E+00	1.813E+00	1.611E+00	1.399E+00	9.314E-01
3.000	4.80E-13	2.510E+00	2.454E+00	2.362E+00	2.248E+00	2.057E+00	1.847E+00	1.627E+00	1.125E+00
3.250	5.20E-13	2.754E+00	2.698E+00	2.606E+00	2.490E+00	2.295E+00	2.078E+00	1.849E+00	1.320E+00
3.500	5.60E-13	3.000E+00	2.943E+00	2.850E+00	2.731E+00	2.531E+00	2.309E+00	2.072E+00	1.520E+00
3.750	6.00E-13	3.249E+00	3.191E+00	3.096E+00	2.974E+00	2.770E+00	2.541E+00	2.296E+00	1.723E+00
4.000	6.40E-13	3.500E+00	3.442E+00	3.344E+00	3.220E+00	3.011E+00	2.775E+00	2.523E+00	1.928E+00
4.500	7.20E-13	3.950E+00	3.894E+00	3.798E+00	3.675E+00	3.464E+00	3.223E+00	2.962E+00	2.332E+00
5.000	8.00E-13	4.400E+00	4.344E+00	4.247E+00	4.121E+00	3.905E+00	3.659E+00	3.390E+00	2.738E+00
5.500	8.80E-13	4.850E+00	4.793E+00	4.695E+00	4.566E+00	4.346E+00	4.093E+00	3.817E+00	3.141E+00
6.000	9.60E-13	5.300E+00	5.243E+00	5.143E+00	5.012E+00	4.787E+00	4.528E+00	4.244E+00	3.545E+00
7.000	1.12E-12	6.150E+00	6.093E+00	5.992E+00	5.859E+00	5.627E+00	5.358E+00	5.062E+00	4.326E+00
8.000	1.28E-12	6.900E+00	6.848E+00	6.753E+00	6.626E+00	6.401E+00	6.138E+00	5.844E+00	5.097E+00
9.000	1.44E-12	7.607E+00	7.555E+00	7.462E+00	7.335E+00	7.112E+00	6.849E+00	6.553E+00	5.801E+00
10.000	1.60E-12	8.300E+00	8.249E+00	8.156E+00	8.029E+00	7.804E+00	7.539E+00	7.241E+00	6.479E+00
15.000	2.40E-12	1.060E+01	1.056E+01	1.049E+01	1.039E+01	1.020E+01	9.981E+00	9.725E+00	9.047E+00
20.000	3.20E-12	1.230E+01	1.227E+01	1.221E+01	1.213E+01	1.197E+01	1.177E+01	1.155E+01	1.095E+01
25.000	4.00E-12	1.360E+01	1.357E+01	1.352E+01	1.344E+01	1.329E+01	1.311E+01	1.290E+01	1.233E+01
30.000	4.80E-12	1.470E+01	1.467E+01	1.462E+01	1.455E+01	1.442E+01	1.425E+01	1.405E+01	1.352E+01
40.000	6.40E-12	1.650E+01	1.648E+01	1.643E+01	1.637E+01	1.625E+01	1.610E+01	1.593E+01	1.544E+01

published similar space electron damage factor curves.[4.6] Barrett also published a similar analysis based on the diffusion length damage coefficient and empirically fitted analytical expressions to the data.[4.5]

The evaluation of ionization dose in solar array materials due to omnidirectional space electron fluences is analogous to that just completed for silicon solar cell degradation. In the case of absorbed dose, the energy deposited by the radiation in the shielding is determined in terms of rads or joules per kilogram. To evaluate this energy deposition at various depths in the shielding, an expression similar to equation (4.3.2) can be used. Equation (4.3.2) is modified to the extent that the electron stopping power together with the flux-to-dose conversion factor (equation 3.1.1)

$$1.6 \times 10^{-8} \left(\frac{1}{\rho} \frac{dE}{dx} \right)_{Collision}$$

replaces $D(E_0,\theta)$, and $D(E,t)$ becomes the absorbed dose per unit fluence. The results of this integration are shown in Figure 4.2 and in Table 4.2. Rosenzweig has published similar curves.[4.6]

The data of Figure 4.2 and Table 4.2 may be used to estimate the energy deposition in coverglasses and their subsequent darkening. The data must be used with caution and somewhat differently than the tabulated solar cell damage coefficient data. For example, an omnidirectional fluence of 0.5 MeV electrons incident on 0.152 cm (0.060 in) thick coverglass material shows no energy deposition at a depth of 0.152 cm. This does not mean, however, that there is no energy deposition in this thick coverglass. Rather, there is a relatively constant energy deposition to a depth of approximately 0.0764 cm (0.030 in) in the glass and it will be darkened fully as much as though it were only 0.0764 cm thick. Thus, for irradiation by monoenergetic electrons, one has a coverglass which is either totally exposed or exposed to some depth relatively uniformly, and a corresponding transmission loss can be easily determined from existing experimental data. In an actual space application, however, the data given in Table 4.2 has to be integrated with the expected electron fluence-energy spectrum to determine the actual

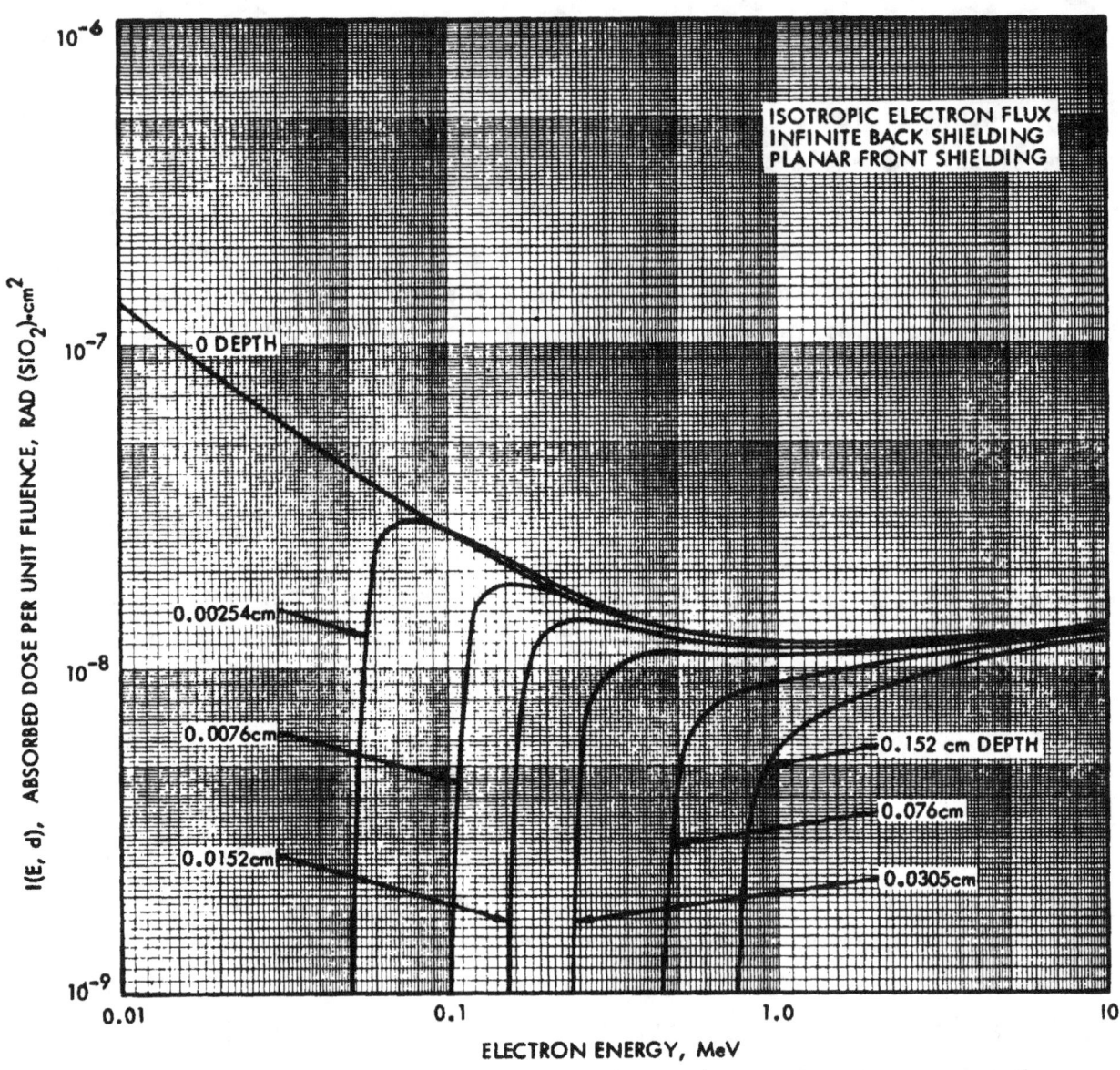

Figure 4.2. Absorbed Dose Per Unit Fluence of Space Electrons for Various Depths in Planar Fused Silica Shielding

Table 4.2. Electron Stopping Power, Rad(SiO_2)/Unit Omnidirectional Flux

ELECTRON STOPPING POWER
RAD(FUSED SILICA)/OMNIDIRECTIONAL FLUX(PARTICLES/CM2)

ENERGY (MEV)	(J)	0. (0.)	5.59E-03 (2.54E-3)	1.68E-02 (7.64E-3)	3.35E-02 (1.52E-2)	6.71E-02 (3.05E-2)	1.12E-01 (5.09E-2)	1.68E-01 (7.64E-2)	3.35E-01 (1.52E-1)
.050	8.00F-15	4.165E-08	0.	0.	0.	0.	0.	0.	0.
.055	8.80E-15	3.696E-08	2.106E-08	0.	0.	0.	0.	0.	0.
.060	9.60E-15	3.665E-08	2.651E-08	0.	0.	0.	0.	0.	0.
.065	1.04F-14	3.474E-08	2.872E-08	0.	0.	0.	0.	0.	0.
.070	1.12E-14	3.307E-08	2.952E-08	0.	0.	0.	0.	0.	0.
.080	1.28F-14	3.025E-08	2.929E-08	0.	0.	0.	0.	0.	0.
.090	1.44E-14	2.809E-08	2.817E-08	0.	0.	0.	0.	0.	0.
.100	1.60E-14	2.629E-08	2.684E-08	9.561E-09	0.	0.	0.	0.	0.
.120	1.92F-14	2.368E-08	2.438E-08	1.751E-08	0.	0.	0.	0.	0.
.140	2.24F-14	2.168E-08	2.238E-08	1.883E-08	0.	0.	0.	0.	0.
.160	2.56F-14	2.018E-08	2.077E-08	1.871E-08	9.739E-09	0.	0.	0.	0.
.180	2.88F-14	1.904E-08	1.949E-08	1.819E-08	1.286E-08	0.	0.	0.	0.
.200	3.20F-14	1.808E-08	1.846E-08	1.756E-08	1.402E-08	0.	0.	0.	0.
.225	3.60F-14	1.724E-08	1.745E-08	1.684E-08	1.447E-08	1.165E-09	0.	0.	0.
.250	4.00F-14	1.653E-08	1.668E-08	1.618E-08	1.447E-08	7.418E-09	0.	0.	0.
.275	4.40F-14	1.591E-08	1.603E-08	1.564E-08	1.431E-08	9.450E-09	0.	0.	0.
.300	4.80F-14	1.536E-08	1.547E-08	1.516E-08	1.409E-08	1.046E-08	0.	0.	0.
.350	5.60F-14	1.465E-08	1.466E-08	1.439E-08	1.365E-08	1.127E-08	5.940E-09	0.	0.
.400	6.40F-14	1.406E-08	1.405E-08	1.383E-08	1.324E-08	1.152E-08	8.133E-09	0.	0.
.450	7.20F-14	1.373E-08	1.365E-08	1.341E-08	1.291E-08	1.155E-08	9.063E-09	4.273E-09	0.
.500	8.00F-14	1.345E-08	1.336E-08	1.312E-08	1.266E-08	1.152E-08	9.559E-09	6.231E-09	0.
.550	8.80E-14	1.319E-08	1.312E-08	1.289E-08	1.247E-08	1.146E-08	9.844E-09	7.281E-09	0.
.600	9.60E-14	1.297E-08	1.290E-08	1.269E-08	1.232E-08	1.142E-08	1.002E-08	7.931E-09	0.
.700	1.12E-13	1.276E-08	1.265E-08	1.244E-08	1.209E-08	1.135E-08	1.023E-08	8.643E-09	1.094E-09
.800	1.28E-13	1.258E-08	1.248E-08	1.229E-08	1.197E-08	1.130E-08	1.035E-08	9.070E-09	4.222E-09
.900	1.44E-13	1.250E-08	1.240E-08	1.220E-08	1.190E-08	1.129E-08	1.045E-08	9.342E-09	5.568E-09
1.000	1.60E-13	1.244E-08	1.234E-08	1.216E-08	1.187E-08	1.131E-08	1.054E-08	9.559E-09	6.320E-09
1.200	1.92F-13	1.246E-08	1.236E-08	1.217E-08	1.190E-08	1.139E-08	1.072E-08	9.883E-09	7.305E-09
1.400	2.24F-13	1.248E-08	1.239E-08	1.222E-08	1.198E-08	1.152E-08	1.091E-08	1.017E-08	7.958E-09
1.600	2.56F-13	1.253E-08	1.244E-08	1.228E-08	1.206E-08	1.164E-08	1.109E-08	1.042E-08	8.446E-09
1.800	2.88E-13	1.262E-08	1.254E-08	1.238E-08	1.217E-08	1.177E-08	1.126E-08	1.064E-08	8.847E-09
2.000	3.20E-13	1.270E-08	1.262E-08	1.248E-08	1.228E-08	1.191E-08	1.143E-08	1.085E-08	9.190E-09
2.500	4.00F-13	1.294E-08	1.287E-08	1.274E-08	1.256E-08	1.223E-08	1.180E-08	1.130E-08	9.872E-09
3.000	4.80E-13	1.318E-08	1.312E-08	1.301E-08	1.284E-08	1.254E-08	1.216E-08	1.170E-08	1.042E-08
3.500	5.60E-13	1.344E-08	1.338E-08	1.327E-08	1.311E-08	1.283E-08	1.247E-08	1.205E-08	1.088E-08
4.000	6.40E-13	1.366E-08	1.360E-08	1.350E-08	1.336E-08	1.310E-08	1.277E-08	1.238E-08	1.129E-08
5.000	8.00E-13	1.414E-08	1.409E-08	1.399E-08	1.386E-08	1.362E-08	1.332E-08	1.296E-08	1.198E-08
6.000	9.60E-13	1.455E-08	1.450E-08	1.442E-08	1.430E-08	1.408E-08	1.381E-08	1.349E-08	1.259E-08
8.000	1.28E-12	1.535E-08	1.531E-08	1.524E-08	1.513E-08	1.493E-08	1.469E-08	1.440E-08	1.359E-08
10.000	1.60E-12	1.614E-08	1.610E-08	1.603E-08	1.592E-08	1.573E-08	1.550E-08	1.522E-08	1.447E-08
15.000	2.40E-12	1.817E-08	1.814E-08	1.806E-08	1.796E-08	1.777E-08	1.754E-08	1.727E-08	1.653E-08

dose-depth profile in the coverglass. For typical trapped electron spectra, this integration will produce a dose-depth profile in which the absorbed dose decreases monotonically through the thickness of the coverglass. If this profile shows that sufficient exposure over a significant depth has occurred, an average energy deposition over that depth may also be estimated. This value may then be used in conjunction with the curves in Figure 3.19 to estimate the transmission loss.

4.4 Proton Space Radiation Effects

For proton space radiation, the evaluation of equation (4.1.2) is more complex than that previously discussed for electrons. Two problems arise in the treatment of space protons with energies less than about 10 MeV, because of their limited penetration and increased damage production. One problem exists because the relative damage constants based on silicon solar cell I_{sc}, V_{oc}, and P_{max} are different and diverge at low proton energies. The second problem is that low energy proton damage has been experimentally characterized only for normal incidence irradiation, and basic considerations indicate that the damage is a strong function of the angle of incidence. The normal incidence proton coefficients for energies of 10 MeV and greater can be assumed to be independent of the angle of radiation incidence for the same reasons discussed for electron irradiation in the previous section.

The physical distribution of low energy proton damage was discussed in Section 3.7. The most significant aspect of the low energy proton damage is the fact most of the displacements are produced at the end of the proton track, as illustrated in Figure 3.11. The high damage concentration near the end of the proton track allows the construction of a simple damage model for the prediction of the effect of angle of incidence on low energy proton damage in silicon solar cells. It is assumed that the effect of a low energy proton, of arbitrary angle of incidence and energy, is roughly equal to that of a normally incident proton with a range equal to the perpendicular penetration of the non-normally incident proton. To partially correct the inaccuracies of this proposed model, a factor is employed which relates the ratio of the total displacements produced by the non-normally incident proton to those of a normally incident proton which would penetrate to the same depth in the cell. The total number of displacements may be computed using the Kinchin and Pease model as discussed in Section (3.1). The low energy proton relative damage coefficient given by the above model can be expressed as follows:

$$D(E_0,\theta) = D(E_n,0) \frac{N_{td}(E_0)}{N_{td}(E_n)} \qquad (4.4.1)$$

where $D(E_0,\theta)$ = relative damage coefficient for protons entering a silicon solar cell with energy E_0 at an angle θ

$D(E_n,0)$ = relative damage coefficient for a proton of normal incidence ($\theta = 0$) with energy E_n (range equal to $R(E_0) \cos\theta$)

$N_{td}(E_0)$ = the total number of silicon displacements created by a proton entering the silicon with energy E_0

$\cos\theta$ = unit cell projected area

$E_n = R^{-1}[R(E_0) \cos\theta]$

E_0 = proton energy as it emerges from the coverglass and enters the solar cell

When the range of a proton of energy E_0 incident on a solar cell at angle θ exceeds (cell thickness)/$\cos\theta$, the proton will penetrate the cell. This case is entirely analogous to the case previously discussed for high energy electrons so that:

$$D(E_0,\theta) = \frac{D(E_0,0)}{\cos\theta} \qquad (4.4.2)$$

Equations (4.4.1) and (4.4.2) allow the evaluation of equation (4.1.2) for infinite backshielding as follows:

$$D(E,t) = \frac{1}{4\pi} \int_0^{\theta_p} D(E_0,0) \, 2\pi \sin\theta \, d\theta$$

$$+ \frac{1}{4\pi} \int_{\theta_p}^{\pi/2} D(E_n,0) \frac{N_{td}(E_0)}{N_{td}(E_n)} \, 2\pi \sin\theta \cos\theta \, d\theta \qquad (4.4.3)$$

where θ_p = the angle of incidence for which a proton of energy E will just penetrate both the coverglass and the solar cell.

The first term in equation (4.4.3) represents the case when the proton completely penetrates the coverglass and the solar cell, while the second term applies when the proton penetrates the coverglass but stops in the cell. This integration has been done by machine using the $D(E_0,0)$ values shown in Figure 3.10. Separate integrations were done for $D(E_0,0)$ values based on I_{sc} and on V_{oc}, P_{max}. $D(E,t)$ values calculated by equation (4.4.3) are unfortunately a function of solar cell thickness. However, evaluation of (4.4.3) for cell thicknesses ranging from 0.0457 cm (0.018 in) to 0.005 cm (0.002 in) has shown that the dependence on cell thickness is very slight, and for practical purposes the results can be considered independent of cell thickness. The results of numerical integration of equation (4.4.3) for several coverglass thicknesses are shown in Figures 4.3 and 4.4. The same data are printed in tabular form in Tables 4.3 and 4.4.

The values of relative damage constants for omnidirectional fluences of protons on shielded solar cells allow a space proton environment to be reduced to an equivalent fluence of normally incident 10 MeV protons on unshielded silicon solar cells. Experimental studies of silicon solar cells have indicated that a fluence of normally incident 10 MeV protons produces damage that can be approximated by a fluence of 1 MeV electrons, which is 3000 times that of the 10 MeV proton fluence.

The evaluation of the absorbed dose in shielding materials due to space protons requires an analysis similar to that done for space electrons. For this evaluation an expression similar to equation (4.3.2) is used. The quantity $D(E_0,0)$ is replaced by the stopping power times the flux-to-dose conversion factor ($-1.6 \times 10^{-8} \frac{1}{\rho} \frac{dE}{dx}$) for protons of energy (E_0), and the quantity $D(E,t)$ becomes the absorbed dose per incident omnidirectional-flux proton of energy E at shielding depth t. The results of this integration for several shielding thicknesses of fused quartz are shown in Figure 4.5 and Table 4.5. The same cautions discussed in section 4.3 regarding use of electron dose calculations apply here also. Rosenzweig has published similar data. [4.6]

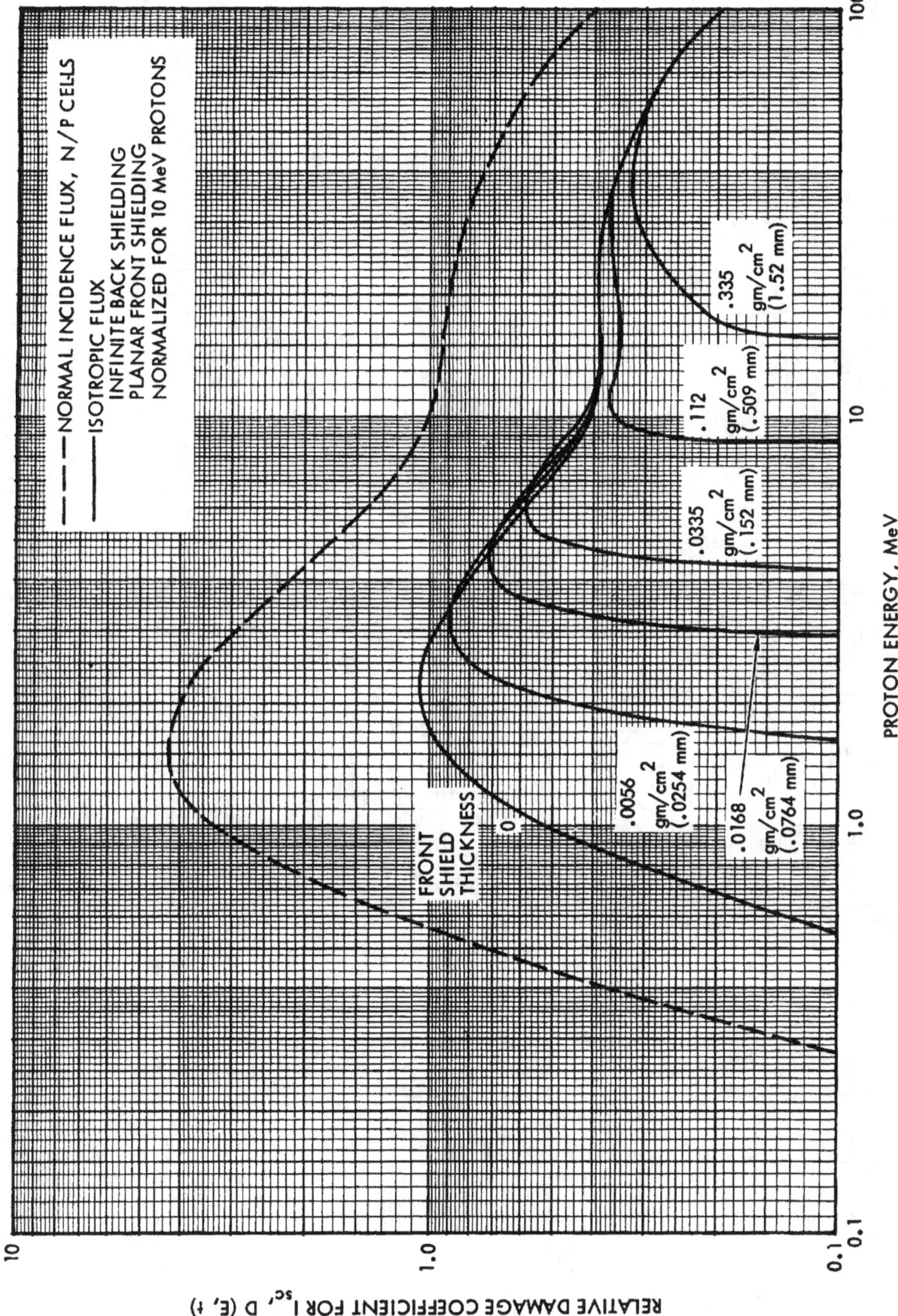

Figure 4.3 Relative Damage Coefficients for Space Proton Irradiation of Shielded n/p Silicon Solar Cells (Based on I_{sc})

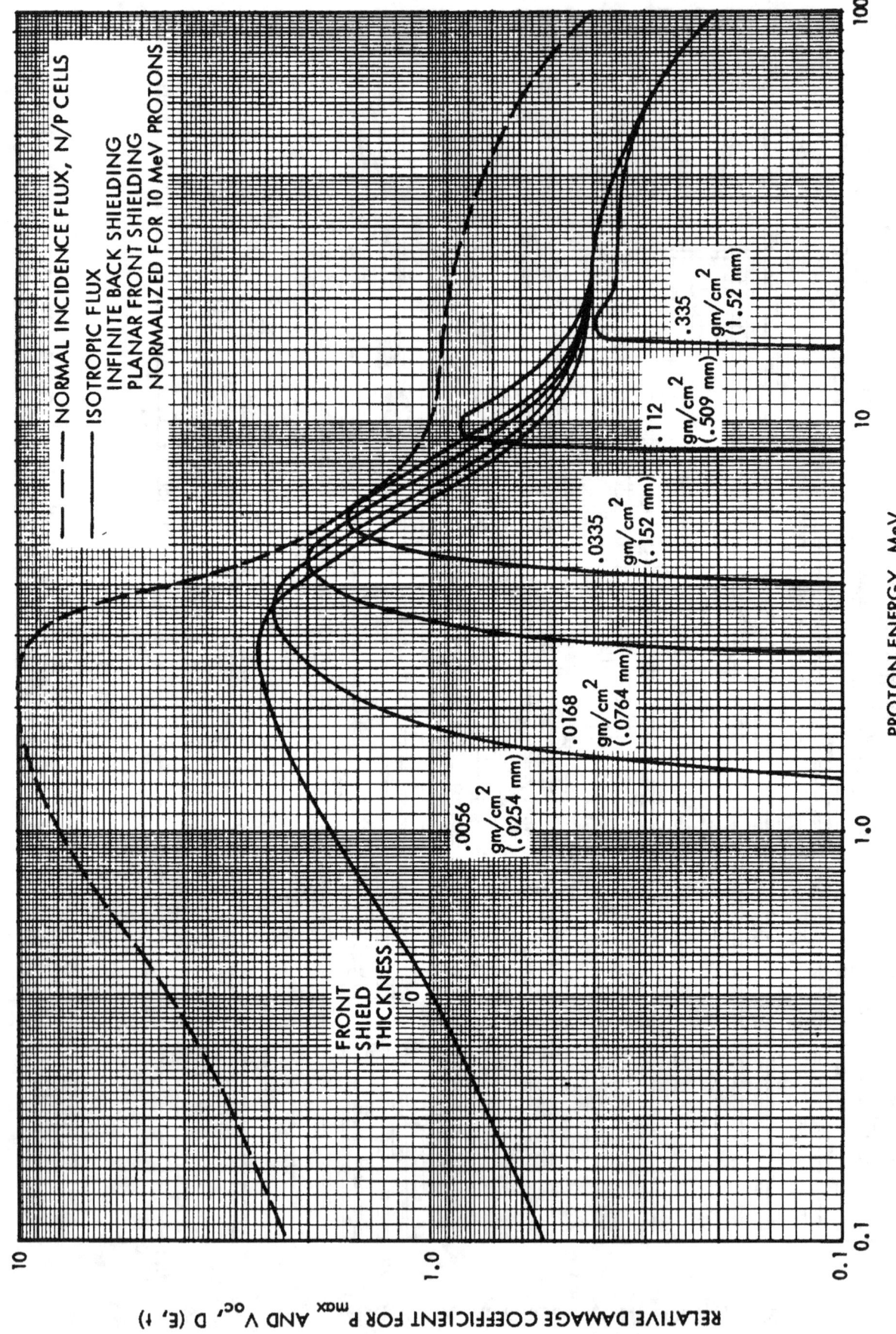

Figure 4.4 Relative Damage Coefficients for Space Proton Irradiation of Shielded n/p Silicon Solar Cells (Based on P_{max} or V_{oc})

Table 4.3. Proton Damage Coefficients for I_{sc}

PROTON DAMAGE COEFFICIENT FOR JSC
OMNIDIRECTIONAL TO EQUIV. 10-MEV UNIDIRECTIONAL NORMAL INCIDENT PROTON FLUX.

ENERGY (MEV)	(J)	\multicolumn{8}{c}{SHIELD THICKNESS, GM/CM2(CM)}
		0. (0.)	5.59E-03 (2.54E-3)	1.68E-02 (7.64E-3)	3.35E-02 (1.52E-2)	6.71E-02 (3.05E-2)	1.12E-01 (5.09E-2)	1.68E-01 (7.64E-2)	3.35E-01 (1.52E-1)
.100	1.60E-14	2.435E-04	0.	0.	0.	0.	0.	0.	0.
.200	3.20E-14	3.047E-03	0.	0.	0.	0.	0.	0.	0.
.300	4.80E-14	1.374E-02	0.	0.	0.	0.	0.	0.	0.
.400	6.40E-14	3.987E-02	0.	0.	0.	0.	0.	0.	0.
.600	9.60E-14	1.502E-01	0.	0.	0.	0.	0.	0.	0.
.800	1.28E-13	3.243E-01	0.	0.	0.	0.	0.	0.	0.
1.000	1.60E-13	5.216E-01	0.	0.	0.	0.	0.	0.	0.
1.200	1.92E-13	7.108E-01	0.	0.	0.	0.	0.	0.	0.
1.300	2.08E-13	7.890E-01	2.322E-05	0.	0.	0.	0.	0.	0.
1.400	2.24E-13	8.549E-01	3.750E-03	0.	0.	0.	0.	0.	0.
1.600	2.56E-13	9.532E-01	8.124E-02	0.	0.	0.	0.	0.	0.
1.800	2.88E-13	1.010E+00	2.525E-01	0.	0.	0.	0.	0.	0.
2.000	3.20E-13	1.039E+00	4.558E-01	0.	0.	0.	0.	0.	0.
2.200	3.52E-13	1.048E+00	6.233E-01	0.	0.	0.	0.	0.	0.
2.400	3.84E-13	1.041E+00	7.426E-01	0.	0.	0.	0.	0.	0.
2.600	4.16E-13	1.023E+00	8.207E-01	1.860E-05	0.	0.	0.	0.	0.
2.800	4.48E-13	9.962E-01	8.680E-01	3.925E-02	0.	0.	0.	0.	0.
3.000	4.80E-13	9.639E-01	8.912E-01	1.794E-01	0.	0.	0.	0.	0.
3.200	5.12E-13	9.286E-01	8.962E-01	3.465E-01	0.	0.	0.	0.	0.
3.400	5.44E-13	8.937E-01	8.871E-01	4.807E-01	0.	0.	0.	0.	0.
3.600	5.76E-13	8.598E-01	8.697E-01	5.787E-01	0.	0.	0.	0.	0.
3.800	6.08E-13	8.273E-01	8.481E-01	6.459E-01	0.	0.	0.	0.	0.
4.000	6.40E-13	7.963E-01	8.243E-01	6.879E-01	1.288E-03	0.	0.	0.	0.
4.200	6.72E-13	7.723E-01	7.989E-01	7.105E-01	7.227E-02	0.	0.	0.	0.
4.400	7.04E-13	7.486E-01	7.734E-01	7.189E-01	2.077E-01	0.	0.	0.	0.
4.600	7.36E-13	7.254E-01	7.499E-01	7.184E-01	3.274E-01	0.	0.	0.	0.
4.800	7.68E-13	7.029E-01	7.280E-01	7.120E-01	4.191E-01	0.	0.	0.	0.
5.200	8.32E-13	6.605E-01	6.866E-01	6.890E-01	5.286E-01	0.	0.	0.	0.
5.600	8.96E-13	6.216E-01	6.479E-01	6.613E-01	5.723E-01	0.	0.	0.	0.
6.000	9.60E-13	5.867E-01	6.119E-01	6.319E-01	5.839E-01	2.142E-03	0.	0.	0.
6.400	1.02E-12	5.585E-01	5.792E-01	6.019E-01	5.793E-01	1.742E-01	0.	0.	0.
6.800	1.09E-12	5.339E-01	5.520E-01	5.731E-01	5.664E-01	3.196E-01	0.	0.	0.
7.200	1.15E-12	5.128E-01	5.285E-01	5.477E-01	5.491E-01	3.945E-01	0.	0.	0.
7.600	1.22E-12	4.947E-01	5.086E-01	5.255E-01	5.299E-01	4.317E-01	0.	0.	0.
8.000	1.28E-12	4.786E-01	4.909E-01	5.058E-01	5.118E-01	4.484E-01	0.	0.	0.
9.000	1.44E-12	4.476E-01	4.565E-01	4.669E-01	4.724E-01	4.478E-01	2.735E-01	0.	0.
10.000	1.60E-12	4.337E-01	4.369E-01	4.401E-01	4.425E-01	4.292E-01	3.537E-01	0.	0.
11.000	1.76E-12	4.232E-01	4.245E-01	4.258E-01	4.226E-01	4.101E-01	3.675E-01	2.061E-01	0.
12.000	1.92E-12	4.196E-01	4.187E-01	4.155E-01	4.110E-01	3.956E-01	3.649E-01	2.839E-01	0.
13.000	2.08E-12	4.185E-01	4.167E-01	4.120E-01	4.040E-01	3.872E-01	3.588E-01	3.062E-01	0.
14.000	2.24E-12	4.181E-01	4.159E-01	4.105E-01	4.020E-01	3.828E-01	3.553E-01	3.131E-01	0.
15.000	2.40E-12	4.194E-01	4.173E-01	4.104E-01	4.010E-01	3.814E-01	3.538E-01	3.159E-01	0.
16.000	2.56E-12	4.214E-01	4.182E-01	4.120E-01	4.025E-01	3.819E-01	3.547E-01	3.187E-01	1.439E-01
18.000	2.88E-12	4.192E-01	4.179E-01	4.133E-01	4.054E-01	3.873E-01	3.606E-01	3.269E-01	2.175E-01
20.000	3.20E-12	4.172E-01	4.159E-01	4.125E-01	4.055E-01	3.900E-01	3.679E-01	3.379E-01	2.441E-01
22.000	3.52E-12	4.144E-01	4.117E-01	4.093E-01	4.047E-01	3.915E-01	3.731E-01	3.473E-01	2.648E-01
24.000	3.84E-12	4.094E-01	4.083E-01	4.059E-01	4.010E-01	3.919E-01	3.757E-01	3.547E-01	2.834E-01
26.000	4.16E-12	4.049E-01	4.039E-01	4.018E-01	3.985E-01	3.898E-01	3.769E-01	3.591E-01	2.984E-01
28.000	4.48E-12	4.000E-01	3.994E-01	3.978E-01	3.939E-01	3.875E-01	3.764E-01	3.613E-01	3.101E-01
30.000	4.80E-12	3.935E-01	3.930E-01	3.918E-01	3.896E-01	3.834E-01	3.753E-01	3.625E-01	3.186E-01
34.000	5.44E-12	3.784E-01	3.782E-01	3.777E-01	3.767E-01	3.739E-01	3.677E-01	3.600E-01	3.291E-01
38.000	6.08E-12	3.664E-01	3.662E-01	3.657E-01	3.650E-01	3.617E-01	3.582E-01	3.529E-01	3.312E-01
42.000	6.72E-12	3.532E-01	3.532E-01	3.532E-01	3.530E-01	3.519E-01	3.484E-01	3.446E-01	3.292E-01
46.000	7.36E-12	3.399E-01	3.399E-01	3.400E-01	3.400E-01	3.396E-01	3.372E-01	3.349E-01	3.245E-01
50.000	8.00E-12	3.272E-01	3.272E-01	3.272E-01	3.273E-01	3.271E-01	3.264E-01	3.250E-01	3.177E-01
55.000	8.80E-12	3.125E-01	3.126E-01	3.128E-01	3.130E-01	3.133E-01	3.132E-01	3.126E-01	3.082E-01
60.000	9.60E-12	2.988E-01	2.989E-01	2.990E-01	2.992E-01	2.995E-01	2.997E-01	2.995E-01	2.969E-01
65.000	1.04E-11	2.844E-01	2.846E-01	2.850E-01	2.855E-01	2.863E-01	2.871E-01	2.875E-01	2.869E-01
70.000	1.12E-11	2.710E-01	2.712E-01	2.715E-01	2.720E-01	2.728E-01	2.736E-01	2.743E-01	2.748E-01
80.000	1.28E-11	2.474E-01	2.476E-01	2.480E-01	2.485E-01	2.494E-01	2.504E-01	2.514E-01	2.531E-01
90.000	1.44E-11	2.245E-01	2.247E-01	2.251E-01	2.256E-01	2.266E-01	2.277E-01	2.289E-01	2.315E-01
100.000	1.60E-11	1.997E-01	1.999E-01	2.004E-01	2.010E-01	2.022E-01	2.037E-01	2.052E-01	2.089E-01
130.000	2.08E-11	1.492E-01	1.493E-01	1.496E-01	1.500E-01	1.509E-01	1.519E-01	1.530E-01	1.560E-01
160.000	2.56E-11	1.183E-01	1.183E-01	1.185E-01	1.188E-01	1.192E-01	1.199E-01	1.206E-01	1.226E-01
200.000	3.20E-11	9.215E-02	9.220E-02	9.229E-02	9.242E-02	9.268E-02	9.302E-02	9.344E-02	9.462E-02

Table 4.4. Proton Damage Coefficients for V_{oc} and P_{max}

PROTON DAMAGE COEFFICIENT FOR VOC
OMNIDIRECTIONAL TO EQUIV. 10-MEV UNIDIRECTIONAL NORMAL INCIDENT PROTON FLUX.

ENERGY (MEV)	(J)	SHIELD THICKNESS, GM/CM2(CM)							
		0. (0.)	5.59E-03 (2.54E-3)	1.68E-02 (7.64E-3)	3.35E-02 (1.52E-2)	6.71E-02 (3.05E-2)	1.12E-01 (5.09E-2)	1.68E-01 (7.64E-2)	3.35E-01 (1.52E-1)
.100	1.60E-14	5.303E-01	0.	0.	0.	0.	0.	0.	0.
.200	3.20E-14	7.150E-01	0.	0.	0.	0.	0.	0.	0.
.300	4.80E-14	8.623E-01	0.	0.	0.	0.	0.	0.	0.
.400	6.40E-14	9.976E-01	0.	0.	0.	0.	0.	0.	0.
.600	9.60E-14	1.271E+00	0.	0.	0.	0.	0.	0.	0.
.800	1.28E-13	1.546E+00	0.	0.	0.	0.	0.	0.	0.
1.000	1.60E-13	1.792E+00	0.	0.	0.	0.	0.	0.	0.
1.200	1.92E-13	1.994E+00	0.	0.	0.	0.	0.	0.	0.
1.300	2.08E-13	2.082E+00	4.303E-02	0.	0.	0.	0.	0.	0.
1.400	2.24E-13	2.160E+00	1.948E-01	0.	0.	0.	0.	0.	0.
1.600	2.56E-13	2.299E+00	5.853E-01	0.	0.	0.	0.	0.	0.
1.800	2.88E-13	2.412E+00	9.827E-01	0.	0.	0.	0.	0.	0.
2.000	3.20E-13	2.502E+00	1.335E+00	0.	0.	0.	0.	0.	0.
2.200	3.52E-13	2.569E+00	1.624E+00	0.	0.	0.	0.	0.	0.
2.400	3.84E-13	2.615E+00	1.860E+00	0.	0.	0.	0.	0.	0.
2.600	4.16E-13	2.645E+00	2.047E+00	1.912E-02	0.	0.	0.	0.	0.
2.800	4.48E-13	2.656E+00	2.191E+00	2.733E-01	0.	0.	0.	0.	0.
3.000	4.80E-13	2.640E+00	2.298E+00	6.092E-01	0.	0.	0.	0.	0.
3.200	5.12E-13	2.597E+00	2.375E+00	9.375E-01	0.	0.	0.	0.	0.
3.400	5.44E-13	2.526E+00	2.416E+00	1.226E+00	0.	0.	0.	0.	0.
3.600	5.76E-13	2.426E+00	2.420E+00	1.468E+00	0.	0.	0.	0.	0.
3.800	6.08E-13	2.302E+00	2.388E+00	1.664E+00	0.	0.	0.	0.	0.
4.000	6.40E-13	2.159E+00	2.320E+00	1.818E+00	4.687E-02	0.	0.	0.	0.
4.200	6.72E-13	2.024E+00	2.219E+00	1.932E+00	2.866E-01	0.	0.	0.	0.
4.400	7.04E-13	1.891E+00	2.093E+00	1.998E+00	5.697E-01	0.	0.	0.	0.
4.600	7.36E-13	1.766E+00	1.962E+00	2.017E+00	8.378E-01	0.	0.	0.	0.
4.800	7.68E-13	1.650E+00	1.839E+00	1.990E+00	1.074E+00	0.	0.	0.	0.
5.200	8.32E-13	1.447E+00	1.616E+00	1.833E+00	1.431E+00	0.	0.	0.	0.
5.600	8.96E-13	1.278E+00	1.428E+00	1.642E+00	1.603E+00	0.	0.	0.	0.
6.000	9.60E-13	1.136E+00	1.268E+00	1.467E+00	1.584E+00	3.741E-02	0.	0.	0.
6.400	1.02E-12	1.020E+00	1.131E+00	1.312E+00	1.468E+00	4.510E-01	0.	0.	0.
6.800	1.09E-12	9.237E-01	1.018E+00	1.178E+00	1.339E+00	8.464E-01	0.	0.	0.
7.200	1.15E-12	8.440E-01	9.252E-01	1.063E+00	1.218E+00	1.101E+00	0.	0.	0.
7.600	1.22E-12	7.775E-01	8.479E-01	9.673E-01	1.109E+00	1.166E+00	0.	0.	0.
8.000	1.28E-12	7.204E-01	7.827E-01	8.867E-01	1.013E+00	1.125E+00	3.696E-03	0.	0.
9.000	1.44E-12	6.134E-01	6.580E-01	7.324E-01	8.256E-01	9.523E-01	7.614E-01	0.	0.
10.000	1.60E-12	5.554E-01	5.834E-01	6.303E-01	6.965E-01	7.998E-01	8.423E-01	0.	0.
11.000	1.76E-12	5.169E-01	5.351E-01	5.691E-01	6.105E-01	6.851E-01	7.463E-01	5.746E-01	0.
12.000	1.92E-12	4.936E-01	5.051E-01	5.264E-01	5.558E-01	6.035E-01	6.552E-01	6.555E-01	0.
13.000	2.08E-12	4.778E-01	4.859E-01	4.997E-01	5.182E-01	5.493E-01	5.830E-01	6.044E-01	0.
14.000	2.24E-12	4.663E-01	4.722E-01	4.815E-01	4.934E-01	5.121E-01	5.320E-01	5.496E-01	0.
15.000	2.40E-12	4.594E-01	4.637E-01	4.688E-01	4.758E-01	4.867E-01	4.961E-01	5.047E-01	0.
16.000	2.56E-12	4.548E-01	4.572E-01	4.606E-01	4.642E-01	4.685E-01	4.713E-01	4.719E-01	3.775E-01
18.000	2.88E-12	4.433E-01	4.458E-01	4.480E-01	4.493E-01	4.478E-01	4.419E-01	4.327E-01	3.956E-01
20.000	3.20E-12	4.352E-01	4.367E-01	4.382E-01	4.379E-01	4.343E-01	4.269E-01	4.142E-01	3.698E-01
22.000	3.52E-12	4.286E-01	4.278E-01	4.290E-01	4.293E-01	4.250E-01	4.172E-01	4.041E-01	3.569E-01
24.000	3.84E-12	4.211E-01	4.211E-01	4.213E-01	4.203E-01	4.176E-01	4.097E-01	3.981E-01	3.530E-01
26.000	4.16E-12	4.146E-01	4.144E-01	4.141E-01	4.137E-01	4.102E-01	4.035E-01	3.931E-01	3.524E-01
28.000	4.48E-12	4.081E-01	4.081E-01	4.079E-01	4.063E-01	4.040E-01	3.978E-01	3.884E-01	3.527E-01
30.000	4.80E-12	4.004E-01	4.005E-01	4.003E-01	3.998E-01	3.969E-01	3.926E-01	3.843E-01	3.530E-01
34.000	5.44E-12	3.836E-01	3.838E-01	3.840E-01	3.839E-01	3.831E-01	3.797E-01	3.749E-01	3.520E-01
38.000	6.08E-12	3.703E-01	3.704E-01	3.704E-01	3.703E-01	3.686E-01	3.669E-01	3.636E-01	3.473E-01
42.000	6.72E-12	3.564E-01	3.565E-01	3.569E-01	3.572E-01	3.571E-01	3.549E-01	3.526E-01	3.409E-01
46.000	7.36E-12	3.426E-01	3.426E-01	3.429E-01	3.433E-01	3.436E-01	3.421E-01	3.409E-01	3.334E-01
50.000	8.00E-12	3.296E-01	3.296E-01	3.298E-01	3.301E-01	3.304E-01	3.303E-01	3.297E-01	3.245E-01
55.000	8.80E-12	3.145E-01	3.145E-01	3.147E-01	3.151E-01	3.158E-01	3.161E-01	3.161E-01	3.132E-01
60.000	9.60E-12	3.005E-01	3.005E-01	3.007E-01	3.010E-01	3.016E-01	3.020E-01	3.022E-01	3.007E-01
65.000	1.04E-11	2.859E-01	2.861E-01	2.865E-01	2.870E-01	2.880E-01	2.890E-01	2.897E-01	2.899E-01
70.000	1.12E-11	2.724E-01	2.726E-01	2.729E-01	2.733E-01	2.743E-01	2.753E-01	2.761E-01	2.773E-01
80.000	1.28E-11	2.481E-01	2.483E-01	2.487E-01	2.492E-01	2.501E-01	2.512E-01	2.523E-01	2.543E-01
90.000	1.44E-11	2.249E-01	2.251E-01	2.255E-01	2.260E-01	2.270E-01	2.281E-01	2.294E-01	2.320E-01
100.000	1.60E-11	1.999E-01	2.001E-01	2.006E-01	2.013E-01	2.025E-01	2.039E-01	2.055E-01	2.092E-01
130.000	2.08E-11	1.492E-01	1.493E-01	1.496E-01	1.500E-01	1.509E-01	1.519E-01	1.530E-01	1.560E-01
160.000	2.56E-11	1.183E-01	1.183E-01	1.185E-01	1.188E-01	1.192E-01	1.199E-01	1.206E-01	1.226E-01
200.000	3.20E-11	9.215E-02	9.220E-02	9.229E-02	9.242E-02	9.268E-02	9.302E-02	9.344E-02	9.462E-02

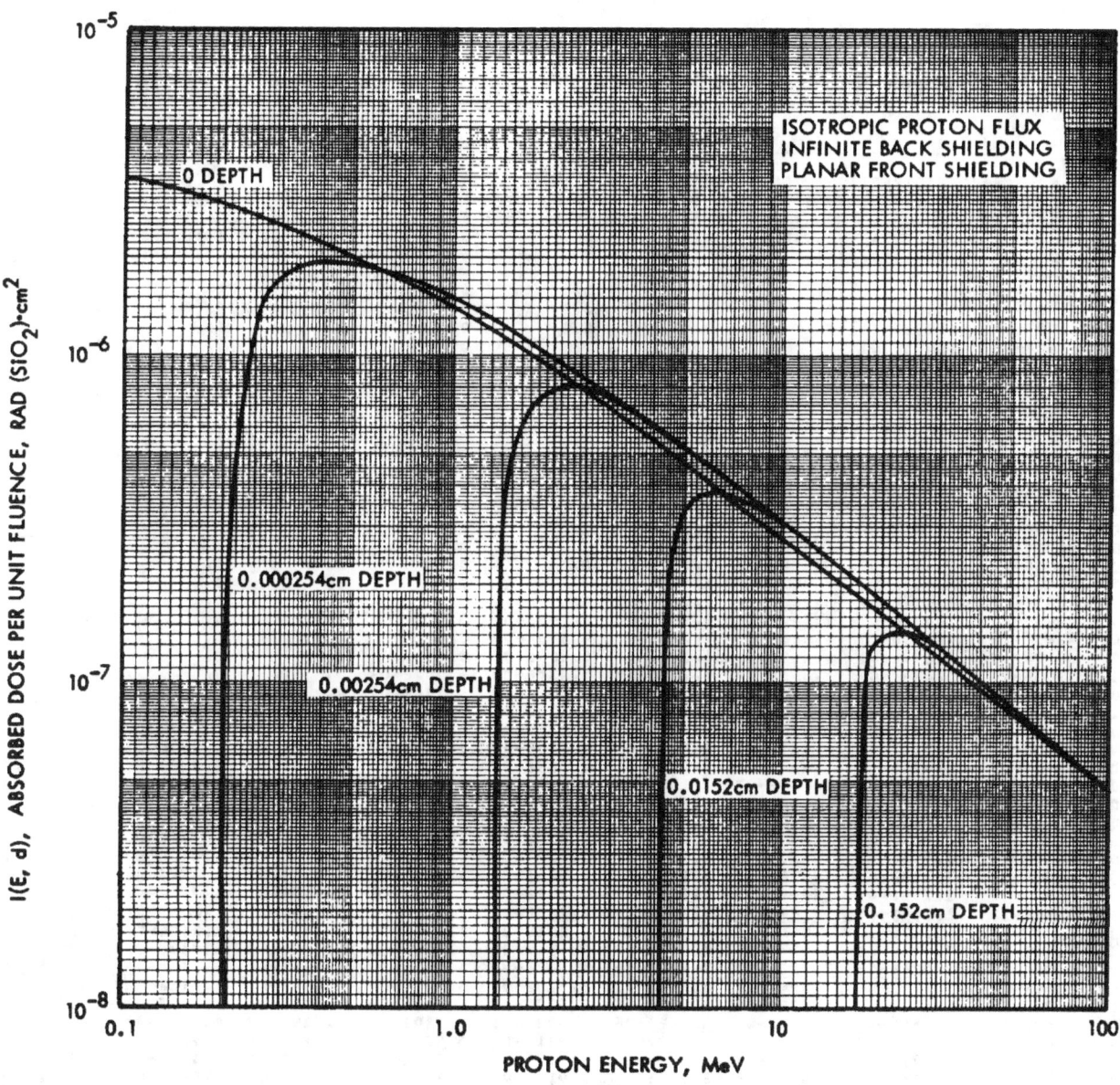

Figure 4.5 Absorbed Dose Per Unit Fluence of Space Proton Irradiation for Various Depths in Planar Fused Silica Shielding

Table 4.5. Proton Stopping Power, Rad(SiO_2)/Unit Omnidirectional Flux

PROTON STOPPING POWER
RAD(FUSED SILICA)/OMNIDIRECTIONAL FLUX(PARTICLES/CM2)

ENERGY (MEV)	(J)	SHIELD THICKNESS, GM/CM2(CM)							
		5.59E-04 (2.54E-4)	2.24E-03 (1.02E-3)	5.59E-03 (2.54E-3)	1.68E-02 (7.64E-3)	3.35E-02 (1.52E-2)	6.71E-02 (3.05E-2)	1.12E-01 (5.09E-2)	1.68E-01 (7.64E-2)
.200	3.20E-14	0.	0.	0.	0.	0.	0.	0.	0.
.225	3.60E-14	9.154E-07	0.	0.	0.	0.	0.	0.	0.
.250	4.00E-14	1.231E-06	0.	0.	0.	0.	0.	0.	0.
.275	4.40E-14	1.465E-06	0.	0.	0.	0.	0.	0.	0.
.300	4.80E-14	1.609E-06	0.	0.	0.	0.	0.	0.	0.
.350	5.60E-14	1.766E-06	0.	0.	0.	0.	0.	0.	0.
.400	6.40E-14	1.825E-06	0.	0.	0.	0.	0.	0.	0.
.500	8.00E-14	1.824E-06	0.	0.	0.	0.	0.	0.	0.
.600	9.60E-14	1.762E-06	0.	0.	0.	0.	0.	0.	0.
.800	1.28E-13	1.564E-06	8.732E-07	0.	0.	0.	0.	0.	0.
1.000	1.60E-13	1.427E-06	1.113E-06	0.	0.	0.	0.	0.	0.
1.200	1.92E-13	1.313E-06	1.144E-06	0.	0.	0.	0.	0.	0.
1.400	2.24E-13	1.199E-06	1.124E-06	4.916E-07	0.	0.	0.	0.	0.
1.600	2.56E-13	1.098E-06	1.077E-06	7.099E-07	0.	0.	0.	0.	0.
1.800	2.88E-13	1.014E-06	1.018E-06	7.937E-07	0.	0.	0.	0.	0.
2.000	3.20E-13	9.414E-07	9.584E-07	8.247E-07	0.	0.	0.	0.	0.
2.250	3.60E-13	8.657E-07	8.900E-07	8.192E-07	0.	0.	0.	0.	0.
2.500	4.00E-13	8.017E-07	8.266E-07	7.933E-07	0.	0.	0.	0.	0.
2.750	4.40E-13	7.472E-07	7.744E-07	7.598E-07	2.788E-07	0.	0.	0.	0.
3.000	4.80E-13	7.005E-07	7.266E-07	7.245E-07	4.193E-07	0.	0.	0.	0.
3.500	5.60E-13	6.237E-07	6.462E-07	6.560E-07	5.204E-07	0.	0.	0.	0.
4.000	6.40E-13	5.644E-07	5.833E-07	5.970E-07	5.337E-07	1.050E-07	0.	0.	0.
5.000	8.00E-13	4.760E-07	4.894E-07	5.033E-07	5.033E-07	3.776E-07	0.	0.	0.
6.000	9.60E-13	4.145E-07	4.242E-07	4.354E-07	4.428E-07	3.995E-07	6.540E-08	0.	0.
8.000	1.28E-12	3.328E-07	3.382E-07	3.459E-07	3.575E-07	3.540E-07	2.944E-07	1.735E-08	0.
10.000	1.60E-12	2.799E-07	2.834E-07	2.886E-07	2.986E-07	3.029E-07	2.880E-07	2.264E-07	0.
12.000	1.92E-12	2.427E-07	2.451E-07	2.487E-07	2.567E-07	2.623E-07	2.606E-07	2.386E-07	1.786E-07
14.000	2.24E-12	2.152E-07	2.169E-07	2.195E-07	2.258E-07	2.311E-07	2.338E-07	2.259E-07	2.005E-07
16.000	2.56E-12	1.938E-07	1.951E-07	1.971E-07	2.020E-07	2.067E-07	2.107E-07	2.088E-07	1.972E-07
18.000	2.88E-12	1.767E-07	1.776E-07	1.792E-07	1.831E-07	1.872E-07	1.914E-07	1.921E-07	1.871E-07
20.000	3.20E-12	1.626E-07	1.634E-07	1.646E-07	1.676E-07	1.713E-07	1.754E-07	1.772E-07	1.754E-07
22.500	3.60E-12	1.482E-07	1.487E-07	1.497E-07	1.522E-07	1.551E-07	1.587E-07	1.611E-07	1.613E-07
25.000	4.00E-12	1.364E-07	1.368E-07	1.375E-07	1.396E-07	1.419E-07	1.452E-07	1.476E-07	1.487E-07
27.500	4.40E-12	1.265E-07	1.268E-07	1.274E-07	1.291E-07	1.311E-07	1.339E-07	1.363E-07	1.377E-07
30.000	4.80E-12	1.181E-07	1.184E-07	1.189E-07	1.203E-07	1.219E-07	1.244E-07	1.266E-07	1.282E-07
35.000	5.60E-12	1.047E-07	1.048E-07	1.052E-07	1.061E-07	1.073E-07	1.092E-07	1.110E-07	1.126E-07
40.000	6.40E-12	9.424E-08	9.437E-08	9.461E-08	9.532E-08	9.623E-08	9.770E-08	9.919E-08	1.006E-07
50.000	8.00E-12	7.925E-08	7.932E-08	7.946E-08	7.988E-08	8.043E-08	8.136E-08	8.237E-08	8.337E-08
60.000	9.60E-12	6.879E-08	6.884E-08	6.893E-08	6.920E-08	6.957E-08	7.020E-08	7.091E-08	7.165E-08
80.000	1.28E-11	5.530E-08	5.532E-08	5.536E-08	5.549E-08	5.568E-08	5.601E-08	5.640E-08	5.682E-08
100.000	1.60E-11	4.688E-08	4.689E-08	4.692E-08	4.699E-08	4.710E-08	4.729E-08	4.753E-08	4.779E-08
130.000	2.08E-11	3.895E-08	3.896E-08	3.897E-08	3.901E-08	3.906E-08	3.916E-08	3.929E-08	3.943E-08
160.000	2.56E-11	3.375E-08	3.375E-08	3.376E-08	3.378E-08	3.381E-08	3.387E-08	3.394E-08	3.403E-08
200.000	3.20E-11	2.915E-08	2.915E-08	2.915E-08	2.916E-08	2.918E-08	2.922E-08	2.926E-08	2.931E-08

4.5 Alpha Particle Space Radiation Effects

Solar flares have been shown to have a component of energetic alpha particles (helium nuclei). The evaluation of the effects of solar flare events on solar arrays requires alpha particle data similar to that for electrons and protons. Smith and Blue compared effects of 10.5 MeV protons and 42 MeV alpha particles on silicon solar cell degradation.[4.7] The results showed that the 42 MeV alpha particle flux degraded the silicon cells 3.8 times as fast as a similar flux of 10.5 MeV protons. These results were in good agreement with a theoretical damage ratio of 4.

Based on the experimental results of Smith and Blue, the proton damage constant curve shown in Figure 4.4 can be translated a factor of 4 higher in energy and a factor of 4 higher in relative damage constant to represent a similar family of relative damage constants for alpha particles in space. Although the relationship found by Smith and Blue may not extend to lower particle energies, a set of effective damage constants for alpha particles, obtained by the above two translations, is shown in Figure 4.6. Data are shown based on P_{max} and V_{oc}. Data based on I_{sc} may be obtained similarly.

The methods for estimating solar cell degradation in space are based on the techniques described in References 4.8 through 4.10. In summary, the omnidirectional space radiation is converted to a damage equivalent unidirectional fluence at a normalized energy and in terms of a specific radiation particle. This equivalent fluence will produce the same damage as that produced by omnidirectional space radiation considered if the relative damage coefficient (RDC) is properly defined to allow the conversion. When the equivalent fluence is determined for a given space environment, the parameter degradation can be evaluated in the laboratory by irradiating the solar cell with the calculated fluence level of unidirectional normally incident flux. The equivalent fluence is normally expressed in terms of 1 MeV electrons or 10 MeV protons. In the presence of a cover shield, angular dependence of both "effective shield thickness" and damage effectiveness (or stopping power for dose calculations) is integrated over 2π for a given energy, assuming semi-infinite planar geometry. As a result,

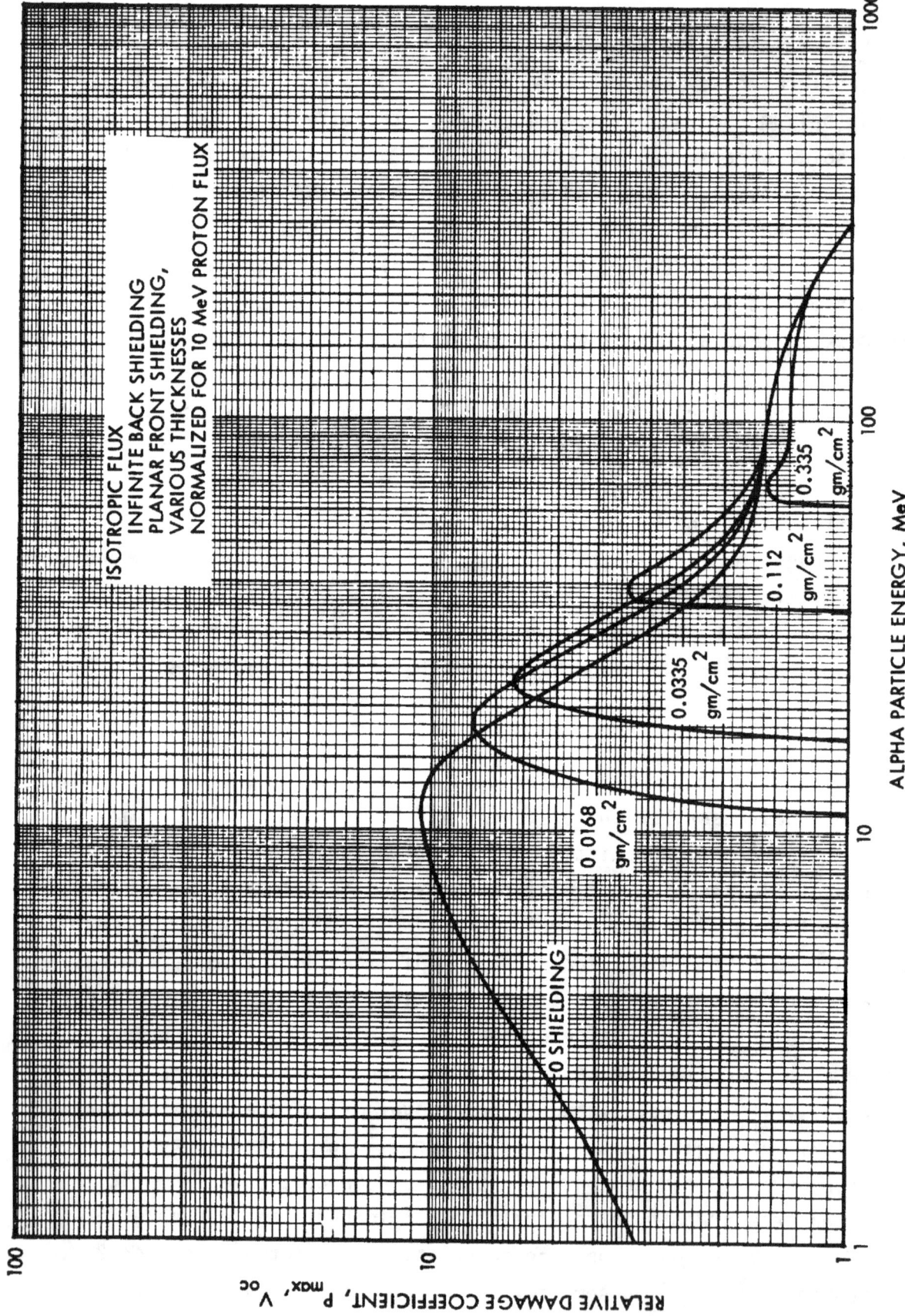

Figure 4.6 Relative Damage Coefficients for Space Alpha Particle Irradiation of Shielded n/p Silicon Solar Cells (Based on P_{max} or V_{oc})

the RDC for a given shield thickness, such as shown in Figures 4.1 through 4.5, is computed only once. Subsequent equivalent fluence calculations simply involve an integration of the omnidirectional fluence times the appropriate damage coefficients as discussed in Chapter 6.

4.6 Alternative Approaches

An alternative approach for estimating solar cell degradation has been proposed by Carosella [4.11] and Picciano and Reitman. [4.12] This method determines the energy spectrum after isotropic space radiation passes through the coverglass (or before entering the solar cell surface), assuming an infinite back shielding. Then the damage coefficients applicable to normally incident particles are applied to determine the damage. There are two drawbacks in this approach: (a) The energy spectrum at the solar cell surface must be recomputed for every change in either the energy spectrum of the space environment or of coverglass thickness. Therefore, the computation is quite repetitive. (b) The calculated energy spectrum at the solar cell surface no longer contains information on angular dependence, and is neither isotropic nor unidirectional. Yet the damage coefficient applied is appropriate only for normally incident radiation. The problems relating to the angular content of the "modified" spectrum emerging from the shielding are of no consequence for some calculations, such as absorbed dose, RDC for electrons, or high-energy protons. It is therefore justified to weigh the "modified" spectrum with RDCs to evaluate electron damage in terms of a damage-equivalent monoenergetic normal incident fluence. In the case of low energy protons, the use of the referenced methods incorrectly assumes that proton damage is independent of the angle of incidence. This shortcoming is particularly serious in the case of many common space environments in which the lower-energy proton damage dominates the solar cell degradation.

Wilkinson and Horne [4.13, 4.14] have proposed an analytical approach based on several computer codes. First a Monte Carlo shielding code (SPARES) [4.15] is used to calculate the modified fluence-energy spectrum inside the solar cell after penetration through a coverglass. Then displacement density profiles are computed as a function of depth inside the solar cell using a theoretical model.[4.16] These profiles are used as an input to the PN code which solves the one-dimensional time-dependent diffusion equation for a diode.[4.17, 4.18] The PN code is capable of handling a number of physical parameters which vary with depth throughout the solar cell including the displacement density profiles. Using experimentally-derived damage constants for minority lifetime correlated with the displacement density profile, the PN code solves the diffusion equation and calculates a solar cell I-V curve. This approach, then, is a basic physics approach and if all the required parameters are accurately known and properly accounted for in the computer codes this would be an ideal technique. The problems with using this approach include the following:

> The code collection is very complex and the capability for using it is available at only one source.

> An enormous amount of data accurately describing the solar cell and its interaction with solar spectrum illumination as well as with electrons and protons must be known. For example a partial list of the parameters which are input to the code as a function of depth are: 1) electron and hole mobility, 2) carrier generation rates, 3) dopant impurity densities (this can account for built-in electric fields in the cell), 4) electron and hole lifetimes, 5) carrier densities in the conduction and valence bands when the Fermi level coincides with the energy level of the recombination centers, 6) intrinsic carrier densities in the conduction and valence bands at the temperature of interest, and 7) radiation defect introduction rates.

> The use of minority carrier damage constants on a microscopic scale may not be correct.

> No cases of in-flight panel degradation are known where the PN code method makes a better prediction than does the equivalent fluence method. Both methods of calculation are based partly on experimental data and partly on theory. The PN code is expected by its authors to make a more accurate calculation for solar panels flying through orbits with a large number of low energy protons, but this has not been verified.

> Assumptions which must be made about the nature of the defects introduced by the radiation and their corresponding energy levels within the bandgap may not be accurate. The relationships between densities and actual minority carrier recombination center kinetics are not yet well established. For example, defect density calculations almost always overestimate the number of defects introduced by factors as large as ten, probably because many dislodged atoms have not been ejected very far from their sites and have a high probability of return. Such defect recombination is likely to be a function of defect density and may cause inaccuracies in the calculation in just those cases where it is thought by its authors to be superior.

REFERENCES

4.1 D. P. LeGalley and A. Rosen (Eds.), Space Physics, John Wiley and Sons, New York, p. 693, 1964.

4.2 J. E. Janni, "Calculations of Energy Loss, Range, Pathlength, Straggling, Multiple Scattering, and the Probability of Inelastic Nuclear Collisions for 0.1 to 1000 MeV Protons," AFWL-TR-65-150, Sept. 1966.

4.3 M. J. Berger and S. M. Seltzer, "Tables of Energy Losses and Ranges of Electrons and Positrons," Paper 10, NASA-NRC Pub. 1133, 1964, also NASA SP-3012, 1964.

4.4 M. J. Berger and S. M. Seltzer, "Additional Stopping Power and Range Tables for Protons, Mesons, and Electrons," NASA SP-3036, 1966.

4.5 M. J. Barrett, "Electron Damage Coefficients in P-Type Silicon," IEEE Trans. on Nuc. Sci., NS-14, 6, 82, 1967.

4.6 W. Rosenzweig, "Space Radiation Effects in Silicon Devices," IEEE Trans. on Nuc. Sci., NS-12, 5, 18, 1965.

4.7 A. B. Smith and J. W. Blue, "A Comparison of Solar Cell Damage by Alpha Particles and Protons," NASA TN D-3427, May 1966.

4.8 W. L. Brown, J. D. Gabbe, and W. Rosenzweig, "Results of the Telstar Radiation Experiments," Bell System Technical J., 42, No. 4, Part 2, 1505, July 1963.

4.9 H. Y. Tada, "Equivalent Fluence and Relative Damage Coefficient - Tools For Space Solar Cell Degradation Estimate," IEEE Trans. on Nuc. Sci., NS-20, 6, 234, December 1973.

4.10 H. Y. Tada, "A New Dimension in Solar Cell Degradation Estimate In-Space RDC Matrix Method," Conf. Rec. of the 10th IEEE Photovoltaic Specialists Conf., 392, 1973.

4.11 C. A. Carosella, "Shielding of Solar Cells Against Van Allen Belt Protons," J. Spacecraft, 5, 7, 878, 1968.

4.12 W. T. Picciano and R. A. Reitman, "Flight Data Analysis of Power Subsystems Degradation at Near Synchronous Altitude," Philco-Ford Report, WDL-TR4223, July 1970.

4.13 M. C. Wilkinson and W. E. Horne, "Limitation of the 1 MeV Equivalent Electron Fluence Method in the Evaluation of Space Radiation Degradation," Conf. Rec. of the 11th IEEE Photovoltaic Specialists Conf., 209, 1975.

4.14 W. E. Horne, R. B. Greegor, M. C. Wilkinson and B. K. Madaras, "Real-Time Space and Nuclear Effects on Solar Cells (Accelerated Evaluation Methods)," Air Force Aero Propulsion Laboratory Publication No. AFAPL-TR-72-69, Vol. III (Final Report), July 1974.

4.15 P. Hahn, "RSIC Computer Code Collection -- Space Radiation Environment Shielding Systems (SPARES)," CCC-148, AS-2807, Oak Ridge National Laboratory Radiation Shielding Information Center, Oak Ridge, Tenn., 1969.

4.16 J. A. Baiker, H. Flicker, and J. Vilms, "Proton Induced Lattice Displacements in Silicon," App. Phys. Letters, $\underline{2}$, 104, 1963.

4.17 R. E. Leadon and M. L. Vaughn, "Short-Pulsed Radiation Effects on Dynamic Electronic Components," Final Report, Contract DASA 01-68-C-0123, prepared for Defense Atomic Support Agency by Gulf General Atomic Inc., June 1969.

4.18 R. E. Leadon, et al., "Radiation Effects on Dynamic Electronic Components," Final Report, Contract DASA 01-69-C-0013, prepared for Defense Atomic Support Agency by Gulf Radiation Technology, June 1970.

CHAPTER 5

5.0 THE SPACE RADIATION ENVIRONMENT

The radiation environment near the Earth consists of electrons and protons trapped in the geomagnetic field, corpuscular radiation associated with large solar flare activity, and to a lesser extent, galactic cosmic ray radiation. Near Jupiter an environment similar to the Earth's trapped particle radiation exists, but the intensity is far greater than that near Earth, due primarily to the large magnetic field. In the following sections each environment is qualitatively described to assist the reader in determining the proper environment for use in making solar cell degradation estimates. Quantitative, or detailed, descriptions of each environment are beyond the scope of this manuscript.

5.1 Geomagnetically Trapped Radiation

The geomagnetic dipole field is responsible for the radiation belts near the Earth, holding the trapped charged particles for long periods of time. It is a plasma confined in an inhomogeneous magnetic field. The understanding of charge transport within the field, loss and capture mechanisms of charged particles has improved considerably over recent years. Models prepared to characterize the trapped radiation are continuously updated and now include solar cycle dependence.

Geomagnetically trapped radiation may be either of natural origin or of artificial origin, such as high-altitude nuclear explosions. Since a particle has to possess a charge to be trapped in a magnetic field, the energetic trapped particles are mainly electrons and protons. Regardless of the origin, particles with just the right momentum and pitch angle can be trapped in the field. The particles will then spiral about a field line with varying pitch angle (angle between magnetic field vector and velocity) and curvature in the inhomogeneous field. They continue the motion until they reach the mirror (or reflection) point where the pitch angle becomes 90°, then turn around and travel back along the field line into the other hemisphere.[5.1] The particles continue to bounce back

and forth between mirror points (latitudinal motion). At the same time
the particles drift in the longitudinal direction as the result of forces
due to the gradient of field strength and the curvature of field lines.
During a quiescent state (periods of normal solar activity) the trapped
particles can be characterized by three periodic motions:
a) circulation about the field line with cyclotron (Larmor) frequency,
b) latitudinal motion between mirror points, and
c) longitudinal drift.

The direction of motion for electrons is eastward, opposite to that of protons (westward) because of their opposite charge. Particles whose mirror points are in the upper atmosphere collide with gas molecules, gradually losing their energy and changing trajectory until they are lost in the lower atmosphere.

At some distance from the Earth the field is distorted by the solar wind as shown in Figure 5.1. The solar wind is a plasma flowing outward from the sun and is dominated by protons with an average energy of approximately one keV and a density on the order of $10/cm^3$. The solar wind interacts with the geomagnetic field, resulting in the formation of a shock wave. This in turn forms and shapes the magnetosphere. As the solar plasma passes the shock wave, the random speeds of the particles increase, producing turbulence in the magnetic field. This turbulent region, the magnetosheath, extends inward from the shock front to the magnetopause, which is the outer boundary of the more regular field region associated with the earth.

The geomagnetic field lines just inside the magnetopause are qualitatively similar to those associated with the simple dipole model and trap corpuscular radiation as described above. During quiescence a relatively steady flow of solar wind blows the field away from the sun, contributing to an asymmetric shape of the radiation belt, compressed on the sun's side and forming the neutral sheet and magnetotail extending away from the Earth in the antisolar direction.

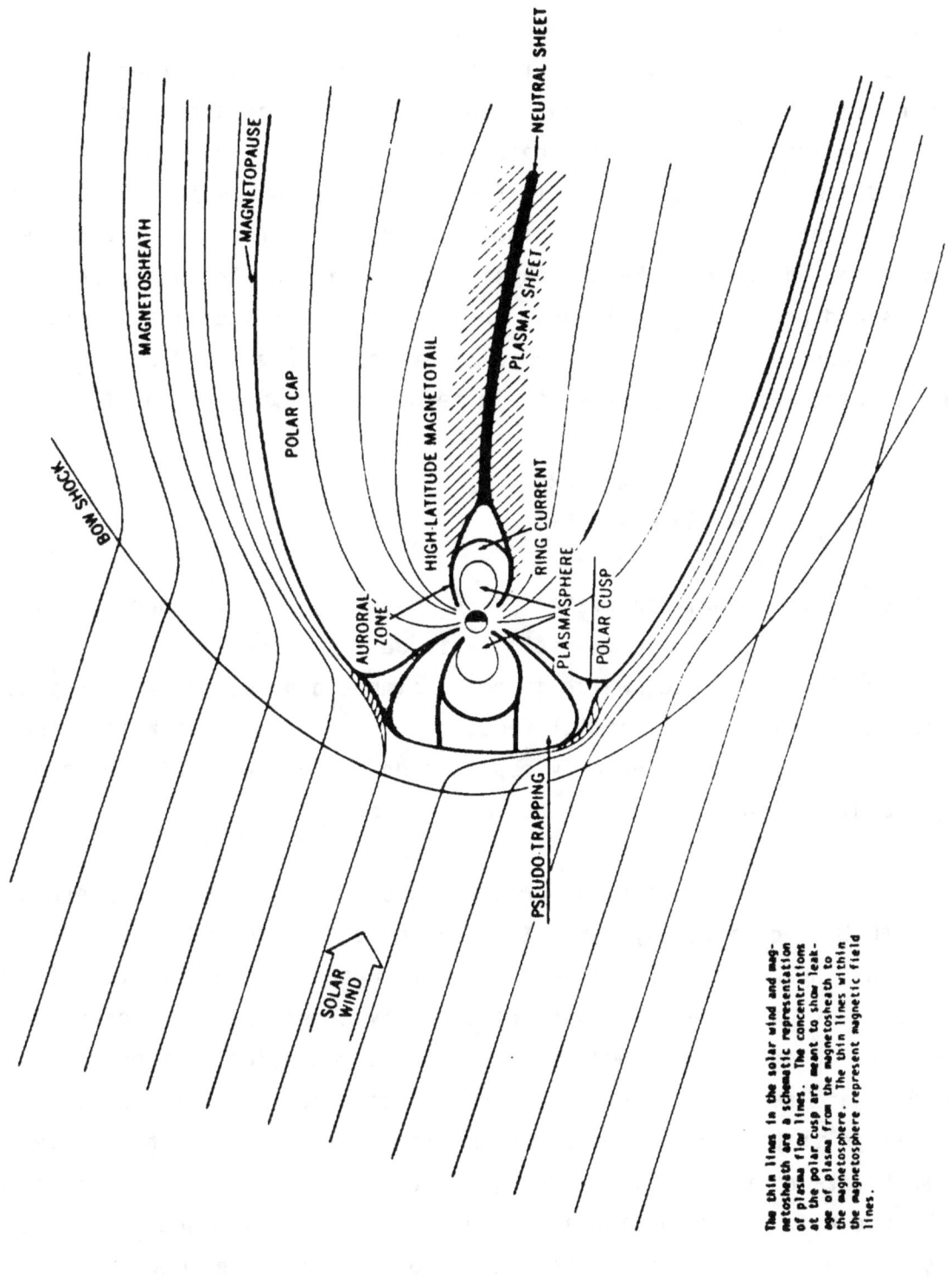

Figure 5.1 Regions of the Magnetosphere Shown in the Noon-Midnight Meridian Plane   5.12

McIlwain [5.2] in 1961 proposed a coordinate system consisting of the magnetic field B and the integral invariant I which can adequately relate measurements made at different geographic locations. The quantity I is the length of the field line between mirror points weighed by a function of the magnetic field along the line and is an adiabatic invariant of the motion. He introduced the magnetic shell parameter L = f(B,I), analogous to a physical distance in a dipole field (which reduces to the equatorial radius of a field line in the case of a dipole field), thus reducing the number of variables needed to describe the physical situation of trapped charged particles and presenting field data in a manner which facilitates its physical interpretation. For a radial distance of R and a dipole moment of M, the transformation using the dipole relation is expressed as follows:

$$B = \frac{M}{R^3}\sqrt{4 - \frac{3R}{L}} \qquad (5.1.1)$$

where the magnetic shell parameter $L = R(\cos\lambda)^{-2}$, M is the geomagnetic dipole moment, and λ is the magnetic latitude. In order to apply this concept to the Earth's field, which is not a simple dipole, McIlwain expanded the parameter L into a polynomial function of a variable which is a function of I, B, and M and elegantly represented trapped particle phenomena using two dimensions (B and L) instead of three.

Since its introduction, numerous particle field data were presented in this (B,L) coordinate system. Stassinopoulos, Vette and co-workers, at the National Space Science Data Center, have concentrated efforts on the compilation of particle field data reported by numerous investigators and continue to construct and update models of the radiation environment. These data are regarded as the best consolidated source of information available on trapped radiation environments and are used as the single source of data on this subject throughout this manuscript. The reader may consult the referenced publications [5.3-5.13] for detailed and quantitative discussions of the trapped electron and proton environment models. The distribution of the charged particles in the magnetosphere is illustrated in Figure 5.2. [5.14]

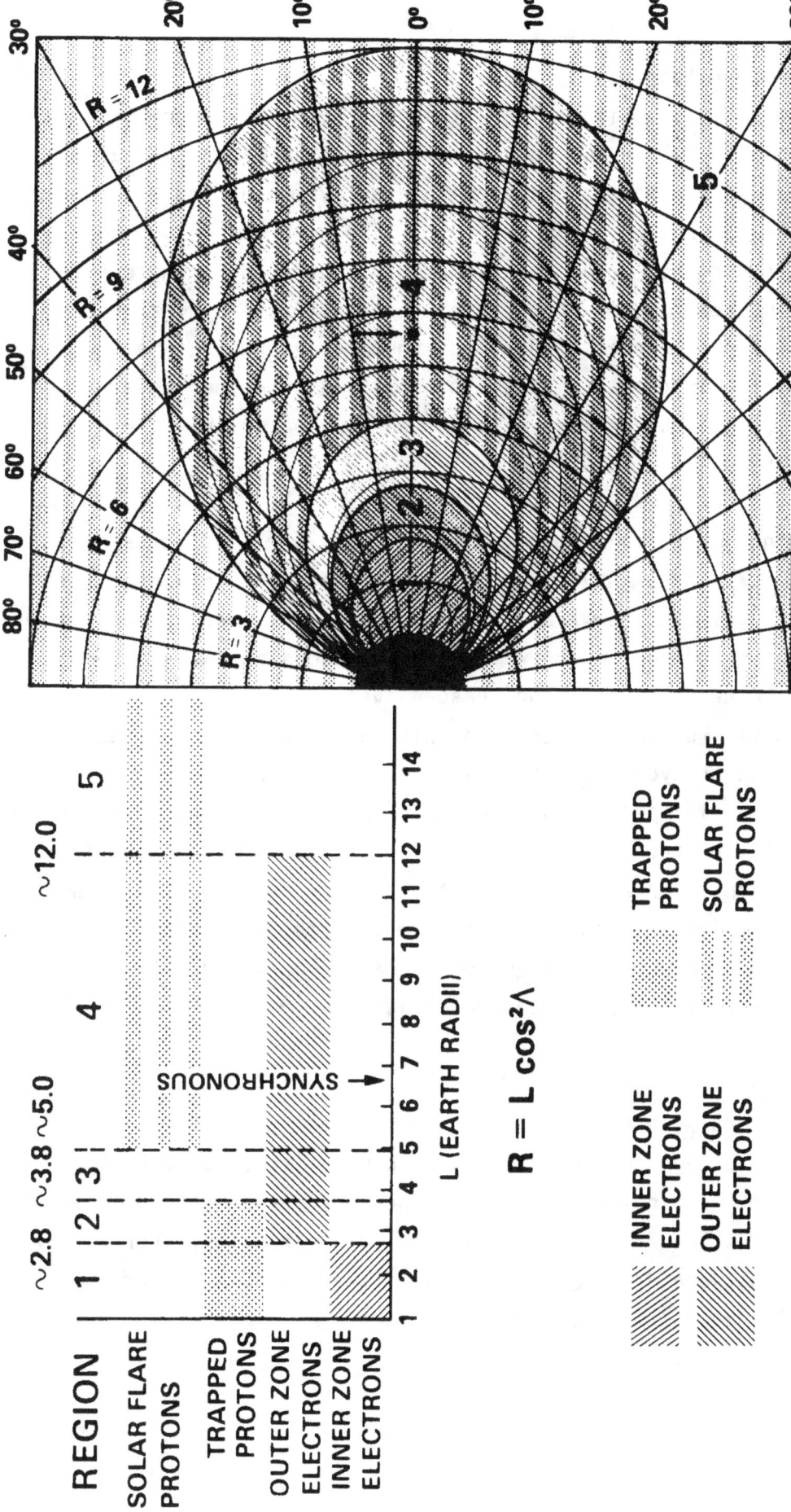

Figure 5.2 Charged Particle Distribution in the Magnetosphere

5.1.1 Trapped Protons

The most recent descriptions of the trapped proton environment are presented in Reference 5.13. The largest concentration of protons at intermediate energies is near the Earth within a L value of four (geocentric) Earth radii, peaked at about two Earth radii. The high energy protons concentrate even closer to Earth, peaking at 1.5 Earth radii, whereas the distribution of the lower energy protons extends nearly to synchronous altitude (L = 6.6 R_e). Generally speaking, the energy spectrum becomes softer as the L-value increases. At synchronous altitude the spectrum is so soft that practically no protons with an energy greater than 2 MeV exist.

In 1976 a single trapped proton model, the AP8 [5.13], replaced several older models (e.g. AP5, AP6, AP7), each of which was valid only over a specific energy range. The new AP8 describes the entire energy spectrum in a coherent, uniform, and continuous way. It was issued in two versions: the AP8-MAX and the AP8-MIN relating, respectively, to average solar maximum and solar minimum conditions. In this case there is good theoretical reasoning for solar cycle effects which have been verified by experimental observations. Trapped protons are affected by solar cycle variations only in the vicinity of the atmospheric cutoff regions. No changes of consequence have been observed in the heart of the proton trapping domain or at synchronous altitudes since the observed temporal variations are in most cases of no greater extremes than the precision obtained between measurements by detectors on different satellites. [5.14]

5.1.2 Trapped Electrons

Trapped electrons with energies of a few hundred keV extend to the outer boundary of the magnetosphere, which fluctuates at 8 to 10 Earth radii. There are two intense regions: an inner one covers the L-values in the range of $1.2 < L < 2.8$ and peaks about 1.4 Earth radii, whereas the outer zone ranges between $3 < L < 11$ and peaks at around 4 to 5 Earth radii with the flux about 10^7 electrons/cm^2-sec for both zones with energies $E \gtrsim 250$ KeV.

The outer zone is a very dynamic region of space where some particles are stably trapped but others are considered to be pseudo-trapped because the lifetimes are shorter than the drift time around the Earth. However, strong external (e.g., galactic and solar) sources supply electrons to this region of space and thus substantial fluxes are always present. In this zone, the flux has large short-term temporal variations related to the local time as well as a long-term change in average flux associated with a solar cycle.

The current models which describe the outer zone electrons ($L \geq 2.8$) are AEI7-HI and AEI7-LO.[5.15] The "HI" version favors Vampola's fits to the enhanced OV1-19 data, while the "LO" version is representative of more quiescent times averaged over longer time periods. These are interim models replacing the older solar min and max versions of AE-4. However, they do not reflect solar cycle variations.[5.16]

Stassinopoulos[5.14] has pointed out that the radiation experienced by a spacecraft in a geosynchronous orbit is dependent on its longitude. This is because the geomagnetic coordinate system is tilted and displaced from the geographic coordinate system on which the satellite orbits are based. Thus, a measurement of L vs. longitude at geosynchronous altitude would yield a periodic curve with two maxima and two minima. The highest maxima occurs at a longitude of 70°W with an L of 7.02 and the lowest minima occurs at 160°W with an L of 6.6. This results in the trapped electron fluence with energies > 3 MeV having an intensity at 160°W about an order of magnitude larger than at 70°W, as shown by Figure 5.3[5.14] Careful assessments of the interaction of solar panels with the trapped electrons at synchronous altitude must take this longitude effect into account. It is not a problem at any other altitude because in general the spacecraft sweeps through all longitude and the effect is averaged out.

In the inner zone, the effect of geomagnetic storms on the average flux is significant at high L values and higher energies. A long-term increase in the inner zone flux is correlated with an increase in solar activity. Past sources of temporal variations include the decay of resi-

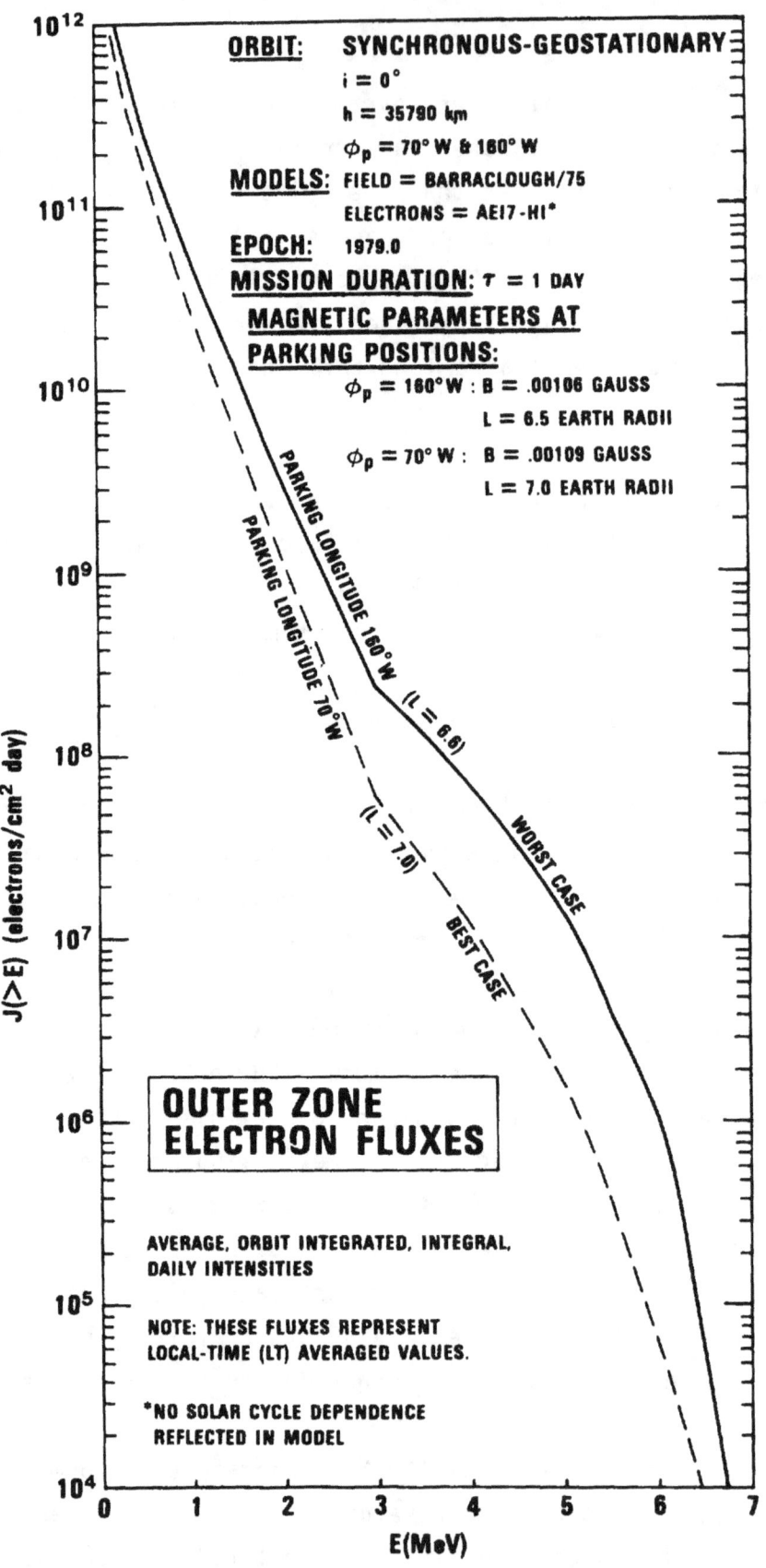

Figure 5.3 Outer Zone Electron Fluxes [5.14]

dual electrons from the high altitude "Starfish" nuclear test, but present data indicate these electrons are no longer present.[5.17] These temporal variations are accommodated in the compilation of data and publications on the AE model sequence by Vette, et al.,[5.5,5.8-5.12] of which the latest versions, AE5 and AE6, are appropriate near solar minimum and solar maximum, respectively.[5.10, 5.11]

5.2 Orbital Integration

Vette and co-workers have time integrated both the trapped proton and electron environments for convenient energy ranges, and have tabulated the average daily fluence for circular orbits having specific altitudes and inclinations. There are two forms of spectra in his data: one is of the form of integral flux, the other of difference flux. (The latter should not be confused with the differential flux.)

If $\phi(E)$ is a differential flux at energy E in MeV, normally expressed in terms of particles/cm^2-sec-MeV, and $\Phi(>E)$ is an integral flux with an energy greater than E, expressed in particles/cm^2-sec, the relationship of these two quantities is

$$\Phi(>E) = \int_\infty^E \phi(E)dE \qquad (5.2.1)$$

$$= \sum_j \phi(E_j) \Delta E_j$$

On the other hand, the difference fluence is simply

$$\Delta \Phi_j = \phi(>E_j) - \phi(>E_j + \Delta E) \qquad (5.2.2)$$

For spacecraft trajectories other than circular orbits tabulated in References 5.3 through 5.13 the radiation environment encountered by the spacecraft must be determined by some other method. Approximate methods may be used within the trajectory may be divided into segments approximating arcs of circular trajectories where the fluence is known and appropriate sums performed. The most accurate way to determine the environment is to make use of the physically significant coordinate system (B,L) so that uncertainties and inaccuracies attributable to the geographic coordinate system are eliminated. A set of state vectors or classical orbital elements can be used to solve Kepler's equation and generate a trajectory with suitable time intervals.[5.15, 5.16, 5.18-5.21] These geographic coordinates are then transformed into geomagnetic shell coordinates (B,L) on which isoflux contour maps are plotted. Computer routines are available from the National Space Science Data Center for conversion of geographic coordinates to (B,L) coordinates[5.18, 5.19] and for performing the time integrals of flux to compute the fluence encountered by the spacecraft.[5.16]

5.3 Cosmic-Ray (Galactic Cosmic-Ray) Radiation

Galactic cosmic rays are a highly penetrating radiation originating beyond the solar system. Many possess energies greater than 1 BeV and are capable of extraordinary interactions with matter in the upper atmosphere such as spallation, fission, fragmentation, and the subsequent secondary processes. The local cosmic-ray radiation in the atmosphere contains protons, neutrons, pi-mesons, mu-mesons, electrons, photons, and strange particles. Near the upper limits of the atmosphere, the primary radiation, consisting of 79 percent protons and 20 percent alpha particles, predominates over the products of nuclear reactions and the decay products; thus the components change with altitude.

One remarkable characteristic of cosmic rays is their isotropy. The average diurnal effect is very small. There is a definite relationship between the fluctuation and solar activity in general; 27-day effects, an 11-year fluctuation cycle, and the Forbush decrease associated with the magnetic storms are examples. Although the energy is very high, the flux is negligibly small compared with other environments considered, and this environment is commonly ignored in solar cell array degradation cases.

5.4 Solar Flare (Solar Cosmic-Ray) Radiation

Solar flares occur in the neighborhood of sunspots, very seldom emit white light, and cause a sudden increase in intensity of the hydrogen alpha line (wavelength 656 nm). After its inception the flare rapidly expands over an area of a few million to a billion km^2 of the solar disk, reaching a peak intensity and gradually decaying and completely disappearing within several minutes to several hours, depending on the size of the flare.

Within half an hour or more following the appearance of large solar flares, energetic particles, consisting mostly of protons, are detected at the Earth, particularly within the auroral zones around the geomagnetic poles. The radiation dies away with a time constant of one to three days. The constituent particles are electrons, protons, alpha particles, and very small numbers of nuclei having intermediate masses (C, N and O). The ratio of protons to alpha particles and of protons to medium nuclei vary considerably between solar events, whereas the ratio of alpha particles to medium nuclei remains relatively constant.

Although the fluctuation in flux intensity is much more severe and random than those of galactic cosmic rays, the following phenomena have been observed: a) there may be an 11-month cycle in the peak number of events; b) there is a semiannual variation which has maxima in March and September, near the equinoxes; c) the maximum number of events occurs on the average near the September equinox and the minimum during December or January; d) the number of flares varies with the 11-year solar cycle; and e) there is a definite tendency for flare events producing a large proton fluence to occur during the increase or decrease of sunspot activity rather than during the maximum. [5.22] Observed sunspot numbers for the previous solar cycles and the predicted numbers for cycle 21, in which a maximum will be reached in 1980, are shown in Figures 5.4 and 5.5. [5.23]

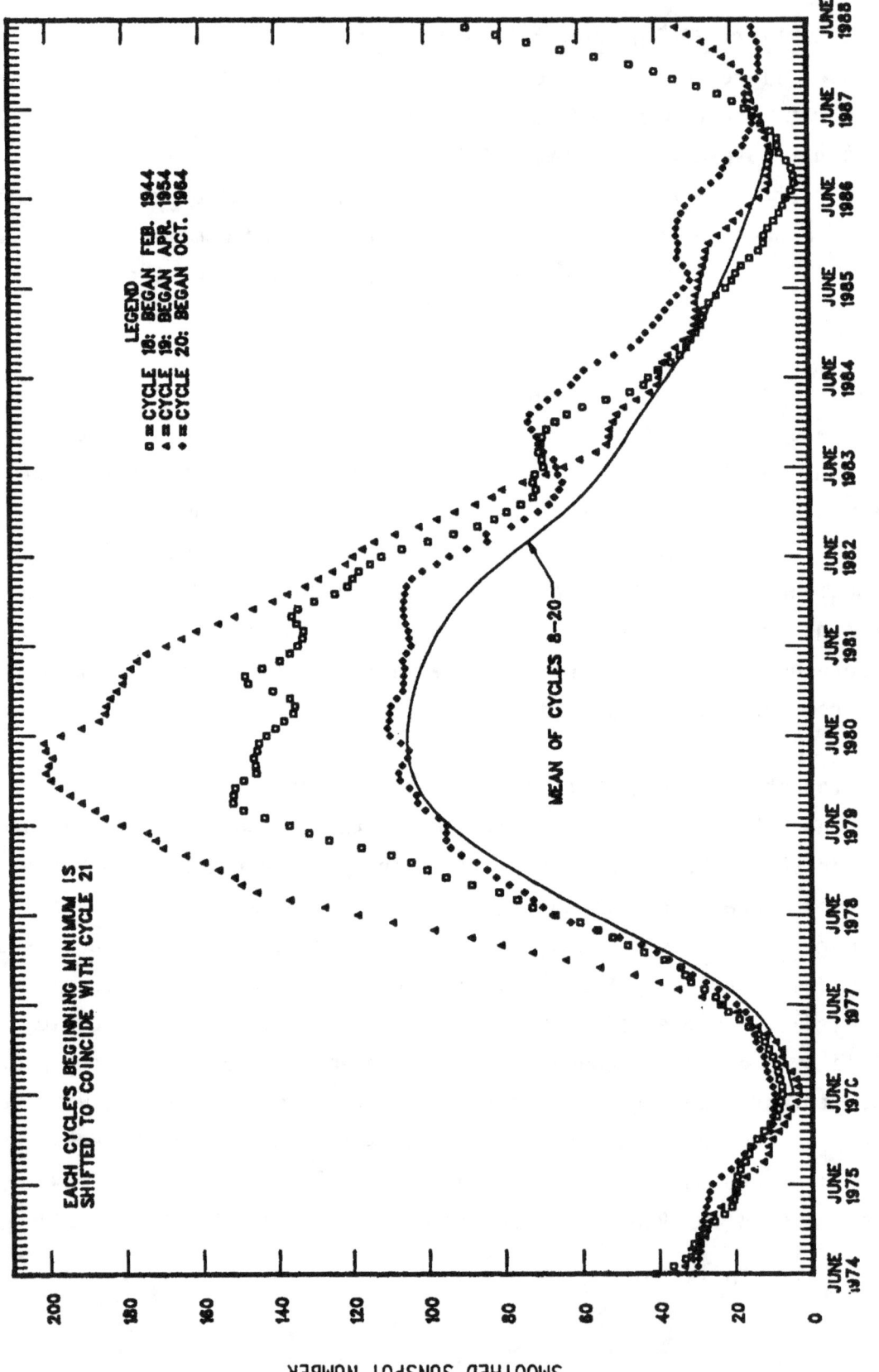

Figure 5.4 Superposition of Cycles 18, 19, and 20

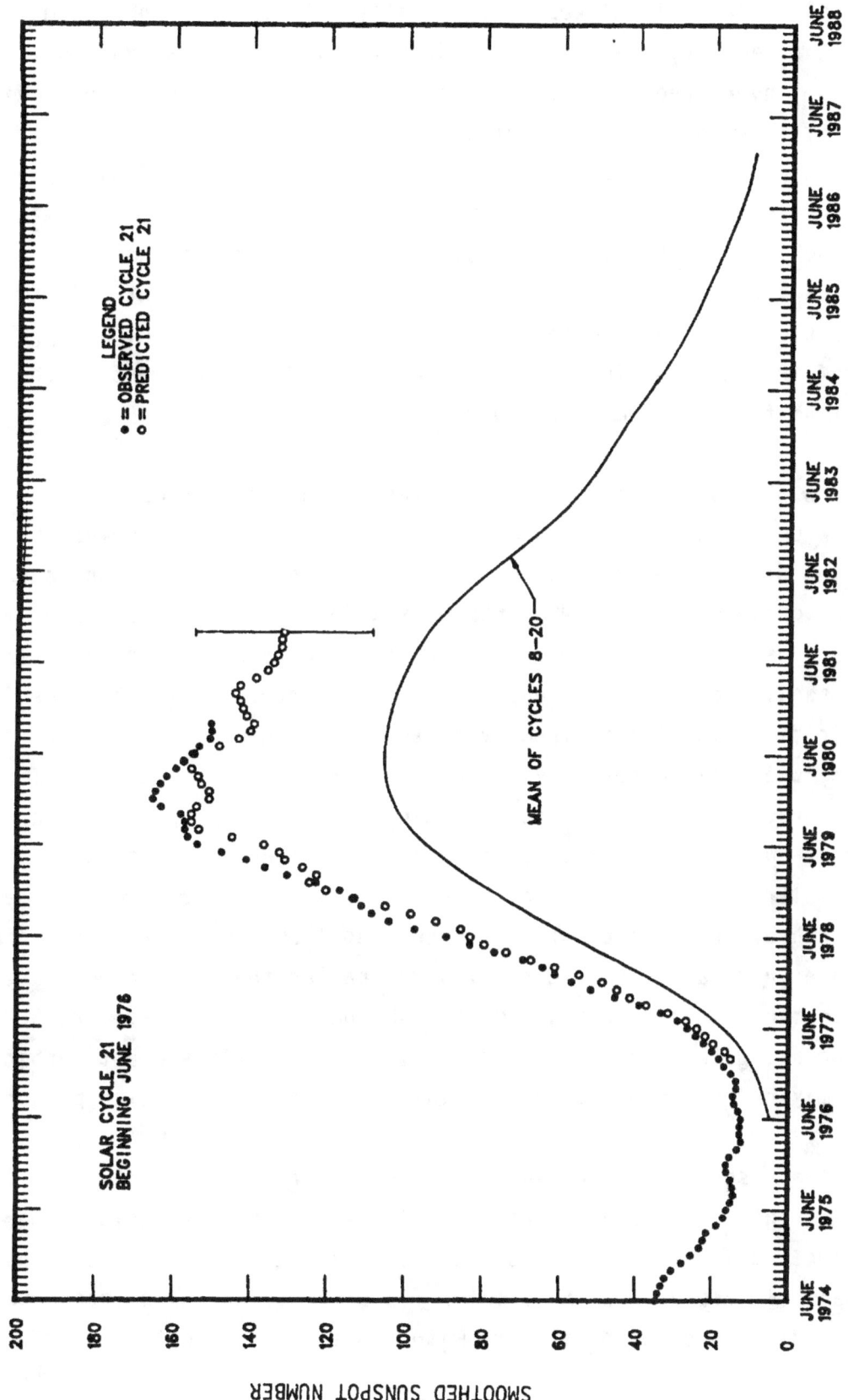

Figure 5.5 Observed and One-Year-Ahead Predicted Smoothed Sunspot Numbers 5.23

5-13

Solar flare particle fluxes arriving at the Earth are highly time-dependent in intensity, spectrum and isotropy. The rise time varies with the individual event and is strongly energy-dependent, reaching the maximum intensity first at higher energies and thus showing a harder spectrum at the beginning. After the peak of radiation, the integral flux decays with time at a rate approximately proportional to t^{-n}, where t is time and n is roughly equal to 3. The particle flux arriving in the upper atmosphere is for the most part isotropic; however, significant anisotropies frequently exist for shorter durations, arriving from a highly preferred and fairly narrow direction in space from 30° to 60° west of the Earth-sun line for a period of a few minutes.

Figure 5.6 is an illustration of a flare event and the magnetic flux associated with the event. The plasma ejected by the flare carries the magnetic line of force of the sunspot with it. The sun's rotation causes the lines to curve westward. When the plasma front impacts the Earth's magnetosphere we experience a magnetic storm. The magnetic fields interact and merge, sending electric currents around the Earth. During this period, the magnetic field is linked directly to the sun. Should another flare event occur, the solar cosmic rays have a direct path to the Earth. The solar cosmic rays travel along the magnetic field lines joining the Earth and the sun. The Earth's magnetic field at this point does not shield the Earth from solar cosmic rays. Galactic cosmic rays, however, are shielded from the Earth by the magnetic fields extending from the sun to the Earth. This decrease in galactic cosmic ray flux is called the Forbush decrease. This will continue until the flare event subsides and the plasma front diffuses, breaking the magnetic field link. During a flare event, orbiting spacecraft may be subjected to substantially high fluxes of solar protons.

Solar flare spectra are often described in terms of a quantity called the magnetic rigidity. Magnetic rigidity is defined by the ratio of momentum to charge and is a measure of the ability of charged particles to penetrate a magnetic field. For many solar flare events the time integrated spectrum exhibits an exponential form with respect to rigidity and is

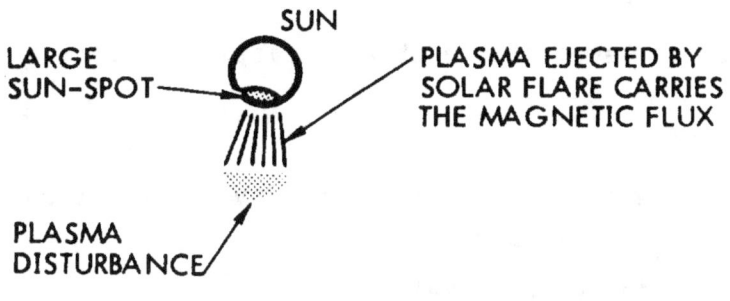
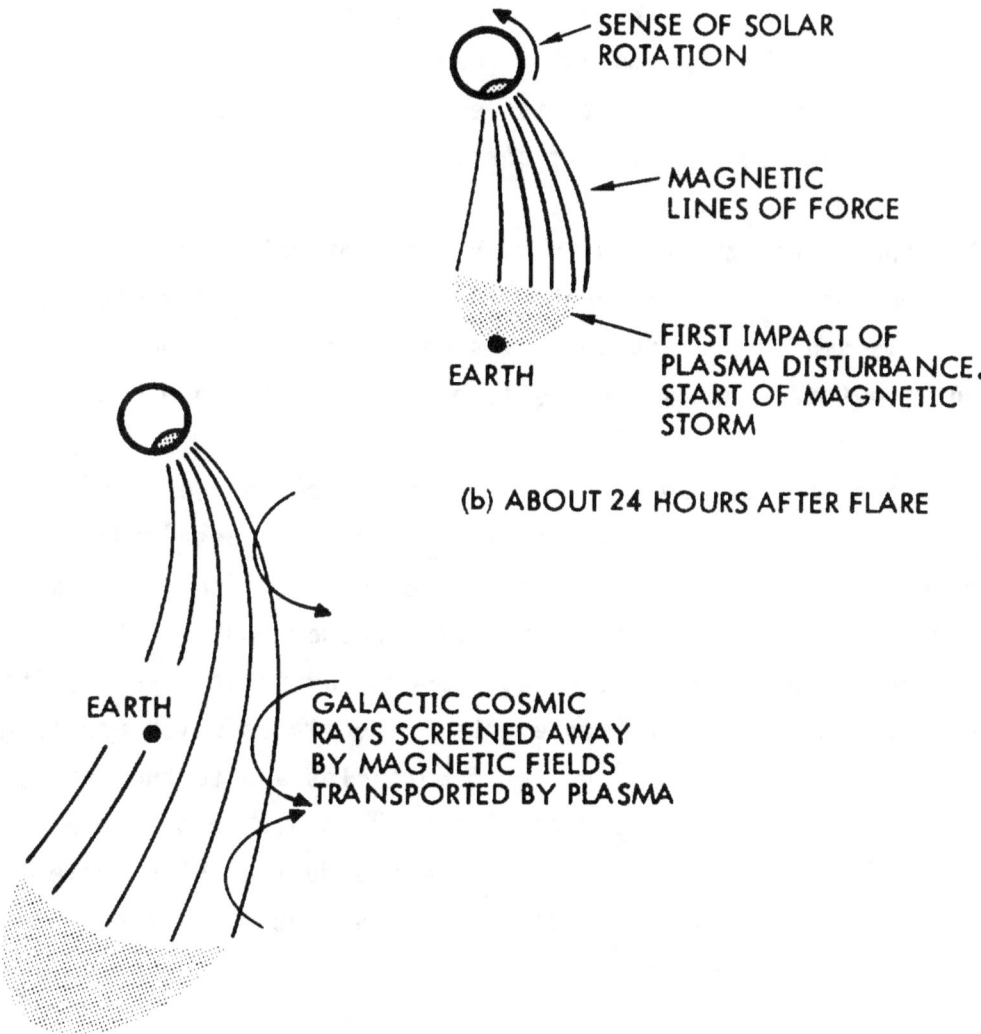

Figure 5.6 Changes of the Interplanetary Magnetic Field Regime Model with Time 5.24

Customarily expressed in terms of the characteristic rigidity R_0 as follows:

$$\Phi(>R) = \text{integral flux having rigidity greater than R}$$
$$= \Phi(>R_0) \cdot (e^{1-R/R_0})$$
$$R = \text{rigidity (v or Mv)}$$
$$= pc/zq = \frac{\sqrt{E^2 - (m_0c^2)^2}}{zq} = \frac{\sqrt{T(T + 2m_0c^2)}}{zq}$$
$$E = \text{total energy (MeV)}$$
$$T = \text{kinetic energy (MeV)}$$
$$p = \text{momentum (MeV/c)}$$
$$m_0c^2 = \text{rest mass energy (938 MeV per nucleon)}$$
$$\text{and } zq = \text{atomic charge}$$

The characteristic rigidity R_0 varies not only with each event but within the spectrum of an event.[5.25, 5.26] The R_0 computed for the annual flux is smaller during the years near sunspot maximum (50 to 70 MV), but the total annual fluence is higher during these years.

For the purpose of predicting the size and spectrum of solar flare proton events, many statistical analyses have been made on proton events observed near or on the Earth. Unfortunately, the correlation between the prediction and observations has been rather poor. A Poisson distribution may be appropriate for sunspot numbers and solar flares on the sun, but not for solar flare proton events. The flares which are large enough to emit a large number of energetic particles and further satisfy the requirements of protons to reach the Earth obviously belong to a special class of solar flare events. Phenomena observed during solar cycle 19 are enumerated below for review, particular empasis be placed on those which appear to be dependent on solar activity.

a) The flares capable of producing large proton events tend to occur when the rate of change in annual sunspot number becomes greater.

b) The characteristic rigidity of solar flare protons is randomly distributed throughout an 11-year cycle, but both the annual expectation value and variance are not. During a period of increasing or decreasing sunspot activity, the R_0 becomes larger on the average than that during the maximum, and the variance becomes smaller during the solar maximum. That is to say, the solar flare proton events are relatively steady and confined in a smaller rigidity range during the solar maximum, whereas the size and spectrum become erratic when the rate of change in sunspot activity becomes severe.

c) The size of each event, as measured by an integral proton flux of energy greater than 30 MeV, is almost randomly distributed over an 11-year cycle, but a line connecting the successive annual fluence plotted against sunspot number is not a single-valued function.

King [5.26] made a probabilistic study of solar proton fluence level based on 1966-1972 data. The probability with which any given solar proton fluence level will be exceeded was computed for the active phase of the current cycle (177-1983) [5.28] The probability is a function of fluence level, proton energy threshold, and mission duration. He assumed that fluences of all anomalously large (AL) events have a spectrum given by the August 1972 event, and fluences of the ordinary (OR) events obey a log normal distribution. The computer code (SOLPRO) developed for this calculation [5.29] is provided to supplement the equivalent fluence calculation code and is listed in Appendix D. The solar flare proton environment of solar cycle 20 is shown in Figure 5.7. The spectrum for an anomalous event [5.26] is also shown in order to compare with the spectrum used in Reference 5.30. A spectrum softer than the August 1972 event is used for the latter model and the annual fluence level is scaled according to the solar activity as measured by smoothed sunspot number (Figure 5.8).

The annual integral flux for solar cycles 19 and 20 from 1956 through 1979 and predicted from 1980 through 1990 using SOLPRO is shown in Table 5.1 [5.25-5.29, 5.31]

Since solar flare particle fluxes are rich in low rigidities, a strong cutoff phenomenon is expected. During the quiescent state, the cutoff rigidity at low latitude is a strong function of direction as well as

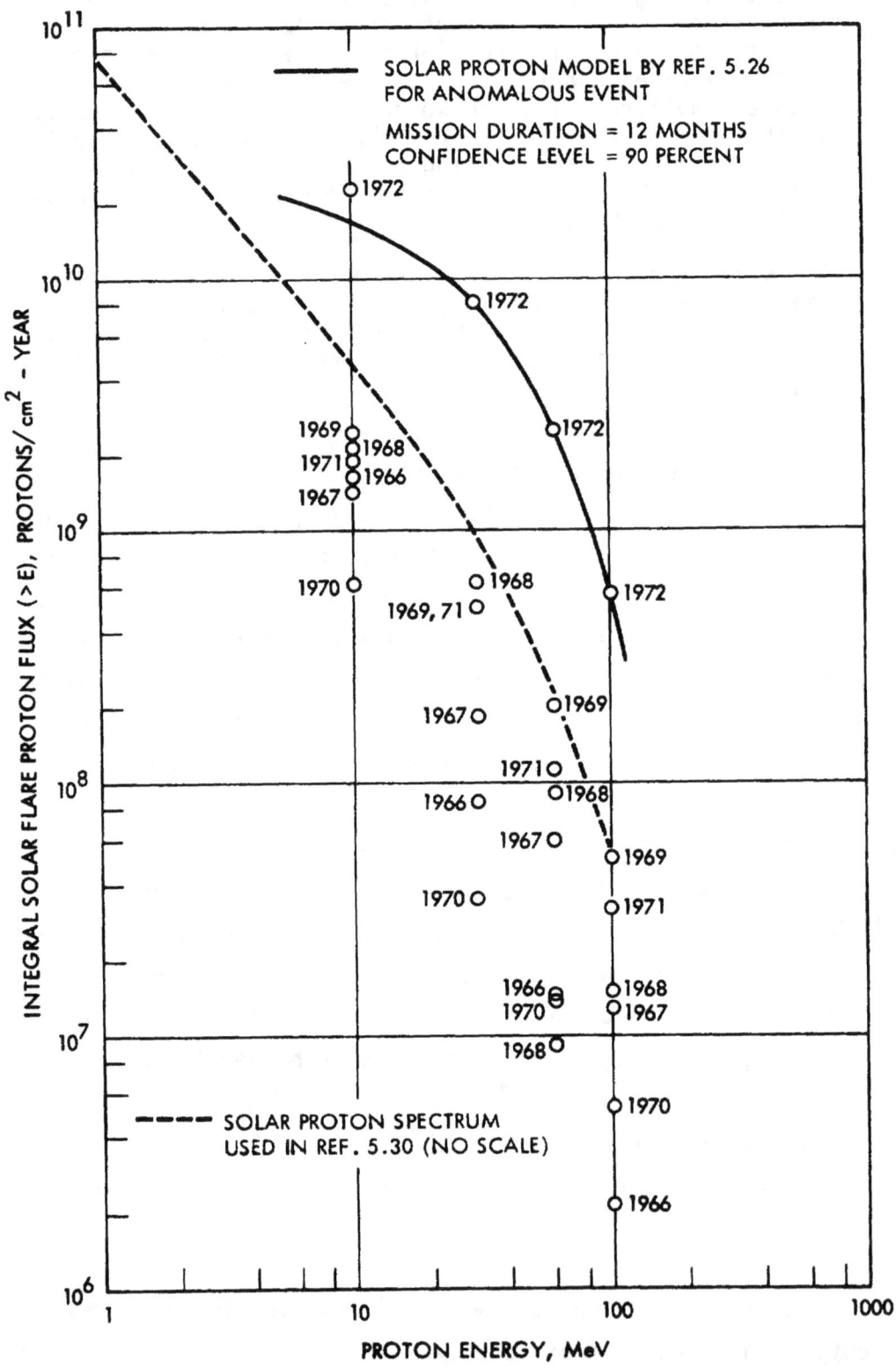

Figure 5.7 Solar Flare Proton Environment of Solar Cycle 20 [5.26]

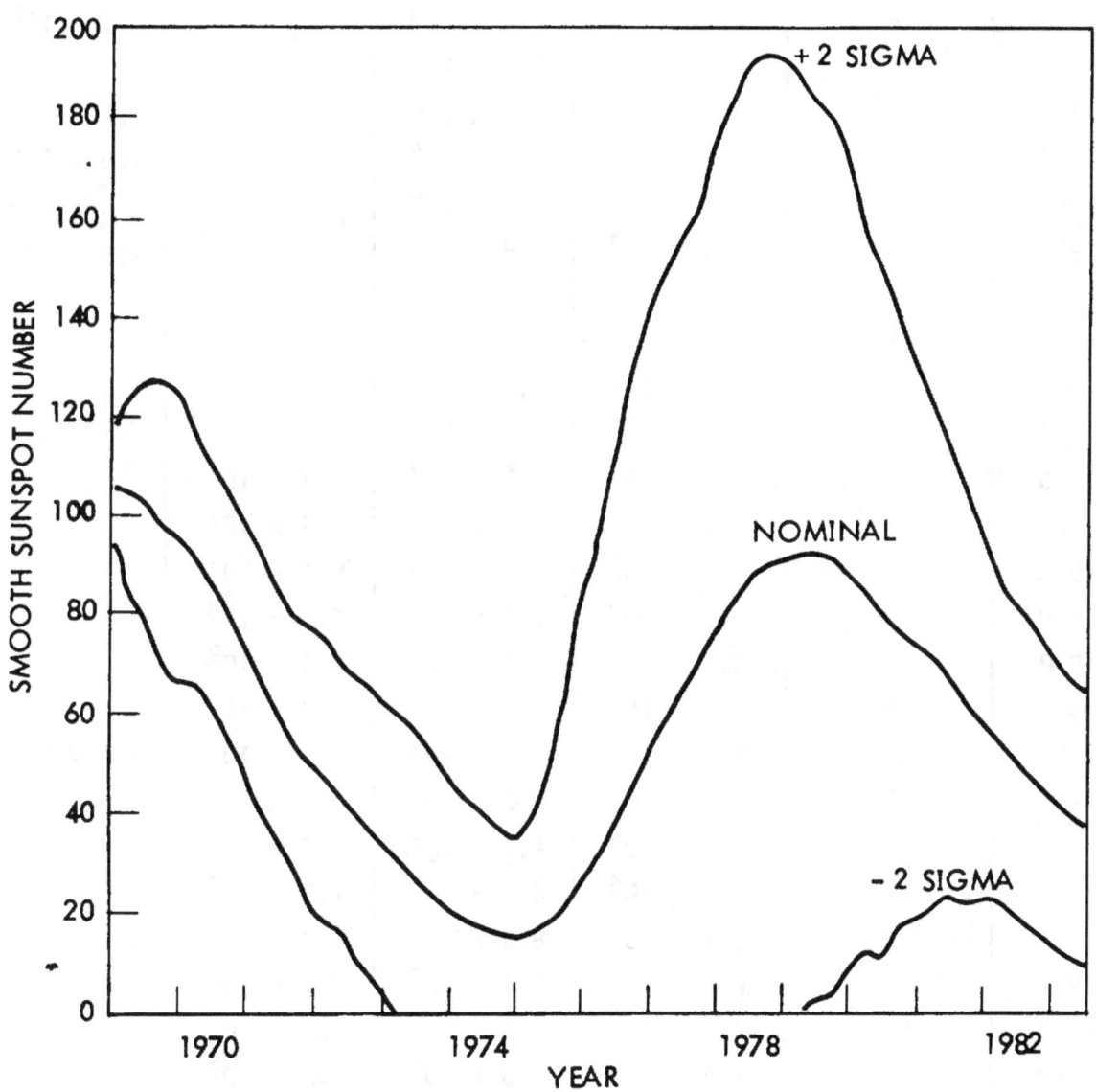

Figure 5.8 Predicted Smoothed Sunspot Number for Solar Cycle 21 [5.30]

Table 5.1. Observed and Predicted Unattenuated Interplanetary Annual Integral Solar Proton Fluence 5.25 - 5.29, 5.31

Year	Number of Events	Integral Fluence (protons/cm^2)			
		$\Phi(>10\text{ MeV})$	$\Phi(>30\text{ MeV})$	$\Phi(>60\text{ MeV})$	$\Phi(>100\text{ MeV})$
1956	4	2.0×10^9	1.0×10^9		3.5×10^8
1957	9	---	4.0×10^8		2.0×10^7
1958	8	7.0×10^9	7.8×10^8		2.4×10^7
1959	6	2.2×10^{10}	4.2×10^9		4.6×10^8
1960	15	6.8×10^9	2.2×10^9		3.8×10^8
1961	6	1.6×10^9	3.5×10^8		4.2×10^7
1962	2				
1963	1				
1964		1.0×10^7	2.4×10^6		
1965	1	2.5×10^7	2.8×10^6		
1966	4	1.7×10^9	8.7×10^7	1.4×10^7	
1967	5	1.5×10^9	1.9×10^8	5.9×10^7	
1968	7	2.2×10^9	2.8×10^8	9.4×10^7	
1969	4	2.5×10^9	5.1×10^8	2.1×10^8	
1970	6	6.3×10^8	3.5×10^7	1.4×10^7	
1971	3	1.9×10^9	5.1×10^8	1.2×10^8	
1972	3	2.3×10^{10}	8.2×10^9	2.5×10^9	5.5×10^8
1973	3	1.4×10^7	6.3×10^6	3.0×10^6	
1974	5	2.8×10^8	4.2×10^7	8.2×10^6	
1975	1	3.3×10^6	2.0×10^6	1.2×10^6	
1976	1	2.1×10^7	8.6×10^6	4.2×10^6	
1977	5	2.6×10^8	8.4×10^7	3.7×10^7	
1978	11	6.3×10^9	7.9×10^8	1.4×10^8	
1979	7	8.4×10^8	1.6×10^8	4.3×10^7	
ANNUAL INTEGRAL FLUENCE PREDICTION FOR CYCLE 21 (1978 - 1984) USING SOLPRO WITH 90% CONFIDENCE LEVEL THAT CALCULATED FLUENCES WILL NOT BE EXCEEDED					
1 AL EVENT		1.7×10^{10}	7.9×10^9	2.5×10^9	5.6×10^8

of latitude (approximately proportional to $\cos^4\lambda$ for large geomagnetic latitudes), and hence of L. Galactic cosmic rays follow this normal Stormer cutoff as do flare particles just before the plasma cloud hits the geomagnetic field. After the impact of the plasma front from the flare event the geomagnetic field is disturbed, resulting in a magnetic storm. This disturbance is in such a manner that a field due to a time-dependent ring current appears to superimpose on the normal geomagnetic dipole field. This causes the disturbed line of force to stretch farther out from the Earth at a given latitude. As a result the particle rigidity necessary to penetrate at a given latitude is greatly reduced and the cutoff energy becomes time-dependent. Satellite observations have indicated that the cutoff energy at synchronous altitude seems to be much less than expected, and flare protons with energy as low as a few hundred keV were observed during the storm. If this is indeed the case, the cutoff energy due to the geomagnetic field becomes insignificant at this altitude, because the cutoff due to solar cell cover shield is normally far greater than the magnetic cutoff during a storm. If both altitude and latitude are low, the field perturbation due to the storm may be insignificantly small compared with that of the quiescent state and the Stormer cutoff approximation may prevail. The geomagnetic shielding phenomena are shown in Figure 5.9 [5.30] for protons from a class three flare on July 18, 1961. It is evident from the figure that orbit inclination is a significant factor for spacecraft at low altitude.

For unmanned missions of a year or longer the ordinary (OR) solar flare fluence may be insignificant because there is a high probability that an anomalously large (AL) solar flare event will occur. This AL event will expend all its fluence in a relatively short time, 2 to 4 days, and totally overshadow all the OR events. Neither the AL event itself nor its time of occurrence can be predicted, but statistically an AL event will occur sometime during a solar cycle. The prediction of solar flare proton fluxes becomes a function of mission duration and a confidence level Q through a modified type of Poisson statistics. Missions of short duration are less subject to AL events and more subject to OR events, but this does not preclude the possibility of an AL event occurring during a short mission. [5.14]

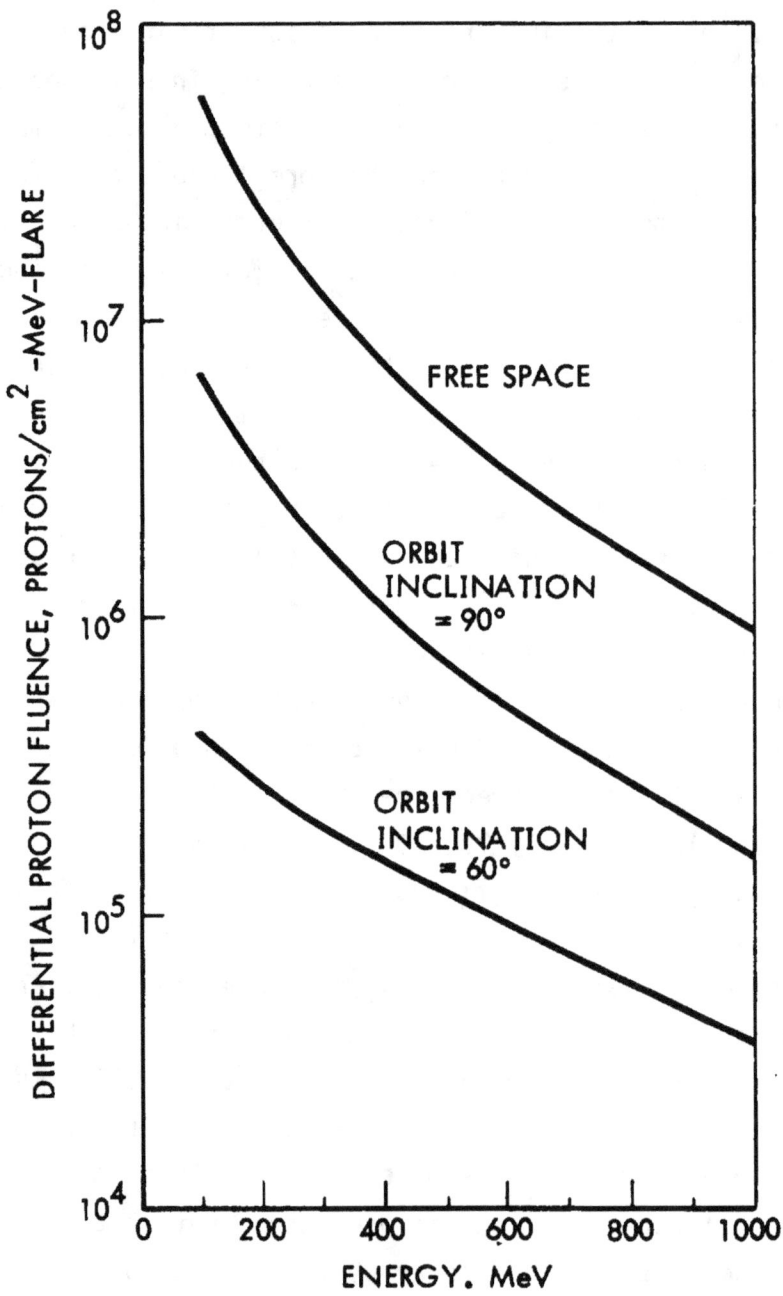

Figure 5.9 Solar Flare Proton Environment in a 200-nmi Circular Orbit Due to a Class Three Flare Event on July 18, 1961 [5.30]

Synchronous orbits do not experience a significant amount of geomagnetic shielding for cosmic rays of solar or galactic origin of energy E>10 MeV. Therefore, spacecraft in geostationary orbit will receive 100% exposure of unattenuated interplanetary solar flare proton intensities of all energies above 10 MeV. To a first approximation this exposure is omnidirectional and isotropic.

REFERENCES

5.1 W. N. Hess, *The Radiation Belt and Magnetosphere*, Blaisdell Publishing Company, Waltham, Mass., 1968.

5.2 C. W. McIlwain, "Coordinates for Mapping the Distribution of Magnetically Trapped Particles," J. Geophys. Res. $\underline{66}$, 3681, 1961.

5.3 J. H. King, "Models of Trapped Radiation Environment, Vol. $\underline{IV}$, Low Energy Protons," NASA SP-3024, 1967.

5.4 J. I. Vette, "Models of the Trapped Radiation Environment, Vol. $\underline{I}$: Inner Zone Protons and Electrons," NASA SP-3024, 1966.

5.5 J. I. Vette, A. B. Lucero, J. A. Wright, "Models of the Trapped Radiation Environment, Vol. $\underline{II}$: Inner and Outer Zone Electrons," NASA SP-3024, 1966.

5.6 J. P. Lavine and J. I. Vette, "Models of the Trapped Radiation Environment, Vol. $\underline{V}$, Inner Belt Protons," NASA SP-3024, 1969.

5.7 J. P. Lavine and J. I. Vette, "Model of the Trapped Radition Environment, Vol. $\underline{VI}$, High Energy Protons," NASA SP-3024, 1970.

5.8 J. I. Vette and A. B. Lucero, "Models of Trapped Radiation Environment, Vol. $\underline{III}$, Electrons at Synchronous Altitudes," NASA SP-3024 1967.

5.9 G. W. Singley and J. I. Vette, "The AE-4 Model of the Outer Radiation Zone Electron Environment," NASA, NSSDC 72-06, 1972.

5.10 M. J. Teague and J. I. Vette, "The Inner Zone Electron Model AE-5," NASA, NSSDC 72-10, 1972.

5.11 M. J. Teague, K. W. Chan, J. I. Vette, "AE6: A Model Environment of Trapped Electrons for Solar Maximum," NASA, NSSDC/WDC-A-R&S 76-04, 1976.

5.12 G. W. Singley and J. I. Vette, "A Model Environment for Outer Zone Electrons," NASA, NSSDC 72-13, 1972.

5.13 D. M. Sawyer and J. I. Vette, "AP8 Trapped Proton Environment for Solar Maximum and Solar Minimum," NSSDC/WDC-A-R&S 76-06, 1976.

5.14 E. G. Stassinopoulos, "The Geostationary Radiation Environment," Journal of Spacecraft and Rockets, $\underline{17}$, 145, 1980.

5.15 H. K. Hills, K. W. Chan, M. J. Teague, J. I. Vette, to be published.

5.16 E. G. Stassinopoulos, J. J. Hebert, E. L. Butler, J. L. Barth, "SOFIP: A Short Orbital Flux Integration Program," NSSDC/WDC-A-R&S 79-01, 1979.

5.17 M. J. Teague, N. J. Schofield, K. W. Chan, J. I. Vette, "A Study of Inner Zone Electron Data and Their Comparison with Trapped Radiation Models," NSSDC/WDC-A-R&S 79-06, 1979.

5.18 A. Hassitt, and C. E. McIlwain, "Computer Programs for the Computation of B and L (May 1966)," NSSDC 67-27, 1967.

5.19 E. G. Stassinopoulos and G. D. Mead, "ALLMAG, GDALMAG, LINTRA: "Computer Programs for Geomagnetic Field and Field-Line Calculations," NSSDC 7212, 1972.

5.20 J. Jensen, G. Townsend, J. Kork, D. Kraft, Design Guide to Orbital Flight, McGraw-Hill, New York, 1962.

5.21 P. R. Escobal, Methods of Orbit Determination, J. Wiley & Sons, New York, 1976

5.22 P. M. Blair, Jr. and H. Y. Tada, "Environments of SYNCOM, Mark II," TM-732, Aerospace Group, Hughes Aircraft Co., 1972.

5.23 Solar Geophysical Data, Prompt report, No. 441-part 1, USDC, NOAA, May 1981.

5.24 F. B. McDonald, (Ed.), Solar Proton Manual, NSAS TR R-169, Dec. 1963.

5.25 W. R. Webber, "An Evaluation of Radiation Hazard Due to Solar Particle Events," The Boeing Co., Report D2-90469, Dec. 1963

5.26 J. H. King, "Solar Proton Fluences for 1977-1983 Space Missions," J. of Spacecraft and Rockets, 11, No. 6, 401, June 1974.

5.27 Courtesy of Gary Heckman of NOAA.

5.28 E. G. Stassinopoulos and J. H. King, "Empirical Solar Proton Model for Orbiting Spacecraft Applications," IEEE Transaction on Aerospace and Electronic Systems, AES-10, No. 4, July 1974.

5.29 E. G. Stassinopoulos, "SOLPRO: A Computer Code to Calculate Probabilistic Energetic Solar Proton Fluences," NASA, NSSDC 75-11, 1975.

5.30 D. K. Weidner, "Natural Space Environment Criteria for 1975-1985 NASA Space Stations," NASA TM X-53865, Aug. 1969, Second Edition, 10 Aug. 1970.

5.31 E. R. Martina and S. A. Yinger, "Energetic Proton Fluxes at Synchronous Altitude 23 Aug. 1977 through 7 Aug. 1979," Aerospace Report No. TOR-0081 (6409-34)-1, March, 1981.

CHAPTER 6

6.0 SOLAR ARRAY DEGRADATION CALCULATIONS

In the previous sections, the three basic input elements necessary to perform degradation calculations were developed. The first of these elements is degradation data for solar cells under normal incidence 1 MeV electron irradiation. The second input element is the effective relative damage coefficients for omnidirectional space electrons and protons of various energies for solar cells with various coverglass thicknesses. The third input element is space radiation environment data for the orbit of interest. This section will cover the use of these data to perform a solar array degradation estimate.

6.1 General Procedure, Equivalent Fluence

The effective relative damage coefficients allow the conversion of various energy spectra of space electrons and protons into equivalent fluences. The equivalent fluences are based on normal-incidence monoenergetic irradiations for which the degradations of the solar cells of interest are characterized. The process of weighting an integral energy spectrum of electrons for a given orbit can be described as follows:

$$\Phi_{1 \text{ MeV e}} = \sum_{E=0}^{\infty} [\Phi(>E) - \Phi(>E + \Delta E)] \cdot D(E,t) \qquad (6.1.1)$$

where $\Phi_{1 \text{ MeV e}}$ = the damage equivalent 1 MeV electron fluence (e/cm^2-year)

$\Phi(>E) - \Phi(>E + \Delta E)$ = the isotropic particle fluence having energies in a small energy increment greater than energy E (e/cm^2-year)

$D(E,t)$ = the relative damage coefficient for isotropic fluences of energy E incident on solar cells shielded by coverglasses of thickness t (dimensionless)

The quantities $\Phi(>E) - \Phi(>E + \Delta E)$ for a range of energies are also known as the difference spectrum. This spectrum can be generated from an integral energy spectrum for any energy increments desired. For the case of space protons, equation (6.1.1) can also be used with the exception that D(E,t) values for protons are based on 10 MeV proton fluences rather than 1 MeV electrons. The calculated equivalent fluence will therefore be a damage equivalent 10 MeV proton fluence. The equivalent 10 MeV proton fluence can be converted to equivalent 1 MeV electron fluence as follows:

$$\Phi(1 \text{ MeV e}) = 3000 \cdot \Phi(10 \text{ MeV p}) \qquad (6.1.2)$$

The above relationship is an approximation which must be made for the purpose of combining electron and proton damage. In Section 3.3, the differences between electron and proton degradation were discussed. Since the slope of the degradation curve (the constant C in equation 3.2.1) is different for 1 MeV electron and 10 MeV proton irradiations, the constant in equation (6.1.2) will differ depending on the level of degraded cell output at which this constant is determined. At present, the best information available indicates a value equal to 3000 when cell output parameters are degraded by 25%. In cases when the cell degradaton is entirely dominated by proton damage, the cell degradation could be estimated more accurately by calculating the equivalent 10 MeV proton fluence, and using 10 MeV proton cell damage data, than by using the equivalent 1 MeV electron fluence and electron data.

An additional problem arises in calculating equivalent fluences for proton environments. The results shown in Figures 4.3 and 4.4 reveal that different values of D(E,t) for proton irradiation are found when this damage constant is based on cell I_{sc} or P_{max} and V_{oc}. The P_{max}, V_{oc} damage coefficients are higher in the low energy region, which accounts for the much higher damage produced in these parameters by low energy protons. This differs with the results of electron irradiation where one value of D(E,t) describes the behavior of all cell output parameters. Because of the two sets of D(E,t) values for proton irradiation, two different equivalent 10 MeV proton fluences must be considered. One of these will describe the variation of solar cell P_{max} and V_{oc}. The other will describe the variation of solar cell I_{sc}.

The values of $D(E,t)$ have been calculated by assuming infinite back shielding. Although this condition is often approached by the body-mounted solar arrays of spinning spacecraft, it is not generally true. The designer must also evaluate the contribution of equivalent fluence resulting from radiation incident on the back side of the solar cells. The result is a front and a back component of equivalent fluence. A question arises as to the values of $D(E,t)$ to be used for back irradiations. In the case of trapped space electron irradiation, it is reasonable to use the same values of $D(E,t)$ for both front and back irradiations. The only problem in this case is to convert the backshielding of the panels, satellite, etc., to an equivalent planar shielding (gm/cm^2).

The case for space protons is considerably more complex because of the nonpenetrating nature of low energy protons. There is an increasing need for a technique to evaluate rear irradiation effects with the increased use of light weight solar panels with negligible back radiation protection. Low-energy proton irradiation from the rear not only increases bulk resistivity, thereby decreasing the fill factor, but also greatly changes the forward dark I-V characteristic curves.[6.1-6.3] These phenomena, peculiar to rear irradiation, must be considered and included in the evaluation of $D(E,t)$. Scarcity of useable data and lack of proper technique prevent the appropriate evaluation of $D(E,t)$ at present. The only alternative is to use the front irradiation data, assuming that both front and back irradiations produce the same result as long as all protons penetrate through the junction. However, the use of the V_{oc}, P_{max} coefficients, which were designed to account for the high junction damage by low energy protons, is not considered proper for protons incident on the rear cell surface. Therefore, we use only the I_{sc} proton damage coefficients for rear incidence protons. To allow for the self-shielding effect for cells irradiated with protons from the rear, the back contact solder thickness (approximately 0.01 to 0.08 mm) plus the thickness of the substrate and the substrate adhesive should be included in the total back shielding.

The various contributions and variations of equivalent fluence which can be encountered in a natural space environment are summarized in Table 6.1. Columns in the right side of the table indicate the contributions from the various radiation components to the two different types of equivalent fluence. Although the most general case can involve all the contributions shown in Table 6.1, except for the P_{max}, V_{oc} coefficients for rear incidence protons, in a typical earth orbit only a few of these contributions may be significant.

Table 6.1
Summary of Equivalent Fluence Contributions

		Contributions	
		I_{sc}	P_{max}, V_{oc}
1.	Trapped electrons, front, (I_{sc}, P_{max}, V_{oc})	X	X
2.	Trapped electrons, back, (I_{sc}, P_{max}, V_{oc})	X	X
3.	Trapped protons, front, (I_{sc})	X	
4.	Trapped protons, back, (I_{sc})	X	X
5.	Trapped protons, front, (P_{max}, V_{oc})		X
6.	Trapped protons, back, (P_{max}, V_{oc})	NA	NA
7.	Flare protons, front, (I_{sc})	X	
8.	Flare protons, back, (I_{sc})	X	X
9.	Flare protons, front, (P_{max}, V_{oc})		X
10.	Flare protons, back, (P_{max}, V_{oc})	NA	NA

Thus, 1, 2, 3, 4, 7, and 8 contribute to the I_{sc} total equivalent fluence and 1, 2, 4, 5, 8, and 9 to the P_{max}, V_{oc} total equivalent fluence. Types 6 and 10 do not contribute.

The calculation of equivalent fluence and subsequent estimation of degraded solar cell output from the data in Figures 3-20 through 3-111 yields data which are valid for temperatures of 28°C and solar illumination power densities of 135 mW/cm^2. When degraded solar cell outputs are desired for temperatures other than 28°C, corrections can be made by use of temperature coefficients discussed in Section 3.11. The evaluation of changes in solar cell response due to reduced light transmission in the coverglass materials will be covered in the next section.

6.2 Effect of Reduced Light Transmission on Solar Cell Response

To use the coverglass darkening data previously presented, a procedure is necessary to evaluate the absorbed dose produced by the various radiation components of the space environment. This can be done by use of data developed in Chapter 4. The procedure is similar to that used for equivalent fluence, with the exception that the absorbed dose is a point function and therefore varies with depth in the cover material. To calculate the absorbed dose at a particular depth in the cover materials, the following expression is used:

$$\text{Dose}(d) = \sum_{E=0}^{\infty} [\Phi(>E) - \Phi(>E + \Delta E)] \cdot I(E,d) \qquad (6.2.1)$$

where Dose(d) = the absorbed dose in the cover material at a depth d

   $I(E,d)$ = the absorbed dose per unit fluence for isotropic space radiation particles of energy E at depth d in the shielding material. See Figures 4.2 and 4.5, Tables 4.2 and 4.5.

The absorbed dose must be calculated at several depths in the cover material, and the electron and proton portions of the environment must be summed to determine the dose-depth profile. The necessity of including contributions from back radiations must also be considered. In practice, the dose deposited will decrease greatly with increasing depth into the cover materials. The greater dose near the surface is due largely to low-energy trapped protons, and contributes little to the average dose deposited in the cover materials. Because of the uncertainties in evaluating cover material transmission loss in space, there is little to be gained in making an extremely accurate evaluation of the surface dose. When the average dose deposited in the cover material is known, the degradation in transmission can be estimated from the data in Section 3.12. These loss factors may then be applied to the estimated solar cell output parameter values.

6.3 Rough Degradation Calculations

For circular orbits around the Earth, Vette, et al. [6.4-6.6] have time-integrated both electron and proton environments for convenient energy ranges, various altitudes, and inclinations of 0°, 30°, 60°, and 90° appropriate for both solar minimum and solar maximum. The average daily omnidirectional integral fluences are presented in the form of carpet plots in References 6.4 through 6.6. For inner zone electrons ($L \leq 2.8$) and at solar maximum, model AE6MAX (epoch 1980) described in Reference 6.4 is used, while at solar minimum, model AE5MIN (epoch 1975 projected) described in Reference 6.6 is used. The carpet plots of circular orbit integrations in References 6.4 and 6.6 use the old AE4 and AE4MIN models to describe the outer zone electrons, but these have both been superseded by the interim AEI7-LO and AEI7-HI models.[6.7] No orbital integration results from NSSDC are yet available using AEI7-LO or AEI7-HI. In the case of protons, the old models, AP1, AP5, AP6, and AP7, have all been replaced by models AP8MIN (epoch 1964) for solar minimum protons and AP8MAX (epoch 1970) for solar maximum protons. Both models are described and carpet plots for circular orbital integrations given in Reference 6.5.

A rough determination of the equivalent fluence can be made by following the procedure described by equations (6.1.1) and (6.1.2). Examples of such rough calculations are shown in Tables 6.2 through 6.4 for a circular orbit of 450 nmi altitude and 90° inclination. Data for the circular integrations are taken from the carpet plots of References 6.4 and 6.5, which are based on the AE6MAX and AP8MAX radiation belt models. As described above, the numbers based on these models are appropriate for periods of maximum solar cycle activity. The $D(E,t)$ values used are taken from Tables 4.1, 4.3 and 4.4 for the average energy value of the energy increment. Calculations shown are for coverglass thicknesses of 6, 12 and 30 mils.

Several observations can be made regarding the calculations in Table 6.2. The largest contribution to the equivalent fluence for the 6 mil cover occurs in the energy increment between 4 and 6 MeV. The equivalent fluence contributions become increasingly less important as the energy increases. The use of the $D(E,6)$ value for 5 MeV (1.253) leads to serious equivalent fluence errors in the most important energy increment of 4 to 6 MeV because $D(E,6)$ changes very rapidly with energy in this region. Similarly, the rapidly changing $D(E,12)$ values in the energy range 6 to 8 MeV and $D(E,30)$ values in the energy range 10 to 15 MeV also lead to serious errors in the calculations for the 12 and 30 mil coverglasses. In each case the region of most rapidly varying $D(E,t)$ is in the energy range just above the energy that will start to penetrate the coverglass. The equivalent 1 MeV electron fluence calculated for 6 mils of shielding by this rough method is 1.14 E13* e/cm^2-year. A similar detailed machine calculation (to be discussed) employing much smaller energy increments yielded an equivalent 1 MeV electron fluence of 1.47 E13 e/cm^2-year. This difference is due to the use of smaller energy increments in the machine calculation and also to differences in the machine calculations of fluence-energy spectra (values from References 6.4 and 6.5 vs. values computed using JPL and NSSDC computer codes). The accuracy of the manual calculation can be improved by using smaller energy increments, but additional values of $\Phi(E_1)$ and $D(E,t)$ must be obtained by interpolation. The omnidirectional fluence should not be reduced by a factor of 1/2 to allow for the assumed infinite rear shielding, because this factor is already included in the $D(E,t)$ term. In Table 6.3, the above calculation is repeated using $D(E,t)$ values based on I_{sc}. If the panel does not have infinite backshielding, similar I_{sc} equivalent fluence contributions based on an appropriate shield thickness would be added to the front surface I_{sc} and V_{oc}, P_{max} contributions. The procedure and problems are identical to those previously discussed.

Table 6.4 displays a rough calculation for trapped electrons in the 450 nmi altitude orbit. The calculation procedure for trapped electrons is exactly the same as that for trapped protons, with the exception that one equivalent fluence value will describe the variation of the solar cell parameters I_{sc}, V_{oc}, and P_{max}. As in the case of the trapped proton evaluations, the major equivalent fluence contributions occur in a few lower-energy increments. For coverglass shielding of 6 mils fused silica, an equivalent fluence of 3.78 E11 e/cm^2-year is determined by these rough methods. A detailed machine calculation of this value indicates 4.59 E11 equivalent 1 MeV electrons/cm^2-year. Such fluences will not produce significant degradation in the performance of current space cells and the electron contribution to the total equivalent fluence may safely be ignored in computing the effect on a solar panel flying in this orbit (450 nmi, 90°).

---

*Throughout this section, the floating point notation will be used to represent exponential quantities. 1.14 E13 = 1.14 x 10^{13}.

The calculation of absorbed dose in shielding materials is very similar to the equivalent fluence calculation and is described mathematically by equation (6.2.1). The I(E,t) value in Figures 4.2 and 4.5 and Tables 4.2 and 4.5 may be used for this purpose. Although the absorbed dose contributed by geomagnetically trapped protons is often very high in the surface layers of shielding, this is usually not a significant contribution to the average absorbed dose in the shielding.

6.4 Computer Calculated Equivalent Fluence

The aforementioned rough calculations can be improved in accuracy and speed with the aid of a computer. Although the quantity computed is exactly the same as before, the selection of difference fluence and the corresponding damage coefficient can be programmed to achieve higher accuracy and more consistent results. The increased accuracy of calculated fluence is achieved mainly by use of finer energy increments for a given environment. A computer program that performs this function for any given electron or proton environment is listed in Appendix D.

Tables 6.6 through 6.38 have been calculated by computing orbital coordinates for various circular orbits, converting the spatial coordinates into B,L coordinates, summing the contributions of electron and proton fluence at each position, then converting the resultant fluence-energy spectra into 1 MeV equivalent electron fluences appropriate for various coverglass thicknesses. The orbital integration is performed by using the equations of Reference 6.8 to compute appropriate trajectories for Earth-orbiting spacecraft. These orbital calculations are estimated [6.8] to give accuracies of generally ± tens of kilometers in altitude, ± one degree in latitude and longitude, and ± one minute in time. The conversion to B,L coordinates uses routines developed at the National Space Science Data Center. [6.9, 6.10] The IGRF model of the Earth's magnetic field was used with epoch 1983. Calculation of the proton and electron fluence at each point in B,L space was performed by the NSSDC program SOFIP. [6.11] The radiation models used in these calculations were AE6MAX and AEI7-LO for electrons and AP8MAX for protons, applicable during periods of solar maximum. Model AP8MIN for solar minimum, will give higher values than AP8MAX at low altitudes, so

AP8MIN was used in calculating proton fluences for a few selected altitudes below 1000 nmi. The program EQFRUX, listed in Appendix D, is used for converting the fluence-energy spectra into 1 MeV equivalent fluence using the $D(E,t)$ values as discussed in Chapter 4.

The annual equivalent 1 MeV electron fluences resulting from geomagnetically trapped particles are tabulated in Tables 6.6 through 6.35 for I_{sc} and for V_{oc} and P_{max}. A summary is given in Table 6.5. The computations are for various circular orbits with inclinations of 0° through 90° using 10° increments in inclination. The final altitude entry in each table is the synchronous altitude for that inclination. Although the damage ratio between 10 MeV protons and 1 MeV electrons varies with degradation level and depends on the solar cell output parameter,[6.12] the ratio of 3000 was assumed throughout the computation.

As discussed in Section 5.1.2, the radiation experienced by a spacecraft in synchronous orbit is dependent on its longitude because the geomagnetic coordinate system is tilted and displaced from the geographic coordinate system. A computation of the equivalent 1 MeV electron fluence as a function of longitude in synchronous orbit is summarized in Tables 6.36 through 6.38. Since the longitude effect is more prominent for higher-energy electrons, the ratio of maximum to minimum equivalent fluence increases with coverglass thickness. The effect is not important for trapped protons because the only protons at this altitude have insufficient energy to penetrate any coverglass of practical thickness.

The assessment of solar-flare proton effects is complicated by several problems:

    a.   the unpredictable nature of future solar flare proton fluxes and energy spectra

    b.   the undefinable nature of geomagnetic cutoff energy during a flare event, and hence, the evaluation of the near-Earth flare environment

    c.   the uncertainty in the isotropy of flare fluxes

The magnetic cutoff energy varies with both altitude and latitude even during quiescent periods, and thus it becomes time-dependent for spacecraft moving with respect to the Earth. Further complications are caused by the plasma disturbance and magnetic field regime sweeping through the Earth, the magnitude of which depends in part on the size and location of flares on the solar disk. Therefore it is impossible to generalize all these conditions. However, there are two distinct cases in which certain assumptions are valid as previously discussed: (a) at high altitude and latitude, the geomagnetic field makes almost negligible contribution to the cutoff phenomena during the storm, and (b) at very low altitude and latitude, the Stormer cutoff approximation may prevail.

The damage coefficients for omnidirectional flux can be confidently used with the following understanding:

a. If the solar flare proton flux is omnidirectional throughout the event, the equivalent fluence computed with the omnidirectional damage coefficients described in Chapter 4 will not result in any error from the directionality of the proton flux.

b. If the flux is unidirectional throughout the event, though such an event is very rare, the computed equivalent fluence based on the omnidirectional damage coefficients will be in error by a factor of two.

Therefore, the uncertainty in flux directionality can be removed by the use of the omnidirectional damage coefficients with the provision that the estimate can be very reasonable for most of the events with a very small probability of underestimating by a factor of two.

The equivalent 1 MeV electron fluences from solar flare protons are calculated for free space and are tabulated in Table 6.39. The environment is for the free space proton fluence-energy spectra listed in Table 5.1, with an extrapolation down to 1 MeV (to be discussed). The values shown are derived from observed values in the years 1972 through 1979, while the 1978-1984 entry is for a computer prediction using SOLPRO. SOLPRO is a computer code available from the National Space Science Data Center (NSSDC) which computes solar proton fluences at 1 A.U. as a function of mission duration and confidence level.$^{6.13}$ The prediction used in this computation was based on a one year mission duration, a confidence level of 90% that the calculated fluences will not be exceeded, and the occurrence of 1 anomalously large (AL) event. The extrapolation from 10 MeV to 1 MeV was performed using a log-log power law matching the slope of the SOLPRO proton spectrum for an AL event (proportional to $E^{-0.7}$) and continuous at 10 MeV. References 6.14 through 6.16 may be consulted for alternate approaches to predicting the solar proton environment for space missions. A similar calculation to the one summarized in Table 6.39 may be performed using these alternate approaches or different assumptions about mission duration and confidence level by using the EQFRUX amd SOLPRO routines listed in Appendix D.

For a trajectory near the Earth, a partial magnetospheric shielding is operative, and a fractional exposure to the flare proton environment has to be calculated if the cutoff energy attributable to coverglass thickness is less than the geomagnetic shielding cutoff energy at various trajectory points. In this case, the determination of a solar flare proton environment requires consideration of both spacecraft trajectory and time dependent flare proton spectra. The computer program SOFIP available from the NSSDC $^{6.11}$ and its Geomagnetic Shielding Module will be found useful for making such calculations.

The damage produced by back radiation is, to first-order, regarded as the same in nature and magnitude as that produced by the front radiation provided only I_{sc} damage coefficients are used. An equivalent fluence attributable to the back radiation can be added to the front contribution by estimating an effective thickness of back shielding. This assumption is not valid when higher order effects are considered. If a composite back-shielding material is similar to the coverglass, only a density correction is required to compute the effective shielding thickness. This is done by comparing shield thicknesses in units of g/cm^2. If the atomic number and/or density of the substrate is vastly different from that of glass, the equivalent fluence should be computed using effective damage coefficients specifically developed for the new shielding material. However, the uncertainty contributed by an improper Z correction is probably much less than the uncertainty introduced by applying these damage coefficients to rear incidence calculations.

6.5 Solar Array Degradation

The process of calculating an equivalent 1 MeV electron fluence reduces the space radiation environment to a laboratory electron environment for which solar cell degradation has been evaluated. When the damage equivalent fluence is known, the estimation of solar array degradation is almost completed. The next step in estimating array degradation is to make use of such variables as base resistivity, cell thickness, front surface treatment (such as AR coating and texturing), and rear surface treatment (such as back surface fields and back surface reflectors) to choose proper solar cell radiation data. The equivalent fluence then allows the estimation of solar cell output parameters through the use of the data in Figures 3-20 through 3-111.

The tabulated equivalence fluence data will be used in two examples to illustrate the calculation of degradation of a solar array. The first example will be for an array in synchronous orbit at 0° inclination with infinite backshielding. The second example will be for a flexible array in a circular orbit at 1500 nmi altitude with 60° inclination which is flying through both trapped protons and electrons.

I. Solar Cell   10 ohm-cm resistivity
                                  0.0203 cm (0.008 in) thick
                                  Dual AR, BSR, No BSF

    Coverglass   0.015 cm (0.006 in) thick
                                  Fused silica, UV filter
                                  antireflecting coating

    Backshielding   Infinite

    Orbit   Synchronous, 0° Inclination

    1 MeV Electron Fluence

        Trapped Electrons   2.48 E13
        Trapped Protons   0
        Total   2.48 E13 e/cm^2-yr.

Solar Cell Output	Absolute	Relative
After 0 Months (Equivalent Fluence = 0)		
I_{sc}	39.4 mA/cm^2	1.00
V_{oc}	538 mV	1.00
P_{max}	16.7 mW/cm^2	1.00
V_{mp}	452 mV	1.00
After 1 Year (Equivalent Fluence = 2.48 E13 e/cm^2)		
I_{sc}	39.0 mA/cm^2	0.990
V_{oc}	535 mV	0.994
P_{max}	16.35 mW/cm^2	0.979
V_{mp}	446 mV	0.987
After 3 Years (Equivalent Fluence = 7.44 E13 e/cm^2)		
I_{sc}	38.3 mA/cm^2	0.972
V_{oc}	529.5 mV	0.984
P_{max}	15.9 mW/cm^2	0.952
V_{mp}	440 mV	0.973

II.
Solar Cell	Same as for Example I
Coverglass	Same as for Example I
Backshielding	0.0076 cm (0.003 in) thick (Equivalent fused silica thickness found in units of g/cm^2)
Orbit	1500 nmi, 60° Inclination

1 MeV Electron Fluence	I_{sc}	V_{oc}, P_{max}
Trapped Electrons, front	1.25 E13	1.25 E13
Trapped Electrons, rear	1.73 E13	1.73 E13
Trapped Protons, front	2.22 E15	4.21 E15
Trapped Protons, rear	<u>3.04 E15</u>	<u>3.04 E15</u>
Total	5.29 E15	7.28 E15 e/cm^2-yr.

Solar Cell Output	Absolute		Relative
After 0 Months (Equivalent Fluence = 0)			
I_{sc}	39.4	mA/cm^2	1.000
V_{oc}	538	mV	1.000
P_{max}	16.7	mW/cm^2	1.000
V_{mp}	452	mV	1.000
After 1 Year (Equivalent Fluence I_{sc} = 5.29 E 15			
V_{oc}, P_{max} = 7.28 E 15)			
I_{sc}	29.6	mA/cm^2	0.751
V_{oc}	465	mV	0.864
P_{max}	10.04	mW/cm^2	0.601
V_{mp}	378	mV	0.836

The effects of coverglass transmission loss due to radiation darkening have been omitted from this estimate. The average absorbed dose due to trapped electrons in the synchronous orbit can be shown (by using the orbital integration data from Reference 6.4 and the data from Table 4.2 in a simple hand calculation similar to those displayed in Tables 6.2 to 6.4) to be approximately 10^7 rad(SiO_2) per year. The data in Figure 3.19 indicate that such a dose would cause a transmission loss of about 0.5%, which would cause 0.5% losses in solar cell I_{sc}, I_{mp} and P_{max}.

An additional factor which must be considered in these calculations is the modification of bare solar cell output by the coverglass. Modern solar cells with Ta_2O_5 AR coatings commonly exhibit an increase in output of approximately 2% due to the mounting of the coverglass. Therefore, if this type of cell/coverglass combination is in use, the values for I_{sc}, I_{mp} and P_{max} should all be multiplied by 1.02 in the above examples. Cells with multiple layer AR coatings show no increase or decrease due to glassing. On older solar cells manufactured with silicon monoxide AR coatings, the addition of the coverglasses caused a decrease of 2 to 6%, but this type of cell need rarely be considered now.

Table 6.2 Manual Calculation of Equivalent Fluence (AP8MAX Protons) V_{oc} and P_{max} Circular Orbit 450 nmi (833 km), 90 Degree Inclination

Energy Increment E_1 E_2 (MeV)	Integral Energy Spectrum $\Phi(>E_1)$ (p/cm²-day)	Difference Spectrum $\Phi(>E_1)-\Phi(>E_2)$ (p/cm²-day)	Shielding Thickness (6 mils) $D(E,6)^2$	Equiv. Fluence (p/cm²-day)	Shielding Thickness (12 mils) $D(12)^2$	Equiv. Fluence (p/cm²-day)	Shielding Thickness (30 mils) $D(E,30)^2$	Equiv. Fluence (p/cm²-day)
4 6	1.90 E7	3.0 E6	1.253	3.76 E6				
6 8	1.60 E7	2.0 E6	1.278	2.56 E6	.980	1.96 E6		
8 10	1.40 E7	1.0 E6	.826	8.26 E5	.952	9.52 E5	.630	7.56 E5
10 15	1.30 E7	1.2 E6	.537	6.44 E5	.576	6.91 E5	.443	4.43 E5
15 20	1.18 E7	1.0 E6	.453	4.53 E5	.453	4.53 E5	.396	4.75 E5
20 30	1.08 E7	1.2 E6	.417	5.00 E5	.414	4.97 E5	.358	5.73 E5
30 50	9.60 E6	1.6 E6	.364	5.82 E5	.363	5.81 E5	.302	4.23 E5
50 70	8.00 E6	1.4 E6	.301	4.21 E5	.302	4.23 E5	.241	3.86 E5
70 100	6.60 E6	1.6 E6	.237	3.79 E5	.239	3.82 E5	.131	2.49 E5
100 150	5.00 E6	1.9 E6	.129	2.45 E5	.130	2.47 E5		
150	3.10 E6							

	6 mils	12 mils	30 mils
Equivalent 10 MeV protons/cm²-day	1.04 E7	6.19 E6	3.30 E6
Days/year	× 365	× 365	× 365
Equivalent 10 MeV protons/cm²-year	3.78 E9	2.26 E9	1.21 E9
Equivalent 1 MeV electrons/10 MeV proton	× 3000	× 3000	× 3000
Equivalent 1 MeV electrons/cm²-year	1.14 E13	6.77 E12	3.62 E12
Equivalent 1 MeV electrons/cm²-year P_{max}, V_{oc} (Machine Calculation: See Table 6.34)	1.47 E13	8.80 E12	4.78 E12

[1] Using AP8MAX from Reference 6.5

[2] From Table 4.4

Table 6.3 Manual Calculation of Equivalent Fluence (AP8MAX Protons) I_{sc} Circular Orbit 450 nmi (833 km), 90 Degree Inclination

Energy Increment E_1 E_2 (MeV)		Integral Energy Spectrum $\Phi(>E_1)$ (p/cm²-day)	Difference Spectrum $\Phi(>E_1)-\Phi(>E_2)$ (p/cm²-day)	Shielding Thickness (6 mils)		Shielding Thickness (12 mils)		Shielding Thickness (30 mils)	
				$D(E,6)^2$	Equiv. Fluence (p/cm²-day)	$D(12)^2$	Equiv. Fluence (p/cm²-day)	$D(E,30)^2$	Equiv. Fluence (p/cm²-day)
4	6	1.90 E7	3.0 E6	.474	1.42 E6				
6	8	1.60 E7	2.0 E6	.558	1.12 E6				
8	10	1.40 E7	1.0 E6	.473	4.73 E5				
10	15	1.30 E7	1.2 E6	.408	4.90 E5				
15	20	1.18 E7	1.0 E6	.405	4.05 E5				
20	30	1.08 E7	1.2 E6	.400	4.80 E5				
30	50	9.60 E6	1.6 E6	.359	5.74 E5	.357	7.14 E5	.295	3.54 E5
50	70	8.00 E6	1.4 E6	.299	4.19 E5	.448	4.48 E5	.325	3.25 E5
70	100	6.60 E6	1.6 E6	.237	3.79 E5	.391	4.69 E5	.357	4.28 E5
100	150	5.00 E6	1.9 E6	.159	3.02 E5	.386	3.86 E5	.349	5.58 E5
150		3.10 E6				.391	4.69 E5	.300	4.20 E5
						.357	5.71 E5	.300	4.20 E5
						.300	4.20 E5	.240	3.84 E5
						.238	3.81 E5	.162	3.08 E5
						.159	3.02 E5		

	6 mils	12 mils	30 mils
Equivalent 10 MeV protons/cm²-day	6.06 E6	4.16 E6	2.78 E6
Days/year	× 365	× 365	× 365
Equivalent 10 MeV protons/cm²-year	2.21 E9	1.52 E9	1.01 E9
Equivalent 1 MeV electrons/10 MeV proton	× 3000	× 3000	× 3000
Equivalent 1 MeV electrons/cm²-year	6.64 E12	4.56 E12	3.04 E12
Equivalent 1 MeV electrons/cm²-year (I_{sc}) (Machine Calculation: See Table 6.35)	8.64 E12	6.03 E12	3.98 E12

[1] Using AP8MAX from Reference 6.5

[2] From Table 4.3

Table 6.4 Manual Calculation of Equivalent Fluence (AE6MAX Electrons)
Circular Orbit 450 nmi (833 km), 90 Degree Inclination

Energy Increment E_1 E_2 (MeV)	Integral Energy Spectrum $\Phi(>E_1)$ (p/cm²-day)	Difference Spectrum $\Phi(>E_1)-\Phi(>E_2)$ (e/cm²-day)	Shielding Thickness (6 mils) $D(E,6)^2$	Shielding Thickness (6 mils) Equiv. Fluence (e/cm²-day)	Shielding Thickness (12 mils) $D(E,12)^2$	Shielding Thickness (12 mils) Equiv. Fluence (e/cm²-day)	Shielding Thickness (30 mils) $D(E,30)^2$	Shielding Thickness (30 mils) Equiv. Fluence (e/cm²-day)
.05 .25	5.5 E10	4.3 E10	8.67 E-5	3.73 E6				
.25 .50	1.2 E10	9.5 E9	1.31 E-2	1.24 E8	2.44 E-3	2.32 E7		
.50 .75	2.5 E9	1.2 E9	.113	1.36 E8	6.45 E-2	7.74 E7	7.74 E-3	9.29 E6
.75 1.00	1.3 E9	6.0 E8	.274	1.64 E8	.197	1.18 E8	6.86 E-2	4.12 E7
1.00 1.25	7.0 E8	2.5 E8	.471	1.18 E8	.374	9.35 E7	.193	4.83 E7
1.25 1.50	4.5 E8	1.6 E8	.655	1.05 E8	.540	8.64 E7	.317	5.07 E7
1.50 1.75	2.9 E8	1.0 E8	.880	8.80 E7	.748	7.48 E7	.479	4.79 E7
1.75 2.00	1.9 E8	6.0 E7	1.119	6.71 E7	.972	5.83 E7	.663	3.98 E7
2.00 2.25	1.3 E8	5.0 E7	1.366	6.83 E7	1.206	6.03 E7	.863	4.32 E7
2.25 2.50	8.0 E7	2.5 E7	1.617	4.04 E7	1.445	3.61 E7	1.071	2.68 E7
2.50 2.75	5.5 E7	2.1 E7	1.871	3.93 E7	1.690	3.55 E7	1.289	2.71 E7
2.75 3.00	3.4 E7	1.3 E7	2.123	2.76 E7	1.935	2.52 E7	1.513	1.97 E7
3.00 3.25	2.1 E7	1.0 E7	2.369	2.37 E7	2.176	2.18 E7	1.738	1.74 E7
3.25 3.50	1.1 E7	5.5 E6	2.611	1.44 E7	2.413	1.33 E7	1.961	1.08 E7
3.50 3.75	5.5 E6	3.6 E6	2.853	1.03 E7	2.651	9.54 E6	2.184	7.86 E6
3.75 4.00	1.9 E6	1.3 E6	3.097	4.03 E6	2.891	3.76 E6	2.410	3.13 E6
4.00 4.25	6.0 E5	5.1 E5	3.334	1.70 E6	3.124	1.59 E6	2.633	1.34 E6
4.25 4.50	9.0 E4	7.6 E4	3.561	2.71 E5	3.351	2.55 E5	2.852	2.17 E5
4.50	1.4 E4							

	6 mils	12 mils	30 mils
Equivalent 1 MeV e/cm²-day	1.04 E9	7.39 E8	3.94 E8
Days/year	x 365	x 365	x 365
Equivalent 1 MeV e/cm²-year	3.78 E11	2.70 E11	1.44 E11
Equivalent 1 MeV e/cm²-year (Machine Calculation: See Table 6.33)	4.59 E11	3.46 E11	1.97 E11

[1] Using AE6MAX from Reference 6.4
[2] From Table 4.1

Table 6.5 Summary of Data in Tables 6.6 Through 6.39

Environment	Reference	Orbital Inclination (Deg.)	Equivalent Fluence for Various Shielding Thicknesses, (I_{sc}) (Table No.)	Equivalent Fluence for Various Shielding Thicknesses, (V_{oc}, P_{max}) (Table No.)
Trapped Electrons	AE6MAX[6.4] AEI7-LO[6.7]	0 10 20 30 40 50 60 70 80 90	6.6 6.9 6.12 6.15 6.18 6.21 6.24 6.27 6.30 6.33	6.6 6.9 6.12 6.15 6.18 6.21 6.24 6.27 6.30 6.33
Trapped Protons	AP8MAX[6.5]	0 10 20 30 40 50 60 70 80 90	* 6.8 6.11 6.14 * 6.17 6.20 6.23 * 6.26 6.29 6.32 * 6.35	* 6.7 6.10 6.13 * 6.16 6.19 6.22 * 6.25 6.28 6.31 * 6.34
Synchronous	AEI7-LO[6.7] AP8MAX[6.5]	0° Various Longitudes	6.36, 6.38	6.36, 6.37
Solar Flare Protons	Table 5.1	Free Space 1 AU	6.39	6.39

* These Tables also include a calculation for AP8MIN [6.5]

Table 6.6. Annual Equivalent 1 MeV Electron Fluence from Trapped Electrons, 0° Inclination (Infinite Backshielding)

ELECTRONS - ISC - CIRCULAR ORBIT
EQUIV. 1 MEV ELECTRON FLUENCE FOR ISC - CIRCULAR ORBIT
DUE TO GEOMAGNETICALLY TRAPPED ELECTRONS - MODELS AE6MAX, AE17LO
ELECTRONS - ISC, VOC, AND PMAX INCLINATION = 0 DEGREES.

ALTITUDE		SHIELD THICKNESS, CM (MILS)							
(N.M.)	(KM)	0 (0)	2.54E-3 (1)	7.64E-3 (3)	1.52E-2 (6)	3.05E-2 (12)	5.09E-2 (20)	7.64E-2 (30)	1.52E-1 (60)
150	277	0.00	0.00	0.00	0.00	0.00	0.00	0.00	0.00
250	463	1.14+07	1.09+07	1.03+07	9.51+06	8.40+06	7.32+06	6.28+06	4.21+06
300	556	4.65+07	4.29+07	3.86+07	3.41+07	2.83+07	2.33+07	1.91+07	1.16+07
450	833	8.03+09	6.16+09	4.37+09	2.96+09	1.64+09	9.63+08	6.03+08	2.36+08
600	1111	4.92+11	3.56+11	2.48+11	1.58+11	7.97+10	4.24+10	2.45+10	8.33+09
800	1481	7.26+12	5.46+12	3.80+12	2.51+12	1.36+12	7.76+11	4.76+11	1.75+11
1000	1852	2.49+13	1.91+13	1.36+13	9.32+12	5.37+12	3.27+12	2.10+12	8.21+11
1250	2315	6.03+13	4.70+13	3.44+13	2.44+13	1.49+13	9.53+12	6.37+12	2.61+12
1500	2778	9.80+13	7.78+13	5.69+13	4.08+13	2.55+13	1.66+13	1.13+13	4.73+12
1750	3241	1.26+14	9.79+13	7.12+13	4.99+13	2.99+13	1.87+13	1.23+13	5.06+12
2000	3704	1.46+14	1.12+14	7.92+13	5.35+13	2.97+13	1.71+13	1.05+13	3.94+12
2250	4167	1.64+14	1.23+14	8.54+13	5.57+13	2.86+13	1.49+13	8.21+12	2.71+12
2500	4630	1.73+14	1.29+14	8.80+13	5.59+13	2.70+13	1.27+13	6.22+12	1.75+12
2750	5093	1.78+14	1.26+14	8.50+13	5.31+13	2.44+13	1.04+13	4.44+12	9.78+11
3000	5556	1.53+14	1.13+14	7.61+13	4.71+13	2.10+13	8.38+12	3.18+12	5.41+11
3500	6482	1.11+14	8.06+13	5.31+13	3.21+13	1.37+13	5.08+12	1.70+12	2.07+11
4000	7408	8.80+13	6.76+13	3.76+13	2.24+13	9.39+12	3.42+12	1.12+12	1.28+11
4500	8334	6.39+13	4.66+13	3.09+13	1.90+13	8.44+12	3.38+12	1.17+12	1.39+11
5000	9260	4.90+13	3.73+13	2.62+13	1.74+13	8.83+12	4.12+12	1.79+12	2.68+11
5500	10186	3.81+13	3.09+13	2.36+13	1.72+13	1.03+13	5.77+12	3.04+12	6.31+11
6000	11112	3.63+13	3.14+13	2.61+13	2.11+13	1.51+13	1.05+13	7.21+12	3.11+12
7000	12964	6.96+13	6.38+13	5.68+13	4.97+13	4.03+13	3.20+13	2.51+13	1.36+13
8000	14816	1.06+14	9.25+13	8.32+13	7.36+13	6.06+13	4.90+13	3.89+13	2.15+13
9000	16668	1.47+14	1.26+14	1.22+14	1.08+14	8.90+13	7.19+13	5.69+13	3.08+13
10000	18520	2.04+14	1.89+14	1.70+14	1.49+14	1.22+14	9.67+13	7.51+13	3.86+13
11000	20372	2.19+14	2.03+14	1.82+14	1.60+14	1.29+14	1.01+14	7.75+13	3.78+13
12000	22224	2.18+14	2.02+14	1.81+14	1.58+14	1.27+14	9.91+13	7.47+13	3.51+13
13000	24076	1.89+14	1.74+14	1.56+14	1.36+14	1.08+14	8.37+13	6.26+13	2.88+13
14000	25928	1.57+14	1.44+14	1.27+14	1.10+14	8.65+13	6.57+13	4.83+13	2.14+13
15000	27780	1.32+14	1.20+14	1.05+14	8.96+13	6.89+13	5.11+13	3.67+13	1.55+13
16000	29632	1.09+14	9.83+13	8.52+13	7.19+13	5.43+13	3.96+13	2.79+13	1.13+13
17000	31484	8.96+13	8.02+13	6.88+13	5.74+13	4.26+13	3.04+13	2.10+13	8.10+12
18000	33336	6.71+13	5.92+13	5.00+13	4.09+13	2.96+13	2.06+13	1.38+13	4.91+12
19327	36794	4.35+13	3.76+13	3.10+13	2.48+13	1.72+13	1.15+13	7.36+12	2.34+12

6-19

Table 6.7. Annual Equivalent 1 MeV Electron Fluence from Trapped Protons (V_{oc}, P_{max}), 0° Inclination (Infinite Backshielding)

PROTONS - VOC AND PMAX

EQUIV. 1 MEV ELECTRON FLUENCE FOR VOC AND PMAX CIRCULAR ORBIT
DUE TO GEOMAGNETICALLY TRAPPED PROTONS, MODEL AP8MAX   INCLINATION = 0 DEGREES.

ALTITUDE		\multicolumn{8}{c}{SHIELD THICKNESS, CM (MILS)}
(N.M.)	(KM)	0 (0)	2.54E-3 (1)	7.64E-3 (3)	1.52E-2 (6)	3.05E-2 (12)	5.09E-2 (20)	7.64E-2 (30)	1.52E-1 (60)
150	277	0.00	0.00	0.00	0.00	0.00	0.00	0.00	0.00
250	463	0.00	0.00	0.00	0.00	0.00	0.00	0.00	0.00
300	555	0.00	0.00	0.00	0.00	0.00	0.00	0.00	0.00
450	833	2.87+12	2.26+12	2.07+12	1.90+12	1.71+12	1.56+12	1.46+12	1.31+12
600	1111	3.88+13	3.30+13	3.03+13	2.79+13	2.48+13	2.18+13	1.96+13	1.62+13
800	1481	3.27+14	2.80+14	2.51+14	2.26+14	1.93+14	1.61+14	1.38+14	1.03+14
1000	1852	1.48+15	1.27+15	1.10+15	9.65+14	7.82+14	5.98+14	4.81+14	3.24+14
1250	2315	6.61+15	5.48+15	4.69+15	3.89+15	2.92+15	1.94+15	1.42+15	8.30+14
1500	2778	2.11+16	1.75+16	1.42+16	1.16+16	8.19+15	4.93+15	3.27+15	1.56+15
1750	3241	5.17+16	4.22+16	3.33+16	2.65+16	1.77+16	9.80+15	5.95+15	2.31+15
2000	3704	1.09+17	8.62+16	6.36+16	4.77+16	2.94+16	1.49+16	8.29+15	2.71+15
2250	4167	2.35+17	1.71+17	1.16+17	7.82+16	4.20+16	1.88+16	9.60+15	2.77+15
2500	4630	4.26+17	2.83+17	1.77+17	1.09+17	5.17+16	2.07+16	9.82+15	2.60+15
2750	5093	7.09+17	4.13+17	2.40+17	1.36+17	5.81+16	2.12+16	9.44+15	2.31+15
3000	5556	1.08+18	5.41+17	2.91+17	1.53+17	5.90+16	2.00+16	8.60+15	1.98+15
3500	6482	2.51+18	8.60+17	3.74+17	1.59+17	4.78+16	1.52+16	6.15+15	1.32+15
4000	7408	4.98+18	1.16+18	4.12+17	1.47+17	3.52+16	1.03+16	3.96+15	8.21+14
4500	8334	8.40+18	1.34+18	3.98+17	1.21+17	2.35+16	6.20+15	2.23+15	4.36+14
5000	9260	1.28+19	1.45+18	3.56+17	9.26+16	1.46+16	3.56+15	1.21+15	2.23+14
5500	10186	1.79+19	1.49+18	2.97+17	6.40+16	8.02+15	1.88+15	6.12+14	1.01+14
6000	11112	2.26+19	1.37+18	2.27+17	4.18+16	4.26+15	8.93+14	2.66+14	3.80+13
7000	12964	2.83+19	8.41+17	9.46+16	1.25+16	8.13+14	1.29+14	3.09+13	3.18+12
8000	14816	2.71+19	3.72+17	2.77+16	2.66+15	1.12+14	1.30+13	2.53+12	2.03+11
9000	16668	1.66+19	1.01+17	4.35+15	2.71+14	6.45+12	5.93+11	1.11+11	1.61+10
10000	18520	1.01+19	2.37+16	6.11+14	2.67+13	7.93+10	1.83+03	0.00	0.00
11000	20372	6.40+18	4.61+15	2.57+13	6.27+11	1.98+09	1.83+03	0.00	0.00
12000	22224	4.57+18	8.67+14	4.37+11	9.63+03	0.00	0.00	0.00	0.00
13000	24076	3.29+18	1.14+14	1.91+07	0.00	0.00	0.00	0.00	0.00
14000	25928	2.22+18	1.44+13	8.82+06	0.00	0.00	0.00	0.00	0.00
15000	27780	1.60+18	3.31+12	4.97+06	0.00	0.00	0.00	0.00	0.00
16000	29632	1.21+18	2.15+10	0.00	0.00	0.00	0.00	0.00	0.00
17000	31484	9.40+17	1.70+10	0.00	0.00	0.00	0.00	0.00	0.00
18000	33336	6.86+17	1.38+10	0.00	0.00	0.00	0.00	0.00	0.00
19327	35794	2.65+17	9.16+09	0.00	0.00	0.00	0.00	0.00	0.00

EQUIV. 1 MEV ELECTRON FLUENCE FOR VOC AND PMAX CIRCULAR ORBIT
DUE TO GEOMAGNETICALLY TRAPPED PROTONS, MODEL AP8MIN   INCLINATION = 0 DEGREES.

150	277	0.00	0.00	0.00	0.00	0.00	0.00	0.00	0.00
250	463	0.00	0.00	0.00	0.00	0.00	0.00	0.00	0.00
300	555	4.54+10	3.04+10	2.41+10	2.02+10	1.63+10	1.30+10	1.05+10	6.80+09
450	833	7.68+12	6.17+12	5.64+12	5.20+12	4.71+12	4.34+12	4.02+12	3.44+12
600	1111	5.82+13	4.95+13	4.53+13	4.18+13	3.74+13	3.35+13	3.04+13	2.51+13
800	1481	3.89+14	3.37+14	3.00+14	2.68+14	2.29+14	1.94+14	1.67+14	1.25+14
1000	1852	1.62+15	1.40+15	1.21+15	1.05+15	8.53+14	6.62+14	5.35+14	3.59+14

6-20

Table 6.8. Annual Equivalent 1 MeV Electron Fluence from Trapped Protons (I_{SC}), 0° Inclination (Infinite Backshielding)

PROTONS - ISC

EQUIV 1 MEV ELECTRON FLUENCE FOR ISC - CIRCULAR ORBIT
DUE TO GEOMAGNETICALLY TRAPPED PROTONS, MODEL AP8MAX   INCLINATION = 0 DEGREES.

ALTITUDE		\multicolumn{8}{c}{SHIELD THICKNESS, CM (MILS)}							
(N.M.)	(KM)	0 (0)	2.54E-3 (1)	7.64E-3 (3)	1.52E-2 (6)	3.05E-2 (12)	6.09E-2 (20)	7.64E-2 (30)	1.52E-1 (60)
150	277	0.00	0.00	0.00	0.00	0.00	0.00	0.00	0.00
250	463	0.00	0.00	0.00	0.00	0.00	0.00	0.00	0.00
300	555	0.00	0.00	0.00	0.00	0.00	0.00	0.00	0.00
450	833	1.91+12	1.78+12	1.70+12	1.61+12	1.51+12	1.42+12	1.36+12	1.24+12
600	1111	2.66+13	2.51+13	2.38+13	2.25+13	2.07+13	1.88+13	1.74+13	1.48+13
800	1481	2.11+14	1.97+14	1.83+14	1.69+14	1.50+14	1.30+14	1.16+14	9.06+13
1000	1852	8.87+14	8.14+14	7.36+14	6.62+14	5.58+14	4.64+14	3.85+14	2.78+14
1250	2316	3.55+15	3.17+15	2.76+15	2.39+15	1.86+15	1.36+15	1.07+15	6.78+14
1500	2778	1.07+16	9.40+15	7.92+15	6.61+15	4.81+15	3.15+15	2.27+15	1.20+15
1750	3241	2.50+16	2.16+16	1.77+16	1.43+16	9.78+15	5.84+15	3.87+15	1.68+15
2000	3704	4.92+16	4.10+16	3.18+16	2.43+16	1.54+16	8.40+15	5.14+15	1.88+15
2250	4167	9.79+16	7.62+16	5.40+16	3.76+16	2.10+16	1.01+16	5.77+15	1.86+15
2500	4630	1.66+17	1.20+17	7.86+16	5.02+16	2.49+16	1.08+16	5.78+15	1.70+15
2750	5093	2.55+17	1.70+17	1.03+17	6.07+16	2.73+16	1.08+16	5.47+15	1.49+15
3000	5556	3.57+17	2.16+17	1.22+17	6.66+16	2.72+16	9.97+15	4.87+15	1.27+15
3500	6482	7.09+17	3.23+17	1.49+17	6.64+16	2.16+16	7.43+15	3.47+15	8.25+14
4000	7408	1.21+18	4.13+17	1.59+17	5.94+16	1.67+16	4.93+15	2.22+15	5.09+14
4500	8334	1.80+18	4.63+17	1.51+17	4.77+16	1.02+16	2.90+15	1.24+15	2.67+14
5000	9260	2.45+18	4.84+17	1.33+17	3.58+16	6.28+15	1.63+15	6.66+14	1.35+14
5500	10186	3.15+18	4.81+17	1.09+17	2.43+16	3.42+15	8.48+14	3.31+14	6.00+13
6000	11112	3.60+18	4.29+17	8.24+16	1.57+16	1.78+15	3.92+14	1.41+14	2.21+13
7000	12964	3.49+18	2.47+17	3.36+16	4.56+15	3.28+14	6.31+13	1.57+13	1.79+12
8000	14816	2.56+18	1.01+17	9.67+15	9.53+14	4.37+13	5.08+12	1.25+12	1.13+11
9000	16668	1.27+18	2.47+16	1.48+15	9.50+13	2.48+12	2.27+11	5.80+10	9.71+09
10000	18520	5.95+17	5.16+15	2.04+14	9.18+12	1.67+10	0.00	0.00	0.00
11000	20372	2.67+17	8.65+14	7.81+12	2.13+11	5.26+08	0.00	0.00	0.00
12000	22224	1.36+17	1.42+13	9.59+10	4.60+04	0.00	0.00	0.00	0.00
13000	24076	6.50+16	1.36+14	1.31+06	0.00	0.00	0.00	0.00	0.00
14000	25928	3.60+16	1.45+12	7.67+05	0.00	0.00	0.00	0.00	0.00
15000	27780	1.74+16	2.77+11	5.10+05	0.00	0.00	0.00	0.00	0.00
16000	29632	1.21+16	4.14+08	0.00	0.00	0.00	0.00	0.00	0.00
17000	31484	8.75+15	3.43+08	0.00	0.00	0.00	0.00	0.00	0.00
18000	33336	6.01+15	2.91+08	0.00	0.00	0.00	0.00	0.00	0.00
19327	35794	2.39+15	2.10+08	0.00	0.00	0.00	0.00	0.00	0.00

EQUIV 1 MEV ELECTRON FLUENCE FOR ISC - CIRCULAR ORBIT
DUE TO GEOMAGNETICALLY TRAPPED PROTONS, MODEL AP8MIN   INCLINATION = 0 DEGREES.

ALTITUDE		0	2.54E-3	7.64E-3	1.52E-2	3.05E-2	6.09E-2	7.64E-2	1.52E-1
(N.M.)	(KM)	(0)	(1)	(3)	(6)	(12)	(20)	(30)	(60)
150	277	0.00	0.00	0.00	0.00	0.00	0.00	0.00	0.00
250	463	0.00	0.00	0.00	0.00	0.00	0.00	0.00	0.00
300	555	2.27+10	1.84+10	1.59+10	1.40+10	1.18+10	9.82+09	8.24+09	5.63+09
450	833	5.17+12	4.85+12	4.61+12	4.39+12	4.11+12	3.86+12	3.63+12	3.20+12
600	1111	4.02+13	3.80+13	3.61+13	3.41+13	3.15+13	2.90+13	2.70+13	2.31+13
800	1481	2.53+14	2.37+14	2.19+14	2.02+14	1.80+14	1.58+14	1.41+14	1.11+14
1000	1852	9.75+14	8.97+14	8.09+14	7.26+14	6.13+14	5.04+14	4.28+14	3.07+14

6-21

Table 6.9. Annual Equivalent 1 MeV Electron Fluence from Trapped Electrons, 10° Inclination (Infinite Backshielding)

ELECTRONS - ISC, VOC, AND PMAX

EQUIV. 1 MEV ELECTRON FLUENCE FOR ISC - CIRCULAR ORBIT
DUE TO GEOMAGNETICALLY TRAPPED ELECTRONS - MODELS AE6MAX, AE17LO

INCLINATION = 10 DEGREES.

ALTITUDE					SHIELD THICKNESS, CM (MILS)			
(N.M.)	(KM)	0 (0)	2.54E-3 (1)	7.64E-3 (3)	1.52E-2 (6)	3.05E-2 (12)	6.09E-2 (20)	7.64E-2 (30)	1.52E-1 (60)
150	277	2.59+06	2.49+06	2.34+06	2.18+06	1.93+06	1.68+06	1.44+06	9.60+05
250	463	4.66+07	4.20+07	3.65+07	3.13+07	2.47+07	1.96+07	1.53+07	8.72+06
300	555	2.16+08	1.84+08	1.51+08	1.20+08	8.61+07	6.26+07	4.63+07	2.37+07
450	833	2.76+10	2.06+10	1.41+10	9.12+09	4.67+09	2.51+09	1.46+09	5.04+08
600	1111	8.04+11	6.98+11	4.07+11	2.60+11	1.32+11	7.09+10	4.15+10	1.44+10
800	1481	7.58+12	5.76+12	4.05+12	2.72+12	1.51+12	8.93+11	5.62+11	2.14+11
1000	1852	2.49+13	1.92+13	1.38+13	9.54+12	5.59+12	3.46+12	2.26+12	8.96+11
1250	2315	5.74+13	4.48+13	3.28+13	2.33+13	1.42+13	9.14+12	6.12+12	2.52+12
1500	2778	8.94+13	6.99+13	5.13+13	3.65+13	2.26+13	1.45+13	9.73+12	4.06+12
1750	3241	1.15+14	8.90+13	6.43+13	4.48+13	2.64+13	1.63+13	1.06+13	4.29+12
2000	3704	1.34+14	1.02+14	7.22+13	4.86+13	2.67+13	1.52+13	9.25+12	3.46+12
2250	4167	1.49+14	1.15+14	7.82+13	5.04+13	2.39+13	1.32+13	7.24+12	2.38+12
2500	4630	1.64+14	1.11+14	7.47+13	4.97+13	2.14+13	1.12+13	5.48+12	1.54+12
2750	5093	1.49+14	1.01+14	6.75+13	4.67+13	1.86+13	9.14+12	3.90+12	8.63+11
3000	5556	1.36+14	7.26+13	4.78+13	4.17+13	1.23+13	7.44+12	2.86+12	5.02+11
3500	6482	9.97+13	6.26+13	3.44+13	2.89+13	8.74+12	4.59+12	1.55+12	1.92+11
4000	7408	7.30+13	4.20+13	2.81+13	2.07+13	3.23+12	3.17+12	1.07+12	1.25+11
4500	8334	6.73+13	3.43+13	2.37+13	1.74+13	7.87+12	3.17+12	1.17+12	1.50+11
5000	9260	4.48+13	3.05+13	2.37+13	1.63+13	8.51+12	4.16+12	1.94+12	3.96+11
5500	10186	3.71+13	3.64+13	2.99+13	1.78+13	1.14+13	7.05+12	4.30+12	1.51+12
6000	11112	4.05+13	6.59+13	5.88+13	2.47+13	1.83+13	1.33+13	9.62+12	4.60+12
7000	12964	7.18+13	9.67+13	8.69+13	5.16+13	4.19+13	3.35+13	2.63+13	1.43+13
8000	14816	1.05+14	1.39+14	1.25+14	7.69+13	6.33+13	5.11+13	4.05+13	2.22+13
9000	16668	1.51+14	1.84+14	1.66+14	1.11+14	9.11+13	7.33+13	5.77+13	3.08+13
10000	18520	1.99+14	1.96+14	1.76+14	1.46+14	1.19+14	9.40+13	7.28+13	3.70+13
11000	20372	2.12+14	1.96+14	1.76+14	1.55+14	1.25+14	9.79+13	7.46+13	3.62+13
12000	22224	2.07+14	1.91+14	1.71+14	1.50+14	1.20+14	9.35+13	7.04+13	3.30+13
13000	24076	1.78+14	1.64+14	1.46+14	1.27+14	1.01+14	7.79+13	5.81+13	2.65+13
14000	26928	1.49+14	1.37+14	1.21+14	1.04+14	8.18+13	6.20+13	4.55+13	2.01+13
15000	27780	1.24+14	1.12+14	9.83+13	8.38+13	6.43+13	4.77+13	3.42+13	1.44+13
16000	29632	9.99+13	8.98+13	7.77+13	6.53+13	4.92+13	3.57+13	2.50+13	1.00+13
17000	31484	7.82+13	6.96+13	5.95+13	4.93+13	3.63+13	2.57+13	1.76+13	6.63+12
18000	33336	5.82+13	5.10+13	4.28+13	3.49+13	2.49+13	1.72+13	1.14+13	3.95+12
19327	36794	3.83+13	3.30+13	2.71+13	2.15+13	1.48+13	9.79+12	6.23+12	1.93+12

6-22

Table 6.10. Annual Equivalent 1 MeV Electron Fluence from Trapped Protons (V_{oc}, P_{max}), 10° Inclination (Infinite Backshielding)

PROTONS - VOC AND PMAX

EQUIV. 1 MEV ELECTRON FLUENCE FOR VOC AND PMAX CIRCULAR ORBIT INCLINATION = 10 DEGREES.
DUE TO GEOMAGNETICALLY TRAPPED PROTONS, MODEL AP8MAX

ALTITUDE		\multicolumn{8}{c}{SHIELD THICKNESS, CM (MILS)}							
(N.M.)	(KM)	0 (0)	2.54E-3 (1)	7.64E-3 (3)	1.52E-2 (6)	3.05E-2 (12)	5.09E-2 (20)	7.64E-2 (30)	1.52E-1 (60)
150	277	0.00	0.00	0.00	0.00	0.00	0.00	0.00	0.00
250	463	8.54+09	6.52+09	6.85+09	5.29+09	4.74+09	4.43+09	4.10+09	3.42+09
300	555	1.36+11	1.07+11	9.53+10	8.50+10	7.49+10	6.92+10	6.50+10	5.82+10
450	833	4.91+12	4.02+12	3.74+12	3.51+12	3.19+12	2.86+12	2.64+12	2.31+12
600	1111	4.42+13	3.75+13	3.43+13	3.16+13	2.79+13	2.43+13	2.16+13	1.74+13
800	1481	3.51+14	3.02+14	2.68+14	2.40+14	2.04+14	1.68+14	1.43+14	1.05+14
1000	1852	1.51+15	1.29+15	1.11+15	9.67+14	7.75+14	5.85+14	4.68+14	3.15+14
1250	2315	6.50+15	5.44+15	4.52+15	3.78+15	2.81+15	1.86+15	1.35+15	7.77+14
1500	2778	2.09+16	1.72+16	1.38+16	1.12+16	7.83+15	4.65+15	3.05+15	1.43+15
1750	3241	5.04+16	4.07+16	3.17+16	2.49+16	1.64+16	8.98+15	5.41+15	2.08+15
2000	3704	1.08+17	8.28+16	6.04+16	4.43+16	2.66+16	1.33+16	7.37+15	2.42+15
2250	4167	2.27+17	1.61+17	1.07+17	7.16+16	3.80+16	1.69+16	8.60+15	2.48+15
2500	4630	4.03+17	2.59+17	1.60+17	9.80+16	4.62+16	1.85+16	8.78+15	2.33+15
2750	5093	6.80+17	3.82+17	2.19+17	1.23+17	5.18+16	1.89+16	8.40+15	2.06+15
3000	5556	1.05+18	4.99+17	2.63+17	1.37+17	5.22+16	1.78+16	7.60+15	1.77+15
3500	6482	2.41+18	7.88+17	3.38+17	1.43+17	4.27+16	1.36+16	5.49+15	1.18+15
4000	7408	4.72+18	1.04+18	3.59+17	1.31+17	3.14+16	9.23+15	3.55+15	7.31+14
4500	8334	7.95+18	1.21+18	3.55+17	1.08+17	2.11+16	5.58+15	2.01+15	3.92+14
5000	9260	1.20+19	1.30+18	3.18+17	8.29+16	1.32+16	3.22+15	1.09+15	2.00+14
5500	10186	1.67+19	1.33+18	2.64+17	5.71+16	7.20+15	1.68+15	5.46+14	8.97+13
6000	11112	2.08+19	1.22+18	2.04+17	3.80+16	3.94+15	8.31+14	2.49+14	3.60+13
7000	12964	2.58+19	7.43+17	8.35+16	1.11+16	7.22+14	1.15+14	2.79+13	2.89+12
8000	14816	2.40+19	3.43+17	2.62+16	2.56+15	1.11+14	1.31+13	2.57+12	2.05+11
9000	16668	1.51+19	9.72+16	4.46+15	2.96+14	7.63+12	7.02+11	1.48+11	3.18+10
10000	18520	8.89+18	2.03+16	5.22+14	2.31+13	7.00+10	1.83-03	0.00	0.00
11000	20372	5.76+18	4.00+15	2.19+13	5.35+11	1.68+09	1.83-03	0.00	0.00
12000	22224	4.18+18	7.88+14	4.03+11	9.63-03	0.00	0.00	0.00	0.00
13000	24076	3.00+18	1.23+14	1.99+07	0.00	0.00	0.00	0.00	0.00
14000	25928	2.08+18	1.78+13	9.67+06	0.00	0.00	0.00	0.00	0.00
15000	27780	1.47+18	4.11+12	5.58+06	0.00	0.00	0.00	0.00	0.00
16000	29632	1.04+18	1.90+10	0.00	0.00	0.00	0.00	0.00	0.00
17000	31484	7.55+17	1.39+10	0.00	0.00	0.00	0.00	0.00	0.00
18000	33336	5.41+17	1.09+10	0.00	0.00	0.00	0.00	0.00	0.00
19327	35794	1.76+17	7.53+09	0.00	0.00	0.00	0.00	0.00	0.00

Table 6.11. Annual Equivalent 1 MeV Electron Fluence from Trapped Protons (I_{sc}), 10° Inclination (Infinite Backshielding)

PROTONS - ISC

EQUIV 1 MEV ELECTRON FLUENCE FOR ISC - CIRCULAR ORBIT
DUE TO GEOMAGNETICALLY TRAPPED PROTONS, MODEL AP8MAX

INCLINATION = 10 DEGREES.

ALTITUDE		SHIELD THICKNESS, CM (MILS)						
(N.M.)	(KM)	0 (0)	2.54E-3 (1)	7.64E-3 (3)	1.52E-2 (6)	3.05E-2 (12)	6.09E-2 (20)	7.64E-2 (30)	1.52E-1 (60)
150	277	0.00	0.00	0.00	0.00	0.00	0.00	0.00	0.00
250	463	5.40+09	5.01+09	4.72+09	4.45+09	4.14+09	3.91+09	3.66+09	3.16+09
300	555	8.81+10	8.16+10	7.64+10	7.17+10	6.67+10	6.31+10	6.03+10	5.48+10
450	833	3.40+12	3.21+12	3.07+12	2.94+12	2.75+12	2.56+12	2.41+12	2.15+12
600	1111	2.99+13	2.81+13	2.66+13	2.51+13	2.29+13	2.06+13	1.89+13	1.58+13
800	1481	2.24+14	2.09+14	1.93+14	1.78+14	1.56+14	1.35+14	1.19+14	9.25+13
1000	1852	8.95+14	8.16+14	7.33+14	6.56+14	5.49+14	4.43+14	3.74+14	2.69+14
1250	2315	3.49+15	3.11+15	2.69+15	2.31+15	1.78+15	1.29+15	1.01+15	6.32+14
1500	2778	1.05+16	9.16+15	7.65+15	6.34+15	4.56+15	2.95+15	2.11+15	1.10+15
1750	3241	2.40+16	2.06+16	1.66+16	1.33+16	9.00+15	5.33+15	3.51+15	1.50+15
2000	3704	4.79+16	3.92+16	2.98+16	2.24+16	1.39+16	7.49+15	4.58+15	1.68+15
2250	4167	9.30+16	7.12+16	4.98+16	3.43+16	1.90+16	9.10+15	5.17+15	1.66+15
2500	4630	1.54+17	1.10+17	7.10+16	4.50+16	2.22+16	9.65+15	5.17+15	1.53+15
2750	5093	2.40+17	1.56+17	9.35+16	5.46+16	2.43+16	9.58+15	4.87+15	1.33+15
3000	5556	3.38+17	1.98+17	1.10+17	5.94+16	2.41+16	8.89+15	4.36+15	1.13+15
3500	6482	6.66+17	2.95+17	1.34+17	5.96+16	1.93+16	6.63+15	3.10+15	7.39+14
4000	7408	1.12+18	3.73+17	1.42+17	5.30+16	1.40+16	4.42+15	1.99+15	4.53+14
4500	8334	1.67+18	4.16+17	1.34+17	4.26+16	9.19+15	2.61+15	1.12+15	2.40+14
5000	9260	2.26+18	4.34+17	1.18+17	3.21+16	5.68+15	1.47+15	6.01+14	1.21+14
5500	10186	2.87+18	4.29+17	9.70+16	2.17+16	3.07+15	7.58+14	2.95+14	5.33+13
6000	11112	3.22+18	3.82+17	7.41+16	1.42+16	1.65+15	3.66+14	1.32+14	2.10+13
7000	12964	3.11+18	2.18+17	2.97+16	4.03+15	2.91+14	4.76+13	1.42+13	1.62+12
8000	14816	2.28+18	9.36+16	9.16+15	9.19+14	4.34+13	5.14+12	1.27+12	1.15+11
9000	16668	1.16+18	2.40+16	1.53+15	1.04+14	2.94+12	2.76+11	8.25+10	1.86+10
10000	18520	5.12+17	4.42+15	1.74+14	7.92+12	1.49+10	0.00	0.00	0.00
11000	20372	2.35+17	7.48+14	6.66+12	1.82+11	4.50+08	0.00	0.00	0.00
12000	22224	1.21+17	1.24+14	8.88+10	4.60+04	0.00	0.00	0.00	0.00
13000	24076	6.03+16	1.57+13	1.34+06	0.00	0.00	0.00	0.00	0.00
14000	25928	2.98+16	1.86+12	8.18+05	0.00	0.00	0.00	0.00	0.00
15000	27780	1.65+16	3.66+11	6.54+05	0.00	0.00	0.00	0.00	0.00
16000	29632	1.02+16	3.76+08	0.00	0.00	0.00	0.00	0.00	0.00
17000	31484	6.68+15	2.91+08	0.00	0.00	0.00	0.00	0.00	0.00
18000	33336	4.43+15	2.41+08	0.00	0.00	0.00	0.00	0.00	0.00
19327	35794	1.61+15	1.79+08	0.00	0.00	0.00	0.00	0.00	0.00

Table 6.12. Annual Equivalent 1 MeV Electron Fluence from Trapped Electrons, 20° Inclination (Infinite Backshielding)

ELECTRONS - ISC, VOC, AND PMAX

EQUIV. 1 MEV ELECTRON FLUENCE FOR ISC - CIRCULAR ORBIT
DUE TO GEOMAGNETICALLY TRAPPED ELECTRONS - MODELS AE6MAX, AE17LO

INCLINATION = 20 DEGREES.

ALTITUDE		\multicolumn{8}{c}{SHIELD THICKNESS, CM (MILS)}							
(N.M.)	(KM)	0 (0)	2.54E-3 (1)	7.64E-3 (3)	1.52E-2 (6)	3.05E-2 (12)	5.09E-2 (20)	7.64E-2 (30)	1.52E-1 (60)
150	277	2.03+07	1.84+07	1.62+07	1.41+07	1.15+07	9.35+06	7.61+06	4.65+06
250	463	8.21+08	6.36+08	4.60+08	3.20+08	1.89+08	1.18+08	7.86+07	3.42+07
300	556	6.44+09	4.94+09	3.52+09	2.42+09	1.40+09	8.63+08	5.53+08	2.22+08
450	833	2.45+11	1.86+11	1.32+11	8.92+10	5.08+10	3.07+10	1.98+10	7.80+09
600	1111	1.60+12	1.22+12	8.58+11	5.79+11	3.26+11	1.95+11	1.25+11	4.86+10
800	1481	7.72+12	5.90+12	4.19+12	2.86+12	1.64+12	9.99+11	6.44+11	2.53+11
1000	1852	2.26+13	1.74+13	1.25+13	8.67+12	5.12+12	3.18+12	2.08+12	8.34+11
1250	2315	4.90+13	3.81+13	2.77+13	1.96+13	1.19+13	7.55+12	5.02+12	2.06+12
1500	2778	7.39+13	5.74+13	4.18+13	2.94+13	1.78+13	1.13+13	7.48+12	3.08+12
1750	3241	9.28+13	7.13+13	5.11+13	3.51+13	2.02+13	1.22+13	7.79+12	3.10+12
2000	3704	1.06+14	8.03+13	5.63+13	3.75+13	2.02+13	1.12+13	6.67+12	2.44+12
2250	4167	1.16+14	8.68+13	5.96+13	3.85+13	1.93+13	9.72+12	5.19+12	1.65+12
2500	4630	1.17+14	8.71+13	5.91+13	3.73+13	1.78+13	8.17+12	3.91+12	1.06+12
2750	5093	1.12+14	8.31+13	5.58+13	3.47+13	1.58+13	6.65+12	2.78+12	6.92+11
3000	5556	1.00+14	7.39+13	4.94+13	3.04+13	1.35+13	5.35+12	2.02+12	3.43+11
3500	6482	7.40+13	5.38+13	3.54+13	2.14+13	9.18+12	3.45+12	1.18+12	1.48+11
4000	7408	5.42+13	3.93+13	2.59+13	1.58+13	6.97+12	2.77+12	1.06+12	2.02+11
4500	8334	4.47+13	3.33+13	2.29+13	1.49+13	7.49+12	3.64+12	1.83+12	5.82+11
5000	9260	3.84+13	3.05+13	2.28+13	1.65+13	1.00+13	6.10+12	3.80+12	1.55+12
5500	10186	3.79+13	3.23+13	2.65+13	2.11+13	1.50+13	1.05+13	7.33+12	3.37+12
6000	11112	4.54+13	4.07+13	3.52+13	3.00+13	2.33+13	1.78+13	1.35+13	6.96+12
7000	12964	7.82+13	7.20+13	6.45+13	5.67+13	4.63+13	3.71+13	2.91+13	1.58+13
8000	14816	1.11+14	1.02+14	9.19+13	8.12+13	6.66+13	5.35+13	4.21+13	2.26+13
9000	16668	1.47+14	1.36+14	1.22+14	1.08+14	8.83+13	7.06+13	5.51+13	2.88+13
10000	18520	1.80+14	1.67+14	1.49+14	1.31+14	1.06+14	8.40+13	6.46+13	3.23+13
11000	20372	1.86+14	1.72+14	1.54+14	1.35+14	1.08+14	8.45+13	6.41+13	3.07+13
12000	22224	1.74+14	1.61+14	1.44+14	1.25+14	1.00+14	7.76+13	5.82+13	2.69+13
13000	24076	1.49+14	1.37+14	1.21+14	1.05+14	8.34+13	6.39+13	4.74+13	2.14+13
14000	25928	1.24+14	1.13+14	9.95+13	8.55+13	6.67+13	5.02+13	3.66+13	1.60+13
15000	27780	9.99+13	9.02+13	7.86+13	6.66+13	5.08+13	3.74+13	2.66+13	1.11+13
16000	29632	7.67+13	6.86+13	5.90+13	4.93+13	3.67+13	2.64+13	1.84+13	7.18+12
17000	31484	5.76+13	5.09+13	4.31+13	3.54+13	2.57+13	1.80+13	1.22+13	4.45+12
18000	33336	4.16+13	3.62+13	3.00+13	2.42+13	1.70+13	1.16+13	7.55+12	2.52+12
19327	35794	2.62+13	2.23+13	1.81+13	1.41+13	9.57+12	6.22+12	3.98+12	1.17+12

Table 6.13. Annual Equivalent 1 MeV Electron Fluence from Trapped Protons (V_{oc}, P_{max}), 20° Inclination (Infinite Backshielding)

PROTONS — VOC AND PMAX

EQUIV. 1 MEV ELECTRON FLUENCE FOR VOC AND PMAX CIRCULAR ORBIT
DUE TO GEOMAGNETICALLY TRAPPED PROTONS, MODEL AP8MAX

INCLINATION = 20 DEGREES.

ALTITUDE		SHIELD THICKNESS, CM (MILS)							
(N.M.)	(KM)	0 (0)	2.54E-3 (1)	7.64E-3 (3)	1.52E-2 (6)	3.05E-2 (12)	5.09E-2 (20)	7.64E-2 (30)	1.52E-1 (60)
150	277	1.22+09	7.88+08	8.93+08	8.48+08	7.93+08	6.76+08	6.73+08	4.81+08
250	463	2.74+11	2.08+11	1.87+11	1.69+11	1.50+11	1.36+11	1.27+11	1.14+11
300	555	9.83+11	7.74+11	7.07+11	6.50+11	5.83+11	5.25+11	4.89+11	4.35+11
450	833	1.25+13	1.06+13	9.70+12	8.97+12	8.00+12	7.00+12	6.31+12	6.25+12
600	1111	6.84+13	5.84+13	5.22+13	4.69+13	4.02+13	3.37+13	2.92+13	2.25+13
800	1481	4.18+14	3.55+14	3.07+14	2.67+14	2.18+14	1.71+14	1.41+14	1.00+14
1000	1852	1.66+15	1.39+15	1.16+15	9.82+14	7.53+14	5.39+14	4.18+14	2.72+14
1250	2315	6.68+15	5.46+15	4.38+15	3.55+15	2.53+15	1.60+15	1.13+15	6.31+14
1500	2778	2.08+16	1.65+16	1.28+16	9.98+15	6.66+15	3.80+15	2.44+15	1.12+15
1750	3241	4.97+16	3.77+16	2.79+16	2.10+16	1.32+16	6.97+15	4.12+15	1.56+15
2000	3704	1.05+17	7.38+16	5.11+16	3.58+16	2.05+16	9.93+15	5.44+15	1.78+15
2250	4167	2.17+17	1.39+17	8.84+16	5.66+16	2.88+16	1.25+16	6.28+15	1.81+15
2500	4630	3.78+17	2.14+17	1.27+17	7.47+16	3.40+16	1.34+16	6.31+15	1.68+15
2750	5093	6.43+17	3.12+17	1.70+17	9.27+16	3.78+16	1.36+16	6.05+15	1.49+15
3000	5556	9.78+17	3.97+17	1.99+17	1.00+17	3.70+16	1.26+16	5.36+15	1.26+15
3500	6482	2.24+18	6.18+17	2.53+17	1.04+17	3.03+16	9.60+15	3.87+15	8.38+14
4000	7408	4.23+18	7.88+17	2.67+17	9.27+16	2.16+16	6.31+15	2.42+15	5.03+14
4500	8334	6.97+18	9.03+17	2.55+17	7.61+16	1.45+16	3.83+15	1.38+15	2.78+14
5000	9260	1.03+19	9.48+17	2.23+17	5.69+16	8.88+15	2.16+15	7.34+14	1.34+14
5500	10186	1.41+19	9.73+17	1.87+17	3.98+16	4.92+15	1.14+15	3.67+14	6.01+13
6000	11112	1.68+19	8.51+17	1.38+17	2.54+16	2.59+15	5.44+14	1.63+14	2.34+13
7000	12964	1.99+19	4.93+17	5.36+16	6.98+15	4.48+14	7.07+13	1.70+13	1.75+12
8000	14816	1.83+19	2.28+17	1.68+16	1.61+15	6.80+13	8.03+12	1.53+12	1.13+11
9000	16668	1.14+19	6.61+16	2.97+15	1.93+14	4.90+12	4.53+11	8.97+10	1.46+10
10000	18520	6.60+18	1.25+16	3.11+14	1.36+13	4.27+10	1.83-03	0.00	0.00
11000	20372	4.40+18	2.52+15	1.58+13	4.20+11	1.40+09	1.83-03	0.00	0.00
12000	22224	3.11+18	4.56+14	2.20+11	9.63-03	0.00	0.00	0.00	0.00
13000	24076	2.23+18	6.37+13	1.62+07	0.00	0.00	0.00	0.00	0.00
14000	25928	1.58+18	1.06+13	8.22+06	0.00	0.00	0.00	0.00	0.00
15000	27780	1.08+18	2.52+12	4.80+06	0.00	0.00	0.00	0.00	0.00
16000	29632	7.22+17	1.45+10	0.00	0.00	0.00	0.00	0.00	0.00
17000	31484	4.79+17	1.03+10	0.00	0.00	0.00	0.00	0.00	0.00
18000	33336	2.92+17	7.83+09	0.00	0.00	0.00	0.00	0.00	0.00
19327	35794	8.64+16	5.22+09	0.00	0.00	0.00	0.00	0.00	0.00

Table 6.14. Annual Equivalent 1 MeV Electron Fluence from Trapped Protons (I_{sc}), 20° Inclination (Infinite Backshielding)

PROTONS - ISC

EQUIV 1 MEV ELECTRON FLUENCE FOR ISC - CIRCULAR ORBIT
DUE TO GEOMAGNETICALLY TRAPPED PROTONS, MODEL AP8MAX

INCLINATION = 20 DEGREES.

ALTITUDE		\multicolumn{8}{c	}{SHIELD THICKNESS, CM (MILS)}
(N.M.)	(KM)	0 (0)	2.54E-3 (1)	7.64E-3 (3)	1.52E-2 (6)	3.05E-2 (12)	5.09E-2 (20)	7.64E-2 (30)	1.52E-1 (60)
150	277	7.96+08	6.70+08	7.17+08	6.85+08	6.40+08	6.57+08	6.14+08	4.45+08
250	463	1.73+11	1.59+11	1.50+11	1.42+11	1.32+11	1.24+11	1.18+11	1.08+11
300	555	6.50+11	6.04+11	5.74+11	5.46+11	6.08+11	4.75+11	4.51+11	4.09+11
450	833	8.55+12	8.06+12	7.65+12	7.24+12	6.06+12	6.06+12	5.62+12	4.82+12
600	1111	4.42+13	4.13+13	3.84+13	3.55+13	3.16+13	2.78+13	2.50+13	2.01+13
800	1481	2.52+14	2.31+14	2.08+14	1.88+14	1.60+14	1.34+14	1.16+14	8.75+13
1000	1852	9.25+14	8.30+14	7.28+14	6.36+14	5.13+14	3.98+14	3.30+14	2.31+14
1250	2315	3.41+15	2.98+15	2.60+15	2.09+15	1.55+15	1.09+15	8.35+14	5.10+14
1500	2778	9.94+15	8.44+15	6.82+15	5.47+15	3.80+15	2.38+15	1.67+15	8.51+14
1750	3241	2.23+16	1.83+16	1.42+16	1.10+16	7.14+15	4.10+15	2.66+15	1.13+15
2000	3704	4.37+16	3.38+16	2.46+16	1.78+16	1.06+16	5.57+15	3.37+15	1.23+15
2250	4167	8.35+16	6.00+16	4.02+16	2.67+16	1.43+16	6.69+15	3.77+15	1.22+15
2500	4630	1.35+17	8.90+16	5.53+16	3.40+16	1.53+16	6.97+15	3.72+15	1.10+15
2750	5093	2.10+17	1.25+17	7.21+16	4.09+16	1.77+16	6.91+15	3.51+15	9.63+14
3000	5556	2.93+17	1.55+17	8.24+16	4.32+16	1.71+16	6.27+15	3.08+15	8.05+14
3500	6482	5.71+17	2.29+17	1.00+17	4.32+16	1.37+16	4.68+15	2.19+15	5.25+14
4000	7408	9.32+17	2.79+17	1.03+17	3.73+16	9.60+15	3.02+15	1.36+15	3.12+14
4500	8334	1.36+18	3.08+17	9.63+16	2.99+16	6.33+15	1.79+15	7.65+14	1.65+14
5000	9260	1.78+18	3.12+17	8.28+16	2.19+16	3.81+15	9.89+14	4.03+14	8.14+13
5500	10186	2.24+18	3.11+17	6.86+16	1.51+16	2.09+15	5.12+14	1.98+14	3.57+13
6000	11112	2.41+18	2.64+17	5.01+16	9.50+15	1.08+15	2.39+14	8.61+13	1.36+13
7000	12964	2.23+18	1.43+17	1.90+16	2.55+15	1.81+14	2.92+13	8.64+12	9.85+11
8000	14816	1.64+18	6.17+16	5.86+15	5.77+14	2.66+13	3.13+12	7.52+11	6.34+10
9000	16668	8.28+17	1.62+16	1.01+15	6.80+13	1.89+12	1.75+11	4.77+10	8.48+09
10000	18520	3.46+17	2.70+15	1.04+14	4.68+12	9.35+09	0.00	0.00	0.00
11000	20372	1.64+17	4.68+14	4.86+12	1.43+11	3.80+08	0.00	0.00	0.00
12000	22224	8.05+16	7.12+13	4.85+10	4.60-04	0.00	0.00	0.00	0.00
13000	24076	3.96+16	7.96+12	1.17+06	0.00	0.00	0.00	0.00	0.00
14000	25928	2.11+16	1.09+12	7.30-05	0.00	0.00	0.00	0.00	0.00
15000	27780	1.16+16	2.22+11	4.97-05	0.00	0.00	0.00	0.00	0.00
16000	29632	6.79+15	3.01+08	0.00	0.00	0.00	0.00	0.00	0.00
17000	31484	4.14+15	2.30+08	0.00	0.00	0.00	0.00	0.00	0.00
18000	33336	2.42+15	1.85+08	0.00	0.00	0.00	0.00	0.00	0.00
19327	35794	8.11+14	1.35+08	0.00	0.00	0.00	0.00	0.00	0.00

Table 6.15. Annual Equivalent 1 MeV Electron Fluence from Trapped Electrons, 30° Inclination (Infinite Backshielding)

ELECTRONS - ISC, VOC, AND PMAX INCLINATION = 30 DEGREES.

EQUIV. 1 MEV ELECTRON FLUENCE FOR ISC - CIRCULAR ORBIT
DUE TO GEOMAGNETICALLY TRAPPED ELECTRONS - MODELS AE6MAX, AE17LO

ALTITUDE		\multicolumn{8}{c	}{SHIELD THICKNESS, CM (MILS)}
(N.M.)	(KM)	0 (0)	2.54E-3 (1)	7.64E-3 (3)	1.52E-2 (6)	3.05E-2 (12)	5.09E-2 (20)	7.64E-2 (30)	1.52E-1 (60)
150	277	3.47+09	2.64+09	1.89+09	1.30+09	7.69+08	4.79+08	3.14+08	1.29+08
250	463	4.34+10	3.29+10	2.33+10	1.60+10	9.27+09	5.68+09	3.69+09	1.48+09
300	555	9.68+10	7.34+10	5.20+10	3.55+10	2.05+10	1.25+10	8.10+09	3.25+09
450	833	5.60+11	4.25+11	3.01+11	2.05+11	1.17+11	7.11+10	4.57+10	1.80+10
600	1111	2.00+12	1.53+12	1.08+12	7.37+11	4.22+11	2.55+11	1.63+11	6.42+10
800	1481	7.23+12	5.54+12	3.96+12	2.73+12	1.59+12	9.79+11	6.36+11	2.54+11
1000	1852	1.81+13	1.39+13	1.00+13	6.93+12	4.09+12	2.54+12	1.66+12	6.68+11
1250	2315	3.70+13	2.86+13	2.07+13	1.46+13	8.71+12	5.49+12	3.63+12	1.48+12
1500	2778	5.44+13	4.20+13	3.03+13	2.11+13	1.25+13	7.74+12	5.05+12	2.05+12
1750	3241	6.78+13	5.17+13	3.67+13	2.49+13	1.40+13	8.26+12	5.17+12	2.01+12
2000	3704	7.64+13	5.76+13	4.01+13	2.64+13	1.40+13	7.61+12	4.41+12	1.57+12
2250	4167	8.23+13	6.14+13	4.20+13	2.69+13	1.33+13	6.56+12	3.42+12	1.06+12
2500	4630	8.19+13	6.07+13	4.11+13	2.59+13	1.22+13	6.57+12	2.63+12	7.06+11
2750	5093	7.79+13	5.75+13	3.87+13	2.41+13	1.10+13	4.69+12	2.01+12	4.71+11
3000	5556	6.96+13	5.13+13	3.44+13	2.13+13	9.68+12	4.06+12	1.71+12	4.18+11
3500	6482	5.30+13	3.90+13	2.63+13	1.66+13	7.94+12	3.72+12	1.88+12	6.85+11
4000	7408	4.18+13	3.15+13	2.20+13	1.47+13	8.04+12	4.54+12	2.80+12	1.25+12
4500	8334	3.88+13	3.07+13	2.30+13	1.68+13	1.06+13	6.89+12	4.70+12	2.23+12
5000	9260	3.85+13	3.24+13	2.62+13	2.07+13	1.47+13	1.05+13	7.58+12	3.70+12
5500	10186	4.36+13	3.85+13	3.29+13	2.76+13	2.11+13	1.59+13	1.19+13	5.95+12
6000	11112	5.35+13	4.86+13	4.28+13	3.71+13	2.95+13	2.31+13	1.77+13	9.18+12
7000	12964	8.13+13	7.49+13	6.71+13	5.91+13	4.82+13	3.85+13	3.01+13	1.60+13
8000	14816	1.06+14	9.76+13	8.76+13	7.73+13	6.31+13	5.05+13	3.95+13	2.08+13
9000	16668	1.31+14	1.21+14	1.09+14	9.55+13	7.77+13	6.17+13	4.78+13	2.45+13
10000	18520	1.50+14	1.38+14	1.24+14	1.09+14	8.76+13	6.88+13	5.25+13	2.58+13
11000	20372	1.48+14	1.36+14	1.22+14	1.06+14	8.52+13	6.62+13	4.99+13	2.36+13
12000	22224	1.33+14	1.22+14	1.09+14	9.48+13	7.53+13	5.79+13	4.31+13	1.97+13
13000	24076	1.14+14	1.05+14	9.26+13	8.00+13	6.29+13	4.78+13	3.52+13	1.57+13
14000	25928	9.37+13	8.51+13	7.46+13	6.38+13	4.94+13	3.69+13	2.67+13	1.15+13
15000	27780	7.17+13	6.45+13	5.59+13	4.71+13	3.56+13	2.60+13	1.84+13	7.49+12
16000	29632	5.34+13	4.75+13	4.07+13	3.38+13	2.50+13	1.79+13	1.24+13	4.77+12
17000	31484	3.93+13	3.46+13	2.92+13	2.39+13	1.73+13	1.21+13	8.17+12	2.97+12
18000	33336	2.77+13	2.40+13	1.99+13	1.60+13	1.13+13	7.63+12	5.00+12	1.68+12
19327	35793	1.65+13	1.40+13	1.13+13	8.80+12	5.94+12	3.86+12	2.42+12	7.27+11

Table 6.16. Annual Equivalent 1 MeV Electron Fluence from Trapped Protons (V_{oc}, P_{max}), 30° Inclination (Infinite Backshielding)

PROTONS - VOC AND PMAX

EQUIV. 1 MEV ELECTRON FLUENCE FOR VOC AND PMAX CIRCULAR ORBIT DUE TO GEOMAGNETICALLY TRAPPED PROTONS, MODEL AP8MAX   INCLINATION = 30 DEGREES.

ALTITUDE		SHIELD THICKNESS, CM (MILS)							
(N.M.)	(KM)	0 (0)	2.54E-3 (1)	7.64E-3 (3)	1.52E-2 (6)	3.05E-2 (12)	5.09E-2 (20)	7.64E-2 (30)	1.52E-1 (60)
150	277	8.78+10	5.86+10	5.12+10	4.54+10	3.82+10	3.19+10	2.87+10	2.52+10
250	463	1.57+12	1.27+12	1.14+12	1.03+12	8.92+11	7.68+11	6.95+11	6.00+11
300	556	4.02+12	3.31+12	2.92+12	2.60+12	2.22+12	1.90+12	1.70+12	1.43+12
450	833	2.96+13	2.41+13	2.06+13	1.78+13	1.45+13	1.16+13	9.91+12	7.72+12
600	1111	1.25+14	1.00+14	8.21+13	6.80+13	5.24+13	3.97+13	3.26+13	2.37+13
800	1481	5.96+14	4.63+14	3.64+14	2.92+14	2.16+14	1.56+14	1.23+14	8.44+13
1000	1852	2.35+15	1.66+15	1.24+15	9.62+14	6.76+14	4.53+14	3.39+14	2.13+14
1250	2315	9.40+15	5.94+15	4.24+15	3.16+15	2.09+15	1.26+15	8.68+14	4.70+14
1500	2778	2.76+16	1.62+16	1.12+16	8.20+15	5.15+15	2.84+15	1.79+15	8.00+14
1750	3241	6.39+16	3.46+16	2.33+16	1.64+16	9.73+15	4.98+15	2.89+15	1.08+15
2000	3704	1.28+17	6.31+16	4.02+16	2.68+16	1.47+16	6.93+15	3.75+15	1.22+15
2250	4167	2.47+17	1.12+17	6.68+16	4.10+16	2.01+16	8.62+15	4.25+15	1.23+15
2500	4630	4.16+17	1.67+17	9.29+16	5.29+16	2.34+16	9.07+15	4.25+15	1.13+15
2750	5093	6.79+17	2.38+17	1.23+17	6.46+16	2.56+16	9.15+15	4.04+15	1.00+15
3000	5556	1.01+18	2.97+17	1.41+17	6.89+16	2.48+16	8.37+15	3.56+15	8.39+14
3500	6482	2.12+18	4.46+17	1.77+17	6.96+16	1.97+16	6.22+15	2.51+15	6.45+14
4000	7408	3.72+18	5.47+17	1.66+17	5.99+16	1.36+16	3.95+15	1.61+15	3.15+14
4500	8334	5.80+18	6.13+17	1.45+17	4.86+16	9.08+15	2.39+15	8.60+14	1.68+14
5000	9260	8.21+18	6.37+17	1.22+17	3.62+16	5.55+15	1.36+15	4.56+14	8.32+13
5500	10186	1.10+19	6.50+17	8.71+16	2.55+16	3.12+15	7.18+14	2.32+14	3.77+13
6000	11112	1.26+19	5.52+17	8.71+16	1.58+16	1.59+15	3.30+14	9.80+13	1.39+13
7000	12964	1.44+19	3.18+17	3.40+16	4.41+15	2.83+14	4.47+13	1.07+13	1.11+12
8000	14816	1.23+19	1.38+17	9.93+15	9.48+14	4.01+13	4.76+12	9.16+11	6.92+10
9000	16668	7.82+18	4.03+16	1.76+15	1.13+14	2.82+12	2.58+11	5.15+10	8.70+09
10000	18520	4.60+18	7.90+15	1.96+14	8.61+12	2.81+10	1.83-03	0.00	0.00
11000	20372	3.08+18	1.60+15	1.15+13	3.21+11	1.11+09	1.83-03	0.00	0.00
12000	22224	2.12+18	2.84+14	1.48+11	9.63-03	0.00	0.00	0.00	0.00
13000	24076	1.56+18	4.09+13	1.42+07	0.00	0.00	0.00	0.00	0.00
14000	25928	1.09+18	6.45+12	7.11+06	0.00	0.00	0.00	0.00	0.00
15000	27780	6.98+17	1.29+12	3.88+06	0.00	0.00	0.00	0.00	0.00
16000	29632	4.68+17	1.09+10	0.00	0.00	0.00	0.00	0.00	0.00
17000	31484	3.18+17	8.11+09	0.00	0.00	0.00	0.00	0.00	0.00
18000	33336	2.02+17	6.29+09	0.00	0.00	0.00	0.00	0.00	0.00
19327	35793	5.60+16	4.05+09	0.00	0.00	0.00	0.00	0.00	0.00

EQUIV. 1 MEV ELECTRON FLUENCE FOR VOC AND PMAX CIRCULAR ORBIT DUE TO GEOMAGNETICALLY TRAPPED PROTONS, MODEL AP8MIN   INCLINATION = 30 DEGREES.

ALTITUDE									
(N.M.)	(KM)	0 (0)	2.54E-3 (1)	7.64E-3 (3)	1.52E-2 (6)	3.05E-2 (12)	5.09E-2 (20)	7.64E-2 (30)	1.52E-1 (60)
150	277	3.70+11	2.85+11	2.62+11	2.42+11	2.18+11	1.94+11	1.80+11	1.61+11
250	463	3.32+12	2.73+12	2.47+12	2.26+12	1.99+12	1.72+12	1.56+12	1.31+12
300	555	7.20+12	6.00+12	5.34+12	4.79+12	4.13+12	3.54+12	3.17+12	2.63+12
450	833	3.95+13	3.27+13	2.80+13	2.42+13	1.99+13	1.62+13	1.39+13	1.08+13
600	1111	1.52+14	1.22+14	9.99+13	8.28+13	6.45+13	4.99+13	4.14+13	3.03+13
800	1481	7.42+14	5.57+14	4.29+14	3.39+14	2.50+14	1.83+14	1.46+14	1.00+14
1000	1852	2.74+15	1.87+15	1.37+15	1.05+15	7.32+14	4.96+14	3.74+14	2.36+14

6-29

Table 6.17. Annual Equivalent 1 MeV Electron Fluence from Trapped Protons (I_{SC}), 30° Inclination (Infinite Backshielding)

PROTONS - ISC

INCLINATION = 30 DEGREES.

EQUIV 1 MEV ELECTRON FLUENCE FOR ISC - CIRCULAR ORBIT
DUE TO GEOMAGNETICALLY TRAPPED PROTONS, MODEL AP8MAX

ALTITUDE					SHIELD THICKNESS, CM (MILS)				
(N.M.)	(KM)	0 (0)	2.54E-3 (1)	7.64E-3 (3)	1.52E-2 (6)	3.05E-2 (12)	6.09E-2 (20)	7.64E-2 (30)	1.52E-1 (60)
150	277	4.72+10	4.11+10	3.80+10	3.51+10	3.14+10	2.81+10	2.63+10	2.37+10
250	463	1.01+12	9.35+11	8.75+11	8.19+11	7.44+11	6.75+11	6.32+11	6.59+11
300	555	2.56+12	2.36+12	2.19+12	2.03+12	1.82+12	1.64+12	1.52+12	1.31+12
450	833	1.75+13	1.68+13	1.43+13	1.29+13	1.11+13	9.51+12	8.52+12	6.94+12
600	1111	6.88+13	6.08+13	5.29+13	4.61+13	3.81+13	3.12+13	2.71+13	2.09+13
800	1481	3.07+14	2.64+14	2.22+14	1.88+14	1.50+14	1.18+14	9.96+13	7.29+13
1000	1852	1.09+15	8.85+14	7.13+14	5.84+14	4.41+14	3.25+14	2.64+14	1.80+14
1250	2315	3.90+15	2.95+15	2.27+15	1.78+15	1.25+15	8.40+14	6.34+14	3.78+14
1500	2778	1.06+16	7.70+15	5.74+15	4.37+15	2.89+15	1.75+15	1.22+15	6.05+14
1750	3241	2.29+16	1.59+16	1.14+16	8.35+15	5.20+15	2.90+15	1.86+15	7.78+14
2000	3704	4.26+16	2.78+16	1.89+16	1.31+16	7.53+15	3.86+15	2.32+15	8.44+14
2250	4167	7.71+16	4.71+16	2.99+16	1.92+16	9.89+15	4.56+15	2.65+15	8.22+14
2500	4630	1.20+17	6.79+16	4.02+16	2.39+16	1.12+16	4.72+15	2.51+15	7.44+14
2750	5093	1.83+17	9.39+16	5.16+16	2.83+16	1.20+16	4.63+15	2.35+15	6.46+14
3000	5556	2.49+17	1.14+17	5.81+16	2.96+16	1.14+16	4.17+15	2.05+15	5.37+14
3500	6482	4.62+17	1.62+17	6.85+16	2.87+16	8.90+15	3.03+15	1.42+15	3.42+14
4000	7408	7.18+17	1.91+17	6.79+16	2.40+16	6.04+15	1.89+15	8.49+14	1.95+14
4500	8334	1.01+18	2.07+17	6.27+16	1.91+16	3.95+15	1.12+15	4.78+14	1.03+14
5000	9260	1.30+18	2.08+17	5.37+16	1.40+16	2.38+15	6.16+14	2.50+14	5.03+13
5500	10186	1.61+18	2.06+17	4.45+16	9.67+15	1.33+15	3.23+14	1.25+14	2.24+13
6000	11112	1.67+18	1.70+17	3.16+16	5.90+15	6.62+14	1.45+14	5.18+13	8.13+12
7000	12964	1.51+18	9.20+16	1.21+16	1.61+15	1.14+14	1.84+13	5.47+12	6.25+11
8000	14816	1.04+18	3.70+16	3.47+15	3.40+14	1.57+13	1.86+12	4.51+11	3.88+10
9000	16668	5.35+17	9.83+15	6.02+14	3.98+13	1.09+12	1.00+11	2.76+10	5.10+09
10000	18520	2.27+17	1.70+15	6.53+13	2.96+12	6.32+09	0.00	0.00	0.00
11000	20372	1.07+17	2.99+14	3.57+12	1.09+11	3.05+08	0.00	0.00	0.00
12000	22224	5.10+16	4.47+13	3.28+10	4.60-04	0.00	0.00	0.00	0.00
13000	24076	2.61+16	5.13+12	1.07+06	0.00	0.00	0.00	0.00	0.00
14000	25928	1.37+16	6.60+11	6.59+05	0.00	0.00	0.00	0.00	0.00
15000	27780	7.07+15	1.11+11	4.25+05	0.00	0.00	0.00	0.00	0.00
16000	29632	4.31+15	2.41+08	0.00	0.00	0.00	0.00	0.00	0.00
17000	31484	2.74+15	1.90+08	0.00	0.00	0.00	0.00	0.00	0.00
18000	33336	1.67+15	1.56+08	0.00	0.00	0.00	0.00	0.00	0.00
19327	35793	5.24+14	1.10+08	0.00	0.00	0.00	0.00	0.00	0.00

EQUIV 1 MEV ELECTRON FLUENCE FOR ISC - CIRCULAR ORBIT
DUE TO GEOMAGNETICALLY TRAPPED PROTONS, MODEL AP8MIN

INCLINATION = 30 DEGREES.

150	277	2.42+11	2.24+11	2.13+11	2.03+11	1.89+11	1.76+11	1.67+11	1.52+11
250	463	2.19+12	2.04+12	1.92+12	1.81+12	1.65+12	1.50+12	1.39+12	1.21+12
300	555	4.66+12	4.33+12	4.03+12	3.75+12	3.39+12	3.05+12	2.82+12	2.42+12
450	833	2.38+13	2.17+13	1.96+13	1.77+13	1.54+13	1.33+13	1.20+13	9.75+12
600	1111	8.43+13	7.48+13	6.52+13	5.70+13	4.75+13	3.95+13	3.45+13	2.68+13
800	1481	3.72+14	3.15+14	2.61+14	2.20+14	1.75+14	1.39+14	1.18+14	8.71+13
1000	1852	1.24+15	9.88+14	7.85+14	6.37+14	4.80+14	3.58+14	2.91+14	1.99+14

6-30

Table 6.18. Annual Equivalent 1 MeV Electron Fluence from Trapped Electrons, 40° Inclination (Infinite Backshielding)

ELECTRONS - ISC, VOC, AND PMAX

INCLINATION = 40 DEGREES.

EQUIV. 1 MEV ELECTRON FLUENCE FOR ISC - CIRCULAR ORBIT
DUE TO GEOMAGNETICALLY TRAPPED ELECTRONS - MODELS AE6MAX, AE17LO

ALTITUDE		\multicolumn{7}{c}{SHIELD THICKNESS, CM (MILS)}						
(N.M.)	(KM)	0 (0)	2.54E-3 (1)	7.64E-3 (3)	1.52E-2 (6)	3.05E-2 (12)	5.09E-2 (20)	7.64E-2 (30)	1.52E-1 (60)
150	277	2.24+10	1.66+10	1.14+10	7.56+09	4.13+09	2.37+09	1.46+09	6.62+08
250	463	8.82+10	6.60+10	4.60+10	3.09+10	1.73+10	1.03+10	6.49+09	2.56+09
300	555	1.65+11	1.17+11	8.19+10	5.53+10	3.15+10	1.89+10	1.21+10	4.84+09
450	833	5.90+11	4.48+11	3.17+11	2.17+11	1.25+11	7.60+10	4.91+10	1.96+10
600	1111	1.65+12	1.25+12	8.86+11	6.04+11	3.46+11	2.10+11	1.36+11	5.44+10
800	1481	5.45+12	4.16+12	2.95+12	2.02+12	1.17+12	7.10+11	4.59+11	1.84+11
1000	1852	1.36+13	1.05+13	7.51+12	5.19+12	3.04+12	1.88+12	1.23+12	4.95+11
1250	2315	2.76+13	2.13+13	1.55+13	1.08+13	6.49+12	4.09+12	2.71+12	1.12+12
1500	2778	4.03+13	3.11+13	2.25+13	1.57+13	9.34+12	5.84+12	3.84+12	1.59+12
1750	3241	5.04+13	3.86+13	2.76+13	1.89+13	1.08+13	6.50+12	4.14+12	1.67+12
2000	3704	5.74+13	4.36+13	3.06+13	2.05+13	1.12+13	6.32+12	3.83+12	1.47+12
2250	4167	6.20+13	4.67+13	3.23+13	2.12+13	1.10+13	5.85+12	3.34+12	1.22+12
2500	4630	6.30+13	4.72+13	3.26+13	2.11+13	1.08+13	5.53+12	3.06+12	1.11+12
2750	5093	6.09+13	4.57+13	3.15+13	2.04+13	1.03+13	5.25+12	2.88+12	1.08+12
3000	5556	5.57+13	4.19+13	2.91+13	1.91+13	9.88+12	5.18+12	2.96+12	1.18+12
3500	6482	4.46+13	3.40+13	2.42+13	1.66+13	9.45+12	5.60+12	3.59+12	1.61+12
4000	7408	3.80+13	3.00+13	2.25+13	1.64+13	1.05+13	7.00+12	4.89+12	2.35+12
4500	8334	3.81+13	3.15+13	2.60+13	1.95+13	1.37+13	9.77+12	7.10+12	3.51+12
5000	9260	4.00+13	3.47+13	2.90+13	2.39+13	1.79+13	1.33+13	9.89+12	4.92+12
5500	10186	4.44+13	3.97+13	3.44+13	2.92+13	2.27+13	1.73+13	1.30+13	6.52+12
6000	11112	5.22+13	4.76+13	4.21+13	3.65+13	2.92+13	2.28+13	1.75+13	8.97+12
7000	12964	7.46+13	6.87+13	6.15+13	5.40+13	4.38+13	3.48+13	2.71+13	1.41+13
8000	14816	9.16+13	8.45+13	7.58+13	6.66+13	5.42+13	4.30+13	3.34+13	1.72+13
9000	16668	1.06+14	9.76+13	8.74+13	7.68+13	6.22+13	4.92+13	3.79+13	1.92+13
10000	18520	1.16+14	1.07+14	9.53+13	8.34+13	6.70+13	5.24+13	3.99+13	1.94+13
11000	20372	1.12+14	1.03+14	9.14+13	7.97+13	6.36+13	4.92+13	3.70+13	1.73+13
12000	22224	9.75+13	8.94+13	7.93+13	6.88+13	5.43+13	4.16+13	3.09+13	1.40+13
13000	24076	8.52+13	7.78+13	6.86+13	5.91+13	4.63+13	3.51+13	2.58+13	1.14+13
14000	25928	6.78+13	6.14+13	5.37+13	4.58+13	3.53+13	2.63+13	1.90+13	8.13+12
15000	27780	4.95+13	4.44+13	3.84+13	3.22+13	2.43+13	1.77+13	1.24+13	5.04+12
16000	29632	3.80+13	3.39+13	2.89+13	2.41+13	1.78+13	1.27+13	8.79+12	3.40+12
17000	31484	2.94+13	2.59+13	2.19+13	1.79+13	1.30+13	9.07+12	6.12+12	2.23+12
18000	33336	2.13+13	1.85+13	1.54+13	1.24+13	8.71+12	5.92+12	3.88+12	1.30+12
19327	35793	1.22+13	1.03+13	8.37+12	6.56+12	4.44+12	2.89+12	1.82+12	5.49+11

Table 6.19. Annual Equivalent 1 MeV Electron Fluence from Trapped Protons (V_{oc}, P_{max}), 40° Inclination (Infinite Backshielding)

EQUIV. 1 MEV ELECTRON FLUENCE FOR VOC AND PMAX CIRCULAR ORBIT
DUE TO GEOMAGNETICALLY TRAPPED PROTONS, MODEL AP8MAX

PROTONS - VOC AND PMAX

INCLINATION = 40 DEGREES.

ALTITUDE		\multicolumn{8}{c	}{SHIELD THICKNESS, CM (MILS)}
(N.M.)	(KM)	0 (0)	2.54E-3 (1)	7.64E-3 (3)	1.62E-2 (6)	3.05E-2 (12)	5.09E-2 (20)	7.64E-2 (30)	1.52E-1 (60)
150	277	6.36+11	4.25+11	3.38+11	2.76+11	2.16+11	1.76+11	1.54+11	1.28+11
250	463	8.62+12	4.91+12	3.47+12	2.57+12	1.81+12	1.37+12	1.15+12	8.95+11
300	555	2.38+13	1.25+13	8.34+12	5.85+12	3.94+12	2.91+12	2.41+12	1.86+12
450	833	1.61+14	6.77+13	4.20+13	2.83+13	1.83+13	1.28+13	1.02+13	7.38+12
600	1111	5.03+14	2.14+14	1.30+14	8.64+13	5.49+13	3.72+13	2.89+13	1.99+13
800	1481	1.95+15	6.69+14	4.17+14	2.90+14	1.91+14	1.30+14	9.94+13	6.56+13
1000	1862	7.89+15	1.95+15	1.22+15	8.52+14	5.63+14	3.55+14	2.61+14	1.61+14
1250	2315	2.67+16	5.96+15	3.72+15	2.59+15	1.62+15	9.54+14	6.49+14	3.48+14
1500	2778	6.14+16	1.48+16	9.27+15	6.40+15	3.85+15	2.09+15	1.30+15	5.80+14
1750	3241	1.17+17	2.96+16	1.84+16	1.25+16	7.22+15	3.67+15	2.13+15	7.94+14
2000	3704	1.96+17	5.19+16	3.12+16	2.02+16	1.09+16	5.13+15	2.77+15	9.01+14
2250	4167	3.22+17	8.89+16	5.06+16	3.04+16	1.47+16	6.25+15	3.12+15	8.98+14
2500	4630	4.90+17	1.30+17	6.98+16	3.93+16	1.72+16	6.69+15	3.14+15	8.33+14
2750	5093	7.28+17	1.83+17	9.13+16	4.76+16	1.88+16	6.72+15	2.97+15	7.33+14
3000	5556	1.01+18	2.26+17	1.05+17	5.07+16	1.82+16	6.16+15	2.61+15	6.15+14
3500	6482	1.91+18	3.28+17	1.26+17	5.03+16	1.42+16	4.49+15	1.81+15	3.93+14
4000	7408	3.17+18	4.01+17	1.28+17	4.35+16	9.90+15	2.88+15	1.10+15	2.28+14
4500	8334	4.70+18	4.44+17	1.20+17	3.51+16	6.59+15	1.74+15	6.24+14	1.22+14
5000	9260	6.46+18	4.60+17	1.04+17	2.61+16	3.99+15	9.68+14	3.28+14	5.99+13
5500	10186	8.49+18	4.75+17	8.92+16	1.88+16	2.30+15	5.32+14	1.72+14	2.80+13
6000	11112	9.50+18	4.03+17	6.37+16	1.16+16	1.16+15	2.43+14	7.22+13	1.03+13
7000	12964	1.06+19	2.35+17	2.54+16	3.31+15	2.14+14	3.39+13	8.18+12	8.48+11
8000	14816	8.87+18	1.00+17	7.31+15	7.03+14	2.99+13	3.55+12	6.87+11	5.26+10
9000	16668	5.74+18	2.93+16	1.28+15	8.26+13	2.06+12	1.88+11	3.76+10	6.44+09
10000	18520	3.43+18	6.01+15	1.51+14	6.64+12	2.22+10	1.83-03	0.00	0.00
11000	20372	2.26+18	1.22+15	9.43+12	2.70+11	9.54+08	1.83-03	0.00	0.00
12000	22224	1.53+18	2.18+14	1.16+11	9.63-03	0.00	0.00	0.00	0.00
13000	24076	1.14+18	3.12+13	1.32+07	0.00	0.00	0.00	0.00	0.00
14000	25928	7.76+17	4.75+12	6.53+06	0.00	0.00	0.00	0.00	0.00
15000	27780	4.75+17	9.48+11	3.58+06	0.00	0.00	0.00	0.00	0.00
16000	29632	3.35+17	9.15+09	0.00	0.00	0.00	0.00	0.00	0.00
17000	31484	2.38+17	6.89+09	0.00	0.00	0.00	0.00	0.00	0.00
18000	33336	1.57+17	5.42+09	0.00	0.00	0.00	0.00	0.00	0.00
19327	35793	4.31+16	3.47+09	0.00	0.00	0.00	0.00	0.00	0.00

Table 6.20. Annual Equivalent 1 MeV Electron Fluence from Trapped Protons (I_{SC}), 40° Inclination (Infinite Backshielding)

PROTONS - ISC

INCLINATION = 40 DEGREES.

EQUIV 1 MEV ELECTRON FLUENCE FOR ISC - CIRCULAR ORBIT
DUE TO GEOMAGNETICALLY TRAPPED PROTONS, MODEL AP8MAX

ALTITUDE		SHIELD THICKNESS, CM (MILS)						
(N.M.)	(KM)	0 (0)	2.54E-3 (1)	7.64E-3 (3)	1.52E-2 (6)	3.05E-2 (12)	5.09E-2 (20)	7.64E-2 (30)	1.52E-1 (60)
150	277	3.20+11	2.63+11	2.28+11	2.01+11	1.71+11	1.50+11	1.37+11	1.18+11
250	463	3.58+12	2.64+12	2.08+12	1.69+12	1.34+12	1.11+12	9.88+11	8.10+11
300	555	9.11+12	6.39+12	4.78+12	3.74+12	2.87+12	2.34+12	2.07+12	1.68+12
450	833	6.16+13	3.25+13	2.29+13	1.72+13	1.26+13	9.85+12	8.45+12	6.53+12
600	1111	1.65+14	1.00+14	6.93+13	5.13+13	3.68+13	2.79+13	2.34+13	1.73+13
800	1481	5.46+14	3.21+14	2.28+14	1.74+14	1.27+14	9.56+13	7.89+13	5.62+13
1000	1852	1.69+15	9.20+14	6.51+14	4.93+14	3.51+14	2.51+14	2.01+14	1.35+14
1250	2315	6.18+15	2.74+15	1.91+15	1.42+15	9.60+14	6.30+14	4.71+14	2.78+14
1500	2778	1.23+16	6.65+15	4.60+15	3.35+15	2.14+15	1.28+15	8.84+14	4.38+14
1750	3241	2.41+16	1.31+16	8.88+15	6.31+15	3.85+15	2.13+15	1.37+15	5.70+14
2000	3704	4.16+16	2.23+16	1.45+16	9.83+15	5.59+15	2.86+15	1.72+15	6.23+14
2250	4167	7.06+16	3.67+16	2.25+16	1.42+16	7.25+15	3.34+15	1.87+15	6.02+14
2500	4630	1.06+17	5.22+16	3.01+16	1.77+16	8.23+15	3.48+15	1.85+15	5.47+14
2750	5093	1.56+17	7.13+16	3.83+16	2.09+16	8.77+15	3.40+15	1.72+15	4.74+14
3000	5556	2.08+17	8.60+16	4.30+16	2.18+16	8.38+15	3.07+15	1.50+15	3.93+14
3500	6482	3.67+17	1.19+17	4.96+16	2.08+16	6.43+15	2.19+15	1.02+15	2.47+14
4000	7408	5.56+17	1.40+17	4.93+16	1.74+16	4.39+15	1.37+15	6.17+14	1.42+14
4500	8334	7.58+17	1.50+17	4.52+16	1.38+16	2.87+15	8.11+14	3.46+14	7.46+13
5000	9260	9.62+17	1.50+17	3.86+16	1.00+16	1.71+16	4.43+14	1.80+14	3.62+13
5500	10186	1.19+18	1.51+17	3.26+16	7.11+15	9.79+14	2.39+14	9.25+13	1.66+13
6000	11112	1.23+18	1.24+17	2.31+16	4.33+15	4.87+13	1.06+14	3.82+13	5.99+12
7000	12964	1.10+18	6.83+16	9.02+15	1.21+15	4.83+13	1.40+13	4.16+12	4.77+11
8000	14816	7.46+17	2.70+16	2.55+15	2.52+14	1.17+13	1.39+12	3.39+11	2.95+10
9000	16668	3.88+17	7.16+15	4.38+14	2.90+13	7.91+11	7.30+10	2.02+10	3.77+09
10000	18520	1.69+17	1.30+15	5.02+13	2.28+12	5.07+09	0.00	0.00	0.00
11000	20372	7.84+16	2.29+14	2.95+12	9.18+10	2.65+08	0.00	0.00	0.00
12000	22224	3.73+16	3.44+13	2.59+10	4.60-04	0.00	0.00	0.00	0.00
13000	24076	1.93+16	3.92+12	1.01+06	0.00	0.00	0.00	0.00	0.00
14000	25928	9.85+15	4.88+11	6.20+05	0.00	0.00	0.00	0.00	0.00
15000	27780	4.86+15	8.19+10	4.00+05	0.00	0.00	0.00	0.00	0.00
16000	29632	3.12+15	2.09+08	0.00	0.00	0.00	0.00	0.00	0.00
17000	31484	2.06+15	1.67+08	0.00	0.00	0.00	0.00	0.00	0.00
18000	33336	1.29+15	1.39+08	0.00	0.00	0.00	0.00	0.00	0.00
19327	35793	4.02+14	9.78+07	0.00	0.00	0.00	0.00	0.00	0.00

Table 6.21. Annual Equivalent 1 MeV Electron Fluence from Trapped Electrons, 50° Inclination (Infinite Backshielding)

ELECTRONS - ISC - CIRCULAR ORBIT
EQUIV. 1 MEV ELECTRON FLUENCE FOR ISC - CIRCULAR ORBIT
DUE TO GEOMAGNETICALLY TRAPPED ELECTRONS - MODELS AE6MAX, AE17L0

ELECTRONS - ISC, VOC, AND PMAX
INCLINATION = 50 DEGREES.

ALTITUDE		SHIELD THICKNESS, CM (MILS)							
(N.M.)	(KM)	0 (0)	2.54E-3 (1)	7.64E-3 (3)	1.52E-2 (6)	3.05E-2 (12)	5.09E-2 (20)	7.64E-2 (30)	1.52E-1 (60)
150	277	8.96+10	7.94+10	6.84+10	5.83+10	4.59+10	3.59+10	2.77+10	1.44+10
250	463	2.04+11	1.76+11	1.47+11	1.22+11	9.36+10	7.18+10	5.47+10	2.79+10
300	555	2.94+11	2.51+11	2.07+11	1.69+11	1.27+11	9.66+10	7.31+10	3.69+10
450	833	7.33+11	6.03+11	4.77+11	3.73+11	2.64+11	1.93+11	1.42+11	6.92+10
600	1111	1.69+12	1.35+12	1.02+12	7.71+11	5.15+11	3.59+11	2.57+11	1.20+11
800	1481	4.91+12	3.85+12	2.84+12	2.05+12	1.29+12	8.54+11	6.89+11	2.62+11
1000	1852	1.17+13	9.13+12	6.69+12	4.76+12	2.94+12	1.91+12	1.29+12	5.56+11
1250	2315	2.31+13	1.81+13	1.33+13	9.50+12	5.90+12	3.84+12	2.60+12	1.11+12
1500	2778	3.31+13	2.58+13	1.89+13	1.34+13	8.26+12	5.32+12	3.58+12	1.53+12
1750	3241	4.19+13	3.24+13	2.34+13	1.63+13	9.68+12	6.01+12	3.94+12	1.65+12
2000	3704	4.83+13	3.70+13	2.64+13	1.80+13	1.03+13	6.03+12	3.80+12	1.53+12
2250	4167	5.22+13	3.97+13	2.80+13	1.87+13	1.02+13	5.78+12	3.50+12	1.38+12
2500	4630	5.36+13	4.07+13	2.85+13	1.90+13	1.02+13	5.62+12	3.35+12	1.33+12
2750	5093	5.22+13	3.97+13	2.79+13	1.86+13	1.00+13	5.54+12	3.31+12	1.36+12
3000	5556	4.86+13	3.71+13	2.64+13	1.79+13	9.96+12	5.73+12	3.57+12	1.55+12
3500	6482	4.06+13	3.17+13	2.34+13	1.67+13	1.03+13	6.68+12	4.57+12	2.17+12
4000	7408	3.64+13	2.95+13	2.28+13	1.73+13	1.18+13	8.24+13	5.95+12	2.92+12
4500	8334	3.65+13	3.07+13	2.49+13	1.98+13	1.43+13	1.04+13	7.68+12	3.79+12
5000	9260	3.77+13	3.29+13	2.78+13	2.31+13	1.74+13	1.31+13	9.72+12	4.79+12
5500	10186	4.03+13	3.61+13	3.13+13	2.67+13	2.08+13	1.59+13	1.19+13	5.90+12
6000	11112	4.55+13	4.14+13	3.66+13	3.18+13	2.53+13	1.97+13	1.51+13	7.59+12
7000	12964	6.05+13	5.57+13	4.97+13	4.36+13	3.53+13	2.79+13	2.16+13	1.11+13
8000	14816	7.14+13	6.58+13	5.89+13	5.17+13	4.19+13	3.31+13	2.56+13	1.31+13
9000	16668	8.30+13	7.65+13	6.84+13	6.00+13	4.85+13	3.83+13	2.95+13	1.49+13
10000	18520	9.10+13	8.38+13	7.48+13	6.54+13	5.25+13	4.11+13	3.13+13	1.53+13
11000	20372	8.69+13	7.99+13	7.12+13	6.20+13	4.95+13	3.83+13	2.88+13	1.35+13
12000	22224	7.53+13	6.91+13	6.13+13	5.32+13	4.22+13	3.23+13	2.40+13	1.10+13
13000	24076	6.77+13	6.19+13	5.47+13	4.72+13	3.70+13	2.81+13	2.07+13	9.18+12
14000	25928	5.33+13	4.84+13	4.24+13	3.62+13	2.80+13	2.09+13	1.51+13	6.50+12
15000	27780	3.76+13	3.38+13	2.92+13	2.46+13	1.86+13	1.36+13	9.58+12	3.91+12
16000	29632	3.04+13	2.71+13	2.32+13	1.93+13	1.43+13	1.02+13	7.10+12	2.75+12
17000	31484	2.40+13	2.12+13	1.79+13	1.47+13	1.06+13	7.42+12	5.02+12	1.83+12
18000	33336	1.78+13	1.55+13	1.29+13	1.03+13	7.29+12	4.95+12	3.24+12	1.09+12
19327	35793	1.00+13	8.52+12	6.90+12	5.41+12	3.67+12	2.39+12	1.50+12	4.54+11

Table 6.22. Annual Equivalent 1 MeV Electron Fluence from Trapped Protons (V_{oc}, P_{max}), 50° Inclination (Infinite Backshielding)

PROTONS - V_{OC} AND P_{MAX}

EQUIV. 1 MEV ELECTRON FLUENCE FOR VOC AND PMAX CIRCULAR ORBIT   INCLINATION = 50 DEGREES.
DUE TO GEOMAGNETICALLY TRAPPED PROTONS, MODEL AP8MAX

ALTITUDE				SHIELD THICKNESS, CM (MILS)				
(N.M.)	(KM)	0 (0)	2.54E-3 (1)	7.64E-3 (3)	1.52E-2 (6)	3.05E-2 (12)	6.09E-2 (20)	7.64E-2 (30)	1.52E-1 (60)
150	277	1.28+13	1.12+12	5.29+11	3.17+11	1.90+11	1.38+11	1.14+11	8.89+10
250	463	7.96+13	8.94+12	4.40+12	2.61+12	1.52+12	1.05+12	8.42+11	6.27+11
300	555	1.63+14	2.04+13	9.85+12	5.71+12	3.25+12	2.22+12	1.77+12	1.31+12
450	833	8.14+14	9.38+13	4.65+13	2.65+13	1.50+13	9.99+12	7.77+12	5.51+12
600	1111	2.28+15	2.53+14	1.26+14	7.57+13	4.40+13	2.88+13	2.20+13	1.50+13
800	1481	6.27+15	6.94+14	3.76+14	2.43+14	1.52+14	1.02+14	7.75+13	5.13+13
1000	1852	1.73+16	1.88+15	1.05+15	6.98+14	4.39+14	2.82+14	2.07+14	1.28+14
1250	2315	4.40+16	5.42+15	3.11+15	2.09+15	1.29+15	7.61+14	5.19+14	2.79+14
1500	2778	8.44+16	1.26+16	7.42+15	5.01+15	3.00+15	1.63+15	1.02+15	4.60+14
1750	3241	1.44+17	2.45+16	1.47+16	9.91+15	5.76+15	2.95+15	1.71+15	6.41+14
2000	3704	2.24+17	4.22+16	2.49+16	1.62+16	8.81+15	4.17+15	2.26+15	7.33+14
2250	4167	3.41+17	7.15+16	4.02+16	2.43+16	1.19+16	5.06+15	2.53+15	7.29+14
2500	4630	4.94+17	1.05+17	5.61+16	3.18+16	1.40+16	5.47+15	2.56+15	6.80+14
2750	5093	7.02+17	1.46+17	7.33+16	3.85+16	1.53+16	5.48+15	2.42+15	5.98+14
3000	5556	9.32+17	1.81+17	8.47+16	4.13+16	1.49+16	5.03+15	2.14+15	5.02+14
3500	6482	1.66+18	2.61+17	1.01+17	4.05+16	1.15+16	3.65+15	1.47+15	3.20+14
4000	7408	2.62+18	3.21+17	1.04+17	3.54+16	8.11+15	2.36+15	9.03+14	1.87+14
4500	8334	3.76+18	3.56+17	9.75+16	2.87+16	5.41+15	1.43+15	5.13+14	1.00+14
5000	9260	5.11+18	3.73+17	8.49+16	2.13+16	3.27+15	7.93+14	2.69+14	4.90+13
5500	10186	6.77+18	3.88+17	7.33+16	1.54+16	1.90+15	4.39+14	1.42+14	2.31+13
6000	11112	7.56+18	3.30+17	5.24+16	9.52+15	9.59+14	2.00+14	6.94+13	8.46+12
7000	12964	8.61+18	1.94+17	2.10+16	2.75+15	1.78+14	2.82+13	6.82+12	7.08+11
8000	14816	7.14+18	8.23+16	6.03+15	5.80+14	2.47+13	2.93+12	5.67+11	4.35+10
9000	16668	4.62+18	2.39+16	1.04+15	6.70+13	1.66+12	1.52+11	3.03+10	5.23+09
10000	18520	2.78+18	5.02+15	1.26+14	5.59+12	1.90+10	1.83-03	0.00	0.00
11000	20372	1.80+18	1.02+15	2.36+13	9.63-03	8.49-08	0.00	0.00	0.00
12000	22224	1.22+18	1.83+14	8.15+12	0.00	0.00	0.00	0.00	0.00
13000	24076	9.23+17	2.59+13	9.90+10	0.00	0.00	0.00	0.00	0.00
14000	25928	6.25+17	3.94+12	1.25+07	0.00	0.00	0.00	0.00	0.00
15000	27780	3.74+17	7.85+11	6.20+06	0.00	0.00	0.00	0.00	0.00
16000	29632	2.74+17	8.18+09	3.41+06	0.00	0.00	0.00	0.00	0.00
17000	31484	1.96+17	6.17+09	0.00	0.00	0.00	0.00	0.00	0.00
18000	33336	1.30+17	4.88+09	0.00	0.00	0.00	0.00	0.00	0.00
19327	35793	3.58+16	3.11+09	0.00	0.00	0.00	0.00	0.00	0.00

Table 6.23. Annual Equivalent 1 MeV Electron Fluence from Trapped Protons (I_{SC}), 50° Inclination (Infinite Backshielding)

PROTONS - ISC

EQUIV 1 MEV ELECTRON FLUENCE FOR ISC - CIRCULAR ORBIT
DUE TO GEOMAGNETICALLY TRAPPED PROTONS, MODEL AP8MAX   INCLINATION = 50 DEGREES.

ALTITUDE				SHIELD THICKNESS, CM (MILS)				
(N.M.)	(KM)	0 (0)	2.54E-3 (1)	7.64E-3 (3)	1.52E-2 (6)	3.05E-2 (12)	5.09E-2 (20)	7.64E-2 (30)	1.52E-1 (60)
150	277	1.63+12	4.73+11	2.76+11	1.92+11	1.38+11	1.11+11	9.84+10	8.07+10
250	463	1.13+13	3.77+12	2.24+12	1.63+12	1.05+12	8.21+11	7.09+11	5.61+11
300	555	2.47+13	8.51+12	4.94+12	3.30+12	2.24+12	1.73+12	1.49+12	1.17+12
450	833	1.14+14	3.91+13	2.26+13	1.50+13	1.01+13	7.57+12	6.39+12	4.85+12
600	1111	3.00+14	1.07+14	6.33+13	4.29+13	2.89+13	2.14+13	1.78+13	1.30+13
800	1481	7.97+14	3.07+14	1.96+14	1.42+14	1.00+14	7.47+13	6.16+13	4.39+13
1000	1852	2.11+15	8.38+14	5.48+14	4.00+14	2.79+14	2.00+14	1.60+14	1.07+14
1250	2315	5.76+15	2.40+15	1.58+15	1.14+15	7.65+14	5.04+14	3.78+14	2.23+14
1500	2778	1.24+16	5.53+15	3.66+15	2.62+15	1.67+15	1.01+15	6.97+14	3.48+14
1750	3241	2.29+16	1.07+16	7.08+15	5.02+15	3.07+15	1.72+15	1.10+15	4.60+14
2000	3704	3.79+16	1.80+16	1.16+16	7.90+15	4.52+15	2.32+15	1.40+15	5.06+14
2250	4167	6.21+16	2.94+16	1.79+16	1.14+16	5.85+15	2.71+15	1.52+15	4.89+14
2500	4630	9.10+16	4.20+16	2.42+16	1.43+16	6.71+15	2.84+15	1.51+15	4.47+14
2750	5093	1.31+17	5.71+16	3.08+16	1.69+16	7.15+15	2.77+15	1.41+15	3.85+14
3000	5556	1.71+17	6.92+16	3.48+16	1.77+16	6.85+15	2.51+15	1.23+15	3.21+14
3500	6482	2.93+17	9.48+16	3.98+16	1.68+16	5.22+15	1.78+15	8.33+14	2.00+14
4000	7408	4.40+17	1.12+17	4.00+16	1.42+16	3.60+15	1.13+15	5.07+14	1.16+14
4500	8334	5.97+17	1.20+17	3.68+16	1.13+16	2.36+15	6.66+14	2.84+14	6.13+13
5000	9260	7.66+17	1.22+17	3.15+16	8.21+15	1.40+15	3.63+14	1.48+14	2.96+13
5500	10186	9.61+17	1.23+17	2.68+16	5.85+15	8.07+14	1.98+14	7.64+13	1.37+13
6000	11112	9.90+17	1.02+17	1.90+16	3.56+15	4.01+14	8.76+13	3.14+13	4.93+12
7000	12964	8.99+17	5.63+16	7.47+15	1.00+15	7.17+13	1.17+13	3.47+12	3.98+11
8000	14816	6.06+17	2.22+16	2.10+15	2.08+14	9.67+12	1.15+12	2.80+11	2.44+10
9000	16668	3.15+17	5.83+15	3.56+14	2.35+13	6.40+11	5.89+10	1.63+10	3.08+09
10000	18520	1.39+17	1.09+15	4.22+13	1.92+12	4.38+09	0.00	0.00	0.00
11000	20372	6.39+16	1.92+14	2.56+12	8.03+10	2.37+08	0.00	0.00	0.00
12000	22224	3.05+16	2.89+13	2.21+10	4.60-04	0.00	0.00	0.00	0.00
13000	24076	1.58+16	3.27+12	9.78+05	0.00	0.00	0.00	0.00	0.00
14000	25928	8.03+15	4.04+11	5.97+05	0.00	0.00	0.00	0.00	0.00
15000	27780	3.88+15	6.81+10	3.86+05	0.00	0.00	0.00	0.00	0.00
16000	29632	2.56+15	1.92+08	0.00	0.00	0.00	0.00	0.00	0.00
17000	31484	1.70+15	1.53+08	0.00	0.00	0.00	0.00	0.00	0.00
18000	33336	1.08+15	1.28+08	0.00	0.00	0.00	0.00	0.00	0.00
19327	35793	3.34+14	8.99+07	0.00	0.00	0.00	0.00	0.00	0.00

Table 6.24. Annual Equivalent 1 MeV Electron Fluence from Trapped Electrons, 60° Inclination (Infinite Backshielding)

ELECTRONS - ISC, VOC, AND PMAX INCLINATION = 60 DEGREES.

EQUIV. 1 MEV ELECTRON FLUENCE FOR ISC - CIRCULAR ORBIT
DUE TO GEOMAGNETICALLY TRAPPED ELECTRONS - MODELS AE6MAX, AEI7LO

ALTITUDE		\multicolumn{7}{c}{SHIELD THICKNESS, CM (MILS)}
(N.M.)	(KM)	0 (0)	2.54E-3 (1)	7.64E-3 (3)	1.52E-2 (6)	3.05E-2 (12)	5.09E-2 (20)	7.64E-2 (30)	1.52E-1 (60)
150	277	2.09+11	1.90+11	1.68+11	1.46+11	1.18+11	9.28+10	7.17+10	3.68+10
250	463	3.45+11	3.10+11	2.70+11	2.32+11	1.84+11	1.44+11	1.11+11	6.66+10
300	555	4.39+11	3.90+11	3.36+11	2.86+11	2.23+11	1.74+11	1.33+11	6.74+10
450	833	8.71+11	7.46+11	6.18+11	5.06+11	3.81+11	2.88+11	2.17+11	1.08+11
600	1111	1.77+12	1.46+12	1.16+12	9.12+11	6.49+11	4.73+11	3.49+11	1.69+11
800	1481	4.69+12	3.75+12	2.85+12	2.12+12	1.40+12	9.68+11	6.88+11	3.18+11
1000	1852	1.08+13	8.48+12	6.33+12	4.60+12	2.94+12	1.96+12	1.36+12	6.05+11
1250	2315	2.09+13	1.64+13	1.22+13	8.87+12	5.63+12	3.74+12	2.58+12	1.13+12
1500	2778	2.97+13	2.33+13	1.73+13	1.25+13	7.85+12	6.16+12	3.53+12	1.54+12
1750	3241	3.77+13	2.94+13	2.15+13	1.52+13	9.25+12	5.89+12	3.94+12	1.69+12
2000	3704	4.37+13	3.37+13	2.43+13	1.69+13	9.85+12	6.02+12	3.90+12	1.63+12
2250	4167	4.72+13	3.62+13	2.58+13	1.76+13	9.97+12	5.87+12	3.69+12	1.53+12
2500	4630	4.87+13	3.73+13	2.66+13	1.80+13	1.00+13	5.79+12	3.59+12	1.49+12
2750	5093	4.76+13	3.64+13	2.59+13	1.76+13	9.85+12	5.68+12	3.53+12	1.49+12
3000	5556	4.40+13	3.38+13	2.43+13	1.67+13	9.57+12	5.67+12	3.61+12	1.57+12
3500	6482	3.62+13	2.84+13	2.11+13	1.52+13	9.46+12	6.14+12	4.21+12	1.96+12
4000	7408	3.18+13	2.57+13	1.98+13	1.50+13	1.02+13	7.05+12	5.04+12	2.41+12
4500	8334	3.08+13	2.58+13	2.08+13	1.64+13	1.17+13	8.46+12	6.16+12	2.98+12
5000	9260	3.03+13	2.63+13	2.21+13	1.82+13	1.36+13	1.01+13	7.45+12	3.61+12
5500	10186	3.23+13	2.89+13	2.49+13	2.11+13	1.63+13	1.24+13	9.28+12	4.55+12
6000	11112	3.57+13	3.24+13	2.86+13	2.47+13	1.96+13	1.53+13	1.16+13	5.86+12
7000	12964	4.88+13	4.49+13	4.01+13	3.51+13	2.84+13	2.25+13	1.74+13	9.03+12
8000	14816	5.74+13	5.29+13	4.73+13	4.16+13	3.37+13	2.68+13	2.08+13	1.07+13
9000	16668	6.95+13	6.40+13	5.73+13	5.03+13	4.08+13	3.22+13	2.49+13	1.26+13
10000	18520	7.74+13	7.13+13	6.37+13	5.57+13	4.48+13	3.51+13	2.68+13	1.31+13
11000	20372	7.35+13	6.76+13	6.03+13	5.26+13	4.20+13	3.26+13	2.45+13	1.16+13
12000	22224	6.42+13	5.90+13	5.24+13	4.55+13	3.61+13	2.78+13	2.07+13	9.45+12
13000	24076	5.84+13	5.33+13	4.71+13	4.07+13	3.19+13	2.42+13	1.78+13	7.94+12
14000	25928	4.59+13	4.16+13	3.65+13	3.12+13	2.41+13	1.80+13	1.31+13	5.62+12
15000	27780	3.21+13	2.89+13	2.50+13	2.11+13	1.60+13	1.17+13	8.25+12	3.38+12
16000	29632	2.64+13	2.36+13	2.02+13	1.68+13	1.25+13	8.93+12	6.19+12	2.40+12
17000	31484	2.09+13	1.85+13	1.56+13	1.28+13	9.28+12	6.49+12	4.39+12	1.60+12
18000	33336	1.57+13	1.36+13	1.13+13	9.10+12	6.41+12	4.35+12	2.85+12	9.57+11
19326	35793	8.89+12	7.57+12	6.13+12	4.81+12	3.26+12	2.13+12	1.34+12	4.05+11

Table 6.25. Annual Equivalent 1 MeV Electron Fluence from Trapped Protons (V_{oc}, P_{max}), 60° Inclination (Infinite Backshielding)

PROTONS - VOC AND PMAX

EQUIV. 1 MeV ELECTRON FLUENCE FOR VOC AND PMAX CIRCULAR ORBIT INCLINATION = 60 DEGREES.
DUE TO GEOMAGNETICALLY TRAPPED PROTONS, MODEL AP8MAX

ALTITUDE		SHIELD THICKNESS, CM (MILS)						
(N.M.)	(KM)	0 (0)	2.54E-3 (1)	7.64E-3 (3)	1.52E-2 (6)	3.05E-2 (12)	5.09E-2 (20)	7.64E-2 (30)	1.52E-1 (60)
150	277	6.23+13	1.28+12	4.12+11	2.19+11	1.30+11	9.64+10	8.15+10	6.52+10
250	463	2.18+14	7.11+12	3.11+12	1.80+12	1.03+12	7.22+11	5.84+11	4.43+11
300	555	3.71+14	1.61+13	6.74+12	4.02+12	2.40+12	1.69+12	1.37+12	1.03+12
450	833	1.40+16	7.75+13	3.55+13	2.04+13	1.16+13	7.75+12	6.07+12	4.35+12
600	1111	3.36+15	2.06+14	9.77+13	5.81+13	3.42+13	2.29+13	1.77+13	1.23+13
800	1481	8.83+15	5.99+14	3.06+14	1.97+14	1.25+14	8.48+13	6.54+13	4.36+13
1000	1852	2.27+16	1.55+15	8.45+14	5.70+14	3.68+14	2.40+14	1.78+14	1.11+14
1250	2315	5.28+16	4.37+15	2.62+15	1.74+15	1.09+15	6.54+14	4.48+14	2.42+14
1500	2778	9.46+16	1.03+16	6.11+15	4.21+15	2.56+15	1.41+15	8.85+14	3.98+14
1750	3241	1.47+17	2.02+16	1.23+16	8.46+15	4.98+15	2.56+15	1.49+15	5.58+14
2000	3704	2.11+17	3.54+16	2.13+16	1.40+16	7.67+15	3.64+15	1.97+15	6.39+14
2250	4167	3.10+17	6.02+16	3.45+16	2.11+16	1.03+16	4.41+15	2.21+15	6.36+14
2500	4630	4.21+17	8.94+16	4.86+16	2.77+16	1.23+16	4.78+15	2.24+15	5.95+14
2750	5093	5.85+17	1.25+17	6.36+16	3.35+16	1.34+16	4.79+15	2.12+15	5.23+14
3000	5556	7.65+17	1.56+17	7.37+16	3.60+16	1.30+16	4.41+15	1.87+15	4.39+14
3500	6482	1.34+18	2.24+17	8.73+16	3.52+16	1.01+16	3.18+15	1.28+15	2.78+14
4000	7408	2.14+18	2.79+17	9.10+16	3.18+16	7.12+15	2.07+15	7.93+14	1.64+14
4500	8334	3.12+18	3.10+17	8.54+16	2.52+16	4.75+15	1.25+15	4.50+14	8.79+13
5000	9260	4.31+18	3.22+17	7.34+16	1.84+16	2.83+15	6.86+14	2.33+14	4.24+13
5500	10186	5.80+18	3.40+17	6.41+16	1.35+16	1.66+15	3.84+14	1.24+14	2.03+13
6000	11112	6.51+18	2.89+17	4.59+16	8.34+15	8.40+14	1.75+14	5.20+13	7.40+12
7000	12964	7.36+18	1.70+17	1.85+16	2.42+15	1.57+14	2.49+13	6.01+12	6.24+11
8000	14816	6.19+18	7.21+16	5.08+15	5.08+14	2.16+13	2.57+12	4.97+11	3.79+10
9000	16668	4.00+18	2.07+16	9.05+14	5.80+13	1.44+12	1.31+11	2.61+10	4.48+09
10000	18520	2.42+18	4.45+15	1.13+14	4.98+12	1.71+10	1.83-03	0.00	0.00
11000	20372	1.56+18	8.99+14	7.48+12	2.19+11	7.96+08	1.83-03	0.00	0.00
12000	22224	1.06+18	1.62+14	8.88+10	9.63-03	0.00	0.00	0.00	0.00
13000	24076	8.01+17	2.29+13	1.21+07	0.00	0.00	0.00	0.00	0.00
14000	25928	5.41+17	3.44+12	5.96+06	0.00	0.00	0.00	0.00	0.00
15000	27780	3.24+17	6.85+11	3.27+06	0.00	0.00	0.00	0.00	0.00
16000	29632	2.39+17	7.58+09	0.00	0.00	0.00	0.00	0.00	0.00
17000	31484	1.72+17	5.73+09	0.00	0.00	0.00	0.00	0.00	0.00
18000	33336	1.15+17	4.54+09	0.00	0.00	0.00	0.00	0.00	0.00
19326	35793	3.21+16	2.92+09	0.00	0.00	0.00	0.00	0.00	0.00

EQUIV. 1 MeV ELECTRON FLUENCE FOR VOC AND PMAX CIRCULAR ORBIT INCLINATION = 60 DEGREES.
DUE TO GEOMAGNETICALLY TRAPPED PROTONS, MODEL AP8MIN

150	277	1.17+14	1.20+13	4.92+12	2.26+12	9.06+11	4.95+11	3.49+11	2.29+11
250	463	3.61+14	3.00+13	1.28+13	6.37+12	2.97+12	1.80+12	1.35+12	9.35+11
300	555	5.74+14	4.89+13	2.09+13	1.06+13	5.18+12	3.27+12	2.51+12	1.80+12
450	833	1.69+15	1.27+14	5.70+13	3.10+13	1.66+13	1.09+13	8.53+12	6.11+12
600	1111	3.73+15	2.79+14	1.30+14	7.52+13	4.32+13	2.91+13	2.27+13	1.59+13
800	1481	9.64+15	7.42+14	3.73+14	2.33+14	1.45+14	9.93+13	7.70+13	5.17+13
1000	1852	2.40+16	1.80+15	9.60+14	6.33+14	4.01+14	2.63+14	1.96+14	1.22+14

6-38

Table 6.26. Annual Equivalent 1 MeV Electron Fluence from Trapped Protons (I_{SC}), 60° Inclination (Infinite Backshielding)

PROTONS - ISC

INCLINATION = 60 DEGREES.

EQUIV 1 MEV ELECTRON FLUENCE FOR ISC - CIRCULAR ORBIT
DUE TO GEOMAGNETICALLY TRAPPED PROTONS, MODEL AP8MAX

ALTITUDE		SHIELD THICKNESS, CM (MILS)						
(N.M.)	(KM)	0 (0)	2.54E-3 (1)	7.64E-3 (3)	1.52E-2 (6)	3.05E-2 (12)	6.09E-2 (20)	7.64E-2 (30)	1.52E-1 (60)
150	277	3.73+12	4.70+11	2.08+11	1.34+11	9.62+10	7.94+10	7.12+10	6.95+10
250	463	1.42+13	2.86+12	1.57+12	1.05+12	7.25+11	6.70+11	4.97+11	3.98+11
300	555	2.87+13	6.18+12	3.48+12	2.39+12	1.69+12	1.33+12	1.16+12	9.22+11
450	833	1.21+14	3.16+13	1.76+13	1.16+13	7.80+12	5.91+12	5.02+12	3.85+12
600	1111	3.02+14	8.56+13	4.92+13	3.34+13	2.28+13	1.72+13	1.44+13	1.07+13
800	1481	8.28+14	2.60+14	1.61+14	1.16+14	8.32+13	6.29+13	5.22+13	3.75+13
1000	1852	2.07+15	6.87+14	4.47+14	3.31+14	2.36+14	1.71+14	1.37+14	9.28+13
1250	2315	6.29+15	1.95+15	1.30+15	9.59+14	6.54+14	4.34+14	3.26+14	1.94+14
1500	2778	1.10+16	4.53+15	3.04+15	2.22+15	1.43+15	8.69+14	6.03+14	3.02+14
1750	3241	1.96+16	8.88+15	6.99+15	4.30+15	2.66+15	1.49+15	9.58+14	4.01+14
2000	3704	3.15+16	1.52+16	9.97+15	6.85+15	3.94+15	2.03+15	1.22+15	4.42+14
2250	4167	5.14+16	2.60+16	1.54+16	9.84+15	5.10+15	2.36+15	1.33+15	4.27+14
2500	4630	7.51+16	3.60+16	2.10+16	1.25+16	5.87+15	2.49+15	1.32+15	3.91+14
2750	5093	1.03+17	4.91+16	2.67+16	1.47+16	6.24+15	2.43+15	1.23+15	3.38+14
3000	5556	1.43+17	5.98+16	3.03+16	1.55+16	5.99+15	2.19+15	1.07+15	2.81+14
3500	6482	2.45+17	8.16+16	3.44+16	1.46+16	4.55+15	1.55+15	7.26+14	1.75+14
4000	7408	3.73+17	9.78+16	3.49+16	1.25+16	3.16+15	9.91+14	4.45+14	1.02+14
4500	8334	5.10+17	1.05+17	3.22+16	9.88+15	2.07+15	6.85+14	2.50+14	5.38+13
5000	9260	6.57+17	1.05+17	2.73+16	7.18+15	1.21+15	3.14+14	1.28+14	2.56+13
5500	10186	8.35+17	1.08+17	2.35+16	5.13+15	7.07+14	1.73+14	6.69+13	1.20+13
6000	11112	8.62+17	8.92+16	1.66+16	3.12+15	3.51+14	7.67+13	2.75+13	4.31+12
7000	12964	7.85+17	4.94+16	6.56+15	8.82+14	6.32+13	1.03+13	3.06+12	3.51+11
8000	14816	5.30+17	1.95+16	1.85+15	1.82+14	8.47+12	1.00+12	2.45+11	2.13+10
9000	16668	2.74+17	6.05+15	3.09+14	2.04+13	6.53+11	5.08+10	1.40+10	2.64+09
10000	18520	1.22+17	9.64+14	3.76+13	1.71+12	3.97+09	0.00	0.00	0.00
11000	20372	5.58+16	1.70+14	2.35+12	7.45+10	2.23+08	0.00	0.00	0.00
12000	22224	2.67+16	2.57+13	1.98+10	4.60-04	0.00	0.00	0.00	0.00
13000	24076	1.38+16	2.88+12	9.54+05	0.00	0.00	0.00	0.00	0.00
14000	25928	6.98+15	3.53+11	5.81+05	0.00	0.00	0.00	0.00	0.00
15000	27780	3.38+15	5.94+10	3.74+05	0.00	0.00	0.00	0.00	0.00
16000	29632	2.24+15	1.80+08	0.00	0.00	0.00	0.00	0.00	0.00
17000	31484	1.49+15	1.45+08	0.00	0.00	0.00	0.00	0.00	0.00
18000	33336	9.47+14	1.21+08	0.00	0.00	0.00	0.00	0.00	0.00
19326	35793	2.99+14	8.66+07	0.00	0.00	0.00	0.00	0.00	0.00

EQUIV 1 MEV ELECTRON FLUENCE FOR ISC - CIRCULAR ORBIT
DUE TO GEOMAGNETICALLY TRAPPED PROTONS, MODEL AP8MIN

INCLINATION = 60 DEGREES.

ALTITUDE									
(N.M.)	(KM)	0 (0)	2.54E-3 (1)	7.64E-3 (3)	1.52E-2 (6)	3.05E-2 (12)	6.09E-2 (20)	7.64E-2 (30)	1.52E-1 (60)
150	277	1.56+13	4.51+12	2.12+12	1.09+12	6.40+11	3.49+11	2.78+11	2.00+11
250	463	4.14+13	1.16+13	5.78+12	3.30+12	1.88+12	1.33+12	1.10+12	8.25+11
300	555	6.76+13	1.91+13	9.65+12	5.66+12	3.39+12	2.47+12	2.08+12	1.60+12
450	833	1.78+14	5.11+13	2.74+13	1.73+13	1.11+13	8.33+12	7.06+12	5.41+12
600	1111	3.82+14	1.15+14	6.49+13	4.30+13	2.90+13	2.20+13	1.86+13	1.39+13
800	1481	9.87+14	3.19+14	1.94+14	1.37+14	9.72+13	7.39+13	6.16+13	4.46+13
1000	1852	2.36+15	7.89+14	5.02+14	3.66+14	2.58+14	1.88+14	1.51+14	1.02+14

Table 6.27. Annual Equivalent 1 MeV Electron Fluence from Trapped Electrons, 70° Inclination (Infinite Backshielding)

ELECTRONS – ISC, VOC, AND PMAX   INCLINATION = 70 DEGREES.

EQUIV. 1 MEV ELECTRON FLUENCE FOR ISC – CIRCULAR ORBIT
DUE TO GEOMAGNETICALLY TRAPPED ELECTRONS – MODELS AE6MAX, AE17LO

ALTITUDE					SHIELD THICKNESS, CM (MILS)				
(N.M.)	(KM)	0 (0)	2.54E-3 (1)	7.64E-3 (3)	1.52E-2 (6)	3.05E-2 (12)	6.09E-2 (20)	7.64E-2 (30)	1.52E-1 (60)
150	277	2.87+11	2.62+11	2.33+11	2.03+11	1.63+11	1.28+11	9.85+10	4.99+10
250	463	4.61+11	4.09+11	3.69+11	3.10+11	2.47+11	1.93+11	1.48+11	7.45+10
300	555	5.50+11	4.94+11	4.30+11	3.69+11	2.91+11	2.27+11	1.73+11	8.71+10
450	833	9.82+11	8.55+11	7.22+11	6.01+11	4.69+11	3.51+11	2.65+11	1.31+11
600	1111	1.86+12	1.65+12	1.26+12	1.00+12	7.29+11	6.38+11	3.99+11	1.93+11
800	1481	4.55+12	3.67+12	2.82+12	2.12+12	1.43+12	9.98+11	7.13+11	3.29+11
1000	1852	1.00+13	7.92+12	6.93+12	4.34+12	2.79+12	1.87+12	1.30+12	6.75+11
1250	2315	1.93+13	1.52+13	1.13+13	8.23+12	5.24+12	3.48+12	2.40+12	1.05+12
1500	2778	2.71+13	2.13+13	1.58+13	1.14+13	7.19+12	4.73+12	3.23+12	1.40+12
1750	3241	3.43+13	2.67+13	1.95+13	1.38+13	8.34+12	5.28+12	3.51+12	1.49+12
2000	3704	3.94+13	3.03+13	2.18+13	1.50+13	8.66+12	5.22+12	3.33+12	1.37+12
2250	4167	4.26+13	3.24+13	2.30+13	1.56+13	8.68+12	5.01+12	3.09+12	1.24+12
2500	4630	4.36+13	3.32+13	2.34+13	1.57+13	8.65+12	4.79+12	2.89+12	1.15+12
2750	5093	4.23+13	3.22+13	2.27+13	1.52+13	8.30+12	4.63+12	2.79+12	1.14+12
3000	5556	3.89+13	2.97+13	2.11+13	1.43+13	7.94+12	4.54+12	2.80+12	1.19+12
3500	6482	3.12+13	2.42+13	1.77+13	1.25+13	7.59+12	4.79+12	3.21+12	1.47+12
4000	7408	2.70+13	2.16+13	1.65+13	1.23+13	8.08+12	5.50+12	3.89+12	1.85+12
4500	8334	2.69+13	2.22+13	1.71+13	1.34+13	9.39+12	6.69+12	4.84+12	2.33+12
5000	9260	2.57+13	2.22+13	1.86+13	1.52+13	1.13+13	8.36+12	6.15+12	2.99+12
5500	10186	2.77+13	2.47+13	2.13+13	1.80+13	1.39+13	1.06+13	7.89+12	3.88+12
6000	11112	3.10+13	2.82+13	2.49+13	2.15+13	1.71+13	1.33+13	1.01+13	5.13+12
7000	12964	4.30+13	3.96+13	3.54+13	3.10+13	2.52+13	1.99+13	1.55+13	8.04+12
8000	14816	5.12+13	4.72+13	4.22+13	3.71+13	3.02+13	2.40+13	1.86+13	9.65+12
9000	16668	6.27+13	5.78+13	5.18+13	4.55+13	3.69+13	2.92+13	2.25+13	1.15+13
10000	18520	7.02+13	6.47+13	5.78+13	5.06+13	4.07+13	3.19+13	2.43+13	1.19+13
11000	20372	6.64+13	6.11+13	5.45+13	4.75+13	3.80+13	2.95+13	2.22+13	1.05+13
12000	22224	5.83+13	6.35+13	4.76+13	4.14+13	3.28+13	2.52+13	1.88+13	8.61+12
13000	24076	5.29+13	4.84+13	4.28+13	3.69+13	2.90+13	2.20+13	1.62+13	7.22+12
14000	25928	4.14+13	3.76+13	3.29+13	2.82+13	2.18+13	1.63+13	1.18+13	5.08+12
15000	27780	2.92+13	2.63+13	2.28+13	1.92+13	1.46+13	1.06+13	7.54+12	3.09+12
16000	29632	2.42+13	2.16+13	1.85+13	1.54+13	1.14+13	8.18+12	5.67+12	2.20+12
17000	31484	1.92+13	1.69+13	1.43+13	1.17+13	8.51+12	5.95+12	4.02+12	1.47+12
18000	33336	1.44+13	1.25+13	1.04+13	8.36+12	5.89+12	4.00+12	2.62+12	8.79+11
19326	35793	8.21+12	6.99+12	5.67+12	4.44+12	3.01+12	1.97+12	1.24+12	3.74+11

Table 6.28. Annual Equivalent 1 MeV Electron Fluence from Trapped Protons (V_{oc}, P_{max}), 70° Inclination (Infinite Backshielding)

PROTONS - VOC AND PMAX

EQUIV. 1 MEV ELECTRON FLUENCE FOR VOC AND PMAX CIRCULAR ORBIT
DUE TO GEOMAGNETICALLY TRAPPED PROTONS. MODEL AP8MAX   INCLINATION = 70 DEGREES.

ALTITUDE		\multicolumn{7}{c}{SHIELD THICKNESS, CM (MILS)}						
(N.M.)	(KM)	0 (0)	2.54E-3 (1)	7.64E-3 (3)	1.52E-2 (6)	3.05E-2 (12)	5.09E-2 (20)	7.64E-2 (30)	1.52E-1 (60)
150	277	2.57+13	7.33+11	2.86+11	1.72+11	1.11+11	8.45+10	7.23+10	6.90+10
250	463	1.45+14	5.64+12	2.53+12	1.51+12	9.08+11	6.49+11	5.30+11	4.04+11
300	555	2.50+14	1.21+13	6.52+12	3.35+12	2.01+12	1.41+12	1.14+12	8.60+11
450	833	1.22+15	6.23+13	2.92+13	1.69+13	9.74+12	6.63+12	5.23+12	3.76+12
600	1111	3.22+15	1.62+14	7.96+13	4.90+13	2.99+13	2.05+13	1.60+13	1.12+13
800	1481	8.74+15	4.77+14	2.55+14	1.71+14	1.11+14	7.65+13	6.93+13	3.97+13
1000	1852	1.94+16	1.28+15	7.33+14	5.07+14	3.32+14	2.18+14	1.62+14	1.01+14
1250	2315	4.17+16	3.75+15	2.24+15	1.57+15	9.96+14	5.97+14	4.09+14	2.21+14
1500	2778	7.52+16	8.90+15	5.44+15	3.78+15	2.32+15	1.28+15	8.05+14	3.63+14
1750	3241	1.17+17	1.79+16	1.11+16	7.69+15	4.54+15	2.34+15	1.36+15	5.11+14
2000	3704	1.67+17	3.19+16	1.94+16	1.28+16	7.02+15	3.33+15	1.80+15	5.86+14
2250	4167	2.50+17	5.44+16	3.14+16	1.92+16	9.46+15	4.04+15	2.02+15	5.82+14
2500	4630	3.48+17	8.13+16	4.45+16	2.54+16	1.12+16	4.38+15	2.06+15	6.45+14
2750	5093	4.91+17	1.14+17	5.81+16	3.07+16	1.22+16	4.39+15	1.94+15	4.79+14
3000	5556	6.58+17	1.42+17	6.74+16	3.30+16	1.19+16	4.04+15	1.71+15	4.02+14
3500	6482	1.16+18	2.02+17	7.92+16	3.30+16	9.16+15	2.90+15	1.17+15	2.54+14
4000	7408	1.90+18	2.55+17	8.35+16	2.85+16	6.54+15	1.90+15	7.29+14	1.51+14
4500	8334	2.79+18	2.84+17	7.83+16	2.31+16	4.36+15	1.15+15	4.13+14	8.06+13
5000	9260	3.90+18	2.95+17	6.74+16	1.69+16	2.60+15	6.30+14	2.14+14	3.89+13
5500	10186	5.28+18	3.11+17	5.88+16	1.24+16	1.52+15	3.52+14	1.14+14	1.86+13
6000	11112	5.93+18	2.65+17	4.21+16	7.65+15	7.70+14	1.60+14	4.76+13	6.78+12
7000	12964	6.72+18	1.56+17	1.70+16	2.22+15	1.44+14	2.29+13	5.53+12	5.75+11
8000	14816	5.66+18	6.62+16	4.84+15	4.66+14	1.98+13	2.35+12	4.55+11	3.48+10
9000	16668	3.65+18	1.89+16	8.23+14	5.26+13	1.30+12	1.18+11	2.36+10	4.07+09
10000	18520	2.22+18	4.12+15	1.04+14	4.62+12	1.60+10	1.83-03	0.00	0.00
11000	20372	1.42+18	8.32+14	7.31+12	2.18+11	7.94+08	1.83-03	0.00	0.00
12000	22224	9.65+17	1.50+14	8.31+10	9.63-03	0.00	0.00	0.00	0.00
13000	24076	7.29+17	2.11+13	1.18+07	0.00	0.00	0.00	0.00	0.00
14000	25928	4.89+17	3.10+12	5.79+06	0.00	0.00	0.00	0.00	0.00
15000	27780	2.97+17	6.31+11	3.20+06	0.00	0.00	0.00	0.00	0.00
16000	29632	2.20+17	7.21+09	0.00	0.00	0.00	0.00	0.00	0.00
17000	31484	1.58+17	5.45+09	0.00	0.00	0.00	0.00	0.00	0.00
18000	33336	1.06+17	4.33+09	0.00	0.00	0.00	0.00	0.00	0.00
19326	35793	2.98+16	2.80+09	0.00	0.00	0.00	0.00	0.00	0.00

Table 6.29. Annual Equivalent 1 MeV Electron Fluence from Trapped Protons (I_{SC}), 70° Inclination (Infinite Backshielding)

PROTONS - ISC

INCLINATION = 70 DEGREES.

EQUIV 1 MEV ELECTRON FLUENCE FOR ISC - CIRCULAR ORBIT
DUE TO GEOMAGNETICALLY TRAPPED PROTONS, MODEL AP8MAX

ALTITUDE		SHIELD THICKNESS, CM (MILS)						
(N.M.)	(KM)	0 (0)	2.54E-3 (1)	7.64E-3 (3)	1.52E-2 (6)	3.05E-2 (12)	5.09E-2 (20)	7.64E-2 (30)	1.52E-1 (60)
150	277	1.81+12	2.93+11	1.55+11	1.11+11	8.49+10	7.05+10	6.38+10	5.42+10
250	463	1.07+13	2.32+12	1.31+12	9.09+11	6.46+11	5.16+11	4.62+11	3.65+11
300	555	2.03+13	6.05+12	2.90+12	2.00+12	1.41+12	1.11+12	9.67+11	7.73+11
450	833	9.51+13	2.57+13	1.46+13	9.76+12	6.63+12	5.08+12	4.33+12	3.32+12
600	1111	2.47+14	6.89+13	4.11+13	2.88+13	2.03+13	1.55+13	1.31+13	9.76+12
800	1481	6.89+14	2.13+14	1.38+14	1.02+14	7.48+13	5.69+13	4.74+13	3.41+13
1000	1852	1.66+15	5.84+14	3.93+14	2.97+14	2.14+14	1.56+14	1.25+14	8.47+13
1250	2315	4.21+15	1.78+15	1.16+15	8.69+14	5.96+14	3.97+14	2.98+14	1.77+14
1500	2778	9.00+15	3.98+15	2.72+15	2.00+15	1.30+15	7.90+14	5.49+14	2.75+14
1750	3241	1.64+16	7.96+15	5.43+15	3.92+15	2.43+15	1.36+15	8.77+14	3.67+14
2000	3704	2.71+16	1.38+16	9.09+15	6.26+15	3.61+15	1.86+15	1.12+15	4.05+14
2250	4167	4.47+16	2.26+16	1.41+16	8.99+15	4.66+15	2.16+15	1.21+15	3.91+14
2500	4630	6.64+16	3.28+16	1.92+16	1.15+16	5.38+15	2.28+15	1.21+15	3.58+14
2750	5093	9.64+16	4.47+16	2.44+16	1.35+16	5.72+15	2.22+15	1.13+15	3.09+14
3000	5556	1.28+17	5.46+16	2.77+16	1.42+16	5.49+15	2.01+15	9.85+14	2.57+14
3500	6482	2.19+17	7.38+16	3.13+16	1.32+16	4.14+15	1.42+15	6.62+14	1.59+14
4000	7408	3.38+17	8.95+16	3.21+16	1.14+16	2.90+15	9.10+14	4.09+14	9.36+13
4500	8334	4.54+17	9.63+16	2.95+16	9.07+15	1.90+15	5.37+14	2.29+14	4.94+13
5000	9260	6.00+17	9.67+16	2.50+16	6.52+15	1.12+15	2.89+14	1.17+14	2.35+13
5500	10186	7.64+17	9.89+16	2.15+16	4.70+15	6.48+14	1.59+14	6.13+13	1.10+13
6000	11112	7.90+17	8.19+16	1.53+16	2.86+15	3.22+14	7.03+13	2.52+13	3.95+12
7000	12964	7.19+17	4.54+16	6.03+15	8.11+14	5.81+13	9.46+12	2.82+12	3.23+11
8000	14816	4.86+17	1.78+16	1.69+15	1.67+14	7.76+12	9.19+11	2.24+11	1.95+10
9000	16668	2.51+17	4.61+15	2.81+14	1.85+13	5.00+11	4.58+10	1.27+10	2.40+09
10000	18520	1.13+17	8.92+14	3.48+13	1.59+12	3.73+09	0.00	0.00	0.00
11000	20372	5.11+16	1.58+14	2.31+12	7.42+10	2.23+08	0.00	0.00	0.00
12000	22224	2.46+16	2.38+13	1.86+10	4.60-04	0.00	0.00	0.00	0.00
13000	24076	1.26+16	2.66+12	9.39+05	0.00	0.00	0.00	0.00	0.00
14000	25928	6.30+15	3.18+11	5.69+05	0.00	0.00	0.00	0.00	0.00
15000	27780	3.10+15	5.48+10	3.68+05	0.00	0.00	0.00	0.00	0.00
16000	29632	2.06+15	1.73+08	0.00	0.00	0.00	0.00	0.00	0.00
17000	31484	1.37+15	1.39+08	0.00	0.00	0.00	0.00	0.00	0.00
18000	33336	8.72+14	1.16+08	0.00	0.00	0.00	0.00	0.00	0.00
19326	35793	2.77+14	8.27+07	0.00	0.00	0.00	0.00	0.00	0.00

6-42

Table 6.30. Annual Equivalent 1 MeV Electron Fluence from Trapped Electrons, 80° Inclination (Infinite Backshielding)

ELECTRONS - ISC, VOC, AND PMAX

EQUIV. 1 MEV ELECTRON FLUENCE FOR ISC - CIRCULAR ORBIT
DUE TO GEOMAGNETICALLY TRAPPED ELECTRONS - MODELS AE6MAX, AEI7LO

INCLINATION = 80 DEGREES.

ALTITUDE		_				SHIELD THICKNESS, CM (MILS)			
(N.M.)	(KM)	0 (0)	2.54E-3 (1)	7.64E-3 (3)	1.52E-2 (6)	3.05E-2 (12)	5.09E-2 (20)	7.64E-2 (30)	1.52E-1 (60)
150	277	2.51+11	2.29+11	2.03+11	1.77+11	1.41+11	1.11+11	8.47+10	4.23+10
250	463	3.80+11	3.44+11	3.01+11	2.60+11	2.06+11	1.61+11	1.23+11	6.12+10
300	555	4.61+11	4.12+11	3.57+11	3.05+11	2.40+11	1.86+11	1.41+11	7.00+10
450	833	8.30+11	7.19+11	6.03+11	4.99+11	3.78+11	2.86+11	2.15+11	1.05+11
600	1111	1.60+12	1.33+12	1.06+12	8.37+11	5.99+11	4.37+11	3.21+11	1.53+11
800	1481	4.08+12	3.26+12	2.48+12	1.84+12	1.22+12	8.36+11	5.91+11	2.68+11
1000	1852	9.22+12	7.26+12	5.40+12	3.91+12	2.48+12	1.64+12	1.13+12	4.95+11
1250	2315	1.79+13	1.41+13	1.04+13	7.53+12	4.75+12	3.13+12	2.14+12	9.26+11
1500	2778	2.53+13	1.98+13	1.46+13	1.05+13	6.56+12	4.28+12	2.91+12	1.25+12
1750	3241	3.21+13	2.49+13	1.81+13	1.27+13	7.65+12	4.80+12	3.17+12	1.34+12
2000	3704	3.69+13	2.84+13	2.03+13	1.39+13	7.96+12	4.75+12	3.01+12	1.22+12
2250	4167	3.98+13	3.10+13	2.14+13	1.44+13	7.94+12	4.52+12	2.76+12	1.09+12
2500	4630	4.09+13	3.10+13	2.18+13	1.45+13	7.82+12	4.32+12	2.57+12	1.01+12
2750	5093	3.96+13	3.01+13	2.11+13	1.41+13	7.56+12	4.15+12	2.46+12	9.90+11
3000	5556	3.64+13	2.78+13	1.97+13	1.32+13	7.26+12	4.09+12	2.50+12	1.05+12
3500	6482	2.91+13	2.25+13	1.64+13	1.15+13	6.87+12	4.28+12	2.85+12	1.30+12
4000	7408	2.51+13	1.99+13	1.51+13	1.12+13	7.30+12	4.93+12	3.47+12	1.65+12
4500	8334	2.39+13	1.97+13	1.56+13	1.22+13	8.48+12	6.01+12	4.34+12	2.09+12
5000	9260	2.38+13	2.06+13	1.71+13	1.40+13	1.04+13	7.67+12	5.65+12	2.74+12
5500	10186	2.57+13	2.29+13	1.98+13	1.67+13	1.29+13	9.76+12	7.29+12	3.59+12
6000	11112	2.90+13	2.64+13	2.32+13	2.01+13	1.60+13	1.24+13	9.47+12	4.80+12
7000	12964	4.04+13	3.72+13	3.32+13	2.91+13	2.36+13	1.87+13	1.45+13	7.66+12
8000	14816	4.83+13	4.46+13	3.99+13	3.51+13	2.85+13	2.27+13	1.76+13	9.14+12
9000	16668	5.93+13	5.46+13	4.89+13	4.30+13	3.49+13	2.76+13	2.13+13	1.08+13
10000	18520	6.65+13	6.13+13	5.48+13	4.79+13	3.86+13	3.02+13	2.31+13	1.13+13
11000	20372	6.26+13	5.77+13	5.14+13	4.49+13	3.59+13	2.78+13	2.10+13	9.90+12
12000	22224	5.52+13	5.07+13	4.51+13	3.92+13	3.11+13	2.39+13	1.78+13	8.16+12
13000	24076	5.00+13	4.57+13	4.04+13	3.49+13	2.74+13	2.08+13	1.53+13	6.83+12
14000	25928	3.91+13	3.55+13	3.12+13	2.66+13	2.06+13	1.54+13	1.12+13	4.81+12
15000	27780	2.78+13	2.50+13	2.17+13	1.83+13	1.38+13	1.01+13	7.17+12	2.94+12
16000	29632	2.30+13	2.05+13	1.76+13	1.46+13	1.09+13	7.78+12	5.39+12	2.10+12
17000	31484	1.83+13	1.61+13	1.36+13	1.12+13	8.11+12	5.67+12	3.83+12	1.40+12
18000	33336	1.37+13	1.19+13	9.91+12	7.97+12	5.62+12	3.81+12	2.50+12	8.39+11
19326	35792	7.88+12	6.71+12	5.44+12	4.27+12	2.89+12	1.89+12	1.19+12	3.60+11

Table 6.31. Annual Equivalent 1 MeV Electron Fluence from Trapped Protons (V_{oc}, P_{max}), 80° Inclination (Infinite Backshielding)

PROTONS — VOC AND PMAX

EQUIV. 1 MEV ELECTRON FLUENCE FOR VOC AND PMAX CIRCULAR ORBIT
DUE TO GEOMAGNETICALLY TRAPPED PROTONS, MODEL AP8MAX

INCLINATION = 80 DEGREES.

ALTITUDE		\multicolumn{8}{c}{SHIELD THICKNESS, CM (MILS)}							
(N.M.)	(KM)	0 (0)	2.54E-3 (1)	7.64E-3 (3)	1.52E-2 (6)	3.05E-2 (12)	5.09E-2 (20)	7.64E-2 (30)	1.52E-1 (60)
150	277	1.85+13	6.39+11	2.68+11	1.55+11	9.68+10	7.16+10	6.00+10	4.75+10
250	463	9.74+13	4.32+12	1.98+12	1.24+12	7.87+11	5.75+11	4.76+11	3.71+11
300	555	1.98+14	1.06+13	6.00+12	2.99+12	1.79+12	1.27+12	1.03+12	7.79+11
450	833	9.44+14	6.21+13	2.49+13	1.48+13	8.76+12	6.01+12	4.75+12	3.43+12
600	1111	2.39+15	1.41+14	7.16+13	4.51+13	2.81+13	1.93+13	1.52+13	1.06+13
800	1481	6.59+15	4.32+14	2.37+14	1.60+14	1.06+14	7.24+13	5.61+13	3.76+13
1000	1852	1.54+16	1.19+15	6.99+14	4.80+14	3.16+14	2.07+14	1.54+14	9.59+13
1250	2315	3.48+16	3.60+15	2.12+15	1.49+15	9.47+14	5.67+14	3.89+14	2.10+14
1500	2778	6.41+16	8.30+15	5.11+15	3.57+15	2.19+15	1.21+15	7.64+14	3.45+14
1750	3241	1.01+17	1.69+16	1.06+16	7.31+15	4.32+15	2.22+15	1.30+15	4.85+14
2000	3704	1.48+17	3.02+16	1.84+16	1.22+16	6.68+15	3.17+15	1.72+15	5.57+14
2250	4167	2.26+17	6.14+16	2.97+16	1.82+16	8.98+15	3.84+15	1.92+15	6.54+14
2500	4630	3.17+17	7.72+16	4.23+16	2.41+16	1.07+16	4.17+15	1.96+15	5.19+14
2750	5093	4.54+17	1.08+17	5.52+16	2.92+16	1.16+16	4.17+15	1.84+15	4.55+14
3000	5556	6.12+17	1.35+17	6.41+16	3.14+16	1.14+16	3.84+15	1.63+15	3.83+14
3500	6482	1.09+18	1.93+17	7.56+16	3.05+16	8.74+15	2.77+15	1.12+15	2.42+14
4000	7408	1.79+18	2.43+17	7.95+16	2.72+16	6.23+16	1.81+15	6.95+14	1.44+14
4500	8334	2.63+18	2.71+17	7.46+16	2.20+16	4.15+15	1.09+15	3.93+14	7.68+13
5000	9260	3.70+18	2.81+17	6.41+16	1.61+16	2.47+15	5.99+14	2.03+14	3.70+13
5500	10186	5.00+18	2.96+17	5.59+16	1.18+16	1.45+15	3.35+14	1.08+14	1.77+13
6000	11112	5.63+18	2.52+17	4.01+16	7.28+15	7.32+14	1.52+14	4.53+13	6.44+12
7000	12964	6.38+18	1.49+17	1.62+16	2.12+15	1.37+14	2.18+13	5.28+12	5.49+11
8000	14816	5.39+18	6.31+16	4.61+15	4.43+14	1.88+13	2.23+12	4.32+11	3.31+10
9000	16668	3.44+18	1.77+16	7.65+14	4.87+13	1.20+12	1.09+11	2.17+10	3.70+09
10000	18520	2.12+18	3.95+15	1.00+14	4.45+12	1.55+10	1.83-03	0.00	0.00
11000	20372	1.34+18	7.98+14	6.89+12	2.04+11	7.49+08	1.83-03	0.00	0.00
12000	22224	9.17+17	1.44+14	8.06+10	9.63-03	0.00	0.00	0.00	0.00
13000	24076	6.91+17	2.01+13	1.17+07	0.00	0.00	0.00	0.00	0.00
14000	25928	4.63+17	2.93+12	5.70+06	0.00	0.00	0.00	0.00	0.00
15000	27780	2.83+17	6.04+11	3.16+06	0.00	0.00	0.00	0.00	0.00
16000	29632	2.09+17	7.02+09	0.00	0.00	0.00	0.00	0.00	0.00
17000	31484	1.51+17	5.31+09	0.00	0.00	0.00	0.00	0.00	0.00
18000	33336	1.01+17	4.22+09	0.00	0.00	0.00	0.00	0.00	0.00
19326	35792	2.86+16	2.73+09	0.00	0.00	0.00	0.00	0.00	0.00

Table 6.32. Annual Equivalent 1 MeV Electron Fluence from Trapped Protons (I_{sc}), 80° Inclination (Infinite Backshielding)

PROTONS – ISC

EQUIV 1 MEV ELECTRON FLUENCE FOR ISC – CIRCULAR ORBIT
DUE TO GEOMAGNETICALLY TRAPPED PROTONS. MODEL AP8MAX

INCLINATION = 80 DEGREES.

ALTITUDE		\multicolumn{8}{c}{SHIELD THICKNESS, CM (MILS)}							
(N.M.)	(KM)	0 (0)	2.54E-3 (1)	7.64E-3 (3)	1.52E-2 (6)	3.05E-2 (12)	6.09E-2 (20)	7.64E-2 (30)	1.52E-1 (60)
150	277	1.42+12	2.57+11	1.37+11	9.65+10	7.12+10	6.86+10	6.22+10	4.33+10
250	463	8.07+12	1.82+12	1.07+12	7.71+11	5.71+11	4.64+11	4.10+11	3.36+11
300	555	1.69+13	4.44+12	2.59+12	1.79+12	1.26+12	1.00+12	8.74+11	6.99+11
450	833	7.83+13	2.18+13	1.27+13	8.66+12	6.00+12	4.62+12	3.94+12	3.03+12
600	1111	2.00+14	6.11+13	3.76+13	2.68+13	1.91+13	1.46+13	1.24+13	9.25+12
800	1481	5.74+14	1.96+14	1.28+14	9.63+13	7.07+13	5.39+13	4.49+13	3.24+13
1000	1852	1.46+15	5.46+14	3.71+14	2.82+14	2.03+14	1.48+14	1.19+14	8.05+13
1250	2315	3.78+15	1.60+15	1.10+15	8.25+14	6.66+14	3.77+14	2.84+14	1.68+14
1500	2778	8.17+15	3.73+15	2.56+15	1.89+15	1.23+15	7.49+14	6.21+14	2.61+14
1750	3241	1.61+16	7.61+15	5.15+15	3.72+15	2.31+15	1.30+15	8.33+14	3.49+14
2000	3704	2.52+16	1.30+16	8.63+15	5.95+15	3.43+15	1.77+15	1.06+15	3.85+14
2250	4167	4.17+16	2.14+16	1.33+16	8.53+15	4.43+15	2.06+15	1.16+15	3.71+14
2500	4630	6.24+16	3.12+16	1.83+16	1.09+16	5.12+15	2.17+15	1.15+15	3.41+14
2750	5093	9.08+16	4.26+16	2.32+16	1.28+16	5.44+15	2.11+15	1.07+15	2.94+14
3000	5556	1.21+17	5.19+16	2.64+16	1.35+16	5.23+15	1.91+15	9.37+14	2.45+14
3500	6482	2.08+17	7.04+16	2.98+16	1.26+16	3.95+15	1.35+15	6.31+14	1.52+14
4000	7408	3.21+17	8.53+16	3.05+16	1.09+16	2.77+15	8.67+14	3.89+14	8.92+13
4500	8334	4.41+17	9.17+16	2.81+16	8.64+15	1.81+15	5.11+14	2.18+14	4.70+13
5000	9260	5.70+17	9.19+16	2.38+16	6.20+15	1.06+15	2.74+14	1.11+14	2.24+13
5500	10186	7.25+17	9.40+16	2.05+16	4.47+15	6.16+14	1.51+14	5.84+13	1.05+13
6000	11112	7.51+17	7.79+16	1.45+16	2.72+15	3.06+14	6.68+13	2.39+13	3.75+12
7000	12964	6.84+17	4.32+16	5.74+15	7.73+14	5.54+13	9.02+12	2.68+12	3.09+11
8000	14816	4.63+17	1.70+16	1.61+15	1.59+14	7.37+12	8.73+11	2.13+11	1.85+10
9000	16668	2.35+17	4.30+15	2.61+14	1.71+13	4.61+11	4.23+10	1.16+10	2.18+09
10000	18520	1.08+17	8.57+14	3.35+13	1.53+12	3.62+09	0.00	0.00	0.00
11000	20372	4.86+16	1.51+14	2.17+12	6.93+10	2.11+08	0.00	0.00	0.00
12000	22224	2.34+16	2.29+13	1.80+10	4.60-04	0.00	0.00	0.00	0.00
13000	24076	1.20+16	2.54+12	9.31+05	0.00	0.00	0.00	0.00	0.00
14000	25928	5.97+15	3.01+11	5.62+05	0.00	0.00	0.00	0.00	0.00
15000	27780	2.96+15	5.25+10	3.65+05	0.00	0.00	0.00	0.00	0.00
16000	29632	1.96+15	1.70+08	0.00	0.00	0.00	0.00	0.00	0.00
17000	31484	1.31+15	1.36+08	0.00	0.00	0.00	0.00	0.00	0.00
18000	33336	8.33+14	1.14+08	0.00	0.00	0.00	0.00	0.00	0.00
19326	35792	2.67+14	8.13+07	0.00	0.00	0.00	0.00	0.00	0.00

Table 6.33. Annual Equivalent 1 MeV Electron Fluence from Trapped Electrons, 90° Inclination (Infinite Backshielding)

ELECTRONS - ISC, VOC, AND PMAX

INCLINATION = 90 DEGREES.

EQUIV. 1 MeV ELECTRON FLUENCE FOR ISC - CIRCULAR ORBIT
DUE TO GEOMAGNETICALLY TRAPPED ELECTRONS - MODELS AE6MAX, AEI7LO

ALTITUDE		SHIELD THICKNESS, CM (MILS)						
(N.M.)	(KM)	0 (0)	2.54E-3 (1)	7.64E-3 (3)	1.52E-2 (6)	3.05E-2 (12)	5.09E-2 (20)	7.64E-2 (30)	1.52E-1 (60)
150	277	2.19+11	2.04+11	1.77+11	1.54+11	1.23+11	9.68+10	7.44+10	3.76+10
250	463	3.42+11	3.08+11	2.69+11	2.31+11	1.83+11	1.43+11	1.09+11	6.47+10
300	556	4.16+11	3.70+11	3.21+11	2.74+11	2.15+11	1.67+11	1.27+11	6.34+10
450	833	7.73+11	6.67+11	5.67+11	4.59+11	3.46+11	2.62+11	1.97+11	9.67+10
600	1111	1.51+12	1.26+12	9.97+11	7.82+11	5.56+11	4.04+11	2.97+11	1.42+11
800	1481	3.97+12	3.17+12	2.40+12	1.78+12	1.17+12	8.01+11	5.65+11	2.56+11
1000	1852	9.02+12	7.10+12	5.27+12	3.81+12	2.41+12	1.59+12	1.10+12	4.78+11
1250	2315	1.75+13	1.38+13	1.02+13	7.37+12	4.63+12	3.05+12	2.09+12	9.01+11
1500	2778	2.48+13	1.94+13	1.43+13	1.03+13	6.40+12	4.17+12	2.84+12	1.22+12
1750	3241	3.15+13	2.44+13	1.78+13	1.29+13	7.47+12	4.68+12	3.09+12	1.30+12
2000	3704	3.52+13	2.78+13	1.99+13	1.36+13	7.78+12	4.63+12	2.93+12	1.19+12
2250	4167	3.90+13	2.97+13	2.10+13	1.41+13	7.76+12	4.41+12	2.68+12	1.05+12
2500	4630	4.02+13	3.05+13	2.14+13	1.42+13	7.64+12	4.21+12	2.50+12	9.79+11
2750	5093	3.88+13	2.94+13	2.07+13	1.38+13	7.36+12	4.03+12	2.38+12	9.63+11
3000	5556	3.67+13	2.72+13	1.92+13	1.29+13	7.08+12	3.98+12	2.42+12	1.01+12
3500	6482	2.85+13	2.20+13	1.60+13	1.12+13	6.68+12	4.15+12	2.76+12	1.26+12
4000	7408	2.45+13	1.96+13	1.47+13	1.09+13	7.10+12	4.78+12	3.36+12	1.60+12
4500	8334	2.32+13	1.92+13	1.52+13	1.18+13	8.20+12	5.80+12	4.18+12	2.01+12
5000	9260	2.33+13	2.01+13	1.68+13	1.37+13	1.02+13	7.50+12	5.52+12	2.68+12
5500	10186	2.52+13	2.24+13	1.93+13	1.63+13	1.26+13	9.55+12	7.14+12	3.51+12
6000	11112	2.83+13	2.57+13	2.27+13	1.96+13	1.56+13	1.21+13	9.25+12	4.69+12
7000	12964	3.96+13	3.64+13	3.26+13	2.86+13	2.32+13	1.84+13	1.42+13	7.41+12
8000	14816	4.76+13	4.39+13	3.93+13	3.46+13	2.81+13	2.24+13	1.74+13	9.01+12
9000	16668	5.84+13	5.39+13	4.82+13	4.24+13	3.44+13	2.72+13	2.10+13	1.07+13
10000	18520	6.64+13	6.02+13	5.39+13	4.71+13	3.80+13	2.97+13	2.27+13	1.11+13
11000	20372	6.14+13	5.65+13	5.04+13	4.40+13	3.62+13	2.73+13	2.06+13	9.71+12
12000	22224	5.41+13	4.97+13	4.42+13	3.84+13	3.05+13	2.35+13	1.75+13	8.01+12
13000	24076	4.89+13	4.47+13	3.95+13	3.41+13	2.68+13	2.04+13	1.50+13	6.68+12
14000	25928	3.86+13	3.50+13	3.07+13	2.63+13	2.03+13	1.52+13	1.10+13	4.74+12
15000	27780	2.74+13	2.46+13	2.13+13	1.80+13	1.36+13	9.98+12	7.06+12	2.89+12
16000	29632	2.27+13	2.02+13	1.73+13	1.44+13	1.07+13	7.67+12	5.32+12	2.07+12
17000	31484	1.80+13	1.59+13	1.34+13	1.10+13	8.00+12	5.60+12	3.78+12	1.38+12
18000	33336	1.36+13	1.18+13	9.79+12	7.88+12	5.55+12	3.77+12	2.47+12	8.30+11
19326	35792	7.84+12	6.67+12	5.41+12	4.24+12	2.88+12	1.88+12	1.18+12	3.58+11

Table 6.34. Annual Equivalent 1 MeV Electron Fluence from Trapped Protons (V_{oc}, P_{max}), 90° Inclination (Infinite Backshielding)

PROTONS - VOC AND PMAX

EQUIV. 1 MEV ELECTRON FLUENCE FOR VOC AND PMAX CIRCULAR ORBIT
DUE TO GEOMAGNETICALLY TRAPPED PROTONS, MODEL AP8MAX   INCLINATION = 90 DEGREES.

ALTITUDE		\multicolumn{7}{c}{SHIELD THICKNESS, CM (MILS)}
(N.M.)	(KM)	0 (0)	2.54E-3 (1)	7.64E-3 (3)	1.52E-2 (6)	3.05E-2 (12)	5.09E-2 (20)	7.64E-2 (30)	1.52E-1 (60)
150	277	3.08+13	7.80+11	2.76+11	1.58+11	9.92+10	7.43+10	6.29+10	5.06+10
250	463	8.59+13	4.70+12	2.27+12	1.38+12	8.32+11	6.99+11	4.91+11	3.77+11
300	555	1.83+14	1.07+13	5.13+12	3.05+12	1.79+12	1.26+12	1.01+12	7.62+11
450	833	8.98+14	6.01+13	2.43+13	1.47+13	8.80+12	6.04+12	4.78+12	3.47+12
600	1111	2.18+15	1.39+14	7.06+13	4.46+13	2.77+13	1.91+13	1.50+13	1.05+13
800	1481	6.05+15	4.16+14	2.32+14	1.57+14	1.04+14	7.16+13	6.56+13	3.73+13
1000	1852	1.47+16	1.16+15	6.78+14	4.72+14	3.11+14	2.04+14	1.51+14	9.44+13
1250	2315	3.31+16	3.43+15	2.08+15	1.46+15	9.33+14	6.59+14	3.83+14	2.07+14
1500	2778	6.15+16	8.12+15	5.02+15	3.51+15	2.16+15	1.19+15	7.52+14	3.39+14
1750	3241	9.79+16	1.66+16	1.04+16	7.18+15	4.25+15	2.19+15	1.27+15	4.78+14
2000	3704	1.45+17	2.97+16	1.81+16	1.26+16	6.57+15	3.11+15	1.69+15	6.48+14
2250	4167	2.19+17	5.04+16	2.92+16	1.79+16	8.83+15	3.78+15	1.89+15	5.45+14
2500	4630	3.11+17	7.60+16	4.16+16	2.38+16	1.05+16	4.11+15	1.93+15	5.10+14
2750	5093	4.43+17	1.06+17	5.43+16	2.87+16	1.14+16	4.10+15	1.81+15	4.47+14
3000	5556	5.99+17	1.33+17	6.31+16	3.09+16	1.12+16	3.78+15	1.60+15	3.77+14
3500	6482	1.07+18	1.90+17	7.43+16	3.00+16	8.60+15	2.72+15	1.10+15	2.38+14
4000	7408	1.75+18	2.39+17	7.83+16	2.68+16	6.14+15	1.79+15	6.84+14	1.42+14
4500	8334	2.59+18	2.67+17	7.35+16	2.17+16	4.09+15	1.08+15	3.87+14	7.56+13
5000	9260	3.63+18	2.77+17	6.32+16	1.59+16	2.44+15	5.91+14	2.00+14	3.64+13
5500	10186	4.92+18	2.91+17	5.50+16	1.16+16	1.43+15	3.29+14	1.05+14	1.74+13
6000	11112	5.64+18	2.48+17	3.94+16	7.16+15	7.20+14	1.50+14	4.45+13	6.32+12
7000	12964	6.28+18	1.46+17	1.59+16	2.08+15	1.35+14	2.15+13	5.19+12	5.40+11
8000	14816	5.32+18	6.21+16	4.54+15	4.35+14	1.85+13	2.19+12	4.23+11	3.24+10
9000	16668	3.39+18	1.74+16	7.51+14	4.78+13	1.17+12	1.07+11	2.12+10	3.58+09
10000	18520	2.09+18	3.92+15	9.99+13	4.44+12	1.55+10	1.83-03	0.00	0.00
11000	20372	1.32+18	7.94+14	7.11+12	2.13+11	7.79+08	1.83-03	0.00	0.00
12000	22224	9.01+17	1.43+14	8.08+10	9.63-03	0.00	0.00	0.00	0.00
13000	24076	6.76+17	1.98+13	1.16+07	0.00	0.00	0.00	0.00	0.00
14000	25928	4.57+17	2.92+12	5.69+06	0.00	0.00	0.00	0.00	0.00
15000	27780	2.79+17	5.94+11	3.15+06	0.00	0.00	0.00	0.00	0.00
16000	29632	2.06+17	6.96+09	0.00	0.00	0.00	0.00	0.00	0.00
17000	31484	1.49+17	5.27+09	0.00	0.00	0.00	0.00	0.00	0.00
18000	33336	9.98+16	4.19+09	0.00	0.00	0.00	0.00	0.00	0.00
19326	35792	2.85+16	2.72+09	0.00	0.00	0.00	0.00	0.00	0.00

EQUIV. 1 MEV ELECTRON FLUENCE FOR VOC AND PMAX CIRCULAR ORBIT
DUE TO GEOMAGNETICALLY TRAPPED PROTONS, MODEL AP8MIN   INCLINATION = 90 DEGREES.

150	277	6.41+13	7.90+12	3.41+12	1.61+12	6.71+11	3.75+11	2.68+11	1.79+11
250	463	1.72+14	1.90+13	8.67+12	4.51+12	2.21+12	1.39+12	1.07+12	7.69+11
300	555	3.00+14	3.09+13	1.39+13	7.34+12	3.74+12	2.42+12	1.88+12	1.36+12
450	833	1.07+15	8.49+13	3.99+13	2.26+13	1.27+13	8.55+12	6.75+12	4.88+12
600	1111	2.41+15	1.83+14	9.17+13	5.63+13	3.45+13	2.40+13	1.90+13	1.35+13
800	1481	6.57+15	5.22+14	2.83+14	1.86+14	1.20+14	8.36+13	6.53+13	4.40+13
1000	1852	1.57+16	1.35+15	7.68+14	5.22+14	3.39+14	2.24+14	1.67+14	1.04+14

6-47

Table 6.35. Annual Equivalent 1 MeV Electron Fluence from Trapped Protons (I_{SC}), 90° Inclination (Infinite Backshielding)

PROTONS - ISC

EQUIV 1 MEV ELECTRON FLUENCE FOR ISC - CIRCULAR ORBIT
DUE TO GEOMAGNETICALLY TRAPPED PROTONS, MODEL AP8MAX   INCLINATION = 90 DEGREES.

SHIELD THICKNESS, CM (MILS)

ALTITUDE (N.M.)	(KM)	0 (0)	2.54E-3 (1)	7.64E-3 (3)	1.52E-2 (6)	3.05E-2 (12)	5.09E-2 (20)	7.64E-2 (30)	1.52E-1 (60)
150	277	2.13+12	2.99+11	1.45+11	9.95+10	7.39+10	6.14+10	5.52+10	4.64+10
250	463	7.19+12	2.00+12	1.19+12	8.33+11	6.96+11	4.78+11	4.20+11	3.40+11
300	555	1.63+13	4.50+12	2.63+12	1.80+12	1.26+12	9.88+11	8.59+11	6.84+11
450	833	7.49+13	2.11+13	1.25+13	8.64+12	6.03+12	4.65+12	3.98+12	3.07+12
600	1111	1.92+14	6.01+13	3.70+13	2.64+13	1.89+13	1.46+13	1.22+13	9.14+12
800	1481	5.41+14	1.90+14	1.26+14	9.50+13	6.99+13	5.33+13	4.45+13	3.21+13
1000	1852	1.41+15	5.36+14	3.65+14	2.77+14	2.00+14	1.46+14	1.17+14	7.92+13
1250	2315	3.68+15	1.57+15	1.08+15	8.12+14	5.58+14	3.71+14	2.79+14	1.66+14
1500	2778	7.95+15	3.66+15	2.52+15	1.86+15	1.21+15	7.38+14	5.13+14	2.57+14
1750	3241	1.48+16	7.38+15	5.06+15	3.66+15	2.27+15	1.28+15	8.20+14	3.43+14
2000	3704	2.48+16	1.28+16	8.50+15	5.85+15	3.37+15	1.74+15	1.04+15	3.78+14
2250	4167	4.08+16	2.10+16	1.31+16	8.39+15	4.36+15	2.02+15	1.14+15	3.65+14
2500	4630	6.14+16	3.07+16	1.80+16	1.07+16	5.04+15	2.13+15	1.13+15	3.35+14
2750	5093	8.91+16	4.17+16	2.28+16	1.26+16	5.34+15	2.08+15	1.05+15	2.89+14
3000	5556	1.19+17	5.11+16	2.60+16	1.33+16	5.14+15	1.88+15	9.22+14	2.41+14
3500	6482	2.04+17	6.92+16	2.93+16	1.24+16	3.89+15	1.33+15	6.21+14	1.49+14
4000	7408	3.16+17	8.40+16	3.01+16	1.07+16	2.73+15	8.55+14	3.84+14	8.79+13
4500	8334	4.34+17	9.03+16	2.77+16	8.51+15	1.78+15	5.03+14	2.15+14	4.63+13
5000	9260	5.61+17	9.06+16	2.34+16	6.11+15	1.04+15	2.70+14	1.10+14	2.20+13
5500	10186	7.14+17	9.25+16	2.01+16	4.39+15	6.06+14	1.48+14	5.74+13	1.03+13
6000	11112	7.39+17	7.67+16	1.43+16	2.68+15	3.01+14	6.56+13	2.35+13	3.68+12
7000	12964	6.73+17	4.25+16	5.65+15	7.61+14	5.45+13	8.88+12	2.64+12	3.04+11
8000	14816	4.57+17	1.67+16	1.58+15	1.56+14	7.23+12	8.56+11	2.09+11	1.82+10
9000	16668	2.32+17	4.23+15	2.56+14	1.68+13	4.51+11	4.14+10	1.13+10	2.11+09
10000	18520	1.07+17	8.51+14	3.34+13	1.53+12	3.61+09	0.00	0.00	0.00
11000	20372	4.80+16	1.61+14	2.24+12	7.24+10	2.19+08	0.00	0.00	0.00
12000	22224	2.31+16	2.28+13	1.81+10	4.60-04	0.00	0.00	0.00	0.00
13000	24076	1.17+16	2.61+12	9.29+05	0.00	0.00	0.00	0.00	0.00
14000	25928	5.91+15	3.00+11	5.62+05	0.00	0.00	0.00	0.00	0.00
15000	27780	2.92+15	5.16+10	3.63+05	0.00	0.00	0.00	0.00	0.00
16000	29632	1.93+15	1.69+08	0.00	0.00	0.00	0.00	0.00	0.00
17000	31484	1.29+15	1.36+08	0.00	0.00	0.00	0.00	0.00	0.00
18000	33336	8.24+14	1.13+08	0.00	0.00	0.00	0.00	0.00	0.00
19326	35792	2.65+14	8.11+07	0.00	0.00	0.00	0.00	0.00	0.00

EQUIV 1 MEV ELECTRON FLUENCE FOR ISC - CIRCULAR ORBIT
DUE TO GEOMAGNETICALLY TRAPPED PROTONS, MODEL AP8MIN   INCLINATION = 90 DEGREES.

ALTITUDE (N.M.)	(KM)	0 (0)	2.54E-3 (1)	7.64E-3 (3)	1.52E-2 (6)	3.05E-2 (12)	5.09E-2 (20)	7.64E-2 (30)	1.52E-1 (60)
150	277	9.24+12	3.03+12	1.49+12	7.92+11	4.06+11	2.67+11	2.15+11	1.57+11
250	463	2.28+13	7.59+12	4.03+12	2.41+12	1.44+12	1.05+12	8.86+11	6.84+11
300	555	3.86+13	1.23+13	6.56+12	4.00+12	2.48+12	1.84+12	1.56+12	1.21+12
450	833	1.12+14	3.50+13	1.97+13	1.29+13	8.62+12	6.57+12	5.61+12	4.33+12
600	1111	2.39+14	7.88+13	4.76+13	3.34+13	2.37+13	1.84+13	1.56+13	1.18+13
800	1481	6.53+14	2.34+14	1.52+14	1.12+14	8.14+13	6.26+13	5.24+13	3.80+13
1000	1852	1.60+15	6.13+14	4.09+14	3.05+14	2.19+14	1.60+14	1.29+14	8.74+13

6-48

Table 6.36. Annual Equivalent 1 MeV Electron Fluence from Trapped Electrons, 0° Inclination at Synchronous Altitude vs. Longitude (Infinite Backshielding)

ELECTRONS - ISC, VOC, AND PMAX INCLINATION = 0 DEGREES.

EQUIV. 1 MEV ELECTRON FLUENCE FOR ISC - CIRCULAR ORBIT
DUE TO GEOMAGNETICALLY TRAPPED ELECTRONS - MODELS AE6MAX, AEI7LO

ALTITUDE		LONGITUDE				SHIELD THICKNESS, CM (MILS)				
(N.M.)	(KM)	(DEG)	0 (0)	2.54E-3 (1)	7.64E-3 (3)	1.52E-2 (6)	3.05E-2 (12)	5.09E-2 (20)	7.64E-2 (30)	1.52E-1 (60)
19327	35794	180 W	4.82+13	4.19+13	3.47+13	2.78+13	1.95+13	1.32+13	8.54+12	2.78+12
19327	35794	170 W	4.96+13	4.30+13	3.57+13	2.87+13	2.02+13	1.36+13	8.85+12	2.90+12
19327	35794	160 W	4.99+13	4.33+13	3.60+13	2.89+13	2.03+13	1.37+13	8.93+12	2.93+12
19327	35794	150 W	4.90+13	4.26+13	3.53+13	2.83+13	1.99+13	1.34+13	8.71+12	2.84+12
19327	35794	140 W	4.70+13	4.08+13	3.38+13	2.71+13	1.89+13	1.27+13	8.24+12	2.66+12
19327	35794	130 W	4.41+13	3.81+13	3.16+13	2.52+13	1.75+13	1.17+13	7.54+12	2.41+12
19327	35794	120 W	4.07+13	3.52+13	2.89+13	2.30+13	1.60+13	1.06+13	6.78+12	2.13+12
19327	35794	110 W	3.74+13	3.22+13	2.64+13	2.09+13	1.44+13	9.50+12	6.03+12	1.86+12
19327	35794	100 W	3.44+13	2.95+13	2.41+13	1.91+13	1.30+13	8.54+12	5.38+12	1.63+12
19327	35794	90 W	3.21+13	2.75+13	2.24+13	1.76+13	1.20+13	7.82+12	4.90+12	1.46+12
19327	35794	80 W	3.06+13	2.62+13	2.13+13	1.67+13	1.13+13	7.36+12	4.59+12	1.36+12
19327	35794	70 W	3.01+13	2.57+13	2.09+13	1.64+13	1.11+13	7.21+12	4.48+12	1.32+12
19327	35794	60 W	3.06+13	2.61+13	2.13+13	1.67+13	1.13+13	7.34+12	4.57+12	1.35+12
19327	35794	50 W	3.19+13	2.73+13	2.23+13	1.75+13	1.19+13	7.75+12	4.84+12	1.44+12
19327	35794	40 W	3.39+13	2.91+13	2.38+13	1.88+13	1.28+13	8.37+12	5.26+12	1.59+12
19327	35794	30 W	3.64+13	3.13+13	2.56+13	2.03+13	1.39+13	9.16+12	5.79+12	1.77+12
19327	35794	20 W	3.91+13	3.36+13	2.76+13	2.19+13	1.51+13	1.00+13	6.36+12	1.97+12
19327	35794	10 W	4.15+13	3.58+13	2.95+13	2.35+13	1.62+13	1.08+13	6.90+12	2.17+12
19327	35794	0	4.35+13	3.76+13	3.10+13	2.48+13	1.72+13	1.15+13	7.36+12	2.34+12
19327	35794	10 E	4.48+13	3.87+13	3.20+13	2.56+13	1.78+13	1.19+13	7.66+12	2.44+12
19327	35794	20 E	4.53+13	3.92+13	3.24+13	2.59+13	1.80+13	1.21+13	7.78+12	2.49+12
19327	35794	30 E	4.50+13	3.90+13	3.22+13	2.57+13	1.79+13	1.20+13	7.73+12	2.47+12
19327	35794	40 E	4.42+13	3.82+13	3.15+13	2.52+13	1.75+13	1.17+13	7.54+12	2.40+12
19327	35794	50 E	4.27+13	3.69+13	3.04+13	2.43+13	1.69+13	1.13+13	7.22+12	2.29+12
19327	35794	60 E	4.12+13	3.56+13	2.93+13	2.33+13	1.62+13	1.08+13	6.88+12	2.17+12
19327	35794	70 E	3.97+13	3.42+13	2.81+13	2.24+13	1.55+13	1.03+13	6.55+12	2.05+12
19327	35794	80 E	3.82+13	3.29+13	2.70+13	2.15+13	1.48+13	9.80+12	6.24+12	1.94+12
19327	35794	90 E	3.71+13	3.20+13	2.62+13	2.08+13	1.43+13	9.46+12	6.01+12	1.86+12
19327	35794	100 E	3.64+13	3.13+13	2.57+13	2.04+13	1.40+13	9.24+12	5.86+12	1.80+12
19327	35794	110 E	3.63+13	3.12+13	2.56+13	2.03+13	1.40+13	9.20+12	5.83+12	1.80+12
19327	35794	120 E	3.67+13	3.16+13	2.59+13	2.06+13	1.41+13	9.33+12	5.92+12	1.83+12
19327	35794	130 E	3.77+13	3.25+13	2.67+13	2.12+13	1.46+13	9.66+12	6.14+12	1.91+12
19327	35794	140 E	3.93+13	3.39+13	2.79+13	2.22+13	1.53+13	1.02+13	6.50+12	2.03+12
19327	35794	150 E	4.14+13	3.57+13	2.95+13	2.35+13	1.63+13	1.09+13	6.96+12	2.20+12
19327	35794	160 E	4.37+13	3.78+13	3.12+13	2.50+13	1.74+13	1.16+13	7.49+12	2.39+12
19327	35794	170 E	4.61+13	3.99+13	3.30+13	2.65+13	1.85+13	1.24+13	8.03+12	2.59+12

Table 6.37. Annual Equivalent 1 MeV Electron Fluence from Trapped Protons (V_{oc}, P_{max}), 0° Inclination at Synchronous Altitude vs. Longitude (Infinite Backshielding)

PROTONS - VOC AND PMAX

EQUIV. 1 MEV ELECTRON FLUENCE FOR VOC AND PMAX CIRCULAR ORBIT INCLINATION = 0 DEGREES.
DUE TO GEOMAGNETICALLY TRAPPED PROTONS, MODEL AP8MAX

ALTITUDE		LONGITUDE		SHIELD THICKNESS, CM (MILS)						
(N.M.)	(KM)	(DEG)	0 (0)	2.54E-3 (1)	7.64E-3 (3)	1.52E-2 (6)	3.05E-2 (12)	6.09E-2 (24)	7.64E-2 (30)	1.52E-1 (60)
19327	35794	180 W	4.02+17	9.95+09	0.00	0.00	0.00	0.00	0.00	0.00
19327	35794	170 W	4.48+17	1.05+10	0.00	0.00	0.00	0.00	0.00	0.00
19327	35794	160 W	4.60+17	1.07+10	0.00	0.00	0.00	0.00	0.00	0.00
19327	35794	150 W	4.29+17	1.05+10	0.00	0.00	0.00	0.00	0.00	0.00
19327	35794	140 W	3.63+17	9.81+09	0.00	0.00	0.00	0.00	0.00	0.00
19327	35794	130 W	2.81+17	8.82+09	0.00	0.00	0.00	0.00	0.00	0.00
19327	35794	120 W	2.05+17	7.70+09	0.00	0.00	0.00	0.00	0.00	0.00
19327	35794	110 W	1.45+17	6.63+09	0.00	0.00	0.00	0.00	0.00	0.00
19327	35794	100 W	1.03+17	5.72+09	0.00	0.00	0.00	0.00	0.00	0.00
19327	35794	90 W	7.76+16	5.14+09	0.00	0.00	0.00	0.00	0.00	0.00
19327	35794	80 W	6.39+16	4.78+09	0.00	0.00	0.00	0.00	0.00	0.00
19327	35794	70 W	5.98+16	4.69+09	0.00	0.00	0.00	0.00	0.00	0.00
19327	35794	60 W	6.36+16	4.83+09	0.00	0.00	0.00	0.00	0.00	0.00
19327	35794	50 W	7.57+16	5.21+09	0.00	0.00	0.00	0.00	0.00	0.00
19327	35794	40 W	9.71+16	5.84+09	0.00	0.00	0.00	0.00	0.00	0.00
19327	35794	30 W	1.29+17	6.67+09	0.00	0.00	0.00	0.00	0.00	0.00
19327	35794	20 W	1.72+17	7.59+09	0.00	0.00	0.00	0.00	0.00	0.00
19327	35794	10 W	2.19+17	8.45+09	0.00	0.00	0.00	0.00	0.00	0.00
19327	35794	0	2.65+17	9.16+09	0.00	0.00	0.00	0.00	0.00	0.00
19327	35794	10 E	2.97+17	9.59+09	0.00	0.00	0.00	0.00	0.00	0.00
19327	35794	20 E	3.10+17	9.69+09	0.00	0.00	0.00	0.00	0.00	0.00
19327	35794	30 E	3.04+17	9.48+09	0.00	0.00	0.00	0.00	0.00	0.00
19327	35794	40 E	2.82+17	9.05+09	0.00	0.00	0.00	0.00	0.00	0.00
19327	35794	50 E	2.47+17	8.42+09	0.00	0.00	0.00	0.00	0.00	0.00
19327	35794	60 E	2.14+17	7.76+09	0.00	0.00	0.00	0.00	0.00	0.00
19327	35794	70 E	1.84+17	7.15+09	0.00	0.00	0.00	0.00	0.00	0.00
19327	35794	80 E	1.59+17	6.58+09	0.00	0.00	0.00	0.00	0.00	0.00
19327	35794	90 E	1.42+17	6.17+09	0.00	0.00	0.00	0.00	0.00	0.00
19327	35794	100 E	1.31+17	5.93+09	0.00	0.00	0.00	0.00	0.00	0.00
19327	35794	110 E	1.29+17	5.86+09	0.00	0.00	0.00	0.00	0.00	0.00
19327	35794	120 E	1.35+17	5.95+09	0.00	0.00	0.00	0.00	0.00	0.00
19327	35794	130 E	1.51+17	6.21+09	0.00	0.00	0.00	0.00	0.00	0.00
19327	35794	140 E	1.78+17	6.71+09	0.00	0.00	0.00	0.00	0.00	0.00
19327	35794	150 E	2.19+17	7.41+09	0.00	0.00	0.00	0.00	0.00	0.00
19327	35794	160 E	2.72+17	8.22+09	0.00	0.00	0.00	0.00	0.00	0.00
19327	35794	170 E	3.36+17	9.10+09	0.00	0.00	0.00	0.00	0.00	0.00

Table 6.38. Annual Equivalent 1 MeV Electron Fluence from Trapped Protons (I_{sc}), 0° Inclination at Synchronous Altitude vs. Longitude (Infinite Backshielding)

PROTONS - ISC

INCLINATION = 0 DEGREES.

EQUIV 1 MEV ELECTRON FLUENCE FOR ISC - CIRCULAR ORBIT
DUE TO GEOMAGNETICALLY TRAPPED PROTONS, MODEL AP8MAX

ALTITUDE		LONGITUDE	B	SHIELD THICKNESS, CM (MILS)						
(N.M.)	(KM)	(DEG)	(0)	2.54E-3 (1)	7.64E-3 (3)	1.52E-2 (6)	3.05E-2 (12)	6.09E-2 (20)	7.64E-2 (30)	1.52E-1 (60)
19327	35794	180 W	3.33+15	2.24+08	0.00	0.00	0.00	0.00	0.00	0.00
19327	35794	170 W	3.70+15	2.34+08	0.00	0.00	0.00	0.00	0.00	0.00
19327	35794	160 W	3.81+15	2.38+08	0.00	0.00	0.00	0.00	0.00	0.00
19327	35794	150 W	3.58+15	2.33+08	0.00	0.00	0.00	0.00	0.00	0.00
19327	35794	140 W	3.08+15	2.21+08	0.00	0.00	0.00	0.00	0.00	0.00
19327	35794	130 W	2.43+15	2.03+08	0.00	0.00	0.00	0.00	0.00	0.00
19327	35794	120 W	1.82+15	1.83+08	0.00	0.00	0.00	0.00	0.00	0.00
19327	35794	110 W	1.32+15	1.62+08	0.00	0.00	0.00	0.00	0.00	0.00
19327	35794	100 W	9.61+14	1.45+08	0.00	0.00	0.00	0.00	0.00	0.00
19327	35794	90 W	7.41+14	1.33+08	0.00	0.00	0.00	0.00	0.00	0.00
19327	35794	80 W	6.24+14	1.26+08	0.00	0.00	0.00	0.00	0.00	0.00
19327	35794	70 W	5.90+14	1.24+08	0.00	0.00	0.00	0.00	0.00	0.00
19327	35794	60 W	6.24+14	1.27+08	0.00	0.00	0.00	0.00	0.00	0.00
19327	35794	50 W	7.38+14	1.34+08	0.00	0.00	0.00	0.00	0.00	0.00
19327	35794	40 W	9.34+14	1.47+08	0.00	0.00	0.00	0.00	0.00	0.00
19327	35794	30 W	1.22+15	1.63+08	0.00	0.00	0.00	0.00	0.00	0.00
19327	35794	20 W	1.60+15	1.81+08	0.00	0.00	0.00	0.00	0.00	0.00
19327	35794	10 W	2.00+15	1.97+08	0.00	0.00	0.00	0.00	0.00	0.00
19327	35794	0	2.39+15	2.10+08	0.00	0.00	0.00	0.00	0.00	0.00
19327	35794	10 E	2.65+15	2.17+08	0.00	0.00	0.00	0.00	0.00	0.00
19327	35794	20 E	2.74+15	2.19+08	0.00	0.00	0.00	0.00	0.00	0.00
19327	35794	30 E	2.67+15	2.15+08	0.00	0.00	0.00	0.00	0.00	0.00
19327	35794	40 E	2.47+15	2.08+08	0.00	0.00	0.00	0.00	0.00	0.00
19327	35794	50 E	2.17+15	1.96+08	0.00	0.00	0.00	0.00	0.00	0.00
19327	35794	60 E	1.88+15	1.84+08	0.00	0.00	0.00	0.00	0.00	0.00
19327	35794	70 E	1.62+15	1.72+08	0.00	0.00	0.00	0.00	0.00	0.00
19327	35794	80 E	1.40+15	1.61+08	0.00	0.00	0.00	0.00	0.00	0.00
19327	35794	90 E	1.25+15	1.53+08	0.00	0.00	0.00	0.00	0.00	0.00
19327	35794	100 E	1.15+15	1.49+08	0.00	0.00	0.00	0.00	0.00	0.00
19327	35794	110 E	1.13+15	1.47+08	0.00	0.00	0.00	0.00	0.00	0.00
19327	35794	120 E	1.18+15	1.49+08	0.00	0.00	0.00	0.00	0.00	0.00
19327	35794	130 E	1.31+15	1.54+08	0.00	0.00	0.00	0.00	0.00	0.00
19327	35794	140 E	1.53+15	1.64+08	0.00	0.00	0.00	0.00	0.00	0.00
19327	35794	150 E	1.86+15	1.77+08	0.00	0.00	0.00	0.00	0.00	0.00
19327	35794	160 E	2.29+15	1.92+08	0.00	0.00	0.00	0.00	0.00	0.00
19327	35794	170 E	2.80+15	2.08+08	0.00	0.00	0.00	0.00	0.00	0.00

Table 6.39 Equivalent 1 MeV Electron Fluence for Solar Flare Protons Based on Fluences in Table 5.1

Years	Cell Parameter	Coverglass Thickness (mils)					
		3	6	12	20	30	60
1972	V_{oc} and P_{max}	1.5 E14	9.1 E13	5.4 E13	3.5 E13	2.5 E13	1.3 E13
	I_{sc}	7.5 E13	5.0 E13	3.3 E13	2.4 E13	1.8 E13	1.0 E13
1973	V_{oc} and P_{max}	9.4 E10	5.5 E10	3.1 E10	1.9 E10	1.3 E10	7.4 E09
	I_{sc}	4.5 E10	2.9 E10	1.9 E10	1.3 E10	9.9 E09	5.9 E09
1974	V_{oc} and P_{max}	2.0 E12	1.2 E12	7.3 E11	5.1 E11	3.4 E11	1.3 E11
	I_{sc}	9.7 E11	6.6 E11	4.4 E11	3.1 E11	2.2 E11	9.3 E10
1975	V_{oc} and P_{max}	2.1 E10	1.2 E10	6.3 E09	3.6 E09	2.5 E09	1.5 E09
	I_{sc}	9.8 E09	6.2 E09	3.8 E09	2.5 E09	1.9 E09	1.3 E09
1976	V_{oc} and P_{max}	1.4 E11	8.1 E10	4.6 E10	2.9 E10	2.1 E10	1.1 E10
	I_{sc}	6.6 E10	4.4 E10	2.8 E10	2.0 E10	1.5 E10	8.5 E09
1977	V_{oc} and P_{max}	1.8 E12	1.1 E12	6.1 E11	3.9 E11	2.8 E11	1.3 E11
	I_{sc}	8.6 E11	5.7 E11	3.7 E11	2.6 E11	1.9 E11	1.0 E11
1978	V_{oc} and P_{max}	4.4 E13	2.7 E13	1.7 E13	1.2 E13	7.6 E12	2.7 E12
	I_{sc}	2.2 E13	1.5 E13	1.0 E13	7.1 E12	4.8 E12	1.9 E12
1979	V_{oc} and P_{max}	5.9 E12	3.6 E12	2.1 E12	1.5 E12	9.9 E11	4.0 E11
	I_{sc}	2.9 E12	2.0 E12	1.3 E12	9.2 E11	6.5 E11	3.0 E11
1978-1984 (SOLPRO)	V_{oc} and P_{max}	1.1 E14	6.8 E13	3.8 E13	2.4 E13	1.8 E13	1.0 E13
	I_{sc}	5.6 E13	3.7 E13	2.4 E13	1.7 E13	1.3 E13	8.5 E12

REFERENCES

6.1 E. Stofel and D. Joslin, "Low Energy Proton Irradiation of Solar Cell Back Contacts," Conf. Rec. of the 8th IEEE Photovoltaic Specialists Conf., 209, 1970.

6.2 C. J. Fischer and A. H. Kalma, "Hardening of Solar Cells Against Low-Energy Rear-Incident Protons," IEEE Trans. Nuc. Sci., NS-22, 6, 2681, 1975.

6.3 A. H. Kalma and C. J. Fischer, "4π Space Radiation of Solar Cells," IEEE Trans. Nuc. Sci., NS-23, 6, 1789, 1976.

6.4 M. J. Teague, K. W. Chan and J. I. Vette, "AE 6: A Model Environment of Trapped Electrons for Solar Maximum," NASA, NSSDC/WDC-A-R&S 76-04, 1976.

6.5 D. M. Sawyer and J. I. Vette, "AP-8 Trapped Proton Environment for Solar Maximum and Solar Minimum," NSSDC/WDC-A-R&S 76-06, 1976.

6.6 M. J. Teague and J. I. Vette, "A Model of the Trapped Electron Population for Solar Minimum," NASA, NSSDC 74-03, 1974.

6.7 H. K. Hills, K. W. Chan, M. J. Teague, and J. I. Vette, To be Published.

6.8 J. A. Barton, B. W. Mar, et al., "Computer Codes for Space Radiation Environment and Shielding," Air Force Weapons Laboratory, Tech. Doc. Report No. WL TDR-64-71, (AD-444602), Vol. I., 1964.

6.9 A. Hassitt and C. E. McIlwain, "Computer Programs for the Computation of B and L (May 1966)," NSSDC 67-27, May 1967.

6.10 E. G. Stassinopoulos and G. D. Mead, "ALLMAG, GDALMG, LINTRA: Computer Programs for Geomagnetic Field and Field-Line Calculations," NSSDC 72-12, 1972.

6.11 E. G. Stassinopoulos, J. J. Hebert, E. L. Butler, and J. L. Barth, "SOFIP: A Short Orbital Flux Integration Program," NSSDC/WDC-A-R&S 79-01, 1979.

6.12 H. Y. Tada, "A New Dimension in Solar Cell Degradation Estimate In Space - RDC Matrix Method," Conf. Rec. of 10th IEEE Photovoltaic Specialists Conf., 393, 1973.

6.13 E. G. Stassinopoulos, "SOLPRO: A Computer Code to Calculate Probabilistic Energetic Solar Proton Fluences," NASA, NSSDC 75-11, 1975.

6.14 D. K. Weidner, "Natural Space Environment Criteria for 1975-1985 NASA Space Stations," NASA TM X53865, August 1969, Second Edition, August 1970.

6.15 J. H. King, "Solar Proton Fluences for 1977 - 1983 Space Missions," J. Spacecraft and Rockets, 11, 6, 401, June 1974.

6.16 J. H. King and E. G. Stassinopoulos, "Energetic Solar Protons vs Terrestrially Trapped Proton Fluxes for the Active Years 1977 - 1983," J. Spacecraft and Rockets, 12, 2, 122, February 1975.

CHAPTER 7

7.0 FLIGHT DATA

Considering the number of satellites in orbit, there is a limited amount of currently usable solar cell radiation degradation data available. Satellite operations have tended to be concentrated in two relatively low-level areas of the geomagnetically trapped radiation belts. The early satellites were placed in low-altitude earth orbits (less than 400 km) where the levels of trapped radiation are very low. Subsequently, as satellite launch capabilities improved, most satellites were placed in synchronous orbit, again avoiding the most intense radiation areas.

The flight data are of two types: (1) the data obtained directly from flight experiments specifically designed for solar cell performance analysis, and (2) the solar array performance data from operational spacecraft. The experiments flown on ATS-1 [7.1], ATS-5, ATS-6, LES-6, NTS-1, and NTS-2 satellites belong to the first category, while examples of the second type are analyses of the IDSCS arrays and Hughes Aircraft Company satellite arrays. It would be reasonable to expect that the data from a well-designed solar cell experiment would be comprehensive and easy to analyze and correlate with laboratory experiments. However, even well designed experiments have experienced unexpected events which make correlation of flight/laboratory data difficult. The following sections discuss the factors affecting data analysis and also comment on the flight data analysis currently available.

For the determination of the radiation environment, the following are required: (a) spacecraft orbital parameters, including launch data and flight duration, (b) the solar panel and surrounding structural configuration, and (c) the most reliable radiation map representing the radiation environment during the flight time span in question or data from on-board radiation spectrometers. Frequently, the information regarding the parking or transfer orbit and its flight duration are neglected in the published flight data. These initial phases of spacecraft flight may be of importance to radiation damage if the trajectory traverses the intense part of the Van Allen Belts for a prolonged period. Launch data and flight duration are also needed to determine the possible occurrence of a solar flare proton event during the flight under consideration.

The most vital part of the environment determination lies in the selection of a reliable, current radiation map matched to the right part of the solar cycle. Yet the models themselves include a factor of 2 intensity uncertainty, not to mention spectral and temporal variations (from solar activity, solar cycle, local time, etc.). The value of equivalent fluence depends entirely on the radiation model on which the calculation is based, and a factor of 10 difference in the resultant equivalent fluence is not uncommon because of the choice of environment.

In comparison with the uncertainties in the radiation environment, the solar panel or surrounding structure geometry is of lesser importance. However, these factors must be considered since variations in solar panel substrates and structure shielding can significantly affect the equivalent fluence. Deficiencies in solar cell/coverglass assembly techniques can lead to unexpected degradation because of the change in radiation environment which is not accounted for in the fluence calculation. In addition, if the sides of the solar cells are not properly protected, especially for the case when a thin substrate is used and the back radiation becomes substantial, damage due to low energy protons can become very important.

The accurate evaluation of solar cell performance data requires not only the cell output parameters but also such factors as (a) solar cell temperature, (b) sun angle, (c) Earth position in terms of seasonal solar irradiance, (d) structural shadowing of the array, (e) identification of "bad" cells, etc. A "shadowed" cell or "bad" cell in a string will become a load instead of a current generator. The spacecraft measurement and telemetry system must be capable of providing the above listed data. In addition, the telemetry resolution and sampling are of importance since sun angle, cell temperature, and shadow problems are usually time dependent.

Many published flight data reports lack detailed solar cell descriptions. The lack of information on the cell manufacturer, for example, may make a substantial difference in the predicted values of solar cell parameters, even for cells with the same physical parameters. The base resistivity, cell thickness, presence of a back surface field or back surface reflector, and coverglass type all influence the output parameter performance, and must be known for accurate degradation predictions to be calculated.

7.1 Flight Data at Synchronous Orbit

7.1.1 Solar Array Performance Data at Synchronous Orbit

The data in Table 7.1, relating to solar array performance in synchronous orbit, were collected by L. A. Gibson of the Aerospace Corporation.[7.2] All the solar cells used in these satellites have 10 ohm-cm base resistivity and coverglass shielding varying from 0.015 cm (0.006 in) microsheet to 0.076 cm (0.030 in) fused silica. No information was reported regarding cell thickness or backshielding. The reported degradation in power in most cases is between 2 to 6% after one year. The power loss estimated in Section 6.5 on the basis of trapped electrons alone was 3% per year for cells with 0.015-cm microsheet shielding. However, this percentage loss does not include an approximate 2% loss usually attributed to UV glass and adhesive darkening effects. In addition, the percentage degradation is estimated from the data in Chapter 3 for currently available commercial cells, not those flown almost a decade ago. Considering the above facts, together with the omission of equivalent fluence contributed by solar flare protons, agreement between satellite performance and the predictions is reasonably good. The omission of solar flare equivalent fluence contributions appears justified in these cases, as flare activity was relatively

Table 7.1 Synchronous Orbit Solar Cell Array Degradation

	TACSAT I	DSP	IDSCS	NATO A/NATO B	INTELSAT 3	INTELSAT 4	ATS-5
Contractor	HAC	TRW	Philco-Ford	Philco-Ford	TRW	HAC	HAC
Launch	9 Feb 69	6 Nov 70	16 Jun 66 18 Jan 67	A: 21 Nov 69 B: 20 Mar 70	3B: 18 Dec 68 3C: 5 Feb 69 3D: 21 May 69	Feb 71	12 Aug 69
Configuration	Drum	Drum & Paddles	24-sided Polygon	Drum	Drum	Drum	Drum
Design Life, yrs.	3	3	5	5	3	7	
Coverglass Thickness, mils	12	6	20	6	12	12	30
Coverglass Material	Fused Silica	Microsheet	Fused Silica	Fused Silica	Fused Silica	Fused Silica	Fused Silica
Solar Cell Resistivity ohm-cm	10	10	10	10	10	10	10
Solar Cell Array:							
Time, years	3	1	5	2 1	1	0.7	1.7
Power, Degradation, %	~6	~5	~17-22	~4 ~2	~6	~3	~4

7-4

low during the time period of the reported flight data. It was reported that the poor performance of IDSCS satellite solar arrays was attributable to excessive ultraviolet transmission loss in the coverglass adhesive due to the use of an improper primer.[7.3]

In a more recent collection of data,[7.4, 7.5] the behavior of 29 Hughes Aircraft synchronous orbit satellites was analyzed. Time in orbit ranged from 3 to 130 months. The arrays were constructed of 0.030 cm (12 mil) thick n/p 10 ohm-cm solar cells from a variety of manufacturers with 0.030 cm (12 mil) thick 7940 fused silica coverglasses. An initial UV degradation of 2% was found to match the data. An equivalent 1 MeV electron fluence of 4.7×10^{13} e/cm^2 was computed for the August 1972 solar flare proton event. It was determined from the flight data that an equivalent 1 MeV electron fluence of 2.1×10^{13} e/cm^2-year would produce the observed array power degradations to within $\pm$ 2% in all cases. Considering all the variables involved, the consistency of these results is remarkable. Further, their deduced equivalent 1 MeV electron fluence agrees well with the value $1.1 - 2.0 \times 10^{13}$ e/cm^2-yr (depending on longitude) predicted in Chapter 6.

An additional analysis of solar array data for spacecraft flying in synchronous orbit was made by Lyons at Comsat.[7.6] He found that the three MARISAT solar arrays (10 ohm-cm, 0.030-cm-thick n/p solar cells with 0.030-cm-thick fused silica coverglasses) experienced power degradation consistent with an equivalent 1 MeV electron fluence of 2.4×10^{13} e/cm^2-yr. This value agrees well with the $1.1 - 2.0 \times 10^{13}$ e/cm^2-yr value predicted in Chapter 6. The INTELSAT V panels utilized 10 ohm-cm n/p cells, 0.025-cm-thick protected by 0.015-cm-thick ceria-doped microsheet coverglasses. Since the panels are built in a wing configuration, they are lightly shielded from the rear. The observed panel power degradation after approximately 2 years in orbit fit an equivalent 1 MeV electron fluence of 4.3×10^{13} e/cm^2-yr. The predicted value from Chapter 6 is $1.6 - 2.9 \times 10^{13}$ e/cm^2-yr for infinite backshielding. The contribution from the rear incidence radiation cannot be estimated from the given data.

An analysis of the FLTSATCOM solar array performance has been performed by Bavaro and Weiner of Aerospace Corp.[7.7] The solar array consists of 10 ohm-cm, 0.020-cm-thick conventional solar cells with 0.015-cm-thick fused silica coverglasses. The Handbook value of 2.5×10^{13} e/cm^2-yr for this panel would predict a power degradation of 4%, whereas the actual array degradation was reported to be 3.9%. A question remains as to how much of the observed degradation was due to UV. Since the reported data is normalized to the initial data taken 1 to 2 months after launch, it is possible the 2% degradation usually attributed to UV exposure had already occurred during that early time period.

Although degradations due to solar flares are often estimated and projected over long satellite missions, the flare events are discrete and their effects occur as rather abrupt degradations. An excellent example of this behavior is shown in Figure 7.1 for two satellites in synchronous orbits during the flare events of August 1972 (also see Figure 7.2). The analysis was provided by H. Riess of TRW.[7.8] The solid line in Figure 7.1 is based on solar cell degradation predictions based on trapped electrons at synchronous altitude. The data indicate that the flares produced an abrupt 2% loss in maximum array current (i.e., short circuit current) in both satellites. It also can be observed that 5 months after the flare, the Flight 3 array current had recovered to within nearly 1% of the value predicted without solar flares. This indicates that considerable annealing of flare radiation damage occurs after termination of the event, matching the results reported in the proton annealing experiments of Reference 7.9.

7.1.2 Flight Experiments at Synchronous Orbits

7.1.2.1 LES-6 [7.10, 7.11]

The sixth Lincoln Laboratory Experiment Satellite (LES-6) was launched into a synchronous orbit on 26 September, 1968. The experiment carried a variety of mostly developmental cells including dendritic cells with and without drift fields, CdS cells, CdTe cells, lithium-doped cells, ion-implanted

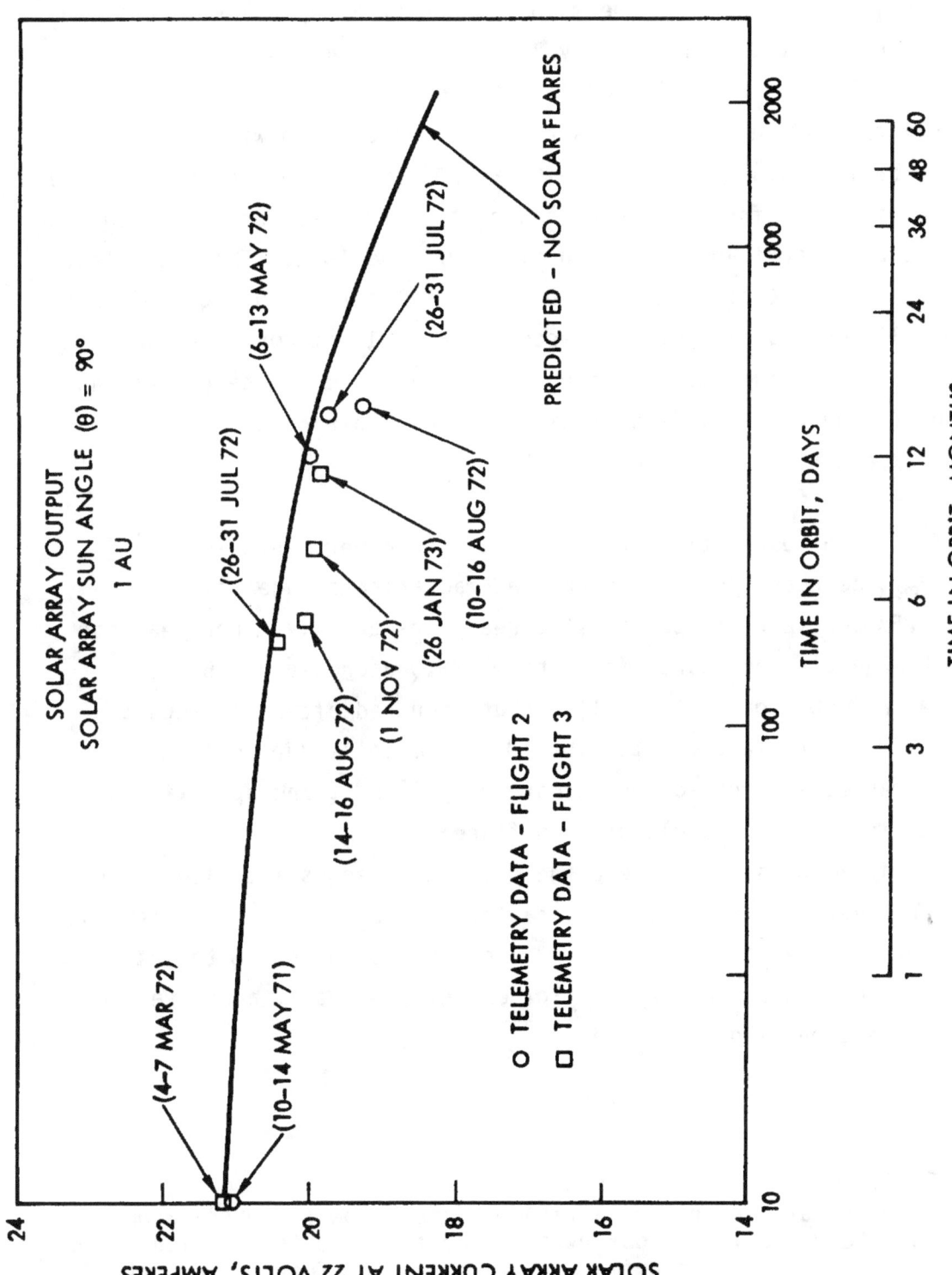

Figure 7.1 Performance of Two Satellite Solar Arrays in Synchronous Orbit During the August 1972 Solar Flares

cells, several cells with 1-2 mil integral coverglasses of sputtered 7940, and cells duplicating those of the main spacecraft array. The latter are 10 ohm-cm, blue-shifted cells made by TI with 0.015 cm 7940 coverglasses. The experimental cells were divided into two groups, one group with deliberately exposed cell/coverglass edges, and one group with the edges and contact bars carefully shielded with a 0.01-cm BeCu window frame. The radiation environment for LES-6 consists of trapped electrons and a substantial fluence of solar flare protons during 1970 and 1972. The equivalent 1 MeV electron fluence attributable to trapped electrons (using the updated AEI7-LO model) for a solar cell protected by a 0.015 cm coverglass after 6-1/2 years in orbit is computed to be 1.6×10^{14} e/cm^2. The equivalent fluence due to the August 1972 solar flare* from Table 6.39 is 5.0×10^{13} e/cm^2 for I_{sc} and 9.1×10^{13} e/cm^2 for V_{oc}, P_{max}.

The following observations were made by the experimenters: 7.11
- P_{max} decrease due to penetrating radiation damage was about 3.5% per year for the first three years and 1.75% per year for the next three years, for a total P_{max} decrease of 15.8%.
- An abrupt drop in the cell output occurred after the August 1972 solar flare event. The 10 ohm-cm cells duplicating the array cells dropped approximately 3% in P_{max} and approximately 2% in I_{sc} as a result of this flare.
- It is important to shield the solar cell edges from low energy protons. A comparison of data from the cells shielded with the BeCu windows and the unshielded cells showed that 8 to 12% degradation can occur from low energy protons if adequate edge protection is not provided.

---

*The cutoff energy at synchronous altitude seems to be somewhere around 5 MeV according to an early ATS observation, contrary to the previous theoretical cutoff energy of approximately 26 MeV. Therefore, no cutoff energy resulting from geomagnetic shielding was assumed for the equivalent fluence calculation. This approximation leads to no appreciable error if the cutoff from coverglass thickness is somewhere around 5 MeV or greater.

In addition to the observations made by the experimenters, the following points are of interest:

- Assuming that the experimenters' value of 15.8% P_{max} degradation (due to radiation only) applies to a 10 ohm-cm 0.036-cm-thick solar cell of mid-1960 vintage protected by a 0.015-cm coverglass, we can use the data of Reference 7.12 to see that this degradation would be caused by an equivalent 1 MeV fluence of 1.5×10^{14} e/cm^2. This compares well with the 1.6×10^{14} e/cm^2 computed for this orbit.
- Using the flight data presented in Figure 7.2 and Reference 7.11 for a 10 ohm-cm cell with a 0.015-cm coverglass, it will be observed that 83% of the initial I_{sc} remains after 6.5 years in orbit. Correcting for a 2% observed loss due to the August 1972 solar flare, a 1% loss in transmission of the 7940 fused silica coverglass (see Figure 3.19) and a 2% UV degradation loss (considered reasonable in conventional solar array design [7.4]) we find that the degradation of I_{sc} due to electrons only is 12%. This degradation would be caused by an equivalent 1 MeV fluence of 3.7×10^{14} e/cm^2, again using Reference 7.12. Correspondingly for P_{max}, we have 77% remaining power at end-of-life. Applying the observed 3% degradation for the solar flare, 1% coverglass transmission loss, and 2% for UV degradation we find a 17% degradation in P_{max} due to trapped electrons. This value implies an equivalent 1 MeV fluence of 1.8×10^{14} e/cm^2.
- Degradation in P_{max} due to the August 1972 solar flare was computed to be 4% as compared to an observed 3%. Similarly, the degradation in I_{sc} was computed to be 1% as compared to an observed 2%.
- Within the accuracies of the assumptions made to compute the above numbers, the experimental values for the 10 ohm-cm cells with 0.015 cm 7940 coverglasses are in reasonable agreement with computed degradations based on the AEI7-L0 environmental model and the relative damage coefficients of Chapter 4.

7.1.2.2 ATS-5 [7.13, 7.14]

The ATS-5 solar cell radiation experiment, consisting of several types of solar cell/coverglass combinations representing 1968 technology,

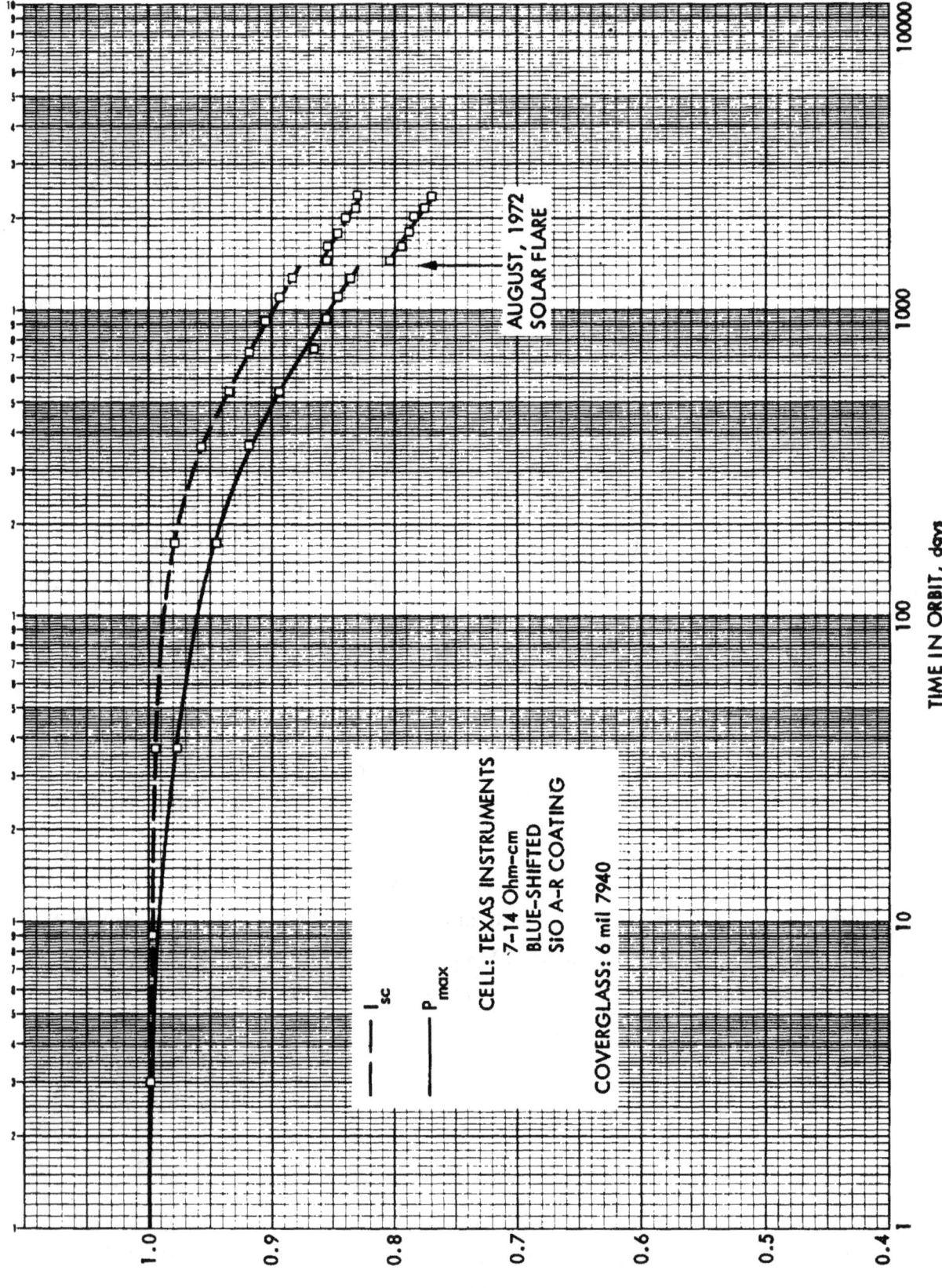

Figure 7.2 LES-6 Flight Data for 10 Ohm-cm Cell No. A3 with 6 mil 7940 Coverglass (From Ref. 7.11)

were mounted on the ATS-5 spacecraft and launched into synchronous orbit in August 1969.

The solar cells were 2 and 10 ohm-cm crucible-grown silicon with thicknesses of 0.02 and 0.03 cm. Coverglasses were 7940 fused silica, ranging in thickness from 0.015 to 0.152 cm. The cells were mounted on two panels, one a rigid aluminum honeycomb structure giving essentially infinite back-shielding, and the other a thin Kapton-fiberglass substrate offering minimal protection to the rear surfaces of the cells.

Cell electrical output was corrected to standard temperature and solar intensity using experimentally derived, radiation-dependent correction factors. The corrected maximum power of a 10 ohm-cm n/p cell is shown for over 6-1/2 years of experimental operation in Figure 7.3.

Some pertinent observations and conclusions drawn from this experiment are:

- The degradation of solar cells mounted on the rigid panel with protected rear surfaces is as predicted using the equivalent 1 MeV electron fluence calculated in Chapter 6. V_{oc} degradation is somewhat less than predicted, but I_{sc} and P_{max} degradations are more than predicted.
- The cells on the flexible panel degrade much more rapidly than predicted, while the rigid panel cells follow the predictions fairly well. Possible causes for the excessive cell degradation on the flexible panels include: deposition of a contaminant on the cell coverglasses, low energy protons entering the edges of the cells or inadequate accounting for the effect of low energy protons incident on the back of the cells through the Kapton-fiberglass substrate.
- An abrupt change in all outputs was observed after the August 1972 solar flare proton event. The equivalent 1 MeV electron fluence for this proton event was used to construct the predicted curve in Figure 7.3. The prediction is within the observation error.

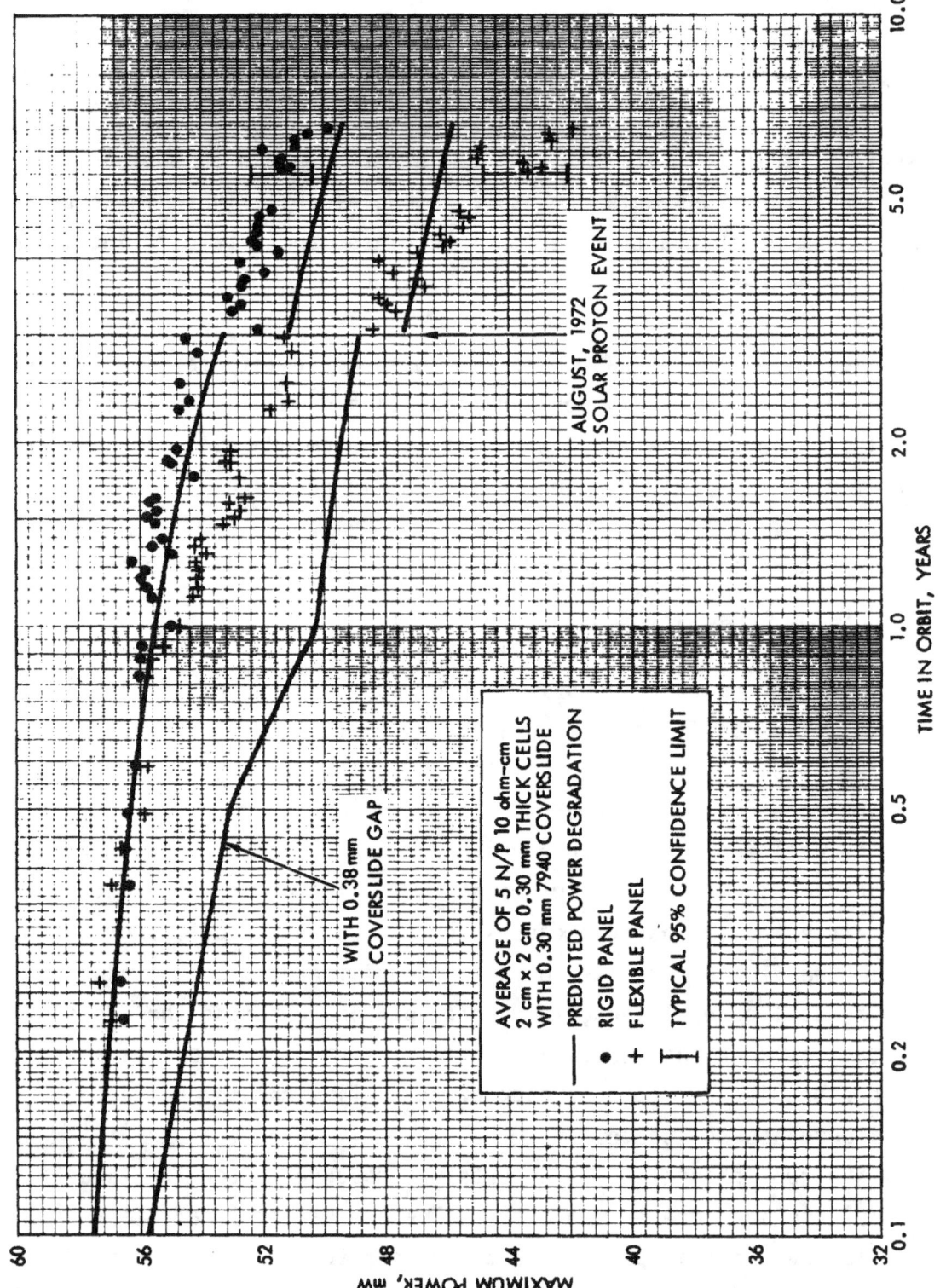

Figure 7.3 Degradation of Solar Cell Maximum Power vs Time in Synchronous Orbit, ATS-5 Experimental Cells. 7.13, 7.14

7.1.2.3 ATS-6 [7.15, 7.16]

The ATS-6 solar cell experiment, with 13 different types of solar cell/coverglass combinations, was launched into synchronous orbit on 30 May 1974. A few comments are made on the data presented in Reference 7.15:

- Soon after orbit insertion, the output of all of the cell configurations on the rigid panel was greater than when measured under the pulsed xenon solar simulator, which was attributed to an electronic offset of the signal processor units. The lack of correlation between simulator data and flight data may very well be due to improper calibration including spectral content of the xenon simulator.
- The temperature of the rigid solar panel ranged from 56°C to 91°C, with outputs reportedly corrected for both temperature and sun angle. However, data inconsistencies prohibit drawing clearcut conclusions. The inconsistency may be attributable to the temperature gradient within the cell itself, inaccurate temperature measurement and sun angle on the flight experiment.
- Despite incomplete flight data, an attempt was made to correlate the prediction with the flight data as shown in Table 7.2. It is assumed that all the cells are conventional. Approximately one third of the predicted values agree with observed values and the observed values varied widely.
- In the final summary report [7.16], the authors conclude that an anomalous and nonuniform loss of optical transmission of the coverglasses produced an additional degradation which in many cases exceeded the expected loss due to penetrating radiation. Although not stated in the final report, this nonuniform degradation suggests the possibility that this loss was caused by deposition of foreign matter on the coverglasses. In the presence of such an anomaly, data analysis and conclusions are tentative at best.

Table 7.2 Percentage Degradation of Predicted and Observed
ATS-6 Solar Cell Experiment Output 7.15

Resistivity	Cell Thickness	Cover-glass Thickness	AVERAGE PERCENTAGE LOSS							
			50 DAYS		247 DAYS					
			I_{sc}		I_{sc}		V_{oc}		P_{max}	
(Ohm-cm)	(cm)	(cm)	Predicted	Observed	Predicted	Observed	Predicted	Observed	Predicted	Observed
10	.030	.0076	1.4	1.1-3.1	5.2	7.6-11.2	3.0	1.2-1.6	9.9	2.3-11.5
10	.030	.015	1.2	1.1-2.1	4.6	9.3-11.1	2.6	1.2-2.0	8.6	5.6-11.2
10	.020	.015	1.2	1.5-2.9	4.6	9.0-10.6	2.6	1.1-1.5	8.6	7.1-11.1
10	.030	.030	.8	1.6-5.0	3.6	3.4-7.2	2.0	.7-1.5	7.1	1.8-7.2
10	.030	.076	.2	1.8-3.2	2.8	8.7-13.1	1.0	.9-1.7	3.9	3.5-12.5
2	.030	.015	2.5	.9-1.3	7.3	6.1-9.5	4.2	.5-.9	12.1	7.5-9.9
2	.030	.015	3.4	.9-3.1	9.4	4.6-9.2	5.0	.3-.7	13.5	6.2-8.6

7.2 Flight Data at Other Than Synchronous Orbits

7.2.1 Solar Array Performance at Other Than Synchronous Orbits

A limited amount of flight data are also available from satellite solar arrays operated in circular orbits at lower altitudes. Data from several such satellites are tabulated in Table 7.3. [7.17-7.19] The approximate equivalent fluence is obtained from tables in Chapter 6 by interpolating both altitude and thicknesses. A density correction was used to convert fused silica data to microsheet coverglasses. For OGO4, the equivalent fluence for 90° inclination is used instead of the actual inclination of 86°. Both electron and proton contributions are shown in Table 7.3. At these altitudes and inclinations, the equivalent fluence is mainly due to protons. The assumptions made are (1) infinite back shielding exists, and (2) cover material darkening losses are negligible. The equivalent fluence values are used to estimate solar cell parameter changes from old radiation data.[7.12] The predicted changes are shown in Table 7.3 along with observed parameter changes from flight data. The predicted degradations are in reasonable agreement with observed values.

7.2.2 Flight Experiments at Other Than Synchronous Orbits

7.2.2.1 ERS 6 [7.18]

The results of experiments on ERS 6 included several observations which have important consequences in array degradation predictions.[7.18] The cells of this satellite were observed to degrade in short circuit current at a rate of 5.5 ± 0.2 mA/cm^2-decade. This value compares well with those reported in Section 3.3 for laboratory proton irradiations in the 10 MeV energy range. It was also observed that cells with adhesively attached coverglasses degraded at the same rate as those with mechanically attached (no adhesive) coverglasses. It was concluded that adhesive darkening effects were either negligible or less than the experimental error. The data also indicated that transmission loss in coverglass is not an important factor in array degradation.

7.2.2.2 NTS-1 (Timation III) [7.20]

The NTS-1 satellite was launched on 14 July 1974 into a nearly circular

Table 7.3 Solar Cell Array Degradation, Various Circular Orbits

Satellite Launch Date	Orbit Altitude, Inclination	Cells and Shielding Data	Equivalent 1 MeV Electron Fluence (Assuming trapped radiation only & infinite back shielding)	Predicted (from equivalent fluence)	Observed
OGO4 28 July 67	930 km (500 nmi) 86°	N/P 10 ohm-cm 0.015 cm micro-sheet	P: 3.7×10^{13} e/cm^2-yr E: 1.1×10^{12} e/cm^2-yr TOTAL: 3.8×10^{13} e/cm^2-yr	$\frac{I_{sc}}{I_{sco}} = 0.96$ @ 1 yr	0.96 @ 1 yr
1963-38C 28 Sept 63	1110 km (600 nmi) 90°	N/P 10 ohm-cm 0.015 cm micro-sheet	P: 7.9×10^{13} e/cm^2-yr E: 2.8×10^{12} e/cm^2-yr TOTAL: 8.2×10^{13} e/cm^2-yr	$\frac{I_{sc}}{I_{sco}} = 0.96$ @ 6 mo	0.95 @ 6 mo
ERS 6 1963-14C 9 May 63	4170 km (2250 nmi) 90°	N/P 1 ohm-cm 0.051 cm fused silica	P: 2.3×10^{15} e/cm^2-yr E: 9.4×10^{12} e/cm^2-yr TOTAL: 2.3×10^{15} e/cm^2-yr	$\frac{I_{sc}}{I_{sco}} = 0.79$ @ 6 mo $\frac{P_{max}}{P_{maxo}} = 0.67$ @ 6 mo	$\frac{P_{max}}{P_{maxo}} =$ 0.70 @ 6 mo
Explorer 38 (RAE 1) 4 July 68	6700 km (3600 nmi) 60°	N/P 10 ohm-cm 0.0473 cm solar cells 0.102 cm fused silica	P: 4.2×10^{14} e/cm^2-yr E: 3.5×10^{12} e/cm^2-yr TOTAL: 4.2×10^{14} e/cm^2-yr	$\frac{P_{max}}{P_{maxo}} = 0.77$ @ 1 yr $\frac{P_{max}}{P_{maxo}} = 0.725$ @ 2 yr	0.72 @ 1 yr 0.65 @ 2 yr
OSO-8 21 Jun 75	556 km (300 nmi) 33°	N/P 2 ohm-cm 0.030 cm 0.015 cm micro-sheet	P: 6.4×10^{12} e/cm^2-yr E: 3.0×10^{10} e/cm^2-yr TOTAL: 6.4×10^{12} e/cm^2-yr	$\frac{P_{max}}{P_{maxo}} = 0.97$ @ 1.7 yr	0.97 @ 1.7 yr

P: Proton contribution
E: Electron contribution

orbit having a perigee of 12,193 km, an apogee of 13,606 km (average of about 7000 nmi), and an inclination of 125.1°. The orbital radiation environment is severe and contains both electrons and protons. A solar cell flight experiment aboard carried conventional silicon Centralab and Heliotek cells, lithium-doped Centralab and Heliotek cells, Comsat violet, and Ferranti float-zone solar cells. Solar cell covers include Corning 7940 fused silica, Pilkington-Perkin Elmer ceria-doped microsheet and Corning 7070 integral coverglasses. Twelve different solar cell/coverglass combinations were represented. Twelve point I-V curves were taken for each of the twelve experiments.

The experimenters reported a number of experimental difficulties and anomalies which render applicable data analysis difficult at best. These effects are summarized below:

- The solar cells are connected into modules consisting of series strings of 5, 23, 47, or 48 cells. Therefore, the resulting data tends to be dominated by the lowest output or most severely degrading cell in each string.
- The initial I_{sc} data deviated from the ground calibration data by as much as 15%.
- The in-flight measured I_{sc} temperature coefficients after only a few months in orbit imply irradiation levels in the 10^{15} to 10^{16} e/cm^2 range (1 MeV), which is inconsistent with the cell output data.
- Spurious I_{sc} currents of 3 to 25 mA were observed in all the experimental modules, leading to uncertainties of as high as 20 to 25% in the telemetry data.
- Centralab lithium-doped cells indicated fill factor losses from an initial 0.75 to 0.47. The Spectrolab lithium-doped cell module failed completely after 261 days.
- Accurate calibration of the experiment during spacecraft integration was not achieved.

Using the techniques of Chapter 6, the 1 MeV equivalent electron fluence for short circuit current degradation of a cell protected by a 0.030-cm-thick coverglass was computed for this orbit. The cumulative effect of both protons and electrons was found to be 7×10^{13} e/cm^2-yr. The averaged observed I_{sc} degradations, however, are best fit with an equivalent 1 MeV electron fluence of 1×10^{15} e/cm^2 the first year and 5.7×10^{15} e/cm^2 the second year. The experimenters offer no explanation for the higher than expected solar cell degradations in this experiment, nor for the variation in annual exposure.

7.2.2.3 NTS-2 [7.21, 7.22]

The NTS-2 solar cell experiment, with 15 different types of solar cell/coverglass combinations, was launched 23 June 1977. The orbit is circular at an altitude of 20,192 km and inclination of 63°. The radiation environment for this orbit consists of trapped electrons. Trapped protons with energies greater than 1 MeV are negligible. The experiments were similar to those previously discussed for NTS-1 and consisted of strings of 5 cells in series for each experiment. Therefore, as previously stated, the possiblility of a single cell dominating the experiment string exists.

After three years in orbit some pertinent observations and conclusions can be made.
- ° The more advanced silicon solar cells utilizing the technologies of texturing, back surface fields, and shallow junctions exhibited higher end-of-life outputs than conventional cells.
- ° The (AlGa)As-GaAs solar cells retained a good power output exceeded only by two other types of cells, the OCLI violet cell and the Comsat textured cell.

There were, however, some discouraging observations which inhibit the drawing of quantitative conclusions. First, apparent UV adhesive degradations of 8 to 15% for experiments containing UV protective filters and 23% for an experiment containing no protective UV filter were reported. These degradations are significantly higher than the existing bulk of both flight and laboratory data on UV degradation. The implication is that some other unidentified factor is influencing the experiment.

Second, after deducting the nonradiation losses, the authors have concluded that the equivalent 1 MeV electron fluence in this orbit required to produce the observed degradations is approximately 3.4×10^{14} electrons/cm^2-yr for an 0.030-cm-thick coverglass. We note that this value is applicable only for fitting the first year data. An additional fluence of 1×10^{15} e/cm^2 e/cm^2-yr is required to fit the second year data and in the third year a fluence of 3×10^{15} e/cm^2-yr is required. These increasingly higher fluences are somewhat unusual.

The radiation environment for this orbit was predicted by Stassinopoulos [7.23] using the same environmental models discussed in Chapters 5 and 6. The environment was also computed with the same program used to calculate the data tabulated in Chapter 6. The two predictions, using the same radiation models but different orbit generation programs, produced fluence-energy spectra which agreed to better than 5%. The equivalent fluence based on these fluence-energy spectra and the damage coefficients of Chapter 4 is 4×10^{13} e/cm^2-yr, which is significantly lower than the apparent equivalent fluences required to fit the data.

Since this orbit is electron dominated just as are the synchronous orbits, it is interesting to compare this case with the synchronous environment. Table 6.6 gives a 1 MeV equivalent fluence of 1.72×10^{13} e/cm^2-yr for cells with 0.030-cm-thick coverglasses in synchronous orbit. This is only a factor of 2 less than calculated for the NTS-2 orbit. A comparison of the fluence-energy spectra for the two orbits reveals that the dominant energy range for solar cell degradation is between 1.0 and 2.0 MeV. In this energy range the NTS-2 orbit has twice as much electron fluence as the synchronous orbit. Fluences at higher energies are so much lower that they are negligible. Therefore, the radiation degradation experienced by the NTS-2 solar cells after 3 years in orbit should be equal to the degradation of equivalent solar cells in synchronous orbit after 6 years. This is clearly not the case.

The following have been offered as possible explanations for this discrepancy:

- The environmental models, AP8MAX and AE6MAX with AEI7-LO, are inaccurate when applied to this particular region in space.
- Temperature nonuniformities over the experimental solar panels together with the problems associated with connecting the cells in series may cause one or more low performing cells to dominate the string degradation. (The panels are running 20° hotter than predicted).
- The individual solar cell temperatures may not be well enough known to permit accurate temperature corrections to the solar cell electrical parameters.
- The AP8MAX proton model does predict a substantial number of protons with energies below 1 MeV. If the solar cell edges in this experiment were exposed, the low energy protons incident on the cell edges may be enhancing the degradation in P_{max} and V_{oc}.
- An unidentified degradation source may be influencing the solar cells in an unknown manner. Contaminant deposition on the cell/coverglass surface is an example of this kind of mechanism.

REFERENCES

7.1 R. C. Waddel, "Solar Cell Radiation Damage on Synchronous Satellite ATS-1," Conf. Rec. of the 7th Photovoltaic Specialists Conf., 195, 1968.

7.2 L. A. Gibson, "Solar Cell Array Degradation at Synchronous Orbit," Aerospace Corp. IOC No. 73.5242.17-1, July 1972.

7.3 W. T. Picciano, R. A. Reitman and R. J. Grant, "Solar Cell and Coverslide Degradations at Near-Synchronous Altitude," Conf. Rec. of the 8th IEEE Photovoltaic Specialists Conf., 221, 1970.

7.4 L. J. Goldhammer and S. W. Gelb, "Synchronous Orbit Performance of Hughes Aircraft Company Solar Arrays," Proc. 11th IECEC, 1379, September 1976.

7.5 L. J. Goldhammer and S. W. Gelb, "Synchronous Orbit Performance of Hughes Aircraft Company Solar Arrays - Update," 17th IECEC, August 1982.

7.6 J. W. Lyons, III, "Comparison of Computer-Predicted and In-Orbit Solar Array Performance for Geosynchronous Communications Satellites," 17th IECEC, August 1982.

7.7 L. T. Bavaro and H. Weiner, "FLTSATCOM Solar Array Degradation," 17th IECEC, August 1982.

7.8 H. Riess, private communication.

7.9 B. E. Anspaugh and R. G. Downing, "Damage Coefficients and Thermal Annealing of Irradiated Silicon and GaAs Solar Cells," Conf. Rec. of the 15th Photovoltaic Specialisits Conf., 499, 1981.

7.10 F. W. Sarles, Jr., A. G. Stanley, and C. Burrowes, "Solar Cell Calibration Experiments on LES-6," Conf. Rec. of the 8th IEEE Photovoltaic Specialists Conf., 262, 1968.

7.11 F. W. Sarles, Jr. "The LES-6 Solar Cell Experiment After Six Years," Conf. Rec. of the 11th IEEE Photovoltaic Specialists Conf., 199, 1975.

7.12 J. R. Carter, Jr. and H. Y. Tada, <u>Solar Cell Radiation Handbook</u>, TRW Report No. 21945-6001-RU-00, June 1973.

7.13 B. E. Anspaugh, "ATS-5 Solar Cell Experiment After 699 Days in Synchronous Orbit," Conf. Rec. of the 9th IEEE Photovoltaic Specialists Conf., 308, 1972.

7.14 B. E. Anspaugh, "The ATS-5 Solar Cell Experiment After 6-1/2 Years in Synchronous Orbit," Conf. Rec. of the 12th IEEE Photovoltaic Specialists Conf., 191, 1976.

7.15 L. J. Goldhammer and L. W. Slifer, Jr., "ATS-6 Solar Cell Flight Experiment Through 2-Years in Orbit," Conf. Rec. of the 12th IEEE Photovoltaic Specialists Conf., 199, 1976.

7.16 L. J. Goldhammer and L. W. Slifer Jr., "Summary Results of the ATS-6 Solar Cell Flight Experiment," Conf. Rec. of 14th IEEE Photovoltaic Specialists Conf., 870, 1980.

7.17 R. E. Fischell, "Solar Cell Power Systems for APL Satellites," Conf. Rec. of the 6th Photovoltaic Specialists Conf., Vol. II, 32, 1967.

7.18 J. M. Denney, "Final Flight Report, Tetrahedral Research Satellites," Vol. II, TRW Systems Report No. 8655-6006-RU-00, 15 February 1964.

7.19 G. R. Brooks, S. W. Gelb, and L. J. Goldhammer, "Orbiting Solar Observatory (OSO-8) Solar Panel Design and In-Orbit Performance," 13th IECEC, August 1982.

7.20 R. L. Statler and D. H. Walker, "The NTS-1 Solar Cell Experiment After Two Years in Orbit," Conf. Rec. of the 12th IEEE Photovoltaic Specialists Conf., 208, 1976.

7.21 R. L. Statler and D. H. Walker, "Three-Year Performance of the NTS-2 Solar Cell Experiment," Space Photovoltaic Research and Technology 1980, NASA Conf. Pub. 2169, NASA Lewis Research Center, 219, 1980.

7.22 D. H. Walker, "Performance of the Solar Cell Experiments Aboard the NTS-2 Satellite After Three Years in Orbit," NRL Memorandum Rept. 4580, July 1981.

7.23 E. Stassinopoulos, "Charged Particle Radiation Environment for NTS-2 and NTS-3 Satellites," NASA Goddard Space Flight Center, X-601-80-1, December 1979.

APPENDIX A

SHIELDING THICKNESS CONVERSION

Table A.1
Shielding Thickness Conversion

Areal Density	Fused Silica 2.2 g/cm^3		Microsheet 2.5 g/cm^3		Aluminum 2.7 g.cm^3	
g/cm^2	cm	in	cm	in	cm	in
0.0168	.00762	.003	0.00671	0.00264	.00621	0.00244
0.0335	.01524	.006	0.0134	0.00528	.01242	0.00489
0.0671	.0305	.012	0.0268	0.0106	.0248	0.00978
0.112	.0508	.020	0.0447	0.0176	.0414	0.0163
0.168	.0762	.030	0.0671	0.0264	.0621	0.0244
0.335	.1524	.060	0.1341	0.0528	.124	0.0489

APPENDIX B
CONSTANTS, PROPERTIES AND VALUES

SILICON

Atomic Weight	28.09
Density	2.33 (g/cm^3)
Crystal Structure	Diamond, 8 atoms/unit cell
Lattice Constant	5.43×10^{-10} m, 5.43 (Å)
Atomic Radius	1.18×10^{-10} m, 1.18 (Å)
Atomic Density	5.00×10^{22} (cm^{-3})
Energy Gap @ 300K	1.78×10^{-19} (J), 1.11 (eV)
Energy Gap @ 0 K	1.91×10^{-19} (J), 1.21 (eV)
Electron Mobility (intrinsic) @ 300K, μ_n	1350 (cm^2/V s)
Hole Mobility (intrinsic) @ 300K, μ_p	480 (cm^2/V s)
Electron Diffusion Constant (intrinsic) @ 300K, D_n	35 (cm^2/s)
Hole Diffusion Constant (intrinsic) @ 300K, D_p	12 (cm^2/s)
n_i @ 300K	1.5×10^{10} (cm^{-3})
Dielectric Constant	11.7
Specific Heat, C_p @ 300K	0.7 (J/g K)
Thermal Conductivity @ 300K	1.5 (W/cm K)
Coefficient of Thermal Expansion, $\frac{\Delta T}{L \Delta T}$	2.5×10^{-6} (K^{-1})
Debye Temperature	658 (K)
Activation Energy, Self Diffusion	7.7×10^{-19} (J), 4.8 (eV)
Energy of Ionization	5.76×10^{-19} (J), 3.6 (eV)
Energy of Sublimation	7.80×10^{-19} (J), 4.9 (eV)
Elastic Moduli	
C_{11}	1.674×10^{11} (N/m^2)
C_{12}	0.652×10^{11} (N/m^2)
C_{44}	0.796×10^{11} (N/m^2)
Index of Refraction	3.5-6.0 (See Figures B-1 and B-2)
Absorption Coefficient	$1-10^5$ (cm^{-1})(See Figure B-3)
Mohs' Hardness	7

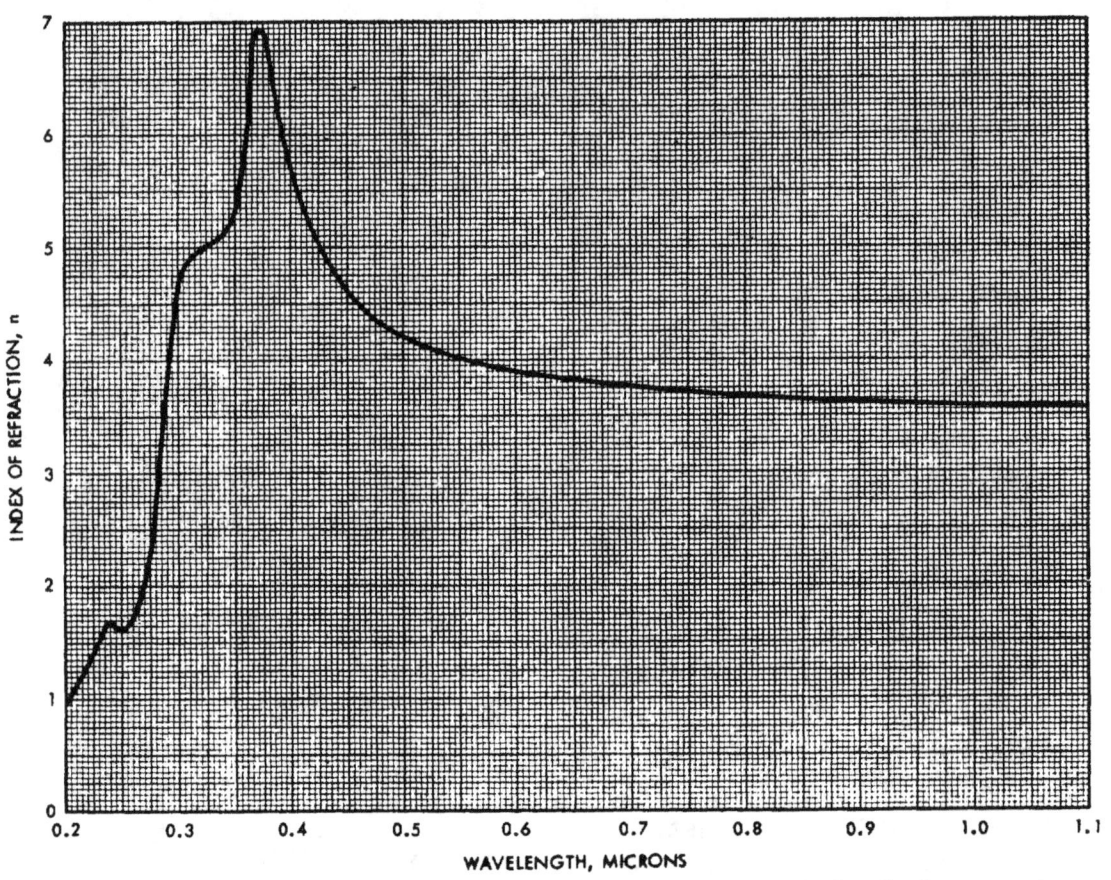

Figure B-1 Refractive Index of Silicon[B-1, B-2]

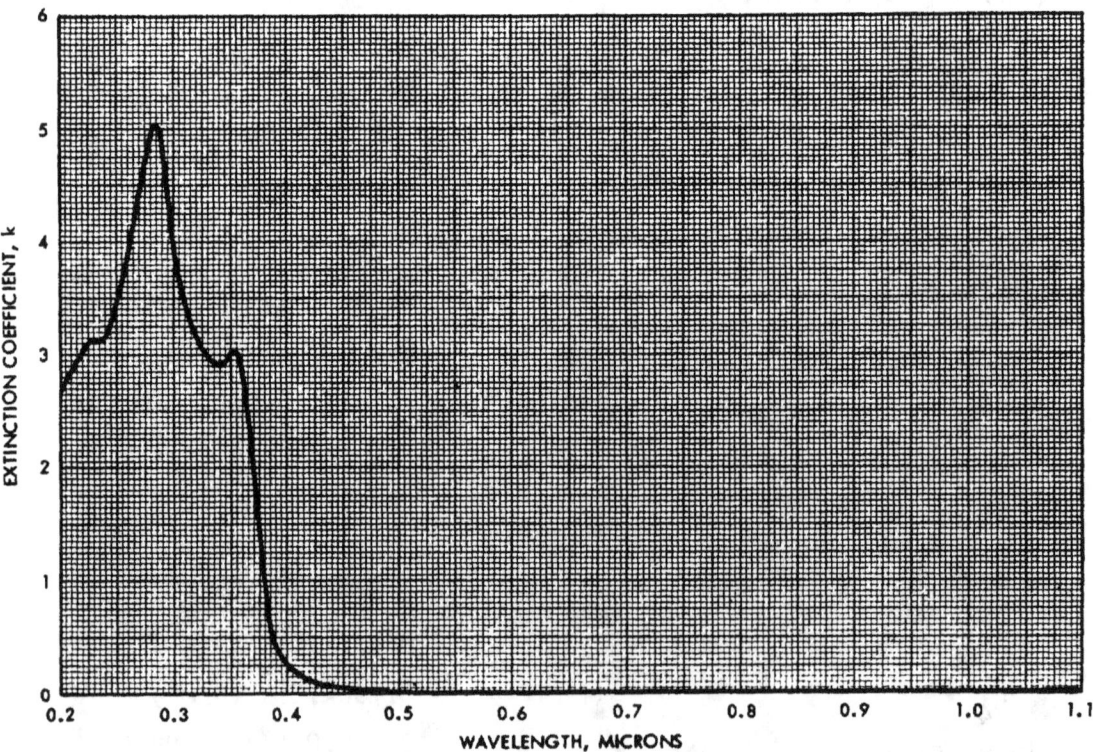

Figure B-2 Extinction Coefficient of Silicon[B-1, B-2]

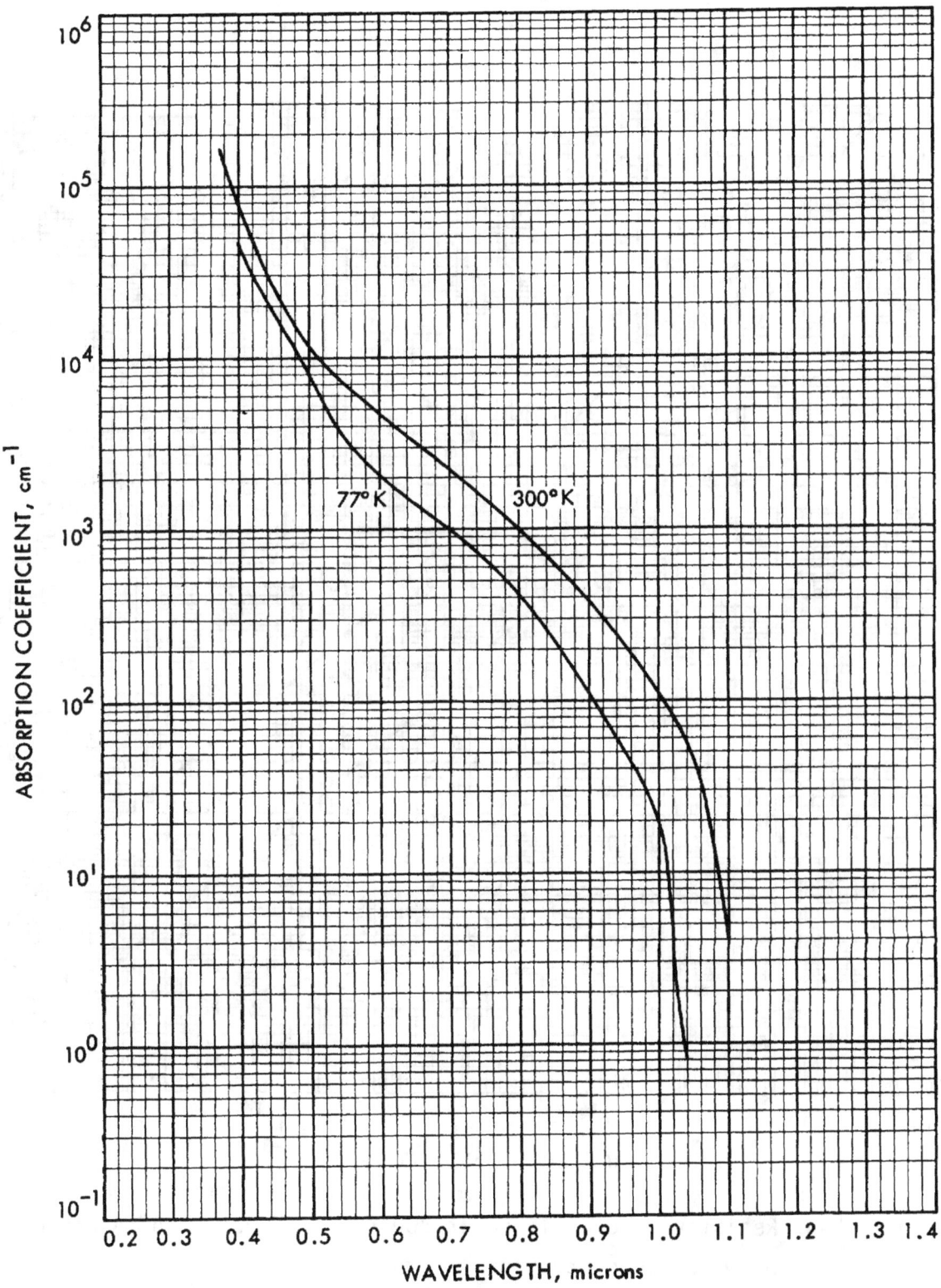

Figure B-3 Absorption Coefficient of Single Crystal Silicon at 77 and 300 K [B-3]

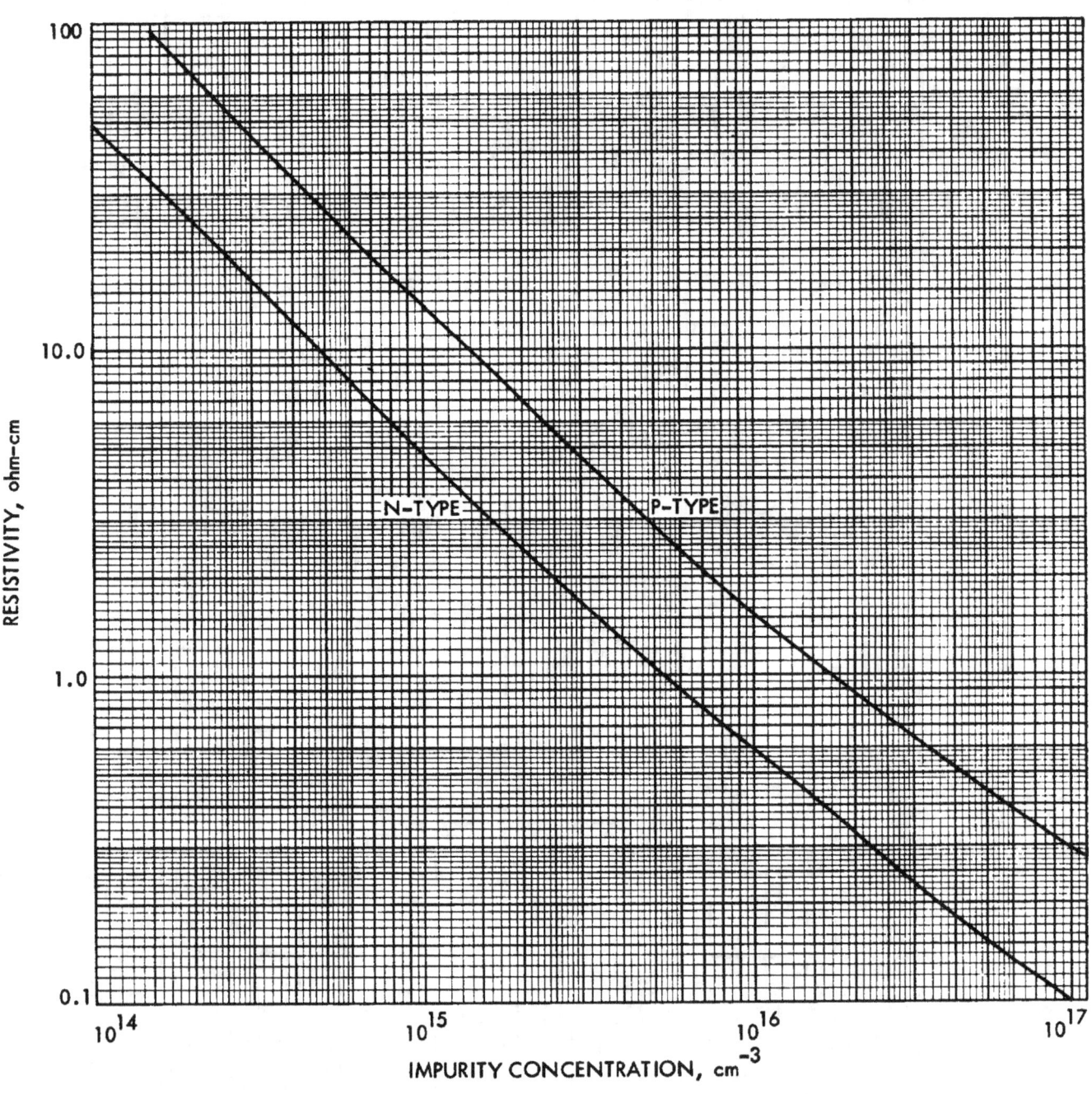

Figure B-4 Resistivity of Silicon at 300 K as a Function of Acceptor or Donor Concentration [B-4]

APPENDIX B (continued)

SILICON (continued)

Solar Absorptance	0.8
Hemispherical Emittance	0.3

QUARTZ GLASS (FUSED SILICA)

Molecular Weight	60.8
Density	2.2 (g/cm^3)
Energy Gap	12.8x10^{-19} (J), ≈8 (eV)
Dielectric Constant	3.5-3.9
Index of Refraction	1.46-1.51
Specific Heat, C_p	1 (J/g K)
Thermal Conductivity	0.014 (W/cm K)
Coefficient of Thermal Expansion, $\frac{\Delta T}{L \Delta T}$	0.55x10^{-6} (K^{-1})
Mohs' Hardness	4.9
Young's Modulus	7.16x10^{10} (N/m^2)
Rigidity Modulus	3.10x10^{10} (N/m^2)
Poisson's Ratio	0.16
Solar Absorptance	0.01
Hemispherical Emittance	0.78
Solar Absorptance (on array)	0.75-0.85
Hemispherical Emittance (on array)	0.78-0.80

SILICONE ELASTOMERS (TYPICAL)

Density	1.1 (g/cm^3)
Index of Refraction	1.41
Coefficient of Thermal Expansion, $\frac{\Delta T}{L \Delta T}$	300 x10^{-6} (K^{-1})
Thermal Conductivity @ 300K	.0017 (W/cm K)
Specific Heat @ 300K	1.0 (J/g K)
Bond Thickness between Coverglass and Solar Cell	75-150 (μm)

APPENDIX B (Continued)

SOME USEFUL PHYSICAL CONSTANTS

Boltzmann's Constant, k	1.3807×10^{-23} (J/K), 8.6171×10^{-5} (eV/K)
Planck's Constant, h	6.6262×10^{-34} (J-s)
Speed of Light, c	2.9979×10^{8} (m/s)
Electron Charge, e	1.6022×10^{-19} (C)
Permittivity of Free Space, ϵ_o	8.8542×10^{-12} (farad/m)
Permeability of Free Space, μ_o	1.2566×10^{-6} (henry/m)
Electron Rest Mass, m_e	9.1095×10^{-31} (kg)
Proton Rest Mass, m_p	1.6726×10^{-27} (kg)
Avogadro's Constant	6.0220×10^{23} (g/mole)
Photon Energy	$E(eV) = 1.23985/\lambda\ (\mu m)$
Thermal Energy	$kT = 0.025$ V at 300 K

SILICON SOLAR CELL DATA

Active Area of 2 cm x 2 cm Solar Cell	≈ 3.8 (cm^2)
Series Resistance	0.05 to 0.1 (ohm)
Shunt Resistance	> 1000 (ohm)

APPENDIX B (Continued)

Table B.1
Solar Cell Parameters for Si, 300 K [B-6]

	n/p Cells. $N_d = 5 \times 10^{19}$, $D_p = 1.295$, $\tau_p = 0.4 \times 10^{-6}$ s				
ρ_{base} (ohm-cm)	N_a (cm^{-3})	μ_n (cm^2/V-sec)	D_n (cm^2/sec)	τ_n (μsec)	L_n (microns)
10	1.25×10^{15}	1390	36	15	232
1	1.50×10^{16}	1040	27	10	164
0.1	$5. \times 10^{17}$	420	10.9	2.5	52
	p/n Cells. $N_a = 5 \times 10^{19}$, $D_n = 2.15$, $\tau_n = 1.1 \times 10^{-6}$ s				
ρ_{base} (ohm-cm)	N_d (cm^{-3})	μ_p (cm^2/V-sec)	D_p (cm^2/sec)	τ_p (μsec)	L_p (microns)
10	4.5×10^{14}	580	15	15	150
1	5.1×10^{15}	500	13	7.5	99
0.1	8.5×10^{16}	350	9	1.5	37

REFERENCES

B-1. H. R. Philipp and E. A. Taft, "Optical Constants of Silicon in the Region of 1 to 10 eV," Phys. Rev. $\underline{120}$, 1, 37, 1960.

B-2. D. E. Gray, Ed., American Institute of Physics Handbook, 6, 3rd Edition, McGraw, Hill, 1972.

B-3. W. C. Dash and R. Newman, "Instrinsic Optical Absorption in Single-Crystal Germanium and Silicon at 77 and 300°K," Phys. Rev., $\underline{99}$, 4, 1151, 1955.

B-4. J. C. Irvin, "Resistivity of Bulk Silicon and of Diffused Layers in Silicon," Bell System Technical Journal, $\underline{41}$, 387, March 1962.

B-5. B. N. Taylor, W. H. Parker, D. N. Langenberg, "Determination of e/h, Using Macroscopic Quantum Phase Coherence in Superconductors: Implications for Quantum Electrodynamics and the Fundamental Physical Constants," Rev. Mod. Phys., $\underline{41}$, 3, 375, July 1969.

B-6. H. J. Hovel, "Semiconductors and Semimetals, Vol. II: Solar Cells," 26, Academic Press, 1975.

APPENDIX C

Table C.1

Solar Cell Types

CELL TYPE	SHALLOW DIFFUSION	BSR	BSF	AR* COATING	TEXTURED	ALTERNATE DESIGNATION
Conventional				SiO		
K4	x			T		Hybrid, Violet
K4 1/2	x	x		T		Hybrid
K4 3/4	x	x		D		Hybrid
K5 3/4	x	x		D	x	
K6	x		x	T		Helios
K6 1/2	x	x	x	T		Helios
K6 3/4	x	x	x	D		Helios
K7	x	x	x	T or D	x	

*AR Coating Nomenclature
 T: Ta_2O_5
 D: Dual AR (TiO_x/Al_2O_3)

Table C.2

Approximate Performance Increases due to
Processing Variable Change

VARIABLE CHANGE	I_{sc} (%)	P_{max} (%)	V_{oc} (mV)
Ta_2O_5 to Dual AR*	5	5	--
No BSF to BSF			
(2 ohm-cm, 4 mil)	15	25	40
(2 ohm-cm, 8 mil)	5	9	25
(10 ohm-cm, 4 mil)	12	25	80
(10 ohm-cm, 8 mil)	5	15	60
(10 ohm-cm, 12 mil)	2	7	40
No BSR to BSR†	3	7	5
Planar to Textured	8	8	--

*Bare cells. Glassed cell increase ≈3%

†Varies with cell thickness

APPENDIX D

COMPUTER PROGRAM, EQFRUX

PROGRAM DESCRIPTION

This program will compute an equivalent fluence for a given space radiation environment for the purpose of estimating solar cell degradation. A geomagnetically trapped electron energy spectrum, or a proton energy spectrum, or both, can be input as the space radiation environment. With a proper choice of input parameter, a free space solar flare proton spectrum will be calculated with the use of a computer code developed by Stassinopoulos,[D-1] (named SOLPRO) based on King's solar proton model [D-2] for the years of 1977 through 1983.

The annual equivalent fluence for a given parameter p, as detailed in Section 6.1, is defined as follows:

$$\Phi^k_{equ}(p,t) = \sum_j D_{kj} \left\{ \sum_i D_j(p,E_i,t) \left[\Phi_j(E_i) - \Phi_j(E_{i+1}) \right] \right\} \tau \qquad (D-1)$$

where

$\Phi^k_{equ}(p,t)$ Equivalent fluence for solar cell output parameter p, normalized to kth particle in the presence of coverglass thickness t.

$\Phi_j(E_i)$ Integral flux of jth radiation particle at energy E_i.

$D_j(p,E_i,t)$ Relative damage coefficient (RDC) for solar cell output parameter p under radiation particle j at energy E_i in the presence of coverglass thickness t.

D_{kj} Radiation damage ratio between jth particle and normalized kth particle by which both particle and energy are normalized (conventionally 1 MeV electrons).

τ Conversion factor for annual fluence. If the integral spectrum is in units of fluence per day, for example, $\tau = 365.2422$.

The space radiation environments should be in a form of integral energy spectra of either electrons, protons or both. The required relative damage coefficients (RDC) for the fluence calculation are provided in BLOCK DATA. The short circuit current RDC's are provided for both electron and proton environments and the open circuit voltage RDC's for the proton environment. The RDC's are evaluated for omnidirectional flux and infinite back shielding.

Basically, all that is needed to run this program is (1) the alphanumeric input to identify the problem (or case run), and (2) either electron, proton, or both energy spectra as the radiation environment input in the namelist format, or proper input to determine the solar proton environment from the subroutine built in the program. In all cases, the equivalent fluence calculation follows.

A time unit of integral flux can be changed as necessary with a proper choice of conversion factor in order to obtain a desired exposure time. For interpolation of both RDC and integral flux, the energy entry of RDC data can be divided into any arbitrary number of points, NSTEP, for accuracy. An NSTEP of 2 to 4 is likely to produce an optimum result. The interpolation scheme is linear on a chosen scale, i.e., if the RDC data are plotted on a log-log scale, the interpolation is linear on this log-log scale; thus the RDC is expressed fragmentally in terms of power of energy. The integration limits of equation (D-1) can be controlled by INTFLG. When INTFLG = 0, the integration proceeds over all energies for which the RDC's are available, and the input spectra will be extrapolated if necessary. When INTFLG = 1, the integration extends only over the input energy range, thus is equivalent to a cutoff RDC at the lowest energy value of the input and an energy cutoff at the highest input value.

Input energy spectra are specified by variable names ESPEC for trapped electrons and PSPEC for trapped protons. The number of input data are indicated by NESPEC and NPSPEC, respectively, and are zero for no calculation. If NPSPEC = 1, a solar flare proton spectrum will be calculated from the subroutine named SOLPRO with two required inputs: (a) mission duration and (b) probabilistic confidence level. Note that the mission duration cannot be greater than 72 months and the confidence limit cannot be less than 80 percent. The damage ratio between 10 MeV protons and 1 MeV electrons for a given solar cell output parameter can also be altered with the use of PEDRI and PEDRV. Print flags, PCKE and PCKP, are provided to check detailed intermediate calculations.

<u>Input Variables</u>

HEADER   Alphanumeric 80 character description which prints as the first line of each output page.

<u>Namelist Variables</u>

NESPEC   Number of input points (maximum of 50) for omnidirectional electron energy spectra. No electron energy spectrum may be input if subroutine SOLPRO is used. If NESPEC = 0 is used, no equivalent fluence calculations for trapped electrons are performed.

NESPEC   Number of input points (maximum of 50) for omnidirectional proton energy spectra. If NESPEC = 1, solar flare proton flux is calculated from subroutine SOLPRO and subsequently the corresponding equivalent fluence. If SOLPRO is used, inputs TAU and IQ are required, otherwise they are disregarded.

No electron spectra may be input if subroutine SOLPRO is being used to calculate solar flare proton fluence.

ESPEC (I,J)   Integral energy spectrum of space electron environment. I ranges from one to NESPEC.
J = 1 for energy in MeV. J = 2 for integral fluence.

PSPEC (I,J)   Integral energy spectrum of space proton environment. I ranges from one to NPSPEC.
J = 1 for energy in MeV. J = 2 for integral fluence.

NSTEP   Number of points between energy entries of relative damage coefficients for interpolation (default value = 2).

TIMIN   12 character Hollerith string which describes the time interval represented by the input spectra. For example if the input spectra are in units of fluence per day
      TIMIN = 12HDAY
If the input spectra are in units of fluence per second
      TIMIN = 12HSECOND

TMULT   Number of "TIMIN" units for which eqivalent fluence is to be computed. For example if input spectra represent fluence per hour and equivalent fluence is to be computed for 24 hours
      TMULT = 24.
(TMULT should be input such that TMULT = TIMOUT/TIMIN)

TIMOUT 12 character Hollerith string which describes total time of exposure and is the product of "TIMIN" units and TMULT. For example if TIMIN = "1 day" and TMULT = 365.2422
TIMOUT = 12H1 Year
If TIMIN = "1 Month" and mission duration is 34.2 months
TIMOUT = 12H34.2 MONTHS
or TIMOUT = 12H1 MISSION
Default values are: TIMOUT = 12H1 YEAR, TMULT = 365.2422
TIMIN = 12HDAY

PEDRI Damage ratio between protons and electrons for I_{sc}. (Default value = 3000).

PEDRV Damage ratio between protons and electons for V_{oc}. (Default value = 3000).

TAU Mission duration in months (used by subroutine SOLPRO).

IQ Confidence level that determines solar flare proton flux (used by subroutine SOLPRO).

PCKE, PCKP Flags to cause printing of differential fluence, damage coefficients, equivalent fluence, etc. for electrons (PCKE) and/or protons (PCKP). 8 values for each variable may be input corresponding to coverglass thicknesses (default value = 0 for no print, set = 1 for print).

IDIAG Flag to print namelist input as a diagnostic aid. (Default value = 0 for no print, set = 1 for print).

INTFLG Flag to establish limits of integration. Proceeds over all energies for which damage coefficients are available and input spectra are extrapolated if necessary. When INTFLG = 1 integration proceeds only over the input energy range, namely, energy intervals ESPEC (1,1) to ESPEC (NESPEC,1) and PSPEC(1,1) to PSPEC(NPSPEC,1) default value = 0. Therefore, the RDC's are regarded as 0 for the energies less than ESPEC(1,1) and PSPEC(1,1), respectively, and an energy cutoff for energies higher than ESPEC(NESPEC,1) and PSPEC(NPSPEC,1), respectively.

REFERENCES

D-1. E. G. Stassinopoulos, "SOLPRO: A Computer Code to Calculate Probabilistic Energetic Solar Proton Fluences," NASA, NSSDC 75-11, 1975.

D-2. J. H. King, "Solar Proton Fluences for 1977-1983 Space Missions," J. Spacecraft and Rockets, 11, 6, 401, June 1974.

```
      EQFLUX PROGRAM
 1    C    ****************************************************************
 2    C    *                                                              *
 3    C    *   PROGRAM FOR COMPUTING EQUIVALENT FLUENCE FROM SPACE ELECTRON *
 4    C    *   AND PROTON ENERGY SPECTRA AND RELATIVE DAMAGE COEFFICIENTS *
 5    C    *   FOR THE PURPOSE OF ESTIMATING SOLAR CELL DEGRADATION.      *
 6    C    *                                                              *
 7    C    *                                                              *
 8    C    * MACHINE / FORTRAN FEATURES NECESSARY                         *
 9    C    *       NAMELIST INPUT/OUTPUT                                  *
10    C    *       INPUT UNIT (CARD READER) IS FORTRAN UNIT 5             *
11    C    *       OUTPUT UNIT (PRINTER) IS FORTRAN UNIT 6                *
12    C    *       BLOCK DATA SUBPROGRAM                                  *
13    C    *       PROGRAM WRITTEN FOR UNIVAC 1108 (FORTRAN 4 COMPATIBLE) *
14    C    *       ALPHANUMERIC INPUT/OUTPUT ( 'A' FORMAT ) ASSUMES       *
15    C    *         6-CHARACTER CAPABILITY.  HOLLERITH STRINGS ARE USED AS *
16    C    *         CHARACTER COUNT ( 5HABCDE ) AND AS QUOTE STRINGS.    *
17    C    ****************************************************************
18    C
19         COMMON/DAMAGE/EMEV(70),EDET(70,8),PMEV(70),PISC(70,8),PVOC(70,8)
20    C
21         DIMENSION TIMIN(2), TIMOUT(2)
22         DIMENSION HEADER(14),THICK(8),COND(2)
23         DIMENSION ED(62),PI(65),PV(65)
24         DIMENSION ESPEC(70,2),PSPEC(70,2)
25         DIMENSION EQUIVE(8),EQV10I(8),EQV10V(8)
26         DIMENSION EMLN(70),PMLN(70)
27         DIMENSION EPTOTV(8),EPTOTI(8)
28         DIMENSION ESPLN(70,2),PSPLN(70,2)
29         DIMENSION TTHICK(8),ITHICK(8)
30    C
31         INTEGER PAGE,PCKE(8),PCKP(8)
32    C
33         DATA THICK/0.,5.69E-3,1.68E-2,3.35E-2,6.71E-2,1.12E-1,1.675E-1,
34        *3.35E-1/
35         DATA NESPEC/0/,NPSPEC/0/,NSTEP/2/,IDIAG/0/
36         DATA PEDRI,PEDRV/3000.,3000./,PCKE,PCKP/16*0/,INTFLG/0/
37         DATA TIMIN/12HDAY         /, TIMOUT/12H1 YEAR      /,
38        * TMULT/365.2422/
39    C
40         NAMELIST /MIKE/ NESPEC,ESPEC,NPSPEC,PSPEC,NSTEP,TIMIN,TIMOUT
41        *,TMULT,PEDRI,PEDRV,TAU,IQ,PCKE,PCKP,IDIAG,INTFLG
42    C
43    C    ****************************************************************
44    C    *   INPUT VARIABLES . . .                                      *
45    C    *                                                              *
46    C    *   HEADER    ALPHANUMERIC RECORD (80 CHARACTERS) TO IDENTIFY CASE.*
47    C    *                                                              *
48    C    *   THE FOLLOWING ARE NAMELIST VARIABLES                       *
49    C    *   PUNCH NAMELIST ITEMS STARTING IN COLUMN 2                  *
50    C    *                                                              *
51    C    *   NESPEC, NPSPEC   NUMBER OF INPUT DATA FOR ELECTRON AND PROTON *
52    C    *          ENERGY SPECTRA.  IF NPSPEC=1 SOLAR FLARE PROTON FLUX IS *
53    C    *          CALCULATED FROM SUBROUTINE SOLPRO AND SUBSEQUENTLY THE *
54    C    *          CORRESPONDING EQUIVALENT FLUENCE, INSTEAD OF CALCULATING *
55    C    *          EQUIVALENT FLUENCE DUE TO AN INPUT PROTON SPECTRUM.  *
56    C    *          IF SOLPRO IS USED, INPUTS TAU AND IQ ARE REQUIRED,  *
57    C    *          OTHERWISE THEY ARE DISREGARDED.  PROGRAM IS CURRENTLY *
```

```
 68    C    *            DIMENSIONED FOR A MAXIMUM OF 50 ELECTRON AND 50 PROTON     *
 69    C    *            SPECTRAL VALUES.                                           *
 60    C    *                 NOTE:  NO ELECTRON SPECTRA MAY BE INPUT IF            *
 61    C    *                        SUBROUTINE SOLPRO IS BEING USED TO             *
 62    C    *                        CALCULATE SOLAR FLARE PROTON FLUENCES.         *
 63    C    *            (SEE NSSDC PUBLICATION 75-11 (STASSINOPOULOS) FOR          *
 64    C    *            DETAILS OF SUBROUTINE SOLPRO.)                             *
 65    C    *     ESPEC(I,J),PSPEC(I,J)   INTEGRAL ENERGY SPECTRUM OF SPACE         *
 66    C    *            ELECTRON AND PROTON ENVIRONMENTS. J=1  ENERGY IN MEV.      *
 67    C    *            J=2  INTEGRAL FLUX IN PARTICLES PER SQUARE CENTIMETER      *
 68    C    *            PER UNIT TIME. INPUT SPECTRAL DATA IN ASCENDING ORDER,     *
 69    C    *            LOWEST ENERGY FIRST, HIGHEST LAST.                         *
 70    C    *     NSTEP   NUMBER OF POINTS BETWEEN ENERGY ENTRIES OF RELATIVE       *
 71    C    *            DAMAGE COEFFICIENTS FOR INTERPOLATION (DEFAULT VALUE = 2)  *
 72    C    *     TIMIN  12 CHARACTER HOLLERITH STRING WHICH DESCRIBES TIME         *
 73    C    *            INTERVAL REPRESENTED BY INPUT SPECTRA.  FOR EXAMPLE IF     *
 74    C    *            INPUT SPECTRA REPRESENT FLUENCES PER DAY                   *
 75    C    *                 TIMIN = 12HDAY                                        *
 76    C    *            IF INPUT SPECTRA REPRESENT FLUENCE PER SECOND              *
 77    C    *                 TIMIN = 12HSECOND                                     *
 78    C    *     TMULT  NUMBER OF 'TIMIN' UNITS FOR WHICH EQUIVALENT FLUENCE       *
 79    C    *            IS TO BE COMPUTED.  FOR EXAMPLE IF INPUT SPECTRA           *
 80    C    *            REPRESENT FLUENCE PER HOUR AND EQUIVALENT FLUENCE IS       *
 81    C    *            TO BE COMPUTED FOR 24 HOURS                                *
 82    C    *                 TMULT = 24.                                           *
 83    C    *            (TMULT SHOULD BE INPUT SUCH THAT TMULT = TIMOUT/TIMIN)     *
 84    C    *     TIMOUT  12 CHARACTER HOLLERITH STRING WHICH DESCRIBES TOTAL       *
 85    C    *            TIME OF EXPOSURE AND IS THE PRODUCT OF 'TIMIN' UNITS       *
 86    C    *            AND TMULT.  FOR EXAMPLE IF TIMIN = '1 DAY' AND             *
 87    C    *            TMULT = 365.2422                                           *
 88    C    *                 TIMOUT = 12H1 YEAR                                    *
 89    C    *            ------------------------------------------------           *
 90    C    *            INCLUDE ALL 12 CHARACTERS IN THE NAMELIST INPUT            *
 91    C    *               INCLUDING TRAILING BLANKS.                              *
 92    C    *            ------------------------------------------------           *
 93    C    *            IF TIMIN = '1 MONTH' AND MISSION DURATION IS 34.2          *
 94    C    *            MONTHS                                                     *
 95    C    *                 TIMOUT = 12H34.2 MONTHS                               *
 96    C    *            OR   TIMOUT = 12H1 MISSION                                 *
 97    C    *            ------------------------------------------------           *
 98    C    *            INCLUDE ALL 12 CHARACTERS IN THE NAMELIST INPUT            *
 99    C    *               INCLUDING TRAILING BLANKS.                              *
100    C    *            ------------------------------------------------           *
101    C    *     **NOTE**  DEFAULT VALUES ARE:                                     *
102    C    *            TIMIN = 12HDAY                                             *
103    C    *            TMULT = 365.2422                                           *
104    C    *            TIMOUT = 12H1 YEAR                                         *
105    C    *     PEDRI   DAMAGE RATIO BETWEEN PROTONS AND ELECTRONS FOR ISC.       *
106    C    *            (DEFAULT VALUE = 3000.)                                    *
107    C    *     PEDRV   DAMAGE RATIO BETWEEN PROTONS AND ELECTRONS FOR VOC        *
108    C    *            (DEFAULT VALUE = 3000.)                                    *
109    C    *     TAU   MISSION DURATION IN MONTHS (USED BY SUBROUTINE SOLPRO).     *
110    C    *     IQ    CONFIDENCE LEVEL THAT DETERMINES SOLAR FLARE PROTON         *
111    C    *            FLUX (USED BY SUBROUTINE SOLPRO).                          *
112    C    *     PCKE, PCKP   FLAGS TO CAUSE PRINTING OF DIFFERENTIAL FLUENCE,     *
113    C    *            DAMAGE COEFFICIENTS, EQUIVALENT FLUENCE, ETC. FOR          *
114    C    *            ELECTRONS (PCKE) AND/OR PROTONS (PCKP).  8 VALUES FOR      *
```

```
115   C      *           EACH VARIABLE MAY BE INPUT CORRESPONDING TO COVER GLASS  *
116   C      *           THICKNESSES (DEFAULT VALUE=0 FOR NO PRINT.               *
117   C      *                          SET=1 FOR PRINT.)                         *
118   C      * IDIAG   FLAG TO PRINT NAMELIST INPUT AS A DIAGNOSTIC AID.          *
119   C      *         (DEFAULT VALUE = 0 FOR NO PRINT.  SET = 1 FOR PRINT.)      *
120   C      * INTFLG  FLAG TO ESTABLISH LIMITS OF INTEGRATION                    *
121   C      *           WHEN INTFLG = 0 INTEGRATION PROCEEDS OVER ALL ENERGIES   *
122   C      *           FOR WHICH DAMAGE COEFICIENTS ARE AVAILABLE AND INPUT     *
123   C      *           SPECTRA ARE EXTRAPOLATED IF NECESSARY.                   *
124   C      *           WHEN INTFLG = 1 INTEGRATION PROCEEDS ONLY OVER THE INPUT *
125   C      *           ENERGY RANGE, NAMELY, ENERGY INTERVALS ESPEC(1,1) TO     *
126   C      *           ESPEC(NESPEC,1) AND PSPEC(1,1) TO PSPEC(NPSPEC,1)        *
127   C      *           DEFAULT VALUE = 0                                        *
128   C      *                                                                    *
129   C      *                                                                    *
130   C      **********************************************************************
131   C
132   C
             PAGE=0
133
134   C
135   C      READ HEADER CARD (IDENTIFIER INFORMATION)
136   C
137       100 READ(5,20,END=9999) HEADER
138   C
139   C      INITIALIZE TOTAL FLUENCE VECTORS
140   C
141          DO 11 I=1,8
142          EPTOTV(I)=0.
143       11 EPTOTI(I)=0.
144   C
145   C      READ INPUT DATA (NAMELIST 'MIKE')
146   C
147          READ(5,MIKE)
148          IF(IDIAG .EQ. 0) GO TO 12
149          PAGE=PAGE+1
150          WRITE(6,25) HEADER,PAGE
151          WRITE(6,MIKE)
152       12 CONTINUE
153          IF(NESPEC .EQ. 0) GO TO 105
154   C
155   C      BYPASS IF NO ELECTRON SPECTRUM
156   C
157          PAGE=PAGE+1
158          WRITE(6,25) HEADER,PAGE
159          WRITE(6,32)TIMIN,((ESPEC(I,J),J=1,2),I=1,NESPEC)
160   C
161   C      TAKE LOGS OF ELECTRON FLUENCES
162   C
163          DO 101 J=1,NESPEC
164      101 ESPLN(J,2) = ALOG(ESPEC(J,2))
165      105 IF(NPSPEC .EQ. 0) GO TO 107
166          IF(NPSPEC .GT. 1) GO TO 104
167   C
168   C      CALCULATE SOLAR FLARE PROTON SPECTRUM BASED ON TAU AND IQ USING
169   C         SUBROUTINE SOLPRO
170   C
171          CALL SOLPRO(TAU,IQ,PSPEC(1,2),PSPEC(1,1),IOR,IERR)
```

```
            IF(IERR .GT. 0) GO TO 100
            PAGE=PAGE+1
            IF(IOR .GT. 0) GO TO 405
            COND(1)='  ORDI'
            COND(2)='NARY'
            WRITE (6,271) COND, PAGE
            GO TO 407
        405 COND(1)=' ANOMA'
            COND(2)='LOUS'
            WRITE (6,27) IOR,COND,PAGE
        407 WRITE(6,272) TAU,IQ
            WRITE(6,37) (PSPEC(I,1),PSPEC(I,2),I=1,10)
            TMULT=1.
            NPSPEC=10
            GO TO 1041
        104 CONTINUE
            PAGE=PAGE+1
            WRITE(6,25) HEADER,PAGE
            WRITE(6,33)TIMIN,((PSPEC(I,J),J=1,2),I=1,NPSPEC)
       1041 CONTINUE
C
C     TAKE LOGS OF PROTON ENERGIES AND FLUENCES
C
            DO 106 J=1,NPSPEC
            PSPLN(J,1) = ALOG(PSPEC(J,1))
        106 PSPLN(J,2) = ALOG(PSPEC(J,2))
        107 DO 9000 L=1,8
C
C     TAKE LOGS OF RELATIVE DAMAGE COEFFICIENTS AND RELATED ENERGIES
C
            IF(NESPEC .EQ. 0) GO TO 190
            DO 187 K=1,47
            IF(L .GT. 1) GO TO 181
            EMLN(K)=ALOG(EMEV(K))
        181 IF(EDET(K,L))183,183,185
        183 ED(K)=-50.
            GO TO 187
        185 ED(K)=ALOG(EDET(K,L))
        187 CONTINUE
        190 IF(NPSPEC .EQ. 0) GO TO 200
            DO 150 K=1,65
            IF(L .GT. 1) GO TO 125
            PMLN(K)=ALOG(PMEV(K))
        125 IF(PISC(K,L))130,130,135
        130 PI(K)=-50.
            GO TO 140
        135 PI(K)=ALOG(PISC(K,L))
        140 IF(PVOC(K,L))145,145,147
        145 PV(K)=-50.
            GO TO 150
        147 PV(K)=ALOG(PVOC(K,L))
        150 CONTINUE
C
C     COMPUTE EQUIVALENT FLUENCE FOR ELECTRON SPECTRUM
C        (BYPASS IF NO ELECTRON SPECTRUM)
C
        200 LINE=1
```

```
229            IF(NESPEC .EQ. 0) GO TO 400
230            EQUIVE(L) = 0.0
231            ELLIM = ESPEC(1,1)
232            EULIM = ESPEC(NESPEC,1)
233     C
234     C      ITERATE OVER ALL ENERGY INCREMENTS
235     C
236            DO 300 K=1,46
237            DIFF=EMLN(K+1)-EMLN(K)
238            DELTA=DIFF/NSTEP
239            DEL2=DELTA/2.
240            DO 300 I=1,NSTEP
241            SPEC1=EMLN(K)+DELTA*(I-1)
242            DSPEC=SPEC1+DEL2
243            EK=EXP(SPEC1)
244            EK1=EXP(SPEC1+DELTA)
245     C
246     C      PERFORM LINEAR INTERPOLATION OF PHI VS. E (SEMI-LOG)
247     C
248            CALL INTP(EK,PHI1,ESPEC(1,1),ESPLN(1,2),NESPEC)
249            CALL INTP(EK1,PHI2,ESPEC(1,1),ESPLN(1,2),NESPEC)
250            PHI1 = EXP(PHI1)
251            PHI2 = EXP(PHI2)
252     C
253     C      DAMAGE COEFFICIENT VS. E (LOG-LOG)
254     C      PERFORM LINEAR INTERPOLATION OF
255     C
256            CALL INTP(DSPEC,D1,EMLN(1),ED(1),47)
257            D=EXP(D1)
258            IF(D .LT. 1.E-4) D=0.0
259     C
260     C      USE RESTRICTED INTEGRATION LIMITS IF INTFLG .GT. 0
261     C
262            IF ( INTFLG .EQ. 0 ) GO TO 201
263            IF(EK .LT. ELLIM .OR. EK1 .GT. EULIM) GO TO 202
264            GO TO 201
265        202 PHI1 = 0.0
266            PHI2 = 0.0
267        201 DPHI = PHI1 - PHI2
268            PROD = DPHI * D
269     C
270     C      SUM PRODUCTS OVER ALL ENERGY INCREMENTS
271     C
272            EQUIVE(L) = EQUIVE(L) + PROD
273            IF(PCKE(L) .EQ. 0) GO TO 300
274     C
275     C      PRINT INTERMEDIATE CALCULATIONS OF DIFFERENTIAL FLUX, RELATIVE
276     C        DAMAGE COEFFICIENT, AND EQUIVALENT FLUENCE
277     C
278            IF(LINE .NE. 1) GO TO 50
279            PAGE=PAGE+1
280            WRITE(6,25) HEADER,PAGE
281            WRITE(6,26) THICK(L)
282            WRITE(6,30)
283         50 DSPEC1=EXP(DSPEC)
284            WRITE(6,10)EK,EK1,PHI1,PHI2,DPHI,D,DSPEC1,PROD,EQUIVE(L)
285            LINE=LINE+1
```

```fortran
286              IF(LINE .GE. 50) LINE=1
287          300 CONTINUE
288       C
289       C     COMPUTE EQUIVALENT FLUENCE FOR PROTRON SPECTRUM
290       C     (BYPASS IF NO PROTON SPECTRUM)
291       C
292          400 IF(NPSPEC .EQ. 0) GO TO 9000
293              LINE=1
294              EQV10I(L) = 0.0
295              EQV10V(L) = 0.0
296              PLLIM = ALOG(PSPEC(1,1))
297              PULIM = ALOG(PSPEC(NPSPEC,1))
298              DO 500 K=1,64
299              DIFF=PMLN(K+1)-PMLN(K)
300              DELTA=DIFF/NSTEP
301              DEL2=DELTA/2.
302              DO 500 I=1,NSTEP
303              SPEC1=PMLN(K)+DELTA*(I-1)
304              SPEC2=SPEC1+DELTA
305              DSPEC=SPEC1+DEL2
306       C
307       C     PERFORM LINEAR INTERPOLATION OF PHI VS. E (LOG-LOG)
308              CALL INTP(SPEC1,PHI1,PSPLN(1,1),PSPLN(1,2),NPSPEC)
309              CALL INTP(SPEC2,PHI2,PSPLN(1,1),PSPLN(1,2),NPSPEC)
310              PHI1 = EXP(PHI1)
311              PHI2 = EXP(PHI2)
312       C
313       C
314       C     PERFORM LINEAR INTERPOLATION OF DAMAGE COEFFICIENT VS. E(LOG-LOG)
315              CALL INTP(DSPEC,DCI,PMLN(1),PI(1),65)
316              CALL INTP(DSPEC,DCV,PMLN(1),PV(1),65)
317              DISC=EXP(DCI)
318              DVOC=EXP(DCV)
319              IF(DISC .LT. 1.E-4) DISC=0.0
320              IF(DVOC .LT. 1.E-4) DVOC=0.0
321              IF(INTFLG .EQ. 0) GO TO 401
322       C
323       C     USE RESTRICTED INTEGRATION LIMITS IF INTFLG .GT. 0
324       C
325              IF(SPEC1 .LT. PLLIM .OR. SPEC2 .GT. PULIM) GO TO 402
326              GO TO 401
327          402 PHI1 = 0.0
328              PHI2 = 0.0
329          401 DPHI = PHI1 - PHI2
330              PROD1=DPHI*DISC
331              EQV10I(L) = EQV10I(L) + PROD1
332              EQV10V(L) = EQV10V(L) + DPHI*DVOC
333              IF(PCKP(L) .EQ. 0) GO TO 500
334              IF(LINE .NE. 1) GO TO 60
335              PAGE=PAGE+1
336              WRITE(6,25) HEADER,PAGE
337              WRITE(6,41) THICK(L)
338              WRITE(6,40)
339           60 EK=EXP(SPEC1)
340              EK1=EXP(SPEC1+DELTA)
341              DFXDCV = DPHI*DVOC
342              DSPEC1=EXP(DSPEC)
```

```
            WRITE(6,10)EK,EK1,PHI1,DPHI,DISC,DVOC,DSPEC1,PROD1,DFXDCV,
           *EQV10I(L),EQV10V(L)
            LINE=LINE+1
            IF(LINE .GE. 50) LINE=1
        600 CONTINUE
       9000 CONTINUE
C
C     PRINT CALCULATION SUMMARY
C
            PAGE=PAGE+1
            WRITE(6,25) HEADER,PAGE
            WRITE(6,2) (THICK(J),J=1,8)
            DO 520 J=1,8
            TTHICK(J)=THICK(J)*178.9008766+.5
            ITHICK(J)=TTHICK(J)
        520 CONTINUE
            WRITE(6,22)(ITHICK(J),J=1,8)
            DO 1000 K=1,8
            EQUIVE(K) = EQUIVE(K) * TMULT
C
C     CONVERT 10 MEV PROTONS TO EQUIVALENT 1 MEV ELECTRONS USING PEDRV
C        AND PEDRI
C
            EQV10I(K) = EQV10I(K) * TMULT * PEDRI
            EQV10V(K) = EQV10V(K) * TMULT * PEDRV
       1000 CONTINUE
            IF(NESPEC .EQ. 0) GO TO 2000
            WRITE(6,3) (EQUIVE(J),J=1,8)
            DO 2001 I=1,8
            EPTOTV(I)=EPTOTV(I)+EQUIVE(I)+EQV10V(I)
            EPTOTI(I)=EPTOTI(I)+EQUIVE(I)+EQV10I(I)
       2001 CONTINUE
       2000 IF(NPSPEC .EQ. 0) GO TO 3000
            WRITE(6,4) (EQV10V(J),J=1,8)
            WRITE(6,5) (EQV10I(J),J=1,8)
            IF(NESPEC .EQ. 0) GO TO 3000
            WRITE(6,28)
            WRITE(6,29) (EPTOTV(J),J=1,8)
            WRITE(6,31) (EPTOTI(J),J=1,8)
       3000 CONTINUE
            WRITE(6,43) TIMOUT,TMULT
            GO TO 100
C
C
          2 FORMAT(1H0,'SHIELD THICKNESS (GM/CM2)',4X8(1PE10.3))
          3 FORMAT(1H0,'ELECTRON FLUENCE'/1H ,2X'EQUIV 1 MEV ELECTRONS/CM2',
           *  2X8(1PE10.3))
          4 FORMAT(1H0,'PROTON FLUENCE'/1H ,2X'EQUIV 1 MEV ELECTRONS/CM2'/
           *  1H ,11X'PMAX VOC',10X8(1PE10.3))
          5 FORMAT(1H ,16X'ISC',10X8(1PE10.3))
         10 FORMAT(11E12.4)
         20 FORMAT(13A6,A2)
         22 FORMAT(1H ,17X,'( MILS ) ',8I10)
         25 FORMAT(1H1,14A6,16X4HPAGE,I4/)
         26 FORMAT(1H ,'(ELECTRON SPECTRUM)',10X'COVER SLIDE THICKNESS =',
           *  F10.5,' GM/CM2'/)
         27 FORMAT(1H1,31HSOLAR FLARE PROTON SPECTRUM FOR,I2,1X,A6,A5,
```

```fortran
      *   'EVENT(S)'51X,4HPAGE,1X,I3/)
  271 FORMAT (1H1,'SOLAR FLARE PROTON SPECTRUM FOR',A6,A5,'EVENT',
      * 51X,'PAGE',1X,I3/)
  272 FORMAT (1H ,5X,17HMISSION DURATION=, F5.1,8H MONTHS.
      * /5X,17HCONFIDENCE LEVEL=, I3, 9H PERCENT.
      * //13X,6HENERGY,10X,13HINTEGRAL FLUX
      * /14X,5H(MEV),7X,20HPROTONS/CM2-MISSION. /)
   28 FORMAT(1H0,'TOTAL FLUENCE (ELECTRONS + PROTONS)'/
      * 1H ,2X'EQUIV 1 MEV ELECTRONS/CM2')
   29 FORMAT(1H ,11X'PMAX VOC',10X8(1PE10.3))
   30 FORMAT(5X,3HEK ,9X,3HEK1,9X,3HFX1,9X,3HFX2,9X,3HDFX,9X
      * ,3HDCI,9X,7HEINTERP,5X,7HDFX*DCI,5X,6HEQFLUX / )
   31 FORMAT(1H ,16X'ISC',10X8(1PE10.3))
   32 FORMAT(1H0,26X,'ELECTRON'/
      *1H ,10X,'ENERGY',10X,'FLUENCE'/
      *1H ,10X, '(MEV)',11X,'(ELECTRONS/CM2-',2A6,')'//
      *(1H ,0PF16.3,1PE18.4))
   33 FORMAT(1H0,26X,'PROTON'/
      *1H ,10X,'ENERGY',10X,'FLUENCE'/
      *1H ,10X, '(MEV)',11X,'(PROTONS/CM2-',2A6,')'//
      *(1H ,0PF16.3,1PE18.4))
   37 FORMAT(0PF20.3,1PE20.4)
   40 FORMAT(5X,2HEK,10X,3HEK1,9X,3HFX1,9X,3HDFX,9X,3HDCI,9X,3HDCV,
      *9X,7HEINTERP,5X,7HDFX*DCI,5X,7HDFX*DCV,5X,4HEQFI,8X,4HEQFV / )
   41 FORMAT(1H ,'(PROTON SPECTRUM)',10X,'COVER SLIDE THICKNESS =',
      * F10.5,' GM/CM2'/)
   43 FORMAT(1H0,'TIME OF EXPOSURE:   ',2A6/3X,'(TMULT = ',1PE12.5,')')
   44 FORMAT(1H1)
 9999 CONTINUE
      WRITE(6,44)
      STOP
      END
```

```
      SUBROUTINE SOLPRO
1           SUBROUTINE SOLPRO(TAU,IQ,F,EF,INALE,IERR)
2     C
3     C     ************************************************************
4     C     *   SUBROUTINE TO COMPUTE INTERPLANETARY SOLAR PROTON FLUX AT  *
5     C     *     1 AU (FROM E>10 TO E>100 MEV)                            *
6     C     *                                                              *
7     C     *   PROGRAM DESIGNED AND TESTED BY E.G. STASSINOPOULOS, CODE 601, *
8     C     *     NASA GODDARD SPACE FLIGHT CENTER, GREENBELT, MARYLAND 20771 *
9     C     *                                                              *
10    C     *   INPUT VARIABLES . . .                                      *
11    C     *     TAU    MISSION DURATION IN MONTHS                        *
12    C     *     IQ     CONFIDENCE LEVEL THAT CALCULATED FLUENCE F(N)     *
13    C     *            WILL NOT BE EXCEEDED                              *
14    C     *                                                              *
15    C     *   OUTPUT: F(N) SPECTRUM OF INTEGRAL SOLAR PROTON FLUENCE FOR *
16    C     *     ENERGIES E>10*N (1=<N=10)                                *
17    C     ************************************************************
18    C
19    C
20          DIMENSION F(1),EF(1),G(10),INDEX(20),ORFLXC(5,9)
21          REAL  NALE,NALECF(7,20)
22    C
23          DATA (NALECF(I),I=1,140)/-.1571,.2707,-.1269E-1,.4428E-3,-.8185E-5
24         *,.7754E-7,-.2939E-9,-.1870,.1951,-.6559E-2,.1990E-3,-.3618E-5,
25         *.3740E-7,-.1599E-9,-.2007,.1497,-.3179E-2,.5730E-4,-.4664E-6,
26         *.1764E-8,0.,-.1882,.1228,-.1936E-2,.2660E-4,-.1022E-6,2*0.,
27         *-.2214,.1149,-.1871E-2,.2695E-4,-.1116E-6,2*0.,-.2470,.1062,
28         *-.1658E-2,.2367E-4,-.9465E-7,2*0.,-.2609,.8710E-1,-.8300E-3,
29         *.8438E-5,3*0.,-.2923,.8932E-1,-.1023E-2,.1029E-4,3*0.,-.3222,
30         *.8648E-1,-.9992E-3,.9935E-5,3*0.,-.3518,.8417E-1,-.1000E-2,
31         *.9956E-5,3*0.,-.3698,.7951E-1,-.8983E-3,.8940E-5,3*0.,-.2771,
32         *.5473E-1,-.1543E-3,4*0.,-.2818,.5072E-1,.2511E-4,4*0.,-.2845,
33         *.4717E-1,.5664E-4,4*0.,-.2947,.4405E-1,.8507E-4,4*0.,-.2923,
34         *.4111E-1,.1106E-3,4*0.,-.2981,.3853E-1,.1312E-3,4*0.,-.3002,
35         *.3585E-1,.1529E-3,4*0.,-.3001,.3312E-1,.1781E-3,4*0.,-.3141,
36         *.3248E-1,.1654E-3,4*0./
37          DATA (ORFLXC(I),I=1,45)/.164047E3,-.522258E4,.714275E5,-.432747E6,
38         *.955315E6,.198004E3,-.448788E4,.438148E5,-.196046E6,.32552E6,
39         *.529120E3,
40         *-.122227E5,.112869E6,-.465084E6,.710572E6,.121141E4,-.266412E5,
41         *.226778E6,-.85728E6,.120444E7,.452062E4,-.103248E6,.896085E6,
42         *-.346028E7,.499852E7,.272028E4,-.499088E5,.35305E6,-.111929E7,
43         *.133386E7,.275697E4,-.469718E5,.314729E6,-.960383E6,.11165E7,
44         *.570997E4,-.799689E5,.381074E6,-.610714E6,0.,.101E3,4*0./
45          DATA (INDEX(I),I=1,20)/2*7,6,3*5,5*4,9*3/
46    C
47        1 FORMAT(' TAU=',F4.0,' IQ=',I3,3X,'PARAMETER(S) EXCEED PROGRAM LIMI
48         *TS')
49        2 FORMAT(2X,'FOR THE COMBINATION OF TAU AND IQ GIVEN, NO SIGNIFICANT
50         * SOLAR PROTON FLUXES ARE TO BE EXPECTED. TAU=',F6.2,' IQ=',I2)
51    C
52    C
53          IERR=0
54          IF(TAU .GT. 72. .OR. IQ .LT. 80) GO TO 500
55          IP=100-IQ
56          M=INDEX(IP)
57          NALE=0.
```

```
              DO 300 J=1,M
  300 NALE=NALE+NALECF(J,IP)*TAU**(J-1)
      INALE=NALE+1.0001
      IF(INALE .GT. 0) GO TO 400
C
C *** CALCULATIONS FOR OR-EVENT CONDITIONS
C
      IT=TAU
      IF(IT .EQ. 1 .AND. IP .GT. 16) GO TO 700
      P=FLOAT(IP)/100.
      OF=0.
      DO 100 J=1,5
  100 OF=OF+ORFLXC(J,IT)* P**(J-1)*1.E7
      E=10.
      DO 200 N=1,10
      G(N)=EXP(.0168*(30.-E))
      F(N)=OF*G(N)
      EF(N)=E
  200 E=E+10.
      RETURN
C
C *** CALCULATIONS FOR AL-EVENT CONDITIONS
C
  400 E=10.
      DO 600 N=1,10
      F(N)=7.9E9*EXP((30.-E)/26.5)*INALE
      EF(N)=E
  600 E=E+10.
      RETURN
C
C     ERROR CONDITIONS - PRINT MESSAGE AND RETURN
C
  700 WRITE(6,2) TAU,IQ
      GO TO 800
  600 WRITE (6,1) TAU,IQ
  800 IERR=1
      RETURN
      END
                INTERPOLATION SUBROUTINE
      SUBROUTINE INTP(XT,YT,X,Y,N)
C
C     **********************************************************************
C     *   LINEAR INTERPOLATION SUBROUTINE                                  *
C     **********************************************************************
C
      DIMENSION X(1),Y(1)
C
      DO 10 I=1,N
      II=I
      IF(XT .LE. X(I)) GO TO 12
   10 CONTINUE
   12 IF(II .EQ. 1) II=2
      IM=II-1
      YT=Y(IM)+(XT-X(IM))*(Y(II)-Y(IM))/(X(II)-X(IM))
      RETURN
      END
```

```
      BLOCK DATA
1           BLOCK DATA
2           COMMON/DAMAGE/EMEV(70),EDET(70,8),PMEV(70),PISC(70,8),PVOC(70,8)
3     C
4     C
5     C     EMEV - ELECTRON ENERGIES FOR DAMAGE COEFFICIENT TABLE EDET
6     C
7           DATA (EMEV(I),I=1,47)
8          *  /1.500E-01,1.600E-01,1.700E-01,1.800E-01,1.900E-01,2.000E-01
9          * ,2.200E-01,2.400E-01,2.600E-01,2.800E-01,3.000E-01,3.200E-01
10         * ,3.600E-01,4.000E-01,4.500E-01,5.000E-01,6.000E-01,7.000E-01
11         * ,8.000E-01,9.000E-01,1.000E+00,1.200E+00,1.400E+00,1.600E+00
12         * ,1.800E+00,2.000E+00,2.250E+00,2.500E+00,2.750E+00,3.000E+00
13         * ,3.250E+00,3.500E+00,3.750E+00,4.000E+00,4.500E+00,5.000E+00
14         * ,5.500E+00,6.000E+00,7.000E+00,8.000E+00,9.000E+00,1.000E+01
15         * ,1.500E+01,2.000E+01,2.500E+01,3.000E+01,4.000E+01/
16    C
17    C     0.0 GM/CM2 COVER GLASS DAMAGE COEFFICIENTS
18    C
19          DATA (EDET(I),I=  1, 47)
20         *  /2.690E-04,5.000E-04,8.951E-04,1.550E-03,2.406E-03,3.650E-03
21         * ,6.750E-03,1.035E-02,1.450E-02,2.010E-02,2.725E-02,3.385E-02
22         * ,5.004E-02,7.000E-02,9.506E-02,1.250E-01,2.000E-01,2.700E-01
23         * ,3.500E-01,4.225E-01,5.000E-01,6.700E-01,8.600E-01,1.060E+00
24         * ,1.260E+00,1.470E+00,1.729E+00,2.000E+00,2.252E+00,2.510E+00
25         * ,2.754E+00,3.000E+00,3.249E+00,3.500E+00,3.950E+00,4.400E+00
26         * ,4.850E+00,5.300E+00,6.150E+00,6.900E+00,7.607E+00,8.300E+00
27         * ,1.060E+01,1.230E+01,1.360E+01,1.470E+01,1.650E+01/
28    C
29    C     0.00559 GM/CM2 COVER GLASS DAMAGE COEFFICIENTS
30    C
31          DATA (EDET(I),I= 71,117)
32         *  /3.687E-05,7.951E-05,1.620E-04,3.168E-04,5.938E-04,1.045E-03
33         * ,2.533E-03,4.924E-03,7.981E-03,1.174E-02,1.668E-02,2.249E-02
34         * ,3.581E-02,5.255E-02,7.562E-02,1.023E-01,1.703E-01,2.400E-01
35         * ,3.166E-01,3.898E-01,4.657E-01,6.303E-01,8.160E-01,1.012E+00
36         * ,1.210E+00,1.418E+00,1.676E+00,1.943E+00,2.197E+00,2.454E+00
37         * ,2.698E+00,2.943E+00,3.191E+00,3.442E+00,3.894E+00,4.344E+00
38         * ,4.793E+00,5.243E+00,6.093E+00,6.848E+00,7.555E+00,8.249E+00
39         * ,1.056E+01,1.227E+01,1.357E+01,1.467E+01,1.648E+01/
40    C
41    C     0.0168 GM/CM2 COVER GLASS DAMAGE COEFFICIENTS
42    C
43          DATA (EDET(I),I=141,187)
44         *  /0.        ,0.        ,0.        ,2.227E-05,5.228E-05,1.143E-04
45         * ,4.375E-04,1.263E-03,2.814E-03,5.052E-03,7.941E-03,1.156E-02
46         * ,2.142E-02,3.423E-02,5.344E-02,7.595E-02,1.343E-01,2.004E-01
47         * ,2.718E-01,3.438E-01,4.169E-01,5.733E-01,7.515E-01,9.405E-01
48         * ,1.136E+00,1.339E+00,1.592E+00,1.854E+00,2.108E+00,2.362E+00
49         * ,2.606E+00,2.850E+00,3.096E+00,3.344E+00,3.798E+00,4.247E+00
50         * ,4.695E+00,5.143E+00,5.992E+00,6.753E+00,7.462E+00,8.156E+00
51         * ,1.049E+01,1.221E+01,1.352E+01,1.462E+01,1.643E+01/
52    C
53    C     0.0335 GM/CM2 COVER GLASS DAMAGE COEFFICIENTS
54    C
55          DATA (EDET(I),I=211,257)
56         *  /0.        ,0.        ,0.        ,0.        ,0.        ,0.
57         * ,1.551E-05,8.667E-05,3.609E-04,1.073E-03,2.400E-03,4.220E-03
```

```
58            *    ,9.858E-03,1.855E-02,3.258E-02,5.059E-02,9.816E-02,1.574E-01
59            *    ,2.225E-01,2.910E-01,3.607E-01,5.072E-01,6.759E-01,8.664E-01
60            *    ,1.045E+00,1.242E+00,1.489E+00,1.744E+00,1.997E+00,2.248E+00
61            *    ,2.490E+00,2.731E+00,2.974E+00,3.220E+00,3.675E+00,4.121E+00
62            *    ,4.566E+00,5.012E+00,5.859E+00,6.626E+00,7.335E+00,8.029E+00
63            *    ,1.039E+01,1.213E+01,1.344E+01,1.455E+01,1.637E+01/
64       C
65       C    0.0671 GM/CM2 COVER GLASS DAMAGE COEFFICIENTS
66       C
67            DATA (EDET(I),I=281,327)
68            *    /0.       ,0.       ,0.       ,0.       ,0.       ,0.
69            *    ,0.       ,0.       ,0.       ,0.       ,2.828E-05,1.481E-04
70            *    ,1.314E-03,4.311E-03,1.106E-02,2.146E-02,5.347E-02,9.769E-02
71            *    ,1.627E-01,2.121E-01,2.759E-01,4.068E-01,5.593E-01,7.256E-01
72            *    ,9.022E-01,1.088E+00,1.323E+00,1.566E+00,1.813E+00,2.057E+00
73            *    ,2.295E+00,2.531E+00,2.770E+00,3.011E+00,3.464E+00,3.905E+00
74            *    ,4.346E+00,4.787E+00,5.627E+00,6.401E+00,7.112E+00,7.804E+00
75            *    ,1.020E+01,1.197E+01,1.329E+01,1.442E+01,1.625E+01/
76       C
77       C    0.112 GM/CM2 COVER GLASS DAMAGE COEFFICIENTS
78       C
79            DATA (EDET(I),I=351,397)
80            *    /0.       ,0.       ,0.       ,0.       ,0.       ,0.
81            *    ,0.       ,0.       ,0.       ,0.       ,0.       ,0.
82            *    ,0.       ,9.075E-05,1.295E-03,4.824E-03,2.158E-02,4.962E-02
83            *    ,9.074E-02,1.385E-01,1.934E-01,3.081E-01,4.419E-01,5.916E-01
84            *    ,7.521E-01,9.245E-01,1.145E+00,1.374E+00,1.611E+00,1.847E+00
85            *    ,2.078E+00,2.309E+00,2.541E+00,2.775E+00,3.223E+00,3.659E+00
86            *    ,4.093E+00,4.528E+00,5.358E+00,6.138E+00,6.848E+00,7.539E+00
87            *    ,9.981E+00,1.177E+01,1.311E+01,1.425E+01,1.610E+01/
88       C
89       C    0.1675 GM/CM2 COVER GLASS DAMAGE COEFFICIENTS
90       C
91            DATA (EDET(I),I=421,467)
92            *    /0.       ,0.       ,0.       ,0.       ,0.       ,0.
93            *    ,0.       ,0.       ,0.       ,0.       ,0.       ,0.
94            *    ,0.       ,0.       ,0.       ,7.759E-05,4.315E-03,1.802E-02
95            *    ,4.262E-02,7.726E-02,1.199E-01,2.172E-01,3.312E-01,4.614E-01
96            *    ,6.040E-01,7.611E-01,9.639E-01,1.178E+00,1.399E+00,1.627E+00
97            *    ,1.849E+00,2.072E+00,2.296E+00,2.523E+00,2.962E+00,3.390E+00
98            *    ,3.817E+00,4.244E+00,5.062E+00,5.844E+00,6.553E+00,7.241E+00
99            *    ,9.725E+00,1.155E+01,1.290E+01,1.405E+01,1.593E+01/
100      C
101      C    0.335 GM/CM2 COVER GLASS DAMAGE COEFFICIENTS
102      C
103           DATA (EDET(I),I=491,537)
104           *    /0.       ,0.       ,0.       ,0.       ,0.       ,0.
105           *    ,0.       ,0.       ,0.       ,0.       ,0.       ,0.
106           *    ,0.       ,0.       ,0.       ,0.       ,0.       ,0.
107           *    ,3.097E-04,4.452E-03,1.566E-02,5.937E-02,1.281E-01,2.120E-01
108           *    ,3.099E-01,4.236E-01,5.793E-01,7.499E-01,9.314E-01,1.125E+00
109           *    ,1.320E+00,1.520E+00,1.723E+00,1.928E+00,2.332E+00,2.738E+00
110           *    ,3.141E+00,3.545E+00,4.326E+00,5.097E+00,5.801E+00,6.479E+00
111           *    ,9.047E+00,1.095E+01,1.233E+01,1.352E+01,1.544E+01/
112      C
113      C    PMEV - PROTON ENERGIES FOR DAMAGE COEFFICIENT TABLES PISC AND PVOC
114      C
```

```
115              DATA (PMEV(I),I=1,65)
116             * /1.000E-01,2.000E-01,3.000E-01,4.000E-01,6.000E-01,8.000E-01
117             * ,1.000E+00,1.200E+00,1.300E+00,1.400E+00,1.600E+00,1.800E+00
118             * ,2.000E+00,2.200E+00,2.400E+00,2.600E+00,2.800E+00,3.000E+00
119             * ,3.200E+00,3.400E+00,3.600E+00,3.800E+00,4.000E+00,4.200E+00
120             * ,4.400E+00,4.600E+00,4.800E+00,5.200E+00,5.600E+00,6.000E+00
121             * ,6.400E+00,6.800E+00,7.200E+00,7.600E+00,8.000E+00,9.000E+00
122             * ,1.000E+01,1.100E+01,1.200E+01,1.300E+01,1.400E+01,1.500E+01
123             * ,1.600E+01,1.800E+01,2.000E+01,2.200E+01,2.400E+01,2.600E+01
124             * ,2.800E+01,3.000E+01,3.400E+01,3.800E+01,4.200E+01,4.600E+01
125             * ,5.000E+01,5.500E+01,6.000E+01,6.500E+01,7.000E+01,8.000E+01
126             * ,9.000E+01,1.000E+02,1.300E+02,1.600E+02,2.000E+02/
127       C
128       C     PISC - PROTON DAMAGE COEFFICIENTS (SHORT-CIRCUIT CURRENT)
129       C
130       C     0.0 GM/CM2 COVER GLASS DAMAGE COEFFICIENTS
131       C
132              DATA (PISC(I),I=  1, 65)
133             * /2.435E-04,3.047E-03,1.374E-02,3.987E-02,1.602E-01,3.243E-01
134             * ,5.216E-01,7.108E-01,7.890E-01,8.549E-01,9.632E-01,1.010E+00
135             * ,1.039E+00,1.048E+00,1.041E+00,1.023E+00,9.962E-01,9.639E-01
136             * ,9.286E-01,8.937E-01,8.598E-01,8.273E-01,7.963E-01,7.723E-01
137             * ,7.486E-01,7.254E-01,7.029E-01,6.605E-01,6.216E-01,5.867E-01
138             * ,5.585E-01,5.339E-01,5.128E-01,4.947E-01,4.786E-01,4.476E-01
139             * ,4.337E-01,4.232E-01,4.196E-01,4.185E-01,4.181E-01,4.194E-01
140             * ,4.214E-01,4.192E-01,4.172E-01,4.144E-01,4.094E-01,4.049E-01
141             * ,4.000E-01,3.935E-01,3.784E-01,3.664E-01,3.532E-01,3.399E-01
142             * ,3.272E-01,3.125E-01,2.988E-01,2.844E-01,2.710E-01,2.474E-01
143             * ,2.245E-01,1.997E-01,1.492E-01,1.183E-01,9.215E-02/
144       C
145       C     0.00559 GM/CM2 COVER GLASS DAMAGE COEFFICIENTS
146       C
147              DATA (PISC(I),I= 71,135)
148             * /0.        ,0.        ,0.        ,0.        ,0.        ,0.
149             * ,0.        ,0.        ,2.322E-05,3.750E-03,8.124E-02,2.525E-01
150             * ,4.558E-01,6.233E-01,7.426E-01,8.207E-01,8.680E-01,8.912E-01
151             * ,8.962E-01,8.874E-01,8.697E-01,8.481E-01,8.243E-01,7.989E-01
152             * ,7.734E-01,7.499E-01,7.280E-01,6.866E-01,6.479E-01,6.119E-01
153             * ,5.792E-01,5.520E-01,5.285E-01,5.086E-01,4.909E-01,4.565E-01
154             * ,4.369E-01,4.245E-01,4.187E-01,4.167E-01,4.159E-01,4.173E-01
155             * ,4.182E-01,4.179E-01,4.159E-01,4.117E-01,4.083E-01,4.039E-01
156             * ,3.994E-01,3.930E-01,3.782E-01,3.662E-01,3.532E-01,3.399E-01
157             * ,3.272E-01,3.126E-01,2.989E-01,2.846E-01,2.712E-01,2.476E-01
158             * ,2.247E-01,1.999E-01,1.493E-01,1.183E-01,9.220E-02/
159       C
160       C     0.0168 GM/CM2 COVER GLASS DAMAGE COEFFICIENTS
161       C
162              DATA (PISC(I),I=141,205)
163             * /0.        ,0.        ,0.        ,0.        ,0.        ,0.
164             * ,0.        ,0.        ,0.        ,0.        ,0.        ,0.
165             * ,0.        ,0.        ,0.        ,1.860E-05,3.925E-02,1.794E-01
166             * ,3.465E-01,4.807E-01,5.787E-01,6.459E-01,6.879E-01,7.105E-01
167             * ,7.189E-01,7.184E-01,7.120E-01,6.890E-01,6.613E-01,6.319E-01
168             * ,6.019E-01,5.731E-01,5.477E-01,5.255E-01,5.058E-01,4.669E-01
169             * ,4.401E-01,4.258E-01,4.155E-01,4.120E-01,4.105E-01,4.104E-01
170             * ,4.120E-01,4.133E-01,4.125E-01,4.093E-01,4.069E-01,4.018E-01
171             * ,3.978E-01,3.918E-01,3.777E-01,3.657E-01,3.532E-01,3.400E-01
```

```
172         *     ,3.272E-01,3.128E-01,2.990E-01,2.850E-01,2.715E-01,2.480E-01
173         *     ,2.251E-01,2.004E-01,1.496E-01,1.185E-01,9.229E-02/
174   C
175   C     0.0335 GM/CM2 COVER GLASS DAMAGE COEFFICIENTS
176   C
177               DATA (PISC(I),I=211,275)
178         *    /0.         ,0.         ,0.         ,0.         ,0.         ,0.
179         *    ,0.         ,0.         ,0.         ,0.         ,0.         ,0.
180         *    ,0.         ,0.         ,0.         ,0.         ,0.         ,0.
181         *    ,0.         ,0.         ,0.         ,0.         ,1.288E-03,7.227E-02
182         *    ,2.077E-01,3.274E-01,4.191E-01,5.286E-01,5.723E-01,5.839E-01
183         *    ,5.793E-01,5.664E-01,5.491E-01,5.299E-01,5.118E-01,4.724E-01
184         *    ,4.425E-01,4.226E-01,4.110E-01,4.040E-01,4.020E-01,4.010E-01
185         *    ,4.025E-01,4.054E-01,4.055E-01,4.047E-01,4.010E-01,3.985E-01
186         *    ,3.939E-01,3.896E-01,3.767E-01,3.650E-01,3.530E-01,3.400E-01
187         *    ,3.273E-01,3.130E-01,2.992E-01,2.855E-01,2.720E-01,2.485E-01
188         *    ,2.256E-01,2.010E-01,1.500E-01,1.188E-01,9.242E-02/
189   C
190   C     0.0671 GM/CM2 COVER GLASS DAMAGE COEFFICIENTS
191   C
192               DATA (PISC(I),I=281,345)
193         *    /0.         ,0.         ,0.         ,0.         ,0.         ,0.
194         *    ,0.         ,0.         ,0.         ,0.         ,0.         ,0.
195         *    ,0.         ,0.         ,0.         ,0.         ,0.         ,0.
196         *    ,0.         ,0.         ,0.         ,0.         ,0.         ,0.
197         *    ,0.         ,0.         ,0.         ,0.         ,0.         ,2.142E-03
198         *    ,1.742E-01,3.196E-01,3.945E-01,4.317E-01,4.484E-01,4.478E-01
199         *    ,4.292E-01,4.101E-01,3.956E-01,3.872E-01,3.828E-01,3.814E-01
200         *    ,3.819E-01,3.873E-01,3.900E-01,3.915E-01,3.919E-01,3.898E-01
201         *    ,3.875E-01,3.834E-01,3.739E-01,3.617E-01,3.519E-01,3.396E-01
202         *    ,3.271E-01,3.133E-01,2.995E-01,2.863E-01,2.728E-01,2.494E-01
203         *    ,2.266E-01,2.022E-01,1.509E-01,1.192E-01,9.268E-02/
204   C
205   C     0.112 GM/CM2 COVER GLASS DAMAGE COEFFICIENTS
206   C
207   C
208               DATA (PISC(I),I=351,415)
209         *    /0.         ,0.         ,0.         ,0.         ,0.         ,0.
210         *    ,0.         ,0.         ,0.         ,0.         ,0.         ,0.
211         *    ,0.         ,0.         ,0.         ,0.         ,0.         ,0.
212         *    ,0.         ,0.         ,0.         ,0.         ,0.         ,0.
213         *    ,0.         ,0.         ,0.         ,0.         ,0.         ,0.
214         *    ,0.         ,0.         ,0.         ,0.         ,0.         ,2.735E-01
215         *    ,3.537E-01,3.675E-01,3.649E-01,3.588E-01,3.553E-01,3.538E-01
216         *    ,3.547E-01,3.606E-01,3.679E-01,3.731E-01,3.757E-01,3.769E-01
217         *    ,3.764E-01,3.753E-01,3.677E-01,3.582E-01,3.484E-01,3.372E-01
218         *    ,3.264E-01,3.132E-01,2.997E-01,2.871E-01,2.736E-01,2.504E-01
219         *    ,2.277E-01,2.037E-01,1.619E-01,1.199E-01,9.302E-02/
220   C
221   C     0.1675 GM/CM2 COVER GLASS DAMAGE COEFFICIENTS
222   C
223               DATA (PISC(I),I=421,485)
224         *    /0.         ,0.         ,0.         ,0.         ,0.         ,0.
225         *    ,0.         ,0.         ,0.         ,0.         ,0.         ,0.
226         *    ,0.         ,0.         ,0.         ,0.         ,0.         ,0.
227         *    ,0.         ,0.         ,0.         ,0.         ,0.         ,0.
228         *    ,0.         ,0.         ,0.         ,0.         ,0.         ,0.
```

```
229             *   ,0.         ,0.         ,0.         ,0.         ,0.         ,0.
230             *   ,0.         ,2.061E-01,2.839E-01,3.062E-01,3.131E-01,3.159E-01
231             *   ,3.187E-01,3.269E-01,3.379E-01,3.473E-01,3.547E-01,3.591E-01
232             *   ,3.613E-01,3.625E-01,3.600E-01,3.529E-01,3.446E-01,3.349E-01
233             *   ,3.250E-01,3.126E-01,2.995E-01,2.875E-01,2.743E-01,2.614E-01
234             *   ,2.289E-01,2.052E-01,1.630E-01,1.206E-01,9.344E-02/
235     C
236     C       0.335 GM/CM2 COVER GLASS DAMAGE COEFFICIENTS
237     C
238             DATA (PISC(I),I=491,555)
239             *   /0.         ,0.         ,0.         ,0.         ,0.         ,0.
240             *   ,0.         ,0.         ,0.         ,0.         ,0.         ,0.
241             *   ,0.         ,0.         ,0.         ,0.         ,0.         ,0.
242             *   ,0.         ,0.         ,0.         ,0.         ,0.         ,0.
243             *   ,0.         ,0.         ,0.         ,0.         ,0.         ,0.
244             *   ,0.         ,0.         ,0.         ,0.         ,0.         ,0.
245             *   ,0.         ,0.         ,0.         ,0.         ,0.         ,0.
246             *   ,1.439E-01,2.175E-01,2.441E-01,2.648E-01,2.834E-01,2.984E-01
247             *   ,3.101E-01,3.186E-01,3.291E-01,3.312E-01,3.292E-01,3.245E-01
248             *   ,3.177E-01,3.082E-01,2.969E-01,2.869E-01,2.748E-01,2.531E-01
249             *   ,2.315E-01,2.089E-01,1.560E-01,1.226E-01,9.462E-02/
250     C
251     C       PVOC - PROTON DAMAGE COEFFICIENTS (OPEN-CIRCUIT VOLTAGE AND P-MAX)
252     C
253     C
254     C       0.0 GM/CM2 COVER GLASS DAMAGE COEFFICIENTS
255     C
256             DATA (PVOC(I),I=  1, 65)
257             *   /5.303E-01,7.150E-01,8.623E-01,9.976E-01,1.271E+00,1.548E+00
258             *   ,1.792E+00,1.994E+00,2.082E+00,2.160E+00,2.299E+00,2.412E+00
259             *   ,2.502E+00,2.569E+00,2.615E+00,2.645E+00,2.656E+00,2.640E+00
260             *   ,2.597E+00,2.526E+00,2.426E+00,2.302E+00,2.159E+00,2.024E+00
261             *   ,1.891E+00,1.766E+00,1.650E+00,1.447E+00,1.278E+00,1.136E+00
262             *   ,1.020E+00,9.237E-01,8.440E-01,7.775E-01,7.204E-01,6.134E-01
263             *   ,5.564E-01,5.169E-01,4.936E-01,4.778E-01,4.663E-01,4.694E-01
264             *   ,4.548E-01,4.433E-01,4.352E-01,4.286E-01,4.211E-01,4.146E-01
265             *   ,4.081E-01,4.004E-01,3.836E-01,3.703E-01,3.564E-01,3.426E-01
266             *   ,3.296E-01,3.145E-01,3.005E-01,2.859E-01,2.724E-01,2.481E-01
267             *   ,2.249E-01,1.999E-01,1.492E-01,1.183E-01,9.215E-02/
268     C
269     C       0.00559 GM/CM2 COVER GLASS DAMAGE COEFFICIENTS
270     C
271             DATA (PVOC(I),I= 71,135)
272             *   /0.         ,0.         ,0.         ,0.         ,0.         ,0.
273             *   ,0.         ,0.         ,4.303E-02,1.948E-01,5.853E-01,9.827E-01
274             *   ,1.335E+00,1.624E+00,1.860E+00,2.047E+00,2.191E+00,2.298E+00
275             *   ,2.375E+00,2.416E+00,2.420E+00,2.388E+00,2.320E+00,2.219E+00
276             *   ,2.093E+00,1.962E+00,1.839E+00,1.616E+00,1.428E+00,1.268E+00
277             *   ,1.131E+00,1.018E+00,9.252E-01,8.479E-01,7.827E-01,6.580E-01
278             *   ,5.834E-01,5.351E-01,5.051E-01,4.869E-01,4.722E-01,4.637E-01
279             *   ,4.572E-01,4.458E-01,4.367E-01,4.278E-01,4.211E-01,4.144E-01
280             *   ,4.081E-01,4.005E-01,3.838E-01,3.704E-01,3.565E-01,3.426E-01
281             *   ,3.296E-01,3.145E-01,3.006E-01,2.861E-01,2.726E-01,2.483E-01
282             *   ,2.251E-01,2.001E-01,1.493E-01,1.183E-01,9.220E-02/
283     C
284     C       0.0168 GM/CM2 COVER GLASS DAMAGE COEFFICIENTS
286     C
```

```
286              DATA (PVOC(I),I=141,205)
287           *  /0.        ,0.        ,0.        ,0.        ,0.        ,0.
288           *  ,0.        ,0.        ,0.        ,0.        ,0.        ,0.
289           *  ,0.        ,0.        ,0.        ,1.912E-02,2.733E-01,6.092E-01
290           *  ,9.375E-01,1.226E+00,1.468E+00,1.664E+00,1.818E+00,1.932E+00
291           *  ,1.998E+00,2.017E+00,1.990E+00,1.833E+00,1.642E+00,1.467E+00
292           *  ,1.312E+00,1.178E+00,1.063E+00,9.673E-01,8.867E-01,7.324E-01
293           *  ,6.303E-01,5.691E-01,5.264E-01,4.997E-01,4.815E-01,4.688E-01
294           *  ,4.606E-01,4.480E-01,4.382E-01,4.290E-01,4.213E-01,4.141E-01
295           *  ,4.079E-01,4.003E-01,3.840E-01,3.704E-01,3.569E-01,3.429E-01
296           *  ,3.298E-01,3.147E-01,3.007E-01,2.865E-01,2.729E-01,2.487E-01
297           *  ,2.255E-01,2.006E-01,1.496E-01,1.185E-01,9.229E-02/
298       C
299       C     0.0335 GM/CM2 COVER GLASS DAMAGE COEFFICIENTS
300       C
301              DATA (PVOC(I),I=211,275)
302           *  /0.        ,0.        ,0.        ,0.        ,0.        ,0.
303           *  ,0.        ,0.        ,0.        ,0.        ,0.        ,0.
304           *  ,0.        ,0.        ,0.        ,0.        ,0.        ,0.
305           *  ,0.        ,0.        ,0.        ,0.        ,4.687E-02,2.866E-01
306           *  ,5.697E-01,8.378E-01,1.074E+00,1.431E+00,1.603E+00,1.584E+00
307           *  ,1.468E+00,1.339E+00,1.218E+00,1.109E+00,1.013E+00,8.256E-01
308           *  ,6.965E-01,6.105E-01,5.558E-01,5.182E-01,4.934E-01,4.758E-01
309           *  ,4.642E-01,4.493E-01,4.379E-01,4.293E-01,4.203E-01,4.137E-01
310           *  ,4.063E-01,3.998E-01,3.839E-01,3.703E-01,3.572E-01,3.433E-01
311           *  ,3.301E-01,3.161E-01,3.010E-01,2.870E-01,2.733E-01,2.492E-01
312           *  ,2.260E-01,2.013E-01,1.500E-01,1.188E-01,9.242E-02/
313       C
314       C     0.0671 GM/CM2 COVER GLASS DAMAGE COEFFICIENTS
315       C
316              DATA (PVOC(I),I=281,345)
317           *  /0.        ,0.        ,0.        ,0.        ,0.        ,0.
318           *  ,0.        ,0.        ,0.        ,0.        ,0.        ,0.
319           *  ,0.        ,0.        ,0.        ,0.        ,0.        ,0.
320           *  ,0.        ,0.        ,0.        ,0.        ,0.        ,0.
321           *  ,0.        ,0.        ,0.        ,0.        ,0.        ,3.741E-02
322           *  ,4.510E-01,8.464E-01,1.101E+00,1.166E+00,1.125E+00,9.523E-01
323           *  ,7.998E-01,6.851E-01,6.035E-01,5.493E-01,5.121E-01,4.867E-01
324           *  ,4.685E-01,4.478E-01,4.343E-01,4.250E-01,4.176E-01,4.102E-01
325           *  ,4.040E-01,3.969E-01,3.831E-01,3.686E-01,3.571E-01,3.436E-01
326           *  ,3.304E-01,3.158E-01,3.016E-01,2.880E-01,2.743E-01,2.501E-01
327           *  ,2.270E-01,2.025E-01,1.509E-01,1.192E-01,9.268E-02/
328       C
329       C     0.112 GM/CM2 COVER GLASS DAMAGE COEFFICIENTS
330       C
331              DATA (PVOC(I),I=351,415)
332           *  /0.        ,0.        ,0.        ,0.        ,0.        ,0.
333           *  ,0.        ,0.        ,0.        ,0.        ,0.        ,0.
334           *  ,0.        ,0.        ,0.        ,0.        ,0.        ,0.
335           *  ,0.        ,0.        ,0.        ,0.        ,0.        ,0.
336           *  ,0.        ,0.        ,0.        ,0.        ,0.        ,0.
337           *  ,0.        ,0.        ,0.        ,0.        ,3.696E-03,7.614E-01
338           *  ,8.423E-01,7.463E-01,6.552E-01,5.830E-01,5.320E-01,4.961E-01
339           *  ,4.713E-01,4.419E-01,4.269E-01,4.172E-01,4.097E-01,4.035E-01
340           *  ,3.978E-01,3.926E-01,3.797E-01,3.669E-01,3.549E-01,3.421E-01
341           *  ,3.303E-01,3.161E-01,3.020E-01,2.890E-01,2.753E-01,2.512E-01
342           *  ,2.281E-01,2.039E-01,1.519E-01,1.199E-01,9.302E-02/
```

```
343       C
344       C     0.1675 GM/CM2 COVER GLASS DAMAGE COEFFICIENTS
345       C
346             DATA (PVOC(I),I=421,485)
347            *    /0.       ,0.       ,0.       ,0.       ,0.       ,0.
348            *    ,0.       ,0.       ,0.       ,0.       ,0.       ,0.
349            *    ,0.       ,0.       ,0.       ,0.       ,0.       ,0.
350            *    ,0.       ,0.       ,0.       ,0.       ,0.       ,0.
351            *    ,0.       ,0.       ,0.       ,0.       ,0.       ,0.
352            *    ,0.       ,0.       ,0.       ,0.       ,0.       ,0.
353            *    ,0.       ,5.746E-01,6.555E-01,6.044E-01,5.496E-01,5.047E-01
354            *    ,4.719E-01,4.327E-01,4.142E-01,4.041E-01,3.981E-01,3.931E-01
355            *    ,3.884E-01,3.843E-01,3.749E-01,3.636E-01,3.526E-01,3.409E-01
356            *    ,3.297E-01,3.161E-01,3.022E-01,2.897E-01,2.761E-01,2.623E-01
357            *    ,2.294E-01,2.055E-01,1.630E-01,1.206E-01,9.344E-02/
358       C
359       C     0.335 GM/CM2 COVER GLASS DAMAGE COEFFICIENTS
360       C
361             DATA (PVOC(I),I=491,555)
362            *    /0.       ,0.       ,0.       ,0.       ,0.       ,0.
363            *    ,0.       ,0.       ,0.       ,0.       ,0.       ,0.
364            *    ,0.       ,0.       ,0.       ,0.       ,0.       ,0.
365            *    ,0.       ,0.       ,0.       ,0.       ,0.       ,0.
366            *    ,0.       ,0.       ,0.       ,0.       ,0.       ,0.
367            *    ,0.       ,0.       ,0.       ,0.       ,0.       ,0.
368            *    ,0.       ,0.       ,0.       ,0.       ,0.       ,0.
369            *    ,3.775E-01,3.956E-01,3.698E-01,3.569E-01,3.530E-01,3.524E-01
370            *    ,3.527E-01,3.530E-01,3.520E-01,3.473E-01,3.409E-01,3.334E-01
371            *    ,3.245E-01,3.132E-01,3.007E-01,2.899E-01,2.773E-01,2.543E-01
372            *    ,2.320E-01,2.092E-01,1.560E-01,1.226E-01,9.462E-02/
373             END
```

```
INPUT STREAM
SIS 42 MONTH HELIOCENTRIC 1 AU MISSION          95 PERCENT PROB.                                    PAGE   1
$MIKE
TMULT=1.0, INTFLG=1, TIMIN=12HMISSION          ,TIMOUT=12HMISSION        ,
NPSPEC=10, PSPEC(1,1)=1.,1.3,4.3,10.,30.,60.,100.,200.,500.,1000.,
PSPEC(1,2)=3.0E11,2.5E11,1.1E11,6.5E10,2.4E10,7.4E9,1.9E9,5.2E8,
3.1E8,2.4E8,
$END
       EXECUTE PROGRAM
   SIS 42 MONTH HELIOCENTRIC 1 AU MISSION       95 PERCENT PROB.

                         PROTON
           ENERGY        FLUENCE
           (MEV)       (PROTONS/CM2-MISSION  )

            1.000        3.000+11
            1.300        2.500+11
            4.300        1.100+11
           10.000        6.500+10
           30.000        2.400+10
           60.000        7.400+09
          100.000        1.900+09
          200.000        5.200+08
          500.000        3.100+08
         1000.000        2.400+08
   SIS 42 MONTH HELIOCENTRIC 1 AU MISSION       95 PERCENT PROB.                                    PAGE   2

SHIELD THICKNESS (GM/CM2)      0.000    5.590-03  1.680-02  3.350-02  6.710-02  1.120-01  1.670-01  3.350-01
              ( MILS )            0         1         3         6        12        20        30        60
PROTON FLUENCE
   EQUIV 1 MEV ELECTRONS/CM2
           PMAX VOC            1.537+15  8.313+14  4.061+14  2.437+14  1.489+14  1.020+14  7.550+13  4.240+13
           ISC                 6.483+14  3.556+14  2.047+14  1.393+14  9.611+13  7.133+13  5.610+13  3.489+13
TIME OF EXPOSURE: MISSION
     (TMULT =  1.00000+00)
```